Advanced Algorithms in Wireless Communication and Internet of Things (IoT)

Advanced Algorithms in Wireless Communication and Internet of Things (IoT)

Guest Editors

Farag M. Sallabi
Mohammad Hayajneh

MDPI

Basel • Beijing • Wuhan • Barcelona • Belgrade • Novi Sad • Cluj • Manchester

Guest Editors
Farag M. Sallabi
Faculty of Information Technology
United Arab Emirates University
Al Ain
United Arab Emirates

Mohammad Hayajneh
Faculty of Information Technology
United Arab Emirates University
Al Ain
United Arab Emirates

Editorial Office
MDPI AG
Grosspeteranlage 5
4052 Basel, Switzerland

This is a reprint of the Special Issue, published open access by the journal *Mathematics* (ISSN 2227-7390), freely accessible at: https://www.mdpi.com/si/mathematics/D4Q85VJREU.

For citation purposes, cite each article independently as indicated on the article page online and as indicated below:

Lastname, A.A.; Lastname, B.B. Article Title. *Journal Name* **Year**, *Volume Number*, Page Range.

ISBN 978-3-7258-5683-1 (Hbk)
ISBN 978-3-7258-5684-8 (PDF)
https://doi.org/10.3390/books978-3-7258-5684-8

Contents

About the Editors

Farag M. Sallabi

Farag M. Sallabi received his B.Sc. and M.Sc. degrees in Electrical and Electronic Engineering from Benghazi University, Libya, in 1987 and 1995, respectively, and a Ph.D. in Electrical and Computer Engineering from the University of Ottawa, Canada, in 2001. He is currently an Associate Professor in the Department of Computer and Network Engineering at the College of Information Technology, United Arab Emirates University (UAEU), which he joined in 2002 after working at Sigpro Wireless in Ottawa, Canada. Farag has served as Principal Investigator and Co-Principal Investigator on several funded research projects at UAEU. His research interests include quality of service provisioning in wired and wireless networks, routing and performance evaluation of wireless sensor networks, network management, modeling and simulation, the Internet of Things (IoT), and smart healthcare systems. He is currently involved in projects on advanced algorithms for wireless communication, IoT-enabled smart city solutions, and AI-driven applications in healthcare and education. To date, Farag has published over 95 refereed journal and conference papers, with more than 1600 citations and an h-index of 20. Around 30% of his most-cited works appear in Q1 journals. He is a member of the IEEE, has served on the technical program committees of numerous international IEEE conferences, and acts as a reviewer for several leading journals in the areas of networking and communications.

Mohammad Hayajneh

Mohammad Hayajneh received his B.Sc. and M.Sc. degrees in electrical engineering, majoring in electronics and communications, from Jordan University of Science and Technology, Irbid, Jordan, in 1995 and 1998, respectively, and his Ph.D. degree in electrical engineering from the University of New Mexico, Albuquerque, NM, USA, in 2004. Before joining the Ph.D. program at the University of New Mexico, he held several positions in academia. He was a Lecturer and the Head of the Department of Telecommunications, Institute Science for Telecom and Technology, Riyadh, Saudi Arabia, from 2000 to 2001. From 1998 to 2000, he was an Instructor at the Princess Sumaya University of Technology (PSUT), Amman, Jordan. In September 2004, he joined the College of Information Technology, United Arab Emirates University, Al Ain, United Arab Emirates, as an Assistant Professor in computer system engineering. His research interests include power control for wireless data networks, WiMAX cross-layer scheduling, resource allocation and pricing in wireless data networks, MIMO systems, performance analysis, and modeling of OFDMA-based wireless networks.

Preface

The Special Issue Reprint "Advanced Algorithms for Wireless Communication and Internet of Things (IoT)" brings together a comprehensive collection of research contributions that address the growing need for efficient, secure, and intelligent communication systems in the era of digital connectivity. With the proliferation of IoT devices, smart environments, and data-intensive applications, the demand for advanced mathematical and computational methods has never been greater. This Reprint highlights innovative approaches that integrate optimization techniques, artificial intelligence, and mathematical modeling to enhance the performance and scalability of next-generation wireless networks and IoT infrastructures.

The studies presented in this volume reflect a balanced combination of theoretical analyses, simulation models, and real-world applications. They cover various topics, including adaptive modulation and coding, intelligent resource allocation, channel estimation, energy-efficient routing, cryptography, and robust network optimization. Several papers also explore the integration of machine learning and deep learning paradigms to enable predictive and self-organizing wireless systems that are capable of responding dynamically to complex network conditions.

This Special Issue Reprint is the result of the dedication of many researchers who have contributed their expertise and creativity to advancing the frontiers of wireless communication and the IoT. The Guest Editors express their sincere gratitude to all authors for their valuable submissions, the reviewers for their constructive and timely evaluations, and the editorial and production teams at MDPI *Mathematics* for their continuous support and professionalism. We hope this Reprint will serve as a valuable reference and inspiration for scientists, engineers, and practitioners working toward the realization of intelligent, reliable, and sustainable communication systems for the connected world.

Farag M. Sallabi and Mohammad Hayajneh
Guest Editors

mathematics

MDPI

Article

CNN-Based End-to-End CPU-AP-UE Power Allocation for Spectral Efficiency Enhancement in Cell-Free Massive MIMO Networks

Yoon-Ju Choi [1,2], Ji-Hee Yu [1,2], Seung-Hwan Seo [1,2], Seong-Gyun Choi [1,2], Hye-Yoon Jeong [1,2], Ja-Eun Kim [1,2], Myung-Sun Baek [3], Young-Hwan You [4] and Hyoung-Kyu Song [1,2,*]

1 Department of Information and Communication Engineering, Sejong University, Seoul 05006, Republic of Korea; jj010513@sju.ac.kr (Y.-J.C.); jeehee59@sju.ac.kr (J.-H.Y.); buffalo1997@sju.ac.kr (S.-H.S.); sk4753611@sju.ac.kr (S.-G.C.); zalfenz@sju.ac.kr (H.-Y.J.); thinkdana@sju.ac.kr (J.-E.K.)
2 Department of Convergence Engineering for Intelligent Drone, Sejong University, Seoul 05006, Republic of Korea
3 Department of Artificial Intelligence and Information Technology, Sejong University, Seoul 05006, Republic of Korea; msbaek@sejong.ac.kr
4 Department of Computer Engineering, Sejong University, Seoul 05006, Republic of Korea; yhyou@sejong.ac.kr
* Correspondence: songhk@sejong.ac.kr

Abstract: Cell-free massive multiple-input multiple-output (MIMO) networks eliminate cell boundaries and enhance uniform quality of service by enabling cooperative transmission among access points (APs). In conventional cellular networks, user equipment located at the cell edge experiences severe interference and unbalanced resource allocation. However, in cell-free massive MIMO networks, multiple access points cooperatively serve user equipment (UEs), effectively mitigating these issues. Beamforming and cooperative transmission among APs are essential in massive MIMO environments, making efficient power allocation a critical factor in determining overall network performance. In particular, considering power allocation from the central processing unit (CPU) to the APs enables optimal power utilization across the entire network. Traditional power allocation methods such as equal power allocation and max–min power allocation fail to fully exploit the cooperative characteristics of APs, leading to suboptimal network performance. To address this limitation, in this study we propose a convolutional neural network (CNN)-based power allocation model that optimizes both CPU-to-AP power allocation and AP-to-UE power distribution. The proposed model learns the optimal power allocation strategy by utilizing the channel state information, AP-UE distance, interference levels, and signal-to-interference-plus-noise ratio as input features. Simulation results demonstrate that the proposed CNN-based power allocation method significantly improves spectral efficiency compared to conventional power allocation techniques while also enhancing energy efficiency. This confirms that deep learning-based power allocation can effectively enhance network performance in cell-free massive MIMO environments.

Keywords: cell-free massive MIMO; power allocation; CNN; CPU-AP-UE optimization; deep learning; spectral efficiency; energy efficiency

MSC: 94A20; 68T07

1. Introduction

Wireless networks demand ultra-high data rates, ultra-low latency, and massive simultaneous user connectivity [1], necessitating the development of cell-free massive multiple-input multiple-output (MIMO) technology. Traditional cellular networks suffer from performance degradation at the cell edges, where user equipment (UE) experiences high interference and unbalanced resource allocation, leading to poor communication quality. To overcome these limitations, cell-free massive MIMO replaces cell boundaries with a cooperative architecture, allowing multiple access points (APs) to jointly serve user equipment across the network. This structure improves uniform quality of service and mitigates performance degradation at the cell edges. The APs are connected to a central processing unit (CPU) by a wired backhaul, enabling centralized coordination and effective interference mitigation [2].

However, the practical deployment of cell-free massive MIMO systems requires efficient power allocation strategies. In particular, power should be allocated both from the CPU to the APs and from the APs to the UEs [3]; both of these stages significantly influence the spectral efficiency (SE) and energy efficiency (EE) [4]. The amount of power assigned by the CPU determines the maximum transmission power of each AP and affects overall interference and power usage. Similarly, the way in which APs distribute power among UEs directly impacts each user's signal quality and data throughput. Therefore, holistic optimization of CPU-AP-UE power allocation is crucial for achieving high system performance with minimal energy wastage. Traditional power allocation methods such as equal power allocation and max–min power allocation rely on heuristic approaches [3]. While computationally simple, these techniques fail to adapt to dynamic network conditions [5]. Equal power allocation ignores spatial variations in channel quality, leading to inefficient resource usage. Max–min power allocation improves fairness by supporting the weakest UEs, but often sacrifices total throughput by underutilizing stronger links [6].

Recently, deep learning-based power control strategies have gained attention thanks to their ability to generalize across complex nonlinear environments. In particular, convolutional neural networks (CNNs) are well suited for capturing spatial features from network topology and channel data. CNNs can process inputs such as channel state information (CSI), AP-UE distance, interference levels, and signal-to-interference-plus-noise ratio (SINR) to learn spatially adaptive power allocation strategies. Based on these inputs, the model learns how to assign power differently depending on user location, channel quality, and interference conditions [7]. This enables flexible and context-aware power control that adapts to dynamic network scenarios instead of relying on fixed allocation rules. This study proposes a CNN-based power allocation model that jointly optimizes both CPU-to-AP and AP-to-UE power distribution. The model is trained end-to-end via supervised learning using data generated from conventional max–min optimization. Simulation results demonstrate that the proposed method significantly improves per-user SE while maintaining EE under realistic network conditions. By integrating both allocation stages into a unified deep learning framework, this work represents a novel approach to hierarchical power control in cell-free networks.

1.1. Related Works

Optimizing power allocation is a key challenge in wireless networks, particularly in terms of balancing SE and EE. In cell-free massive MIMO environments, a CPU coordinates power control for multiple distributed APs, which subsequently allocate power to UEs. This hierarchical structure highlights the importance of efficient CPU-to-AP and AP-to-UE power management in achieving system-wide performance and fairness. Prior research has investigated various power allocation schemes, with heuristic-based and AI-based approaches being most commonly explored for practical implementation. While heuristic methods offer simplicity, they often lack adaptability. In contrast, deep learning-based approaches demonstrate promising flexibility and spatial awareness, particularly in complex and dynamic network environments. Recent deep learning-based studies have applied hybrid architectures combining CNNs with long short-term memory (LSTM) networks to address power control and beamforming in massive and ultra-massive MIMO systems [8–10]. These works leverage temporal dependencies via LSTM and spatial features via CNN; however, they typically target only AP-to-UE power allocation or beamforming, and do not perform full-chain hierarchical optimization. In contrast, our proposed method performs joint end-to-end CPU-to-AP and AP-to-UE power allocation using a lightweight CNN structure designed for real-time inference and system-wide optimization.

Table 1 summarizes the architectural and functional differences between our model and recent deep learning-based methods.

Table 1. Comparison with recent deep learning-based power control methods.

Feature	CNN–LSTM [8]	CNN–LSTM [9]	CNN–LSTM [10]	Proposed CNN
Target Application	CF-mMIMO IoT	Massive MIMO	UM-MIMO Beamforming	CF-mMIMO
CPU-to-AP Optimization	No	No	No	Yes
AP-to-UE Optimization	Yes	Yes	No	Yes
Temporal Modeling	Yes (LSTM)	Yes (LSTM)	Yes (LSTM)	No
Input Features	CSI, SE/EE, time series	Location, CSI	CSI sequences	CSI, distance, SINR, interference
Model Complexity	High	Moderate	High	Low
Real-Time Suitability	No	No	No	Yes
End-to-End Control	Partial	Partial	–	Yes

1.2. Heuristic-Based Power Allocation Methods

Equal power allocation and max–min power allocation are representative heuristic power control methods frequently used as baseline comparisons in cell-free network studies. The equal power allocation strategy assigns the same amount of power to all APs, which then uniformly distribute it among their served UEs. Although this method is simple and computationally efficient, it fails to account for spatial variations in channel quality and user distribution, resulting in inefficient resource utilization and reduced SE. The max–min power allocation method enhances fairness by maximizing the minimum SINR across users, ensuring that even the weakest UEs maintain basic service quality; however, this fairness-driven approach can limit overall SE by underutilizing UEs with stronger channels. While max–min allocation is mathematically formulated as an optimization problem, it is often used as a practical benchmark due to its interpretability and fairness-centric allocation. Overall, heuristic-based methods are easy to implement but cannot dynamically adapt to channel conditions or user density, motivating the development of more flexible data-driven strategies.

1.3. CNN-Based Power Allocation

Convolutional neural networks have recently emerged as an effective alternative for wireless power control. Unlike traditional fully connected neural networks, CNNs are capable of extracting spatial dependencies among APs and UEs. By leveraging input features such as channel state information, AP-UE distance, interference level, and SINR, CNNs can learn context-aware and fine-grained power allocation strategies that adapt to real-time network conditions. However, most existing CNN-based models focus solely on AP-to-UE power allocation and assume fixed or uniform CPU-to-AP power distribution [11,12]. In these stage-wise architectures, the CPU allocates power equally to all APs regardless of their spatial location or connectivity to UEs, and only the AP-to-UE power split is learned by the model. This decoupled approach limits the ability to perform holistic optimization, especially in scenarios where certain APs require more power due to user density or channel conditions. Consequently, network performance may be suboptimal due to misaligned resource allocation at the AP level.

In contrast, our proposed CNN-based model performs end-to-end learning that jointly optimizes both CPU-to-AP and AP-to-UE power allocation within a unified framework. This allows the model to dynamically adjust the total transmit power at each AP as well as its distribution across UEs in a spatially adaptive manner. The proposed joint learning strategy captures global spatial correlations and enables system-wide optimization based on both topological and signal conditions. In addition to CNN-based methods, other learning-based approaches such as deep reinforcement learning (DRL) and graph neural networks have also been investigated. DRL methods such as deep q-networks and deep deterministic policy gradient offer long-term optimization under time-varying conditions, but require complex reward design and extensive training [13]. Graph neural network-based models are suitable for distributed scenarios and can represent AP-UE connectivity as graphs, but often decouple the hierarchical power control stages and may not be suitable for centralized architectures [14]. In comparison, our proposed CNN model provides an efficient and interpretable solution for centrally coordinated systems, achieving a strong balance between throughput, fairness, and energy efficiency while offering significantly lower complexity than alternatives such as GNNs that are better suited to decentralized architectures. In contrast to GNNs, CNNs excel in capturing spatial locality in centralized cell-free MIMO systems.

This study evaluates the proposed method against traditional heuristic schemes such as equal power allocation and max–min power allocation as well as against stage-wise CNN models. Simulation results demonstrate that the integrated CNN architecture consistently improves per-user SE while enhancing EE, confirming the effectiveness of joint spatially-aware learning in cell-free massive MIMO environments.

2. System Model

2.1. Network Architecture

This study considers a downlink power allocation model based on a cell-free massive MIMO network. In conventional cellular networks, each base station (BS) independently transmits signals and serves its own user equipment within a fixed coverage area. However, UEs located near the cell edge often experience poor communication quality due to strong inter-cell interference and imbalanced resource allocation. To address these issues, cell-free massive MIMO eliminates the concept of fixed cells and allows multiple distributed access points to cooperatively serve all UEs. Each AP is connected to a CPU by a wired backhaul, enabling centralized coordination and efficient signal processing (Figure 1).

The network model considered in this study consists of M APs and K UEs, with all APs operating under the control of the CPU. Unlike cellular systems, in which each BS serves its own UEs independently, cell-free massive MIMO adopts a cooperative downlink transmission strategy in which multiple APs jointly serve each UE. This coordination enhances signal quality while mitigating interference. Instead of independent transmission, APs follow centrally optimized transmission parameters determined by the CPU, improving the overall SINR and resulting in better system performance. Unlike idealized models that assume a uniform UE distribution, this study considers a more realistic scenario in which UEs are densely clustered in certain regions. In practice, UEs are often concentrated in urban centers, stadiums, or business districts, creating localized traffic hotspots. Within these clusters, the UEs are randomly distributed to reflect the stochastic nature of user mobility and spatial behavior. These nonuniform yet random user distributions significantly impact network load, and should be considered when designing power allocation strategies. The CNN-based model proposed in this paper explicitly accounts for such spatial patterns in order to ensure efficient power utilization under real-world deployment conditions [15].

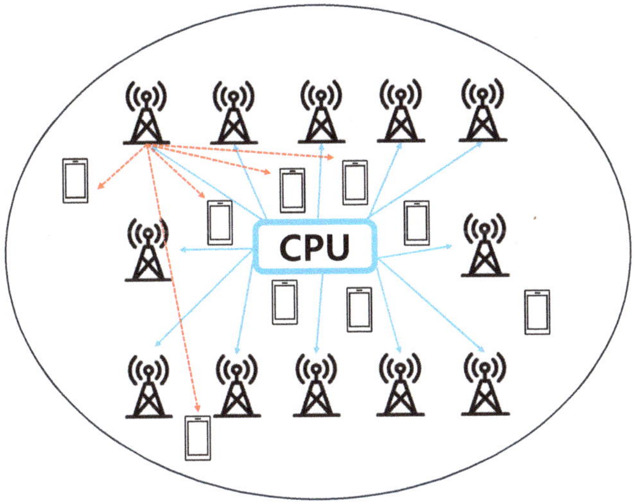

Figure 1. A cell-free massive MIMO downlink architecture in which a CPU coordinates power allocation across distributed APs.

In a cell-free massive MIMO network, AP cooperation plays a vital role in minimizing interference and enhancing both SE and EE. Unlike conventional systems where each BS handles power control independently, the CPU in cell-free networks can coordinate global power allocation across all APs. Moreover, UEs benefit from constructive signal combination from multiple APs through joint downlink beamforming, which improves the effective SINR and data rate. To address the challenge of downlink power allocation under this cooperative architecture, a CNN-based model is proposed that jointly optimizes CPU-to-AP and AP-to-UE power distribution. By enabling centralized power control, the proposed method can flexibly distribute power budgets across APs depending on their spatial demand and channel conditions. This joint optimization of CPU-to-AP and AP-to-UE stages enhances power efficiency and overall system performance compared to traditional rule-based schemes such as equal power allocation and max–min power allocation [16].

2.2. Channel Model

In the proposed cell-free massive MIMO system, the downlink wireless channel between APs and UEs is modeled as a Rayleigh fading channel. Each AP is equipped with a single antenna, and all APs are coordinated by a CPU responsible for power allocation. The small-scale fading component of the AP–UE channel follows a Rayleigh distribution, which is commonly used in rich scattering environments. The complex baseband channel coefficient between AP i and UE j is modeled as follows:

$$h_{i,j} \sim \mathcal{CN}(0, \beta_{i,j}) \tag{1}$$

where $\beta_{i,j}$ is the large-scale fading coefficient that accounts for path loss and shadowing. This coefficient is defined using a distance-dependent path loss model, as follows:

$$\beta_{i,j} = \frac{1}{1 + d_{i,j}^{\alpha}} \tag{2}$$

where $d_{i,j}$ is the distance between AP i and UE j and where α is the path loss exponent. Each AP transmits with an individual power level, and each UE receives the superposition of signals from multiple APs. The received signal at UE j can be expressed as follows:

$$y_j = \sum_{i=1}^{M} h_{i,j}\sqrt{p_{i,j}}x_i + n_j \tag{3}$$

where $p_{i,j}$ denotes the transmit power from AP i to UE j, x_i is the signal transmitted by AP i, and n_j represents additive white Gaussian noise (AWGN) at UE j, modeled as $n_j \sim \mathcal{CN}(0, \sigma^2)$. The downlink SINR at UE j is provided by [3]

$$\text{SINR}_j = \frac{\sum_{i=1}^{M} |h_{i,j}|^2 p_{i,j}}{\sum_{k\neq j}^{K} \sum_{i=1}^{M} |h_{i,k}|^2 p_{i,k} + \sigma^2}. \tag{4}$$

Using the SINR, the achievable rate for UE j is calculated based on the Shannon capacity formula as follows [17]:

$$R_j = B \log_2(1 + \text{SINR}_j) \tag{5}$$

where B denotes the available system bandwidth. As a system-level performance indicator, the average downlink SE is obtained by averaging the achievable rates across all UEs as follows:

$$R_{\text{avg}} = \frac{1}{K}\sum_{j=1}^{K} R_j. \tag{6}$$

While cellular networks can improve the summed SE by allocating more power to users with favorable channel conditions, this often results in unfair resource distribution and degraded performance for users at the cell edges. In contrast, cell-free massive MIMO systems aim to provide uniformly good service to all users through cooperative transmission and centralized power coordination. Therefore, the average SE serves as a more appropriate and fair metric for evaluating the overall performance of cell-free architectures.

2.3. Problem Formulation

In cell-free massive MIMO networks, efficient downlink power allocation plays a crucial role in achieving high SE and EE. The power control process involves two stages: power allocation from a CPU to each of the M APs, followed by power distribution from each AP to the K UEs that it serves. Let $P_{\text{CPU-AP}} = [p_1, p_2, \ldots, p_M]$ denote the CPU-to-AP power allocation vector and let $P_{\text{AP-UE}} = [p_{i,j}] \in \mathbb{R}^{M\times K}$ represent the AP-to-UE power allocation matrix, where $p_{i,j}$ is the power allocated from AP i to UE j. Max–min power allocation has traditionally been used as a fairness-oriented benchmark, aiming to maximize the minimum SINR across all UEs. However, this approach may limit performance for UEs with favorable channels. To overcome this limitation, a CNN-based power allocation framework is proposed which learns from max–min optimized data while incorporating additional spatial and contextual information such as channel state information, AP–UE distance, and interference conditions [18]. While trained on fairness-centric max–min results, the model learns to generalize beyond static optimization by inferring context-aware patterns. This allows it to adaptively allocate power in ways that improve both fairness and efficiency under realistic dynamic network conditions.

During training, the CNN minimizes the error between its predicted power allocations and the optimal values obtained from max–min optimization [11]. The loss function is defined as follows:

$$\min_{\theta} \frac{1}{N} \sum_{i=1}^{N} \left(\| P^{\text{CNN}}_{\text{CPU-AP},i} - P^{\text{opt}}_{\text{CPU-AP},i} \|^2 + \| P^{\text{CNN}}_{\text{AP-UE},i} - P^{\text{opt}}_{\text{AP-UE},i} \|^2 \right) \tag{7}$$

where $P^{\text{CNN}}_{\text{CPU-AP},i}$ and $P^{\text{CNN}}_{\text{AP-UE},i}$ are the CNN-predicted values and $P^{\text{opt}}_{\text{CPU-AP},i}$, $P^{\text{opt}}_{\text{AP-UE},i}$ are the max–min optimized references [3,7]. To generate predictions, the trained CNN takes as input a tensor consisting of spatial and signal features such as large-scale fading coefficients, AP–UE distances, channel gain magnitudes, and estimated interference levels. These features are stacked into a multi-channel matrix of size $C \times M \times K$, where C denotes the number of input feature types, M is the number of APs, and K is the number of UEs. The network outputs the CPU–AP and AP–UE power allocation values through two separate fully connected branches.

The optimization is subject to practical constraints. The power allocated from the CPU to each AP must lie within a valid range:

$$0 \leq p_i \leq P_{\text{CPU-AP,max}}, \quad \forall i \in \{1, \ldots, M\}. \tag{8}$$

In addition, the total transmit power from each AP to its served UEs should not exceed the power it receives from the CPU:

$$\sum_{j=1}^{K} p_{i,j} \leq p_i, \quad \forall i \in \{1, \ldots, M\}. \tag{9}$$

To ensure reliable communication quality, each UE should also satisfy a minimum SINR requirement as follows:

$$\text{SINR}_j \geq \gamma_{\min}, \quad \forall j \in \{1, \ldots, K\} \tag{10}$$

where $\gamma_{\min}$ denotes the minimum SINR threshold to guarantee acceptable service quality. In this study, $\gamma_{\min}$ is set as a design parameter based on typical quality of service requirements in downlink communication. This formulation establishes the power allocation problem as a constrained optimization task. In the next section, a learning-based solution using a CNN is presented to approximate this solution in a data-driven and scalable manner.

3. Proposed CNN-Based Power Allocation Method

This study proposes a CNN-based power allocation framework that jointly optimizes both CPU-to-AP and AP-to-UE power distribution in cell-free massive MIMO networks. Unlike conventional methods that allocate power in two disjoint stages, the proposed model performs end-to-end learning to dynamically assign power based on real-time network conditions and spatial context.

3.1. Motivation for Using CNN

Traditional power allocation schemes such as equal power allocation and max–min power allocation either lack adaptability or sacrifice efficiency. Heuristic methods do not account for spatial variations and channel diversity, while fairness-oriented strategies such as max–min improve service guarantees for weak users at the cost of degrading high-quality links. Moreover, many learning-based approaches focus only on AP-to-UE power allocation, treating CPU-to-AP power as fixed or uniform. This simplification neglects the reality that AP-UE link quality, distance, and interference vary significantly across the network. Consequently, uniform CPU-to-AP allocation can result in resource misalignment, as some APs may require more power to maintain reliable communication but be under-provisioned due to their static assignment.

To validate the design choice of using a CNN-only architecture, a CNN–LSTM variant was constructed by appending a recurrent layer to the convolutional feature extractor. Although the CNN–LSTM model has a smaller parameter count, it incurs significantly higher inference latency due to the sequential nature of LSTM operations [19].

Figure 2 presents a radar chart that compares the two models in terms of five normalized criteria, namely, inverse latency, inverse parameter count, real-time suitability, training stability, and architectural simplicity. To enable fair comparison, all scores are normalized to the range [0, 1]. The inference times for the CNN and CNN–LSTM models were measured as 9.68 ms and 18.29 ms, respectively, and the corresponding parameter counts were 34.7 million and 9.8 million. These values were normalized using inverse scaling, with each score computed as follows: $\text{Score}(x) = \min(x)/x$. The qualitative metrics were scored based on architectural depth, convergence behavior, and observed inference consistency. The CNN model demonstrates strong advantages in four out of five dimensions, particularly in inference latency and deployment efficiency, while the CNN–LSTM model benefits from a smaller parameter count. Table 2 provides the corresponding numerical results.

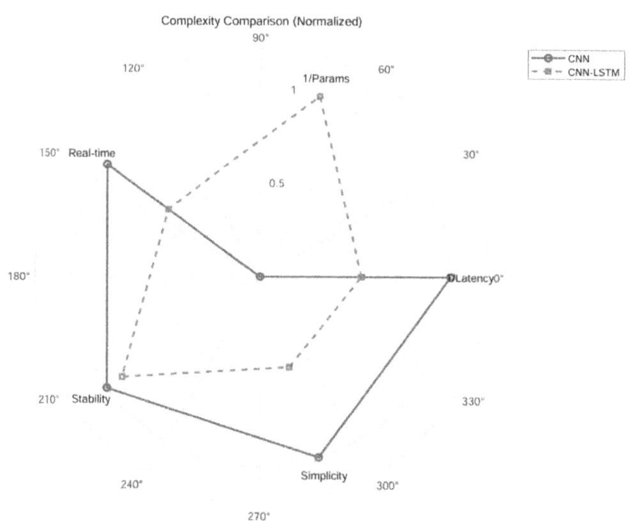

Figure 2. Radar chart comparing CNN and CNN–LSTM models.

Table 2. Comparison of parameter count and inference time

Model	Parameter Count	Inference Time (ms)
CNN	34.7M	9.68
CNN–LSTM	9.8M	18.29

These results indicate that the CNN model achieves a superior tradeoff between architectural simplicity and inference speed, making it more appropriate for real-time power allocation in per-snapshot cell-free massive MIMO systems.

CNNs are particularly well suited to wireless network applications due to their ability to extract local spatial correlations through shared weights and small receptive fields. In cell-free massive MIMO systems, where the spatial arrangement of APs and UEs directly impacts interference and link performance, CNNs can learn how geometry and environmental conditions affect power demand. This enables flexible and adaptive power allocation tailored to network topology. Building on these strengths, this study proposes a CNN model that simultaneously learns both stages of power allocation, i.e., CPU-to-AP and AP-to-UE, within a unified framework. By learning spatial relationships and channel patterns, the model determines power allocations based not only on instantaneous channel gain but also on the level of power that each UE requires to meet its SINR target under adverse conditions.

3.2. CNN Model Architecture and Feature Definition

The proposed CNN maps input network features to hierarchical power allocation outputs as follows:

$$[P_{\text{CPU-AP}}, P_{\text{AP-UE}}] = f_\theta(\mathbf{X}) \tag{11}$$

where $\mathbf{X} \in \mathbb{R}^{C\times M\times K}$ is the input tensor, θ denotes the CNN parameters, $P_{\text{CPU-AP}} \in \mathbb{R}^{M}$ is the CPU-to-AP power vector, and $P_{\text{AP-UE}} \in \mathbb{R}^{M\times K}$ is the AP-to-UE power matrix. The input tensor $\mathbf{X}$ consists of $C = 4$ feature channels, capturing spatial and contextual characteristics of the wireless environment. Each channel corresponds to a specific feature describing the AP–UE relationship, including the channel gain $|h_{i,j}|^2$, distance $d_{i,j}$, estimated interference at each link, and initial SINR estimated under uniform power allocation. These four feature maps are stacked into a three-dimensional tensor of size $4 \times M \times K$, enabling the CNN to process spatially correlated patterns across all AP–UE pairs. By feeding this structured input to the network, the model is able to capture both local channel conditions and global interference characteristics, which are essential for making informed power allocation decisions.

The CNN architecture consists of three convolutional layers, each with 32 filters and a kernel size of 3×3, followed by ReLU activation and batch normalization. This configuration was selected in order to balance computational efficiency and representation capacity. The 3×3 kernel size enables the model to capture fine-grained spatial features while maintaining a low parameter count, and 32 filters provides a practical baseline that offers sufficient expressiveness without incurring excessive complexity [20,21]. Pooling layers are omitted to preserve the spatial resolution of the AP–UE grid, which is critical for fine-grained link-specific power control. The final convolutional output is flattened and passed through two separate fully-connected branches; one branch generates the CPU-to-AP power vector while the other produces the AP-to-UE power matrix, both of which are reshaped to match their target dimensions. This dual-head architecture allows the model to jointly learn global and local power control strategies within a single end-to-end framework. A detailed description of the CNN model architecture and training procedure is provided in Appendix A to support reproducibility (Figure 3).

For comparison, a separate CNN was also trained using a stage-wise configuration in which the CPU-to-AP power is fixed equally across all APs and only the AP-to-UE allocation is learned. While this approach permits partial adaptability, it fails to account for heterogeneous AP conditions and spatial variations in user demand. In contrast, the proposed integrated CNN learns both allocation layers jointly, enabling globally optimized and context-aware power assignment tailored to dynamic network environments.

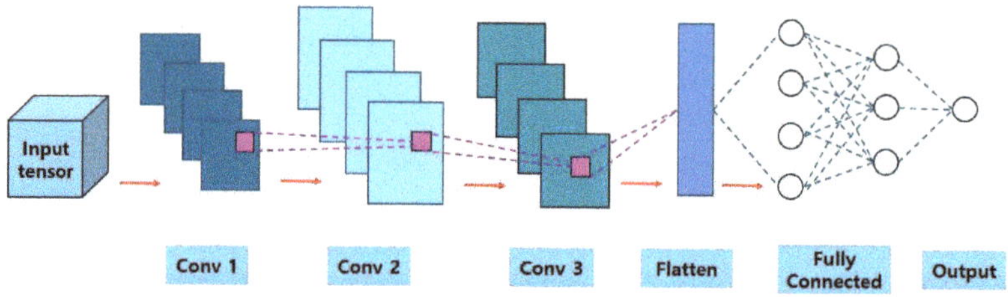

Figure 3. Overall architecture of the proposed CNN model.

3.3. CNN Training and Optimization Objective

To train the proposed CNN model, a supervised learning approach was adopted using data generated from traditional max–min power optimization. Specifically, the method introduced in [3] is followed, in which the downlink power control problem is solved under a max–min fairness criterion to maximize the minimum achievable SINR across all UEs under per-AP power constraints. This is achieved by iteratively transforming the problem into a sequence of second-order cone programs (SOCPs), which are globally optimal under convex feasibility. The resulting power allocation matrices provide fairness-centric ground truth for supervised training.

To simulate practical network environments, the dataset included diverse network realizations with user distributions clustered around high-traffic regions such as urban centers or event areas. Each training sample consisted of a spatial layout of APs and UEs, large-scale fading coefficients, and corresponding input features. These features are used to form the input tensor to the CNN, while the optimal power values from the max–min solution form the output supervision targets. The CNN model was trained on 10,000 samples generated under Rayleigh fading channels with a path loss exponent of $\alpha = 3.76$. Each sample encodes spatial features such as clustered UE distributions, AP-UE distances, channel gain magnitudes, estimated interference levels, and initial SINR values. Training was conducted for 50 epochs using the Adam optimizer, with a batch size of 32 and a learning rate of 10^{-3}. A validation split of 20% was considered to apply early stopping if necessary.

Figure 4 illustrates the training loss curve of the CNN model. The loss rapidly decreases within the first ten epochs and converges smoothly, remaining stable with minimal fluctuations throughout the remaining training process. This confirms that the model learns efficiently and is not prone to overfitting.

The CNN was trained to minimize the discrepancy between its predicted power allocations and the max–min references while ensuring compliance with power constraints. By learning from optimal yet computationally expensive power control solutions, the CNN generalizes to new scenarios with far lower inference complexity and supports end-to-end hierarchical power control beyond the original optimization formulation. The loss function combines two mean squared error (MSE) terms, corresponding to the CPU-to-AP and AP-to-UE power variables, along with a regularization term used to ensure that the sum of AP-to-UE power does not exceed each AP's CPU-assigned power [11]:

$$\mathcal{L}(\theta) = \frac{1}{N}\sum_{i=1}^{N}\left(\|P^{\text{CNN}}_{\text{CPU-AP},i} - P^{\text{opt}}_{\text{CPU-AP},i}\|_2^2 + \|P^{\text{CNN}}_{\text{AP-UE},i} - P^{\text{opt}}_{\text{AP-UE},i}\|_F^2\right) + \lambda\sum_{i=1}^{M}\max\left(0, \sum_{j=1}^{K}\hat{p}_{i,j} - \hat{p}_i\right) \quad (12)$$

where $\|\cdot\|_F$ is the Frobenius norm and λ is a hyperparameter controlling the penalty for violating AP power constraints.

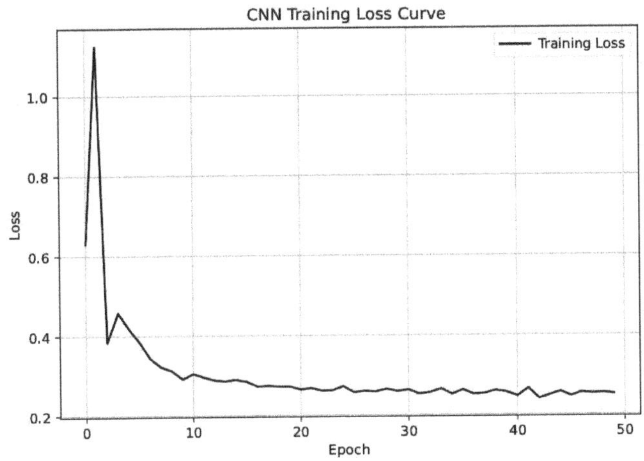

Figure 4. Training loss curve of the CNN model over 50 epochs.

3.4. Context-Aware Power Inference via CNN

The proposed CNN enables spatially adaptive power inference by interpreting complex network features in real time. Instead of simply reinforcing strong links, the model learns to identify UEs that require greater power due to unfavorable conditions such as weak channels, far distances from APs, or high interference levels. This behavior emerges as a result of supervised training based on max–min optimized power allocations, which inherently prioritize fairness by allocating more power to disadvantaged users. During training, the CNN receives input tensors representing spatial and signal features and minimizes the discrepancy between its predicted allocations and the max–min references. Through this process, the model implicitly learns the relationship between environmental conditions and required power levels to meet fairness-driven SINR targets. After the AP-to-UE power matrix is determined, the required CPU-to-AP power is derived by aggregating the per-AP transmission demands. This hierarchical inference allows the network to flexibly and efficiently allocate power while respecting hardware limitations and service constraints. In this way, the proposed design ensures improved user-level performance and EE through globally optimized and context-aware decision making.

4. Simulation Setup

To evaluate the effectiveness of the proposed CNN-based power allocation method, we conducted simulations in a downlink cell-free massive MIMO network under realistic deployment conditions. UEs were densely clustered in localized high-traffic regions, creating spatial imbalance and heterogeneous interference. The goal of the simulation was to verify how well the proposed method handles per-user performance in terms of SE and EE while minimizing power wastage and improving fairness. To emulate realistic network dynamics, each simulation snapshot assumed fixed user positions within a coherence time, with UE locations randomly redrawn across samples based on clustered spatial distributions. This structure reflects pseudo-mobility and variable load conditions, as is commonly observed in urban deployments, while also preserving quasi-static channel assumptions at the per-snapshot level.

Performance was evaluated using per-user metrics and visualized through cumulative distribution functions (CDF). These plots highlight not only the average performance but also how power allocation affects users at different percentiles. The per-user average SE and EE serve as key performance indicators, as the proposed model aims to improve not only total system throughput but also user-level quality of service under nonuniform and interference-limited conditions [4]. These metrics provide more meaningful insights into spatially adaptive power allocation behavior across the entire network.

4.1. Simulation Parameters

The simulated network consists of multiple APs connected to a CPU, serving multiple UEs distributed across the coverage area. The simulation parameters are summarized in Table 3 [3,6,16].

Table 3. Simulation parameters.

Parameter	Value
Number of APs (M)	64
Number of UEs (K)	20
System Bandwidth (B)	10 KHz
Noise Power Density (N_0)	-174 dBm/Hz
Path Loss Model	$\beta_{i,j} = \frac{1}{1+d_{i,j}^{\alpha}}$, where $\alpha = 3.76$
Channel Model	Rayleigh fading
UE Distribution	Clustered in high-density regions
AP Transmission Power ($P_{\mathrm{AP,max}}$)	200 mW
CPU Transmission Power ($P_{\mathrm{CPU,max}}$)	10 W
Minimum SINR Threshold ($\gamma_{\min}$)	0 dB
CNN Training Epochs	1000
CNN Learning Rate	1×10^{-3}

The CNN model was trained using the Adam optimizer with a batch size of 64 and a learning rate of 1×10^{-3}. A validation set comprising 20% of the training data was used for early stopping, and training was continued for up to 1000 epochs. All training samples were generated under random user distributions with nonuniform clustering to enhance generalization.

4.2. Comparison Methods

The performance of the proposed CNN-based method was compared with three baseline schemes: equal power allocation (EPA) assigns uniform power from the CPU to all APs and from APs to UEs regardless of their spatial or channel conditions; max–min power allocation is a fairness-oriented approach that allocates power to maximize the minimum SINR among all users; finally, the third method, referred to as "CPU–AP Equal + AP–UE CNN", applies equal power from the CPU to each AP while using a CNN to determine the AP-to-UE power distribution.

4.3. Evaluation Metrics

Per-user SE is defined as follows:

$$\mathrm{SE}_j = \log_2(1 + \mathrm{SINR}_j).$$

SE is measured in bits per second per Hertz (bps/Hz), based on the Shannon capacity formula; accordingly, the achievable data rate for user j is provided as follows:

$$R_j = B \cdot \mathrm{SE}_j$$

where B is the system bandwidth. This metric quantifies how efficiently each user's channel can carry information under the given SINR conditions.

EE is calculated for each user as follows:

$$\mathrm{EE}_j = \frac{R_j}{P_{\mathrm{total},j}}$$

where $P_{\mathrm{total},j}$ represents the total power consumed to serve user j, including both the CPU-to-AP and AP-to-UE power contributions [4]. This metric reflects how effectively power is used to support communication performance on a per-user basis. All metrics are evaluated at the per-user level and averaged across multiple randomized UE distributions to ensure statistically meaningful comparisons.

5. Performance Analysis

This section presents simulation results comparing the proposed CNN-based power allocation method with three baselines: EPA, max–min power allocation, and a stage-wise CNN approach in which the CPU-to-AP power is fixed equally and only the AP-to-UE power allocation is learned. Performance is evaluated in terms of per-user SE and EE and visualized using CDF plots. These CDFs illustrate both average performance and distribution across users, allowing for a detailed evaluation of fairness, adaptivity, and energy usage.

Figure 5 shows the CDF plot of the per-user SE. The CNN integrated method achieves the best overall performance, with a right-shifted curve compared to the other methods. A larger portion of users attain higher SE values, and the per-user average SE is also higher. However, the difference in SE between the two CNN-based methods is relatively small, as both effectively learn user-specific allocation patterns. The EPA method performs the worst, as it does not account for spatial or channel variability. The max–min method improves fairness but sacrifices overall throughput. The stage-wise CNN approach provides notable improvement over traditional baselines but is limited by its fixed CPU-to-AP allocation, which restricts full spatial optimization.

Figure 6 illustrates the CDF plot of the per-user EE. While SE and EE often trade off against each other, the CNN integrated model shows strong performance on both fronts. It achieves the most right-shifted EE curve, demonstrating that users can maintain high data rates with reduced power consumption. This is because the model learns to avoid unnecessary transmission by APs with poor link conditions, and allocates more power only where it is most effective. The stage-wise CNN model achieves similar SE but lower EE, as it cannot control how much total power is delivered to each AP. Even when some APs are underutilized, they still consume fixed power, reducing the overall EE. In contrast, the integrated model dynamically adjusts both AP-to-UE and CPU-to-AP allocations, enabling global energy savings without sacrificing throughput. EPA has the worst EE performance due to its indiscriminate power usage. Max–min guarantees minimum service levels, but is not energy-aware and often allocates excessive power to weak links.

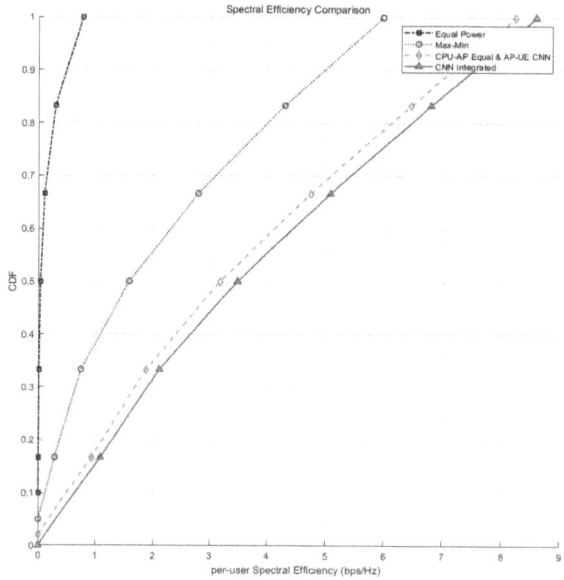

Figure 5. CDF plot of per-user spectral efficiency for four power allocation methods.

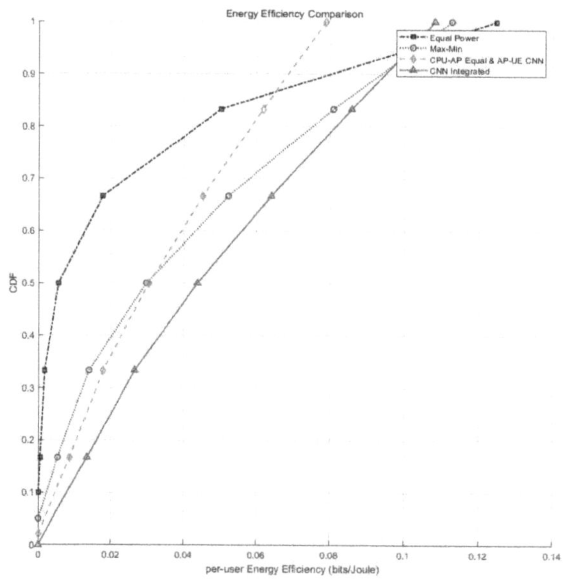

Figure 6. CDF plot of per-user energy efficiency for four power allocation methods.

Table 4 summarizes the per-user average and standard deviation for SE and EE across all methods. The proposed CNN model outperforms all baselines in terms of average SE and EE while also achieving the lowest standard deviation, indicating improved fairness and consistency across users.

Table 4. Per-user average and standard deviation of SE and EE.

Method	Avg SE (bps/Hz)	Std SE	Avg EE (bits/J)	Std EE
Equal Power	2.93	0.67	0.042	0.009
Max-Min	3.58	0.54	0.050	0.007
Stage-wise CNN	4.06	0.49	0.061	0.005
Proposed CNN	4.28	0.46	0.075	0.004

These results clearly demonstrate that although both CNN-based methods perform similarly in terms of SE, the CNN integrated model offers a clear advantage in terms of EE. By jointly learning both levels of power allocation, it avoids wasteful power assignments and adapts better to user and channel conditions across the network. The ability to optimize energy usage without degrading throughput confirms the effectiveness of the integrated approach, particularly in scenarios where EE is critical.

6. Conclusions

This paper proposed a CNN-based power allocation framework for cell-free massive MIMO networks. The proposed method jointly learns CPU-to-AP and AP-to-UE power allocation using spatially distributed input features such as CSI, interference levels, and AP–UE distances. Unlike traditional schemes such as equal power allocation and max–min power allocation, the proposed method performs end-to-end optimization that adapts power allocation dynamically across the entire transmission hierarchy. Simulation results under realistic UE clustering scenarios demonstrate that the CNN integrated model consistently achieves the highest performance in both SE and EE. While the SE performance of the two CNN-based methods is similar, the integrated model provides significantly better EE by avoiding unnecessary power use at underutilized APs. This shows that the proposed method not only improves throughput but also minimizes energy waste through globally coordinated power control.

Future work may explore the integration of CNN-based power control with user-centric clustering strategies in cell-free massive MIMO systems. As both power control and clustering decisions are centrally managed at the CPU level, their joint optimization can enhance overall system efficiency. Dynamically grouping APs based on user location and channel conditions may reduce unnecessary power consumption while maintaining quality of service, ultimately improving scalability and fairness in dense network deployments.

Author Contributions: Conceptualization, Y.-J.C. and J.-H.Y.; methodology, Y.-J.C. and S.-H.S.; software, Y.-J.C. and S.-G.C.; validation, Y.-J.C. and H.-K.S.; formal analysis, Y.-J.C.; investigation, Y.-J.C. and H.-Y.J.; resources, Y.-J.C. and J.-E.K.; data curation, Y.-J.C. and M.-S.B.; writing—original draft preparation, Y.-J.C.; writing—review and editing, Y.-J.C. and H.-K.S.; visualization, Y.-J.C. and Y.-H.Y.; supervision, Y.-H.Y. and H.-K.S.; project administration, M.-S.B. and H.-K.S.; funding acquisition, H.-K.S. All authors have read and agreed to the published version of the manuscript.

Funding: This research was supported by the Basic Science Research Program through the National Research Foundation of Korea (NRF) funded by the Ministry of Education (2020R1A6A1A03038540). This work was supported by Institute of Information & Communications Technology Planning & Evaluation (IITP) under the metaverse support program to nurture the best talents (IITP-2025-RS-2023-00254529) grant funded by the Korean Government (MSIT). This work was supported by an IITP (Institute of Information & Communications Technology Planning & Evaluation)–ITRC (Information

Technology Research Center) grant funded by the Korean Government (Ministry of Science and ICT) (IITP-2025-RS-2021-II211816).

Data Availability Statement: The original contributions presented in this study are included in the article. Further inquiries can be directed to the corresponding author.

Conflicts of Interest: The authors declare no conflicts of interest; the funders had no role in the design of the paper, in the collection, analysis, or interpretation of the data, in the writing of the manuscript, or in the decision to publish the results.

Appendix A. Detailed CNN Model Description

The proposed CNN-based power allocation model is implemented in PyTorch (version 1.10.0) and follows a dual-output architecture that simultaneously predicts both CPU-to-AP and AP-to-UE power distributions.

Input: Each sample is represented as a four-channel input tensor of size $M \times K$, where M and K denote the number of APs and UEs, respectively. The four input channels include:

- Channel gain magnitudes (from APs to UEs)
- AP–UE distances
- Interference estimates
- Initial SINR values.

Architecture:

- **Conv1:** $3 \rightarrow 16$ channels, 3×3 kernel, ReLU activation, batch normalization
- **Conv2:** $16 \rightarrow 32$ channels, 3×3 kernel, ReLU activation, batch normalization
- **Flatten:** Converts the 3D feature map to 1D vector of size $32 \times M \times K$
- **FC1:** Fully connected layer with 1024 neurons, ReLU
- **Dual output heads:**
  - **Head 1:** $1024 \rightarrow M \rightarrow$ CPU-to-AP power vector
  - **Head 2:** $1024 \rightarrow M \times K \rightarrow$ AP-to-UE power matrix.

Training Setup:

- Optimizer: Adam
- Learning rate: 10^{-3}
- Batch size: 32
- Epochs: 50
- Loss function: Dual MSE loss with regularization (see Equation (7))
- Supervision: Optimal power values derived from max–min optimization.

Output: The first output head predicts the CPU-to-AP power vector (M-dimensional), while the second predicts the AP-to-UE power matrix ($M \times K$). These outputs are constrained during postprocessing to meet per-AP power limits.

References

1. Ghosh, A.; Maeder, A.; Baker, M.; Chandramouli, D. 5G Evolution: A View on 5G Cellular Technology Beyond 3GPP Release 15. *IEEE Access* **2019**, *7*, 127639–127651. [CrossRef]
2. Ngo, H.Q.; Ashikhmin, A.; Yang, H.; Larsson, E.G.; Marzetta, T.L. Cell-Free Massive MIMO: Uniformly great service for everyone. In Proceedings of the 2015 IEEE 16th International Workshop on Signal Processing Advances in Wireless Communications (SPAWC), Stockholm, Sweden, 28 June–1 July 2015 ; pp. 201–205. [CrossRef]
3. Ngo, H.Q.; Ashikhmin, A.; Yang, H.; Larsson, E.G.; Marzetta, T.L. Cell-Free Massive MIMO Versus Small Cells. *IEEE Trans. Wirel. Commun.* **2017**, *16*, 1834–1850. [CrossRef]
4. Imoize, A.L.; Obakhena, H.I.; Anyasi, F.I.; Sur, S.N. A Review of Energy Efficiency and Power Control Schemes in Ultra-Dense Cell-Free Massive MIMO Systems for Sustainable 6G Wireless Communication. *Sustainability* **2022**, *14*, 11100. [CrossRef]

5. Chen, L.; Sun, F.; Li, K.; Chen, R.; Yang, Y.; Wang, J. Deep Reinforcement Learning for Resource Allocation in Massive MIMO. In Proceedings of the 2021 29th European Signal Processing Conference (EUSIPCO), Virtual, 27–30 September 2021; pp. 1611–1615. [CrossRef]
6. Björnson, E.; Hoydis, J.; Sanguinetti, L. Massive MIMO Networks: Spectral, Energy, and Hardware Efficiency. *Found. Trends Signal Process.* **2017**, *11*, 154–655. [CrossRef]
7. Kim, D.; Lee, I.H. Deep Learning-Based Power Control Scheme for Perfect Fairness in Device-to-Device Communication Systems. *Electronics* **2020**, *9*, 1606. [CrossRef]
8. Vaziri, A.; Moghaddam, P.S.; Shoeibi, M.; Kaveh, M. Energy-Efficient Secure Cell-Free Massive MIMO for Internet of Things: A Hybrid CNN–LSTM-Based Deep-Learning Approach. *Future Internet* **2025**, *17*, 169. [CrossRef]
9. Cui, Z.; Tan, F. Power Allocation Based on CNN and Long Short-Term Memory Networks in Massive MIMO Systems. In Proceedings of the 12th International Conference on Computer Engineering and Networks, Haikou, China, 4–7 November 2022; Liu, Q., Liu, X., Cheng, J., Shen, T., Tian, Y., Eds.; Springer Nature: Singapore, 2022; pp. 441–447.
10. Murshed, R.U.; Ashraf, Z.B.; Hridhon, A.H.; Munasinghe, K.; Jamalipour, A.; Hossain, M.F. A CNN-LSTM-Based Fusion Separation Deep Neural Network for 6G Ultra-Massive MIMO Hybrid Beamforming. *IEEE Access* **2023**, *11*, 38614–38630. [CrossRef]
11. Van Chien, T.; Nguyen Canh, T.; Björnson, E.; Larsson, E.G. Power Control in Cellular Massive MIMO with Varying User Activity: A Deep Learning Solution. *IEEE Trans. Wirel. Commun.* **2020**, *19*, 5732–5748. [CrossRef]
12. Perdana, R.H.Y.; Nguyen, T.V.; An, B. Deep Learning-based Power Allocation in Massive MIMO Systems with SLNR and SINR Criterions. In Proceedings of the 2021 Twelfth International Conference on Ubiquitous and Future Networks (ICUFN), Jeju Island, Republic of Korea, 17–20 August 2021; pp. 87–92. [CrossRef]
13. Malhotra, S.; Yashu, F.; Saqib, M.; Mehta, D.; Jangid, J.; Dixit, S. Deep Reinforcement Learning for Dynamic Resource Allocation in Wireless Networks. *arXiv* **2025**, arXiv:2502.01129.
14. Shen, Y.; Shi, Y.; Zhang, J.; Letaief, K.B. Graph Neural Networks for Scalable Radio Resource Management: Architecture Design and Theoretical Analysis. *IEEE J. Sel. A Commun.* **2021**, *39*, 101–115. [CrossRef]
15. Interdonato, G.; Björnson, E.; Ngo, H.Q.; Frenger, P.; Larsson, E.G. Ubiquitous Cell-Free Massive MIMO Communications. *EURASIP J. Wirel. Commun. Netw.* **2019**, *2019*, 197. [CrossRef]
16. Hamdi, R.; Driouch, E.; Ajib, W. Resource Allocation in Downlink Large-Scale MIMO Systems. *IEEE Access* **2016**, *4*, 8303–8316. [CrossRef]
17. Tse, D.; Viswanath, P. *Fundamentals of Wireless Communication*; Cambridge University Press: Cambridge, UK, 2005.
18. Sun, H.; Chen, X.; Shi, Q.; Hong, M.; Fu, X.; Sidiropoulos, N.D. Learning to optimize: Training deep neural networks for wireless resource management. In Proceedings of the 2017 IEEE 18th International Workshop on Signal Processing Advances in Wireless Communications (SPAWC), Sapporo, Japan, 3–6 July 2017; pp. 1–6. [CrossRef]
19. Patil, V.T.; Deore, S.S. Deep Learning-Driven IoT Defence: Comparative Analysis of CNN and LSTM for DDoS Detection and Mitigation. *J. Inf. Secur. Appl.* **2023**, *10*, 951. [CrossRef]
20. He, H.; Wen, C.K.; Jin, S.; Li, G.Y. Deep Learning-Based Channel Estimation for Beamspace mmWave Massive MIMO Systems. *IEEE Wirel. Commun. Lett.* **2018**, *7*, 852–855. [CrossRef]
21. Goodfellow, I.; Bengio, Y.; Courville, A. *Deep Learning*; MIT Press: Cambridge, MA, USA, 2016.

mathematics

MDPI

Review

Smart Healthcare Network Management: A Comprehensive Review

Farag M. Sallabi *, Heba M. Khater †, Asadullah Tariq †, Mohammad Hayajneh, Khaled Shuaib and Ezedin S. Barka

College of Information Technology, United Arab Emirates University,
Al Ain P.O. Box 15551, Abu Dhabi, United Arab Emirates
* Correspondence: f.sallabi@uaeu.ac.ae
† These authors contributed equally to this work.

Abstract: Recent developments in sensors, wireless communications, and data processing technologies are the main drivers for adopting the Internet of Things (IoT) in healthcare systems. IoT-based healthcare systems can enhance the quality of life significantly and help prevent the occurrence of health problems and epidemics. Deploying IoT-based healthcare on a massive scale raises several issues and challenges. One of the main challenges is the management of the end-to-end network connections of the IoT-based healthcare system. This paper presents a comprehensive survey of smart network management protocols that improve IoT-based healthcare efficiency, ensuring real-time monitoring, secure data transmission, and effective device management. Moreover, a reference architecture has been proposed for the network management of IoT-based smart healthcare systems to ensure the sustainability of service delivery to patients and caregivers. The architecture avoids health-related risks and anomalies by incorporating proper network management techniques and operational requirements pertaining to smart healthcare systems. This paper also discusses architectural implementation insights supported by new technologies such as software-defined networking (SDN) and deep learning (DL). Finally, this paper explores emerging paradigms to advance next-generation network management protocols for future smart healthcare systems.

Keywords: wireless body area network; Internet of Things; patient monitoring; smart healthcare; location contexts; network management; wearable devices

MSC: 68-02

1. Introduction

The widespread development of Internet of Things (IoT)-based smart systems and the integration of different infrastructure technologies has inspired several academic and industrial institutions to adopt IoT-based research and development and consider it in their strategic plans. Professionals estimate that the number of connected IoT objects will reach 500 billion by 2030 [1]. The development of IoT technologies stimulated the massive deployment of a wide range of intelligent small-form-factor sensing devices in various vertical applications. Sensing devices are characterized by constrained energy resources, limited processing capability, and limited Radiofrequency (RF) communication capabilities [2]. Sensing devices and sensor networks are being implemented and deployed in several domains, like homes, hospitals, factories, mobile phones, shopping malls, roadsides, and vehicles. Despite their benefits, some barriers still limit their social acceptance. Yang et al. [3] discussed these barriers and their solutions. The core concept of IoT is to enable smart sensors to interact autonomously, eliminating the need for human intervention and facilitating a new range of applications. IoT is anticipated to integrate various technologies, allowing physical objects to connect and support intelligent decision-making [4].

One of the key application domains of IoT is healthcare, where medical sensors track vital health indicators such as body temperature, blood glucose levels, and blood pressure [5]. By 2030, the healthcare sector is projected to have the most substantial global economic influence among IoT applications, with an estimated market value between USD 1.4 and 3.3 trillion annually [4]. This contribution is particularly notable given that the overall economic impact of IoT is expected to range from USD 5.5 to 12.6 trillion per year [6]. Advances in sensor technology, wireless communication, and data processing play a crucial role in enabling IoT-driven healthcare solutions. These sensors facilitate patient monitoring, daily activity tracking, and elderly care. The adoption of IoT in healthcare is expected to significantly improve the quality of life and aid in the prevention of health complications.

Smart systems are predictive and operate autonomously, interacting with the environment to make critical decisions based on the collected data. These systems are a cornerstone of Healthcare 4.0 [7], an evolution driven by the convergence of advanced technologies to enhance healthcare services and patient outcomes. Healthcare 4.0 leverages cutting-edge concepts like the Internet of Things (IoT), artificial intelligence (AI), big data analytics, cloud computing, edge computing, and cyber–physical systems to revolutionize healthcare delivery. Incorporating sensors and actuators into medical and environmental products provides healthcare professionals with precise tools to monitor and treat patients at the hospital and home. These smart systems improve patient care and integrate seamlessly with telemetric and tele-diagnostic infrastructure to significantly reduce decision response time. This integration is particularly valuable in containing epidemic diseases and optimizing healthcare expenditures.

Managing devices and networks in IoT-enabled smart healthcare systems differs significantly from traditional networks. Unlike general-purpose systems, healthcare systems must manage patient-related and environmental data with high precision and reliability. Fault management is critical to these systems, as failures can lead to severe consequences for patient health. Unlike traditional wireless sensor networks that rely on reactive fault-tolerant techniques such as redundancy, proactive network management leverages predictive analytics, machine learning, and real-time monitoring to prevent failures before they occur. By continuously analyzing sensor data, AI-driven fault detection systems can identify early indicators of potential device malfunctions and trigger preemptive maintenance or load balancing. This proactive approach ensures continuous system operation, minimizes downtime, and enhances the reliability of healthcare services [8]. Additionally, self-healing network architectures that dynamically adjust routing paths or reallocate computational resources improve resilience against unexpected failures, particularly in environments with life-critical applications.

Healthcare 4.0 brings about a paradigm shift in how healthcare systems operate. It emphasizes interoperability, scalability, and intelligence across all healthcare processes. Integrating cloud and edge computing enables distributed processing and real-time decision-making, reducing latency and improving efficiency. Additionally, advanced analytics and machine learning algorithms enhance the predictive capabilities of healthcare systems, enabling early intervention and personalized care. This holistic approach redefines healthcare into a smart, connected, and adaptive ecosystem [9].

Despite numerous research studies that have addressed IoT-based network management, there are several open issues, especially in the healthcare field. Smart healthcare systems' device and network management needs a well-defined design strategy before considering their technical problems. The strategy should consider all organizations involved in providing end-to-end services to end-users. For instance, if a healthcare provider remotely monitors a patient at home, the end-to-end connection includes different segments of network authorities. The first segment is located on the patient's premises, which usually consists of sensors, actuators, gateways, and Wi-Fi networks. The connection between the patient's premises and the healthcare provider may go through several Internet Service Providers (ISPs). Finally, the healthcare provider segment may include network elements, wireless and wired connections, and backend servers. Therefore, the entities involved should cooperate and support the end-to-end solution to guarantee a reliable and secure service. The design strategy should articulate all technical and medical requirements that should be handled in the implementation to cope with the constraints of IoT objects and the sensitivity of smart healthcare applications. In addition, deploying IoT-based smart healthcare systems on a massive scale would raise other issues. Patients and elderly individuals will rely on the service to monitor their health and daily wellness, while caregivers should provide reliable, secure, effective, and well-performed services. If an IoT-based smart healthcare system is not effective, reliable, and secure, patients' and elderly individuals' health becomes prone to risk. In addition, if it does not provide a reasonable return on investment, the caregiver's business model will be unsatisfactory. This research includes a literature review of IoT-based smart healthcare systems and management solutions referenced in [10]. We comprehensively reviewed network management protocols that suit IoT-based smart healthcare systems based on telecommunications management standards. Moreover, we propose a reference architecture to support network management protocols in smart healthcare network management. The main contributions of the paper are as follows:

1. This paper explores the integration of various smart network management protocols to enhance the efficiency and reliability of IoT-based healthcare systems. It evaluates protocols that facilitate real-time monitoring, secure data transmission, and efficient device management in resource-constrained environments.
2. This paper introduces a reference architecture designed to manage the end-to-end network of smart healthcare systems. It ensures sustainable service delivery for both patients and caregivers while addressing challenges such as fault management, security, and scalability. The architecture integrates emerging technologies like software-defined networking (SDN) and deep learning (DL) to enhance network adaptability and predictive management.
3. Finally, this paper explores emerging paradigms to advance next-generation network management protocols for future smart healthcare systems.

This paper is organized as follows: Section 2 presents the literature review, while Section 3 explores smart healthcare remote monitoring contexts. Section 4 discusses network management protocols, and Section 5 introduces the reference architecture for smart healthcare systems. Section 6 outlines the architecture implementation plan, detailing deployment strategies. Section 7 delves into Advanced Futuristic Network Management Protocols, addressing emerging techniques for enhancing network efficiency and reliability. The paper concludes with key findings and future research directions.

2. Literature Review

The vast deployment of the Internet of Things, Wireless Body Area Networks (WBANs), and the associated standards (e.g., IEEE 802.15.4 [11] and IEEE 802.15.6 [12]) have boosted the emergence of smart healthcare systems. Despite their acceptance in the healthcare sector, smart healthcare systems still face challenges that hinder their large-scale deployment. This section surveys scholarly research studies that tackle the large-scale deployment of smart healthcare systems. This section starts with a literature review of IoT and its architectures, applications, and protocols. Then, it reviews topics related to Wireless Body Area Networks, as these networks play an essential role in the context of smart healthcare systems. Finally, a discussion of the network management of IoT systems is presented.

2.1. Internet of Things

Interest in IoT has increased among industrial and academic institutions, which has necessitated the development of testbeds and platforms to evaluate IoT solutions. Gluhak et al. identified some requirements for IoT experimental facilities [13]. They surveyed currently available research testbeds and suggested new directions. The development of IoT-based systems is an incremental process starting from existing technologies and applications. IoT-based systems should comply at a minimum with these non-functional requirements: heterogeneity, scalability, ubiquitous data exchange through proximity wireless technologies, energy-optimized solutions, localization and tracking capabilities, self-organization capabilities, semantic interoperability as well as data management, and embedded security and privacy-preserving mechanisms [14].

Even though IoT is envisioned to augment the global Internet and make a unified mega-network that connects the physical and digital worlds, current Internet protocols may not apply to IoT due to its resource-constrained devices. Therefore, the IETF is developing new protocol suites tailored to the IoT paradigm [2]. Al-Fuqaha et al. [4] summarized the most relevant protocols and discussed some application issues related to IoT. They also discussed relations between IoT and emerging technologies like big data and cloud computing. Whitmore et al. [15] presented the current state of research on IoT through an intensive literature review. They identified current trends and described challenges that threaten IoT diffusion. Interested researchers in this area may consult the same paper for open research questions, future directions, and a comprehensive reference list.

As of today, IoT consists of heterogeneous devices, technologies, and protocols, making it difficult for hardware and software developers to develop applications that seamlessly integrate. This heterogeneity necessitates the involvement of middleware that supports standard functional and non-functional IoT requirements [16]. In addition, middleware provides hardware and software abstractions that facilitate developers' work so that they focus more on their applications. IoT-based applications generate considerable data (i.e., big data) that require storage, analysis, and ease of access. These requirements pose challenges for systems deployed within enterprises. Therefore, there is a need to integrate cloud computing and IoT [17]. This work provided a literature survey on integrating the cloud, IoT, and applications. They also discussed the available platforms for cloud and IoT integration.

Mobility management is an inevitable requirement for the IoT paradigm. Objects attached to people, vehicles, animals, etc., move with their hosts' bodies. Mobility represents a significant challenge for IoT, which requires extensive research to develop suitable algorithms and protocols to handle it. Ghaleb et al. [18] reviewed and discussed algorithms to address IP integration challenges with resource-constrained IoT devices. They also provided a comprehensive review encompassing the mechanisms, advantages, and disadvantages of related works on IPv6 mobility management. For areas with poor social infrastructures and information communication infrastructures, IoT can be deployed to provide an IoT-based mobility information network [19]. An IoT-based mobility management information network consists of fixed wireless nodes deployed on many roads (semi-electrostatic field sensor, acceleration sensor, gyro sensor, temperature sensor, humidity sensor, infrared sensor, and sensor server) and mobile nodes. Mobile nodes collect data captured by road sensors while moving along the roads and transmit them via the Internet to residents. Such transmitted data are intended to help in disaster prevention, sightseeing information, and shopping.

Fifth-generation (5G) mobile networks have become the preferred choice for connecting IoT devices to backend systems, offering capabilities far beyond those of fourth-generation (4G) networks [20]. While 4G supported the initial wave of IoT adoption, the rapid growth in IoT devices and applications soon exceeded its performance limits, particularly in bandwidth, latency, and overall network efficiency. The transition to 5G has significantly expanded the scope of IoT, offering remarkable speed and ultra-low latency. It provides tailored solutions to meet the diverse demands of IoT applications, particularly in healthcare, where real-time data transmission, remote monitoring, and telemedicine services are critical for improving patient outcomes and operational efficiency. Mahdi et al. [21] reviewed ongoing advancements in IoT, highlighting the transformative impact of 5G while exploring the potential of sixth-generation (6G) networks to meet future demands for ubiquitous IoT connectivity and enhanced service capabilities.

Stankovic [22] identified eight key research areas that tackle existing challenges and encourage further investigation. These areas include large-scale system expansion, architectural design, interdependencies in knowledge generation and big data, system resilience, openness, security, privacy, and human-in-the-loop considerations. Despite these advancements, numerous unresolved research challenges remain, primarily due to the complex nature of system deployment and the strict requirements imposed by various services utilizing these complex frameworks. Therefore, it is essential to examine existing standardization efforts, explore ways to enhance them, and identify research opportunities that will contribute to the advancement of the IoT field.

IoT influenced enterprises' and governments' digital strategies with disruptive business models and fast-changing markets. Many vertical industrial sectors, like transportation, healthcare, and logistics, have incorporated IoT into their digital strategy. IoT still experiences heterogeneity issues in devices, technologies, and protocols. These heterogeneity issues impose challenges on new IT strategies. Zimmermann et al. [23] described a new meta-model-based approach for integrating IoT architectural objects semi-automatically federated into a holistic digital enterprise architectural environment. The large-scale deployment of IoT in various vertical applications is still a challenge. The most important is the diversity of protocols developed for IoT and WSNs [24]. For a recent survey of IoT architectures, protocols, and challenges, the reader may refer to [25].

IoT provides several services that can be utilized to create many vertical applications. IoT services can be divided into four categories: identity-related services, information aggregation services, collaborative-aware services, and ubiquitous services [26]. IoT applications may use the services of one or more of these categories. Moreover, IoT services are essential to application developers to help them focus on their tasks, while the services are provided through some middleware technologies.

Cook et al. [27] presented a full-fledged smart home application based on the ZigBee WSN. This is a product in a box that can be easily installed in homes to monitor specific parameters. The product utilizes recent application protocols developed initially for WSNs and later adopted for IoT. Smart homes can be connected through their gateways and form a smart community [28]. An innovative community could provide useful services to its residents, specifically the elderly and residents with disabilities, by incorporating smart healthcare and connecting it to a community call center. Another exciting possible application of IoT is smart cities that allow municipalities to incorporate IoT into their cities [29].

2.2. Smart Healthcare

Smart IoT objects fulfill the requirements of ubiquitous communication, pervasive computing, and ambient intelligence [30]. These characteristics have encouraged scientific and industrial developers to propose services and applications for the healthcare industry. A direct and useful application of IoT in healthcare is the Ambient Assisted Living (AAL) application, which aims to help older people and people with disabilities and chronic diseases live their everyday lives without being restricted by conventional monitoring and assistive tools. Also, people can stay in their homes and receive treatment similar to that which they would receive if admitted to a hospital [31]. Bui et al. [32] inferred the non-functional requirements for smart healthcare applications through the storyline of a person with diabetes. They highlighted the requirements for interoperability, bounded latency, reliability, and security as the main ones for successful monitoring and supervision. IoT can be used to manage the administration of diabetic therapy by developing a personal device to assist and manage the insulin therapy dosage calculation [33]. The authors provided an overall architecture based on new communications protocols developed by the IETF that enable the connectivity of personal devices with the global Internet. A new concept called 'Health Internet of Things (HIoT)' was proposed to exploit sensor technologies and wireless networks for monitoring medical conditions [34]. This work presents an interconnection framework for mobile Health (mHealth) based on IoT.

Jung [35] explored how wearable and motion sensors, in the context of a smart home, can be used to monitor the movement of senior citizens by continuously logging and reporting their vital signs and the surrounding environmental conditions. It is worth mentioning that pervasive computing, as part of the IoT paradigm, is becoming an essential part of consumers' electronics and can be integrated with a smart home environment to adopt a smart-home-based healthcare model [36]. Collected data from various sensors deployed throughout the home can generate a smart home visualization system that provides different analyses to help detect and understand abnormal behaviors of targeted household residents [37]. Integrating mobile health as part of smart cities is becoming inevitable [38]. The authors provided an overview of how smart health can complement mobile health as part of a smart city initiative. They discussed the various aspects needed to achieve a comprehensive strategy for providing smart healthcare applications as part of a smart city environment. Ghose et al. [39] proposed a mobile software platform for ubiquitous healthcare support in homes and small clinic areas. The IoT platform replaces the PC gateway used in traditional telemedicine with a mobile device to ensure mobility support. Santos et al. [40] proposed an IoT-based mobile health and patient monitoring application. The proposed application is a set of health monitoring services deployed in a mobile gateway device. It collects real-time information about the user's heart rate, location, and environmental conditions and forwards them to an intelligent personal assistant platform called AMBRO. Pereira et al. [41] studied the performance of IoT-based mobile e-health applications and services.

Khattak et al. [42] conducted a comprehensive survey on the technologies and methodologies used for connecting and monitoring medical sensors. They introduced a testbed architecture specifically designed for healthcare applications. Similarly, Rahmani et al. [43] proposed the Smart e-Health Gateway, a system that facilitates the connection between sensor networks and the Internet. This gateway effectively addresses several challenges in ubiquitous healthcare systems, including energy efficiency, scalability, and reliability. Implementing Smart e-Health Gateways on a large scale can significantly enhance ubiquitous health monitoring, particularly in clinical settings. Additionally, they introduced a framework designed for various healthcare environments, including hospitals, specialized clinical settings, patient homes, senior citizen residences, gyms, and mobile healthcare services such as ambulances, mobile clinics, and travel health services.

Islam et al. [44] conducted a review of research on advancements in IoT-driven healthcare technologies. They analyzed cutting-edge network architectures, platforms, applications, and industry developments within IoT-based healthcare solutions. Furthermore, they proposed an intelligent collaborative security framework aimed at mitigating security threats. Their study also explored how emerging technologies such as big data, ambient intelligence, and wearable devices can enhance healthcare applications. Additionally, they examined global IoT and eHealth policies and regulations, assessing their economic and societal impact on sustainable development.

Farahani et al. [45] presented an architecture for applying IoT in healthcare and medicine. The architecture is patient-centric, where all constituents, such as the hospital, patients, and services, are seamlessly connected. The architecture consists of three layers, the device, fog computing, and the cloud, to empower handling complex data in terms of variety, speed, and latency. They also addressed the challenges of IoT eHealth, such as data management, scalability, regulations, interoperability, device–network–human interfaces, security, and privacy. Smart healthcare systems may proactively manage the patient's health—prevent disease, detect disease early, and improve healthcare outcomes [46]. Adame et al. [47] developed a platform for monitoring healthcare environments that integrates RFID and WSN technologies, providing the location, status, and tracking of patients and assets.

2.3. *WBAN*

The IEEE 802.15.4 standard, published in 2006, specifies the PHY and the MAC layers for short-range, low-power, low-cost, and low-bit-rate networks. These capabilities make it suitable for general WSN applications and IoT applications. However, IEEE 802.15.4 was not intended to support wireless communications in the vicinity of or inside the human body [12]. Therefore, IEEE 802.15 Task Group 6 initiated WBAN standardization activities in November 2007. The group recognized that existing standards did not fully meet the medical (i.e., proximity to human tissues) and relevant communication regulations for some application environments. The first draft was released in May 2010, and the final version was published in February 2012. IEEE 802.15.6 is targeted to serve a range of medical and non-medical applications.

Since its initial release, the IEEE 802.15.6 standard has undergone revisions to accommodate emerging use cases. The IEEE 802.15.6ma project is updating the standard to improve dependability for Human Body Area Networks (HBANs) and expand its applicability to Vehicle Body Area Networks (VBANs) [48]. Additionally, an amendment has been introduced to enhance the Ultra-Wideband (UWB) physical layer (PHY) and medium access control (MAC) to support more reliable communication in HBANs [49].

These advancements reflect ongoing efforts to refine WBAN standards to address modern medical and non-medical requirements better.

Cavallari et al. [50] surveyed the main WBAN applications, technologies, and standards, as well as issues related to WBAN design and evolution. They also presented a comparative study of the WBAN standards with simulation and experimental results. In another work [51], the authors surveyed the current state-of-the-art WBANs based on the latest standards and publications. They also discussed some open issues and challenges with WBANs. Healthcare systems in residential environments use WBANs for remote monitoring and health management. Ghamari et al. [52] provided an architecture for residential environment healthcare systems and stated the technologies and networks needed in such an environment.

2.4. IoT Network Management Protocols

The Simple Network Management Protocol (SNMP) is the standard management protocol for traditional TCP/IP networks. However, several issues make SNMP ineffective for IoT. Firstly, the communication overhead associated with SNMP is significant for low-bandwidth and low-power wireless IoT links. Secondly, SNMP follows a centralized management approach where the management station pulls the managed nodes at specified intervals. Thirdly, Management Agent and Management Information Base (MIB) object variables should be configured at each sensor node, which often has limited storage. Finally, SNMP may not adequately handle sensor-specific failures, which is common in IoT [53].

Sheng et al. [54] conducted a review of IoT network management and highlighted that, despite extensive research on various aspects of sensor networks, the management of sensor devices remains largely unexplored. To bridge this gap, a lightweight web service grounded in REST (Representational State Transfer) principles was introduced to streamline wireless sensor device management. Leveraging IPv6-based open standards, the service enables access to resource-restricted wireless networks. Specifically, it incorporates essential IETF protocols intended for such environments, including IPv6 over Low-Power Wireless Personal Area Networks (6LoWPANs), the Routing Protocol for Low-Power and Lossy Networks (RPL), and the Constrained Application Protocol (CoAP). A CoAP-driven device management method further simplifies the administration and accessibility of IPv6-capable sensor devices.

Given that end-to-end connectivity in IoT environments involves multiple networks with varying performance and connectivity requirements, sectional monitoring is suggested as a more effective management strategy [55]. Furthermore, a layered fault management scheme is recommended for end-to-end transmission to accommodate the heterogeneous nature of IoT networks.

Lindholm et al. [56] examined the feasibility of deploying SNMP (Simple Network Management Protocol) in IoT environments and found it to be highly challenging, despite its established usefulness as an Internet management protocol. By leveraging the CoAP protocol, they explored techniques that enable SNMP-based tools and applications to coexist and interact with CoAP-enabled endpoints. While SNMP has the potential to manage sensors in IoT networks [57], its implementation is impractical due to excessive memory and power consumption, as well as increased communication overhead.

Fault tolerance is a critical yet challenging non-functional requirement in IoT systems, particularly in healthcare applications. While fault tolerance mechanisms can be effectively applied to IoT applications such as smart homes [58], they may not be suitable for certain smart healthcare scenarios, such as implanted sensors. A study in [59] explores the opportunities and challenges of healthcare monitoring and management within the

IoT framework, incorporating cloud-based processing. The same study also reviews the current state and future directions for integrating remote health monitoring technologies into clinical medical practice.

Gia et al. [60] propose an IoT-based architecture that supports scalability and fault tolerance in healthcare applications. However, their approach primarily addresses faults in a reactive manner, meaning it lacks proactive mechanisms to assess the condition of sensor nodes and prevent failures in advance.

The widespread adoption of smart applications and services is expected to generate vast amounts of data, including text, audio, and video, necessitating batch, interactive, or real-time processing. One of the key challenges is the integration of heterogeneous data from multiple sources [61]. To address this, researchers have developed a flexible and cost-effective infrastructure that incorporates cloud and fog computing, blockchain technology, and message brokers to enable secure and private IoT deployment for smart healthcare applications and services. Another major challenge is efficiently managing this large volume of data throughout its lifecycle.

Gharaibeh et al. [62] take a data-centric approach, outlining key data management techniques designed to ensure the consistency, interoperability, granularity, and reusability of IoT-generated data in smart cities. Additionally, they identify security and privacy strategies while discussing the networking and computing technologies that support smart city implementations [63].

The Open Mobile Alliance (OMA) SpecWorks [64] developed a device management protocol called LightweightM2M (LwM2M). The protocol is designed for sensor networks and supports a machine-to-machine (M2M) environment. It has also been developed to manage lightweight and low-power devices on various networks, which is necessary to realize the potential of IoT. The LwM2M protocol features an architectural design based on REST. It defines an extensible resource and data model and builds on the CoAP standard, making it suitable for the remote management of M2M devices and related service enablement.

The Message Queuing Telemetry Transport (MQTT) protocol, widely used in IoT systems, is vulnerable to various cyberattacks, including replay and Man-in-the-Middle (MITM) attacks. A novel cyber range system was proposed in [65] to analyze vulnerabilities and implement countermeasures. This system provides an experimental environment for real-world attack scenarios, supporting new research to enhance MQTT protocol security and ensure robust communication in IoHT systems.

To address network congestion and reliability issues in IoHT, an adaptive data communication model was introduced in [66]. The model ensures efficient and reliable data transmission in high-traffic scenarios by dynamically switching between MQTT-SN and CoAP protocols based on network conditions.

A hybrid network management system combining the NETCONF protocol and decision tree algorithms was designed to improve configuration efficiency and fault diagnosis in IoHT networks [67]. This system achieves a fault diagnosis accuracy of 96%, outperforming traditional methods and demonstrating the potential of AI in enhancing network management scalability and adaptability.

A comparative analysis of the Lightweight Machine to Machine (LwM2M) and Extensible Messaging and Presence Protocol (XMPP) highlights their suitability for IoHT applications in [68]. While LwM2M offers efficient device management and low power consumption, XMPP supports robust messaging in constrained systems. The study recommends protocol selection based on specific IoHT application needs.

3. Smart Healthcare Remote Monitoring Contexts

The Internet of Things (IoT) extends Internet connectivity beyond conventional devices such as desktops, laptops, tablets, and smartphones, enabling a broader range of computing and communication-enabled objects. These objects, equipped with sensing, computing, and communication capabilities, can be integrated with both living and non-living entities. In the context of smart healthcare, such devices facilitate the collection of vital signs and physiological parameters, enabling continuous patient monitoring and timely medical interventions.

Smart healthcare is poised to transform the healthcare industry by offering personalized and adaptive healthcare environments. However, for the effective deployment and management of a robust smart healthcare system, it is crucial to understand the various contexts in which a patient may exist. This section categorizes the different contextual scenarios that a patient might encounter.

Most smart healthcare solutions advocate the use of a gateway between sensors and the backend server. This gateway can be either a dedicated device or a smartphone equipped with IoT and Wireless Sensor Network (WSN) protocols. By performing preliminary data processing and filtering, these gateways help minimize the volume of data transmitted to the backend. Rahmani et al. [43] proposed a framework designed for various healthcare settings, including hospitals, specialized clinical facilities, home care, senior citizen residences, gyms, and mobile healthcare environments such as ambulances, mobile clinics, and travel health services.

While the primary settings for smart healthcare systems are typically homes, nursing homes, and hospitals, additional contexts can be considered to provide patients with greater mobility and enable them to fully benefit from the IoT framework. Figure 1 presents both conventional and unconventional contexts where smart healthcare systems can be utilized. These contexts are categorized into three main groups, as outlined below.

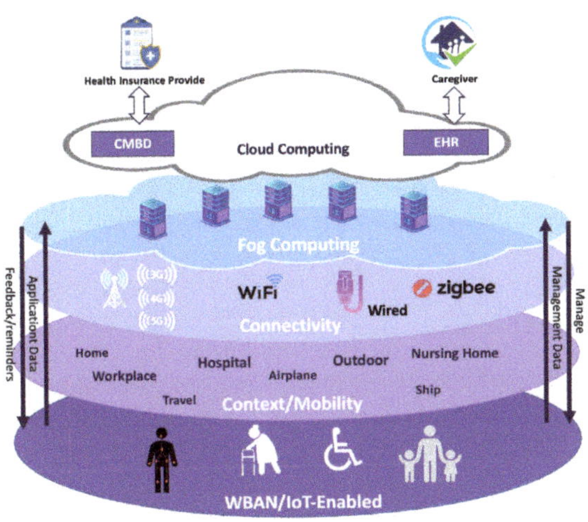

Figure 1. Typical and atypical location contexts.

3.1. Dominant Contexts (Typical)

These contexts refer to locations where patients spend the majority of their time, including homes, hospitals, and nursing homes. Such environments are characterized by a fixed IoT infrastructure. According to Khattak et al. [42], a sensor network deployed in a hospital—where the infrastructure is stable and continuously powered—differs significantly from a network designed for home-based patient monitoring, where the patient is in a remote setting. Home-based monitoring requires a more pervasive and unobtrusive approach to effectively manage patient care.

Various studies have explored patient monitoring systems in hospital settings [69] and personal monitoring at home [70]. In most IoT-driven patient monitoring solutions, a gateway serves as an intermediary between the sensor network and the Internet. This gateway typically handles essential tasks such as protocol translation between Internet-based systems and sensor networks.

3.2. Considered Contexts (Typical)

These contexts encompass, but are not limited to, the workplace and outdoor environments. Patients can continue their daily activities, including going to work, while being monitored for specific health conditions. Workplace monitoring can be facilitated through a dedicated gateway or by utilizing the patient's smartphone. When patients travel outside their homes but remain within the same country, connectivity to the backend system can be maintained through their mobile device or a specialized device provided by the service provider.

3.3. Possible Contexts (Atypical)

These contexts should be determined based on the patient's preferences and health condition. They may include travel-related locations such as airplanes, ships, or international destinations. When a patient travels outside their home country, maintaining connectivity to the healthcare system becomes essential. To address this, the service broker should facilitate seamless integration by connecting the patient to a local service provider in the visiting country. This connection can be established using the patient's mobile phone or a dedicated device that links them to the home network via the local provider.

4. Smart Healthcare Network Management Protocols

Network management protocols play a crucial role in healthcare systems, enabling the efficient monitoring, configuration, and maintenance of vast device networks. These protocols form the backbone of healthcare system management, ensuring reliable operation and seamless communication between devices and management systems. This section examines the primary protocols that facilitate IoT network management and their specific applications.

4.1. Simple Network Management Protocol (SNMP) in IoT for Healthcare

SNMP operates in a manager–agent architecture, where the manager queries or modifies the state of agents. In healthcare IoT, SNMP monitors devices like wearables and medical equipment, ensuring real-time tracking of device health, usage, and alerts. SNMP enables timely responses to device failures, improving patient safety. The protocol flow is described in Algorithm 1.

Algorithm 1 SNMP protocol flow

Initialize SNMP Manager and Agents with MIB definitions.
Establish a secure communication channel between Manager and Agents.
Manager sends a GET request to retrieve device status; Agent responds with data.
while Network is operational **do**
    Manager periodically polls Agents with GET requests.
    Agents respond with updated MIB values.
    if the threshold is exceeded or an event occurs **then**
        Agent sends a Trap notification to the Manager.
        Manager logs and processes the event.
    end if
    if configuration changes are required **then**
        Manager sends SET requests to modify MIB variables.
        Agent applies changes and sends an acknowledgment.
    end if
    if request fails or authentication is invalid **then**
        Agent responds with an error; Manager retries or logs the issue.
    end if
end while
Close the session and log activities for auditing.

Let M represent the manager, $A = \{A_1, A_2, \ldots, A_n\}$ the agents (IoT devices), and $D_i(t)$ the monitored data for agent A_i at time t. Here, f is a function representing the SNMP GET request operation, through which the manager M obtains monitored data $D_i(t)$ from an agent A_i at time t. The SNMP operations are defined as follows:

$$\text{GET operation: } D_i(t) = f(M, A_i, t) \tag{1}$$

$$\text{SET operation: } D_i^{\text{new}}(t) = g(M, A_i, V(t)) \tag{2}$$

where $D_i^{\text{new}}(t)$ explicitly represents the updated monitored data after applying the new parameters $V(t)$. Here, $V(t)$ represents the set of parameters or configuration values provided by the manager M at time t. These parameters are used in the SNMP SET operation to update or modify the settings or monitored data of the agent device A_i.

The Simple Network Management Protocol (SNMP) is widely used for network device monitoring and management in centralized healthcare settings, such as hospitals and medical data centers. It facilitates the remote tracking of IoT-enabled medical equipment like ventilators and infusion pumps, allowing healthcare providers to monitor their status and performance in wired, high-bandwidth hospital networks. Moreover, SNMP enhances hospital safety by generating automated alerts for critical device failures, enabling timely maintenance and reducing the risk of equipment malfunctions. This proactive management approach improves healthcare efficiency and ensures reliable medical device operation.

However, SNMP is not well suited for resource-constrained IoT healthcare environments, such as wearable health devices, Wireless Body Area Networks (WBANs), and home-based patient monitoring systems. Its centralized architecture and polling-based communication mechanism result in high bandwidth consumption, increased latency, and excessive energy overhead, making it inefficient for real-time monitoring in power-constrained medical IoT networks. Furthermore, frequent polling mechanisms introduce latency and unnecessary energy consumption in battery-powered medical sensors, limiting SNMP's applicability in ultra-low-power healthcare scenarios [71].

4.2. Message Queuing Telemetry Transport (MQTT)

MQTT [72] is a lightweight messaging protocol designed for low-bandwidth and resource-constrained environments. It operates on a publish/subscribe messaging model, where a central broker manages communication between publishers (data producers) and subscribers (data consumers), as shown in Figure 2. Initially developed for satellite communication, MQTT has become a cornerstone of IoT communications, offering efficiency and scalability. It offers several key features that make it highly suitable for IoT-based healthcare network management [73]. One of its most significant advantages is minimal packet overhead, with packet sizes as small as 2 bytes, ensuring efficient data transmission over low-bandwidth networks. Another crucial feature is its support for different quality of service (QoS) levels, allowing for flexible message delivery reliability. QoS 0 (at most once) sends messages without requiring acknowledgment, making it ideal for non-critical data. QoS 1 (at least once) ensures message delivery with acknowledgment, preventing data loss while avoiding duplicates. QoS 2 (exactly once) guarantees that messages are received and processed only once, ensuring the highest level of reliability for critical applications. Additionally, MQTT includes a Last Will and Testament mechanism, which notifies the broker in case of an unexpected device disconnection. This feature enhances fault detection and allows for timely interventions. Another essential capability is persistent sessions and queuing, which ensures that messages are stored and delivered even when devices experience intermittent connectivity. These features collectively enhance MQTT's reliability, making it a robust and efficient protocol for managing healthcare IoT networks.

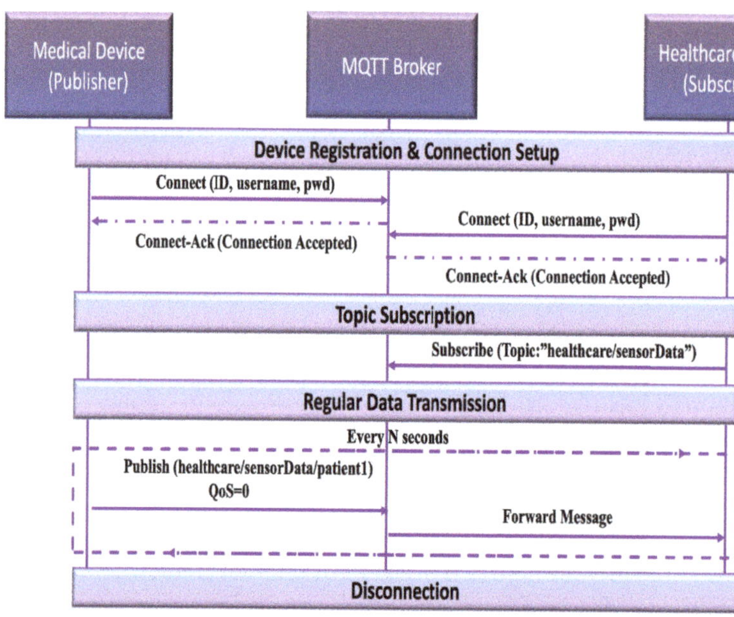

Figure 2. MQTT protocol.

In network management, MQTT facilitates real-time monitoring by allowing devices to publish telemetry data, enabling network managers to track device status, health, and performance continuously. The protocol also ensures reliable command execution, as messages sent by network managers to devices are delivered based on QoS levels, providing appropriate reliability for various applications. Furthermore, MQTT's Last Will and Testament feature allows the network to detect disconnected devices and alert administrators for prompt remedial action. Its scalability is another major advantage, efficiently handling communication in large-scale IoT networks with thousands of connected devices while maintaining minimal overhead. These features make MQTT an essential protocol for managing healthcare networks, ensuring reliable and efficient communication across medical devices and systems. However, it lacks robust security features, making it vulnerable to cyber threats [72].

4.3. NETCONF/RESTCONF in IoT for Healthcare

NETCONF [74] and RESTCONF protocols offer effective mechanisms for configuring and monitoring healthcare IoT networks. Utilizing structured data formats such as XML and JSON, they automate network updates, thus ensuring interoperability among diverse healthcare devices.

Configuration updates across devices in the network can be represented by a set:

$$C = \{X_1, X_2, \ldots, X_n\}, \quad \text{where } X_i \text{ is the configuration parameter for device } A_i. \tag{3}$$

The manager performs updates to a device by employing a remote procedure call (RPC) as follows:

$$C_i' = f(M, X_i, \text{Update}), \tag{4}$$

where $f(\cdot)$ is a function describing the update process, M denotes the managing entity, and X_i specifies the current configuration parameters of device A_i. Typically, $f(\cdot)$ is implemented as an integral operation defined over time or update steps, represented as

$$C_i' = \int_{t_0}^{t_f} g(M, X_i(t), \text{Update}(t))\, dt, \tag{5}$$

where $g(\cdot)$ describes the instantaneous rate or mechanism of configuration change, and t_0 and t_f are the initial and final limits indicating the duration of the update process.

The NETCONF and RESTCONF protocols streamline remote configurations, enhancing the seamless integration and interoperability of IoT diagnostic devices in healthcare settings. Additionally, they facilitate secure firmware updates, thus bolstering both security and operational performance. These protocols also provide flexibility in dynamically adjusting configurations during emergencies, ensuring continuous connectivity and efficient operation within healthcare facilities. However, these protocols are relatively complex and may not be ideal for low-power healthcare IoT devices, as they require higher computational resources and introduce parsing overhead due to their reliance on structured data formats like XML and JSON [75].

4.4. gRPC in IoT for Healthcare

gRPC [76] is a communication framework built on top of HTTP/2 that enables high-throughput, bidirectional data exchange between healthcare IoT devices and backend systems. This approach supports both request–response interactions and continuous data streaming, ensuring minimal transmission delays for critical patient information. The Request–Response Model gRPC is defined as follows:

$$R(t) = \{\text{Req}(t), \text{Resp}(t)\}, \tag{6}$$

where $R(t)$ captures a discrete request at time t and its corresponding response. In a healthcare context, each $\mathrm{Req}(t)$ may represent a specific query for clinical data, while $\mathrm{Resp}(t)$ holds the information returned by the device. The Streaming Model is defined below:

$$S(t) = \int_{t_0}^{t_1} f(\text{Metrics}, t)\, dt, \tag{7}$$

where $f(\text{Metrics}, t)$ denotes a time-varying function of health-related data (e.g., heart rate or ECG signals). The integration from t_0 to t_1 aggregates these continuous measurements, which are then delivered in near real time via gRPC streams. Because healthcare applications often require immediate insights, we focus on end-to-end latency. Let T_i^s be the sending time for a data chunk i and T_i^r its receiving time. Then, the latency L_i for chunk i can be defined as

$$L_i = T_i^r - T_i^s. \tag{8}$$

$$\overline{L} = \frac{1}{N} \sum_{i=1}^{N} L_i \tag{9}$$

Minimizing the average latency is crucial to ensure fast feedback and timely clinical decision-making.

Algorithm 2 NETCONF-Light

Algorithm.

Require: Set of capabilities $C = \{c_1, c_2, \ldots, c_n\}$, client request $R = \{r_1, r_2, \ldots, r_m\}$
Ensure: Processed operations P, unsupported operations E

Initialize supported capabilities:

$$S \leftarrow \{c \in C \mid \text{device supports } c\} \tag{10}$$

Advertise S during `<hello>` message exchange.
Receive client request:

$$R \leftarrow \{r \in C \mid \text{requested by client}\} \tag{11}$$

Validate requests:

$$P \leftarrow R \cap S \quad \text{(supported operations)} \tag{12}$$

$$E \leftarrow R - S \quad \text{(unsupported operations)} \tag{13}$$

for all $p \in P$ **do**
    Execute operation $f(p)$
end for
for all $e \in E$ **do**
    Notify client: `"Operation` e `not supported."`
end for
if device capabilities change **then**
    Update S:

$$S \leftarrow S' \quad \text{(new capabilities set)} \tag{14}$$

    Re-advertise S during the next `<hello>` exchange.
end if

4.5. NETCONF-Light: A Lightweight Protocol for Resource-Constrained Devices

NETCONF-Light [77] is a lightweight version of the NETCONF protocol specifically designed for resource-constrained devices. It provides a streamlined subset of NETCONF operations to facilitate configuration management, including the ability to install, manipulate, and delete network configurations as described in Algorithm 2 . The key operations of NETCONF-Light (O) can be expressed as

$$O_{NETCONF} = \{\text{install}, \text{modify}, \text{delete}\} \tag{15}$$

These operations are applied to device configurations (D), and their effect is to transition the device to a desired configuration state:

$$O_{NETCONF} : D \rightarrow \text{State} \tag{16}$$

where

- D: the set of device configurations.
- State: the target state after applying the operations.

4.6. 6LowPAN-SNMP: Lightweight Management for Resource-Constrained Networks

6LowPAN-SNMP [78] adapts the Simple Network Management Protocol (SNMP) for IPv6 Low-Power Wireless Personal Area Networks (6LowPANs). It introduces mechanisms such as SNMP header compression, as described in Algorithm 3 , to reduce message overhead, making it suitable for resource-constrained networks. Compatibility with standard SNMP is maintained through a proxy forwarder that translates between SNMP and 6LowPAN-SNMP messages. The header compression in 6LowPAN-SNMP can be mathematically modeled as

$$C_{SNMP} = \frac{H_{SNMP}}{R_{6LowPAN}} \tag{17}$$

where

- C_{SNMP}: compression ratio of the SNMP header.
- H_{SNMP}: original header size.
- $R_{6LowPAN}$: reduced header size after compression.

Applications of 6LoWPAN-SNMP include enabling the lightweight monitoring and management of sensor nodes, facilitating efficient data collection, and detecting sensor failures, transmission errors, and battery levels to ensure reliable operation.

4.7. CoAP Management Interface (CoMI)

The CoAP Management Interface (CoMI) [79] is a specialized protocol for managing IoT devices and networks with constrained resources. It uses CoAP (Constrained Application Protocol) as a transport layer to perform operations on device resources defined in YANG or converted from SMIv2 to YANG. CoMI focuses on lightweight device management and supports actions such as configuring, monitoring, and updating devices.

Algorithm 3 6LowPAN-SNMP
compression and message handling.

Require: SNMP message M_{SNMP}, header size H_{SNMP}
Ensure: Compressed message $M_{6LowPAN}$

Input SNMP message M_{SNMP} with header size H_{SNMP}
Apply header compression:

$$R_{6LowPAN} \leftarrow H_{SNMP} \times \text{compression factor} \tag{18}$$

Calculate compression ratio using Equation (15):

$$C_{SNMP} \leftarrow \frac{H_{SNMP}}{R_{6LowPAN}}$$

Generate compressed message $M_{6LowPAN}$ with reduced header size $R_{6LowPAN}$
Send $M_{6LowPAN}$ to the destination over the 6LowPAN network.
if proxy translation is required **then**
    Forward $M_{6LowPAN}$ to the SNMP proxy.
    Translate $M_{6LowPAN}$ back to M_{SNMP} for compatibility with standard SNMP devices.
end if
return Compressed message $M_{6LowPAN}$

In CoMI, resource management can be represented using a mapping between YANG-modeled resources $\mathcal{R}_y$ and their CoAP representations:

$$\mathcal{R}_y = \{\text{res}_1, \text{res}_2, \ldots, \text{res}_m\}, \quad \text{res}_i \mapsto \text{URI}_{\text{CoAP}} \tag{19}$$

where each resource is identified by a URI. Operations such as `GET`, `POST`, `PUT`, and `DELETE` are performed over CoAP to manage resources.

CoMI is widely used in *smart home systems*, *environmental monitoring*, and *wearable device configurations*, where lightweight communication is vital. However, despite its efficiency, CoMI faces scalability challenges in complex healthcare IoT networks, where diverse medical devices require high levels of security, fault tolerance, and real-time interoperability [80].

4.8. Lightweight Machine to Machine (LWM2M)

LWM2M (Lightweight Machine to Machine) [81] is a client-server protocol specifically designed for managing IoT (Internet of Things) devices with low power and constrained resources. Developed by the Open Mobile Alliance (OMA), it builds upon protocol and security standards from the Internet Engineering Task Force (IETF). The protocol supports functionalities such as *connectivity monitoring*, *resource monitoring*, and *firmware upgrades*, which are essential for managing IoT ecosystems efficiently.

The LWM2M architecture follows a hierarchical structure. The *LWM2M server* resides on the network manager device, while *LWM2M clients* operate on managed IoT devices. These devices organize their resources into objects. For instance, a *location object* encompasses resources that provide geolocation information. This modular structure facilitates the efficient management of resources.

A mathematical representation of LWM2M operations can be modeled as resource mappings:

$$R = \{r_1, r_2, \ldots, r_n\} \tag{20}$$

where R represents the set of IoT resources organized into objects. The protocol supports operations like read (O_r), write (O_w), and execute (O_e):

$$O_r : R \rightarrow V, \quad O_w : V \rightarrow R, \quad O_e : R \rightarrow \text{Action} \tag{21}$$

Here, V denotes the value space of a resource.

Applications of LWM2M in healthcare include the remote management of medical devices, ensuring timely software updates, continuous performance monitoring, and secure communication. While LwM2M is designed to manage constrained IoT devices and provide efficient remote device management, it may face performance limitations in large-scale deployments [82].

4.9. Technical Report 069 (TR-069)

TR-069 [83], also known as CPE WAN Management Protocol (CWMP), is a protocol designed for the remote management of customer premises equipment (CPE), including IoT devices. It provides a comprehensive framework for device provisioning, diagnostics, and maintenance in IoT networks. The protocol features auto-configuration capabilities, allowing IoT devices to be remotely configured and updated without manual intervention. Its diagnostic features enable administrators to monitor device health and troubleshoot issues efficiently. TR-069's secure communication framework ensures data integrity and confidentiality during remote management. It uses a hierarchical management model where a centralized Auto-Configuration Server (ACS) communicates with devices to perform operations such as firmware updates, configuration changes, and performance monitoring. The protocol is particularly suitable for large-scale IoT deployments due to its scalable architecture, enabling the efficient management of thousands of connected devices.

TR-069 plays a crucial role in network management by enabling remote device configuration and provisioning, ensuring consistency and scalability across large IoT deployments. With its auto-configuration capabilities, administrators can remotely set up devices without requiring manual intervention, streamlining operations in dynamic network environments. One of its key advantages is support for over-the-air firmware updates, which help maintain device security and functionality by ensuring that all connected devices run the latest software versions. Additionally, TR-069 provides real-time device health monitoring and remote diagnostics, allowing network administrators to quickly identify and troubleshoot issues, minimizing downtime and enhancing overall network efficiency. Its scalable architecture makes it particularly well suited for managing large IoT deployments in geographically distributed networks, ensuring seamless and centralized control over numerous connected devices.

In healthcare, TR-069 is widely used to manage the connectivity and firmware updates of medical devices, ensuring they remain functional and secure. Automated device updates play a crucial role in maintaining compliance with the latest security protocols, safeguarding sensitive patient data and protecting against cyber threats. Furthermore, TR-069 facilitates the real-time monitoring of device health and usage patterns in hospital environments, helping healthcare providers optimize device performance and proactively address potential failures. By ensuring the reliable and efficient management of medical IoT devices, TR-069 contributes to improved patient care and operational efficiency in healthcare networks.

4.10. Open Mobile Alliance Device Management (OMA DM)

OMA DM [84] has evolved from its mobile device management roots to address IoT device management needs. The protocol's standardized management objects and session-based communication provide a robust framework for remote device administration. OMA DM's strong security features and extensible framework enable comprehensive device configuration management, software updates, and security policy enforcement in IoT environments. The protocol flow, as shown in Figure 3, begins with mutual authentication between the server and the mobile device, ensuring a secure connection is established. After authentication, the device transmits its capabilities and authentication details to the server, allowing the server to determine the appropriate management operations required for the device. The server then issues management commands, which the device processes and executes. These commands may include updates to network configurations, the implementation of security policy changes, or software and firmware updates. Following the execution of these tasks, the device reports the results back to the server. Finally, the session concludes with the proper cleanup of resources to maintain operational efficiency.

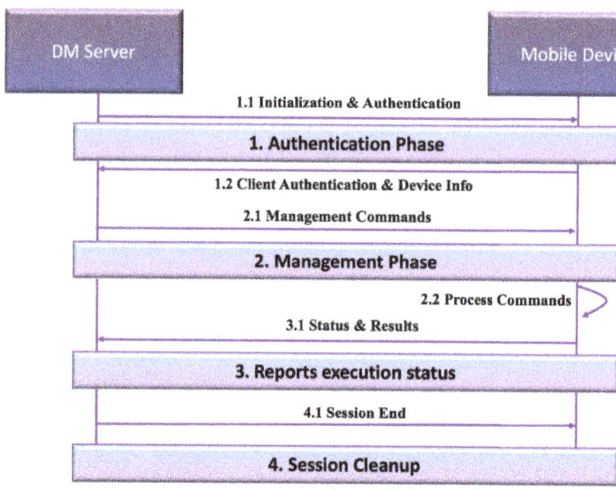

Figure 3. OMA DM protocol.

OMA DM plays a vital role in IoT network management by enabling centralized remote administration, significantly reducing the need for on-site interventions. Through its efficient device configuration capabilities, the protocol ensures consistent network operations by remotely pushing updates and enforcing security policies. One of its key features is the ability to deliver firmware updates over the air, keeping connected devices secure and ensuring they operate with the latest software. Additionally, OMA DM facilitates task monitoring, where devices report execution results back to the server, allowing network administrators to track performance, identify potential issues, and streamline troubleshooting. This remote management capability enhances operational efficiency while ensuring the security and reliability of IoT networks.

In the healthcare sector, OMA DM is widely used for the remote management of wearable healthcare devices, such as glucose monitors and heart rate sensors, ensuring they function optimally and securely. The protocol also enables the secure configuration and monitoring of critical medical equipment, including infusion pumps and patient monitors, which require continuous oversight to maintain accuracy and reliability. Furthermore, OMA DM supports remote diagnostics for advanced medical equipment, such as MRI machines and ventilators, allowing healthcare providers to detect and resolve potential issues before they impact patient care. By ensuring seamless device management and security enforcement, OMA DM enhances the efficiency and reliability of healthcare IoT networks.

4.11. Representational State Transfer (RESTful)

RESTful (Representational State Transfer) [85] is an architectural style widely adopted for IoT network management and device communication. Built on top of the HTTP protocol, RESTful APIs offer a lightweight, stateless, cacheable, and uniform interface for interacting with IoT devices. The protocol's resource-oriented design uses standard HTTP methods—GET (retrieve resources), POST (create resources), PUT (update resources), and DELETE (remove resources)—to enable seamless device management, configuration, and monitoring.

Figure 4 illustrates how RESTful APIs enable efficient communication between a medical device and a WBAN (Wireless Body Area Network) controller. Initially, the medical device undergoes initialization, where it gathers its metadata (e.g., ID, type, and network details). The device then registers with the WBAN controller through a POST request, which is validated and forwarded to the API server. Configuration updates, such as network adjustments, are handled via PUT requests, while GET requests retrieve performance metrics for bottleneck detection and troubleshooting. This flow highlights the stateless and resource-oriented nature of RESTful APIs, which are particularly suited for lightweight, scalable IoT environments like healthcare networks. Their simplicity and widespread adoption make RESTful APIs an ideal choice for cloud-based IoT management systems.

RESTful APIs have become a fundamental component of modern network management, providing a flexible and efficient approach to device communication. One of their key applications is device monitoring, where GET requests enable real-time retrieval of critical information. Additionally, RESTful APIs support configuration management by allowing administrators to remotely update device settings using PUT or POST requests, ensuring seamless adjustments without manual intervention. They also facilitate data collection, enabling scalable telemetry data retrieval from IoT devices for performance analysis and optimization. Moreover, RESTful APIs enhance cloud integration by simplifying communication between IoT devices and cloud platforms, allowing centralized control and analytics. Their stateless operation ensures lightweight and efficient communication, making them particularly suitable for resource-constrained devices by eliminating the need for maintaining session states. These capabilities contribute to a more scalable, responsive, and efficient approach to managing IoT networks.

They are widely utilized in healthcare network management to enhance performance monitoring, device connectivity, and fault detection. One key application is the ability to monitor healthcare network performance, allowing administrators to identify bottlenecks and troubleshoot issues in real time. By continuously retrieving network metrics and analyzing device data, RESTful APIs help maintain optimal system functionality. Additionally, they play a crucial role in managing the connectivity of medical devices in Wireless Body Area Networks (WBANs) by enabling the registration of new medical devices and updating their network configurations remotely. This ensures seamless integration and reliable communication within healthcare environments. Furthermore, RESTful APIs contribute to fault detection by identifying connectivity failures and triggering automated recovery mechanisms, ensuring continuous operation and minimizing disruptions in healthcare networks.

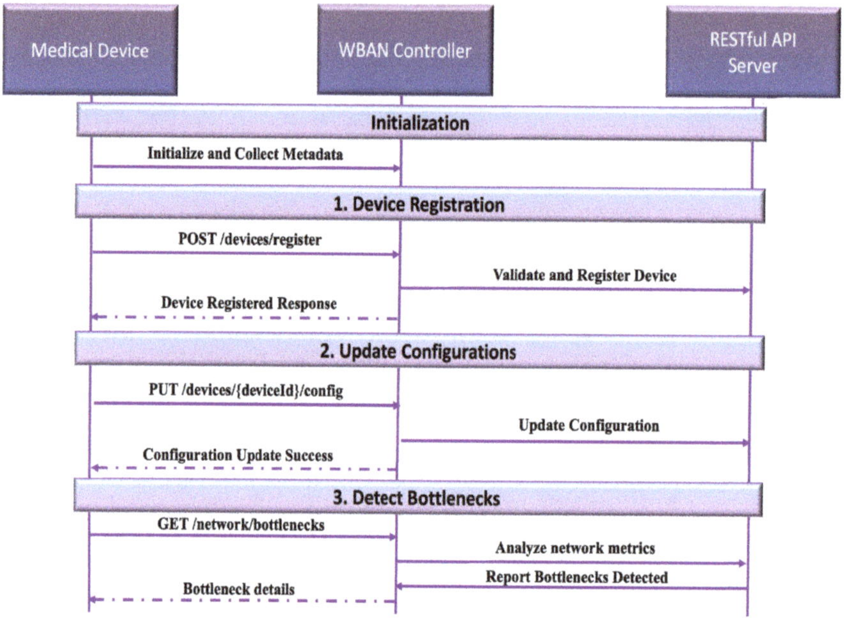

Figure 4. RESTful protocol.

4.12. OpenFlow in IoT for Healthcare

OpenFlow [86] is a foundational SDN protocol that dynamically updates flow tables in switches, allowing fine-grained traffic management. In healthcare IoT, it ensures optimal routing and prioritization of critical health data traffic, reducing latency and ensuring reliability. The flow table F is defined as

$$F(i) = \{M_i, A_i, P_i, S_i\}, \quad \text{where } M_i = \text{Match fields}, A_i = \text{Actions}, P_i = \text{Priority}, S_i = \text{Statistics}. \tag{22}$$

The controller dynamically programs flows:

$$F'(i) = h(M, \text{Traffic Type}, \text{Device Priority}). \tag{23}$$

One of the critical applications of OpenFlow in healthcare IoT networks is prioritizing emergency health data from IoT devices, ensuring that vital patient information, such as alerts from heart rate monitors or emergency notifications, is transmitted with minimal latency. This prioritization helps healthcare providers respond promptly to critical situations. Additionally, OpenFlow plays a crucial role in managing traffic in hospital IoT networks, enabling dynamic resource allocation and optimizing network performance to prevent congestion. By intelligently routing data flows, it ensures efficient communication between connected medical devices, hospital systems, and cloud platforms. Furthermore, OpenFlow facilitates the dynamic control of video streaming for remote patient monitoring, ensuring that bandwidth is allocated efficiently to support high-quality video feeds for telemedicine applications. This capability is essential for continuous patient observation and remote consultations, improving accessibility to healthcare services.

4.13. NetFlow/IPFIX in IoT for Healthcare

NetFlow/IPFIX [87] collects and analyzes traffic flows in healthcare IoT networks. It monitors patterns, detects anomalies, and ensures the uninterrupted flow of critical health data, such as those from wearable devices or connected medical equipment. In NetFlow/IPFIX, the flow record R is

$$R = \{S_{IP}, D_{IP}, S_{Port}, D_{Port}, \text{Protocol}, T_{Start}, T_{End}, \text{Bytes}\}. \tag{24}$$

Traffic utilization is calculated as

$$\text{Utilization: } U(t) = \frac{\text{Total Bytes Transferred at } t}{\text{Network Capacity}}. \tag{25}$$

NetFlow/IPFIX supports key healthcare IoT applications by monitoring data flows from wearable devices, ensuring seamless transmission for real-time health assessments. It enhances cybersecurity by detecting and mitigating abnormal traffic, preventing threats such as unauthorized access and data breaches. Additionally, it ensures uninterrupted data transfer for remote surgery devices, maintaining stable connectivity for latency-sensitive procedures. These capabilities make NetFlow/IPFIX essential for secure, efficient, and reliable healthcare network management. A detailed comparison of different IoT network management protocols is provided in Table 1. Moreover, Table 2 offers a wide-ranging synthesis of heterogeneous IoT devices, core technologies, protocols, and essential requirements, establishing a robust foundation for analyzing and developing cutting-edge solutions across diverse application scenarios.

Table 1. Comparison of different network management protocols in IoT.

Protocol	Resource Efficiency	Scalability	Security Features	Implementation Complexity	Message Overhead	Primary Use Case
SNMP [71]	Moderate	High	Weak (community strings, SNMPv3 improved)	Low	High	Network management
MQTT [72]	High	High	TLS, authentication support	Moderate	Low	IoT messaging
TR-069 [83]	Moderate	Moderate	TLS, ACS authentication	High	High	Remote device management
OMA DM [84]	Moderate	High	Strong (TLS, DTLS, authentication)	High	Moderate	Mobile device management
RESTful [85]	High	High	HTTPS-based security	Moderate	Moderate	Web services, API communication
OpenFlow [86]	Moderate	High	TLS support, secure channel	High	High	SDN (software-defined networking)
NETflow [87]	Moderate	High	Secure export options available	Moderate	High	Network traffic monitoring
NETCONF [74]	Moderate	High	SSH, TLS for secure communication	High	Moderate	Network configuration
gRPC [76]	High	High	TLS, authentication, encryption	Moderate	Low	High-performance RPC
NETCONF Light [77]	High	High	SSH, TLS	Moderate	Low	Lightweight network configuration
6LowPAN-SNMP [78]	High	Moderate	DTLS, AES encryption	Low	Low	Low-power IoT network management

Table 1. *Cont.*

Protocol	Resource Efficiency	Scalability	Security Features	Implementation Complexity	Message Overhead	Primary Use Case
LWM2M [81]	High	High	DTLS, authentication	Moderate	Low	IoT device management
CoAP [79]	Very High	High	DTLS for security	Low	Very Low	Constrained IoT environments

Table 2. Comparative review of heterogeneous devices, technologies, and specified protocols in IoT topic areas.

Topic Area	Devices	Technologies	Protocols	IoT Requirements (Functional and Non-Functional)
Massive Scaling	IoT Sensors, Smart Meters	Cloud Platforms, Edge/Fog Computing	6LoWPAN, NetFlow/IPFIX	*Functional:*High data throughput, real-time monitoring. *Non-Functional:* Scalability, load balancing, fault tolerance.
Architecture	Gateways, Controllers	Microservices, Distributed Systems	OpenFlow, NETCONF, NETCONF-light, LWM2M	*Functional:* Modular data flow, flexible routing. *Non-Functional:* Maintainability, easy integration, clear separation of concerns.
Knowledge-Edge and Big Data	Cameras, Environmental Sensors	Machine Learning, Big Data Analytics	MQTT, TR069	*Functional:* On-node analytics, large-scale data processing. *Non-Functional:* High performance, adaptive resource usage, minimal overhead.
Robustness	Industrial Controllers, Actuators	Redundant Network Designs, Fault-Tolerant Systems	SNMP, CoAP	*Functional:* Reliable data transfer, continuous operation in harsh conditions. *Non-Functional:* High availability, resiliency, error detection, and recovery.
Openness	Open-Source Boards (Arduino, Raspberry Pi)	Open Standards, Interoperability Frameworks	NETCONF, NETCONF-light, REST-based APIs	*Functional:* Cross-platform compatibility, flexible integration. *Non-Functional:* Vendor independence, extensibility, community-driven support.
Security and Privacy	Biometric Devices, Secure Gateways	Encryption, Authentication, IDS/IPS	SNMP (v3), CoAP (DTLS), MQTT (TLS)	*Functional:* Encryption of sensitive data, strong access controls. *Non-Functional:* Regulatory compliance, data integrity, user trust.
Human in the Loop	Wearables, Haptic Interfaces	Interactive Dashboards, Real-Time Alerts	MQTT, 6LoWPAN, Bluetooth Low Energy (BLE)	*Functional:* Real-time user feedback, intuitive control loops. *Non-Functional:* Usability, user-centric design, responsiveness.
Role of Middleware	Smart Routers, Network Hubs	IoT Platforms, Orchestration Tools	TR069, NETCONF, CoAP, LWM2M	*Functional:* Protocol translation, unified data handling, device orchestration. *Non-Functional:* Scalability, resource efficiency, simplified management.

5. Reference Architecture for Smart Healthcare System Network Management

In this work, architecture is defined as the structure of components that are involved in the network management practice of smart healthcare systems, the relationships between them, and the policies, requirements, compliance, and standards that govern the operations, administration, maintenance, and provisioning of networked systems. Network management architecture consists of three primary functions: (1) the transport of management information across the system, (2) the management of network elements, networks, services, and business, and (3) the storage of configuration management data. Even though the reference architecture deals with end-to-end network management, the focus of this work is on the segment that constitutes the patient's context (i.e., the sensors, actuators, gateways, and Internet interfaces). The rest of the network consists of traditional networks and elements managed by existing protocols. The requirements for smart healthcare system network management architecture are outlined in Table 3, detailing the key design considerations necessary for ensuring efficient, secure, and scalable network management in smart healthcare environments [88].

Table 3. Smart healthcare system network management architecture requirements.

Requirement	Description	Measurable Criteria	Importance	Management Function
Fault management	Deal with faults that occur in the network, such as device or software failures.	Network monitoring, alarm management, and trouble ticketing.	Critical	Fault
Scalability	Expand to serve the growing demand for smart healthcare services.	The number of users, sensors per user, data volume.	Important	Configuration
Portability	Move data, components, applications, and resources efficiently.	The number of platforms the system supports.	Useful	Configuration
Extensibility	Support new functionalities and new environments.	Design-dependent.	Important	Configuration
Human Tissue Heating Avoidance	Limit radio transmission power to prevent heating effects.	Temperature threshold, time at threshold.	Critical	Configuration
Energy Saving	Efficient monitoring and configuration to save energy.	Energy dissipation metrics.	Important	Configuration
Asset Management (Inventory)	Implement inventory tracking as part of network management.	Asset management system.	Important	Configuration
Centralized Monitoring	Monitor the entire system from a single location.	Configuration management support.	Useful	Configuration
Compliance	Ensure compliance with SLAs and policies.	SLA adherence.	Important	Account
Real-Time Communication Services	Support real-time WBAN services and applications.	QoS metrics.	Important	Performance
Availability	Ensure resource availability for use.	MTBF/(MTBF+MTTR).	Critical	Performance
Reliability	Proactive failure detection and management.	MTBF measurement.	Critical	Performance
Recoverability	Restore the system after failure or interruptions.	MTTR measurement.	Critical	Performance
Security Services	Protect infrastructure, data, and applications.	Security compliance measures.	Important	Security
Mobility Awareness	Support patient independence with mobility features.	Tracking mobility using protocols.	Critical	Mobility Management
Object Identification	Identify objects like healthcare recipients, sensors, and PDAs.	The uniqueness and correctness of object IDs.	Critical	Identity Management

Smart healthcare systems integrate sensing, actuation, and control functions to assess and analyze situations, enabling predictive or adaptive decision-making that leads to intelligent actions. The "smart" nature of these systems is primarily attributed to their autonomous operation, which relies on closed-loop control and networking capabilities.

Managing large-scale smart healthcare systems is a complex task that demands meticulous coordination and systematic oversight. A structured management approach emphasizes the management process itself rather than just the outcomes. To enhance efficiency, we propose a management system reference architecture, which aims to define the key components of smart healthcare system network management, their interrelationships, and their roles. This framework ensures that operational processes remain cost-effective, precise, and well organized while maintaining an optimal inventory.

The architecture must accommodate the diverse restoration requirements of biomedical sensors, as determined by healthcare professionals. The key architectural requirements include the following:

- Provide reactive and proactive fault management.
- Enable objects to cooperate in providing the desired service.
- Provide a reliable end-to-end connection and service.
- Implement high-level asset management.
- Provide scalability to accommodate new patients and services.
- Support patient mobility.
- Enforce security measures at all levels.
- Identify network management protocols.

Figure 5 shows a network management reference architecture for a smart healthcare system. The architecture consists of five layers, and each layer consists of different components. The architecture considers the stakeholders' requirements and policies. It consolidates existing industry frameworks and standards to maximize systems' interoperability. The system architecture should also consider requirements from external regulators for compliance.

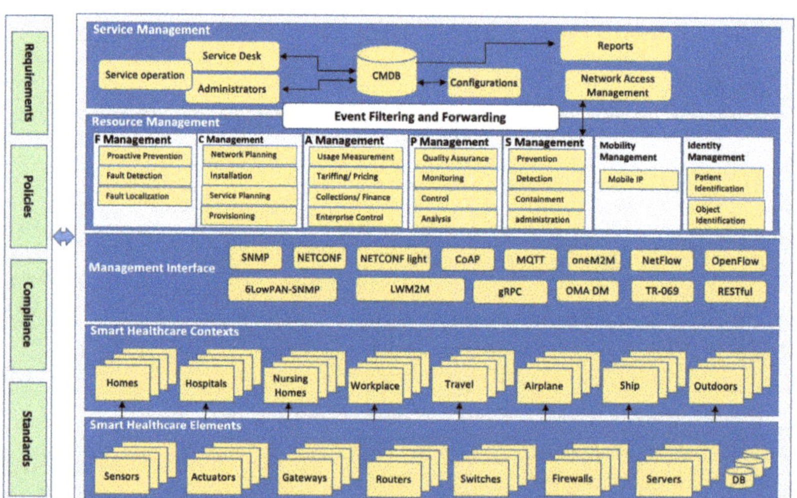

Figure 5. Smart healthcare system management reference architecture.

5.1. Smart Healthcare Element Layer

Network devices such as wearables, implants, IoT devices, gateways, routers, switches, and backend devices are individual components of the network that run across one or more of the network protocol layers [89]. Network device performance can be evaluated using multiple categories: end-to-end, per-link, per-network, and per-element, each corresponding to a different scope. End-to-end measurements look at the collective behavior across multiple network devices or the entire network path for one or more traffic flows. They may include metrics such as availability, capacity, delay, jitter, throughput, error rates, and utilization. In contrast, per-link, per-network, and per-element assessments concentrate on particular types of hardware or individual segments of the network. Per-link metrics often highlight factors such as propagation delay and link utilization, while per-element metrics (for example, for an IP router) might include IP forwarding rates (like packets per second), buffer usage, and authentication failure logs. Furthermore, these lower-level characteristics can be combined to provide an aggregated end-to-end performance profile.

Wearables, implants, and IoT devices are the main elements of the smart healthcare system as they connect directly to the patient [90]. Once installed, the broker should configure, manage, and keep them operational and updated. The network management system can poll individual components for status updates, and these components can either respond with the requested information or send automatic alerts in the event of abnormal behavior. Monitoring needs depend on the type of physiological data being measured. Some devices, such as spirometers and glucometers, only require attention when a reading is captured (discrete monitoring). Others require continuous data sampling and transmission, as is the case with electrocardiogram (ECG) devices, which constantly track cardiac signals [91].

5.2. Smart Healthcare Context Layer

Smart healthcare contexts are configured using the components in the first layer. The broker configures and maintains the contexts of the patient, for example, home, hospital, or workplace. A smart healthcare network must support patient mobility and switch between different location contexts [59]. Details of this layer are explained in Section 4.

5.3. Management Interface Layer

This layer defines the network management interfaces, usually referred to as the southbound interface, needed to connect the management station with the management agents (i.e., managed devices). In the traditional Internet, several protocols can be used for this purpose, such as SNMP. For the WBAN and IoT segments, there is no specific interface protocol defined for network management. Existing management protocols are not appropriate for resource-constrained WBAN/IoT devices. In the literature, there are several attempts to define these interfaces. Some of these attempts suggested using SNMP to manage sensor nodes [57], while other work suggested using SNMP over traditional IP networks and using other protocols over WBAN/IoT network segments and then linking the two segments through a proxy [56]. Sheng et al. [54] proposed a lightweight RESTful web service approach to enable device management of wireless sensor devices, motivated by the recent development of 6LoWPAN and CoAP protocols for resource-constrained devices. They developed a prototype and demonstrated that the proposed solution was efficient and effective for managing wireless sensor devices. Another work [92] investigated how existing IP-based network management protocols, SNMP and NETCONF, can be implemented on resource-constrained devices. They conducted a resource requirement analysis of SNMP and NETCONF on an 8-bit AVR-based device. They implemented light versions of the SNMP and NETCONF protocols and identified some challenges in message processing, session establishment and maintenance, and security. In their implementation, the authors did not mention how they implemented MIBs and how they could be used in resource-constrained devices.

Other widely used lightweight machine-to-machine IoT protocols are MQTT (Message Queuing Telemetry Transport) [93] and oneM2M [94]. MQTT is a publish/subscribe messaging protocol that is extremely simple and lightweight. It is designed for constrained devices and low-bandwidth, high-latency, or unreliable networks. It can be used in sensor networks and bridged to SNMP-enabled IP networks [95]. The oneM2M protocol is a global initiative to standardize a common M2M service layer platform for globally applicable and access-independent M2M services. The current M2M solutions are proprietary systems, which make them difficult to extend to support new services, integrate new data, and interoperate with other M2M systems [96]. Woo et al. [97] proposed a oneM2M-based IoT system for Personal Healthcare Devices (PHDs). They implemented a protocol to convert between ISO/IEEE 11073 [98] and oneM2M protocol messages. The conversion protocols are deployed in gateways located between the PHD and the management server. Bedside monitoring systems in hospitals use the ISO/IEEE 11073 standards. With minor modifications, the ISO/IEEE 11073 standards can be successfully applied to wearable, wireless, point-of-care systems in home healthcare [99].

5.4. Resource Management Layer

The resource management layer abstracts all smart healthcare elements as reachable resources that can be managed remotely by the network management system. This layer consists of the network management functional areas, including Fault, Configuration, Accounting, Performance, and Security (FCAPS), mobility, and identity management [100]. The structure of this layer aligns with the Telecommunications Management Network (TMN) reference model, which provides a hierarchical framework for managing open systems. As illustrated in Figure 6, the TMN's layered approach enables efficient resource monitoring and control, ensuring the seamless operation of smart healthcare networks.

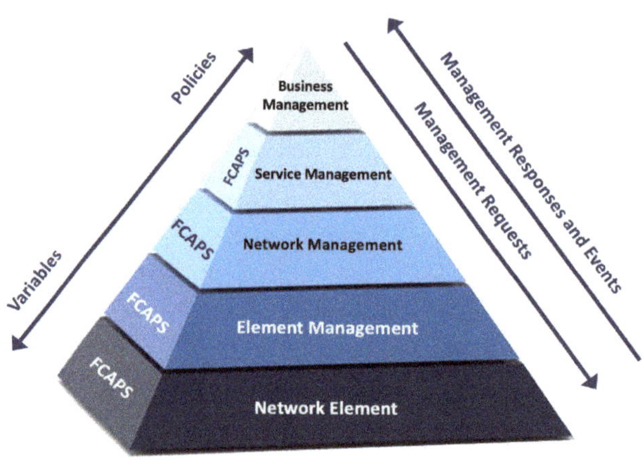

Figure 6. TMN model for managing open systems.

5.4.1. Fault Management

Fault Management is a set of functions, which enables the detection, localization, isolation, and correction of abnormal operations of the telecommunication network and its environment. Various types of faults can arise in IoT-based healthcare networks, affecting data integrity, system performance, and network reliability. These faults can be categorized into data-centric, system-centric, and network-centric issues, each with distinct causes, durations, and impacts. A detailed classification and analysis of these fault types are provided in Table 4, highlighting their characteristics and potential consequences for system operations [101].

5.4.2. Configuration Management

It encompasses the functions needed to control network elements, identify and gather their data, and provide them with required information. It also facilitates network planning, engineering, installation, service planning and negotiation, provisioning, status checks, and control.

5.4.3. Accounting Management

This measures how network services are used and determines the associated costs for providers, as well as charges for customers. It also assists in calculating appropriate pricing for these services.

5.4.4. Performance Management

It evaluates and reports on the behavior of telecommunications devices and the overall network, measuring their effectiveness. It additionally gathers and examines statistical data to both monitor and optimize the performance of the network and its individual components.

5.4.5. Security Management

It enables the management of security, which is required for all management functional areas and TMN transactions.

5.4.6. Mobility Management

It provides the necessary functions that allow the patient to switch between different contexts on different IP networks.

5.4.7. Identity Management

It provides the necessary functions to identify patients and objects and the relationship between them.

5.5. Service Management Layer

The uppermost layer in the architecture is the service management layer. This layer manages the end-to-end services provided to end-users. It receives filtered events from the resource management layer. Within this layer lies the architecture repository, the Configuration Management Database (CMDB), that saves all inventory, configurations, and filtered events related to all devices. Service delivery and management represent the organizational strategic objectives that require effective and efficient support. Therefore, the services should be planned in a systematic and standard way to guarantee service continuity and interoperability. In this architecture, we recommend using the ITILv3 framework. ITILv3 best practices define five core processes, namely, service strategy, service design, service transition, service operation, and continual service improvement. Service operation management further defines five service operation processes, namely, incident management, problem management, event management, service request fulfillment, and access management. The ITILv3 framework can be adapted for smart healthcare systems' service management by incorporating smart healthcare's technical and medical requirements. Defining a service management layer requires comprehensive strategy development and implementation. ITILv3 defines a priority framework to prioritize incidents based on impact and urgency. It also defines the IT support roles and levels that handle different incident complexities and time constraints.

Table 4. Fault type analysis.

Fault Type	Description	Cause of Fault	Duration of Fault	Impact of Fault
	Data-Centric			
Outlier	A data point that deviates from other observations.	Noise, errors, events, or malicious attacks.	Randomly occurring and instantaneous.	It may skew the mean, variance, and gradient.
Spike	The rate of change in the gradient of data over a period of time is higher than expected.	Frequent causes include battery failure and other hardware failures or loose wire connections.	The fault consists of more than one point. May or may not be temporary.	Spikes do not hold any informational value and should be discarded.
Stuck-at	Sensor values experience zero variation for an unexpected length of time.	Frequently, the cause of this fault is a sensor hardware malfunction.	The fault consists of more than one point. May or may not be temporary.	If the cause is environment out of range or clipping, the data still hold value and can be interpreted at lower fidelity. Otherwise, discard the data.
High Noise or Variance	Sensor values experience unexpectedly high variation or noise.	This may be due to hardware failure, environment out of range, or low battery supply.	The fault consists of more than one point. Usually not temporary.	If the noisy data track other sensors, then the data still offer value and should not be discarded.
	System-Centric			
Calibration	The sensor reports values that are offset from the ground truth.	Calibration error and sensor drift are the primary causes of this fault.	The fault remains throughout deployment.	Data should not be discarded. Uncalibrated data can still provide insights.
Hardware	A malfunction in the sensor hardware that causes inaccurate data.	Environmental disturbances and sensor age may indicate higher probabilities of failure. Other causes include a short circuit or a loose wire.	Permanent once it occurs.	Data are meaningless, as the sensor is not performing as designed. Should be discarded.
Low Battery	Battery voltage drops to a point where the sensor cannot confidently report data.	The battery state is an indicator of system performance.	Permanent until battery replacement or recharge.	Results in useless data. The exception is if sensor behavior at a low voltage gives added noise, and then there may be informational value.
Environment Out of Range	The environment exceeds the sensitivity range of the transducer.	There may be much higher noise or a flattening of the data. It may also be a sign of improper calibration.	The sensor returns to normal after the environment returns within range.	Holds some information. At a minimum, it indicates that the environment exceeds the sensor sensitivity range.
Clipping	The sensor maxes out at the limits of the ADC.	This is due to the environment exceeding the range of the ADC.	The sensor returns to normal after the environment returns within range.	Holds some information. Indicates that data exceed the upper or lower ADC values.
	Network-Centric			
Lost Wireless Connection	The connection to the gateway is lost.	Due to interference, noise, low battery, mobility, or sensor out of range.	Randomly occurring. May be temporary or permanent.	Loss of data.
Average Packet Delay	The packet arrives late to the final destination.	Due to congestion along the way of transmission.	Randomly occurring and instantaneous.	Delayed or dropped packets.
No Internet Connection	Connection to the Internet is lost.	Due to mobility.	This occurs when changing the home network. May be temporary or permanent.	No connection to the backend servers.

6. Architecture Implementation Plan

This section discusses the implementation approach of the management entities that are under the administrative domain of a third party accountable for managing the smart healthcare system.

6.1. Supporting the Architecture with Software-Defined Networking and Deep Learning

Providing reliable, timely, secure, cost-effective, and pervasive smart healthcare is an essential requirement of the healthcare industry nowadays. These requirements are driven by the increasing number of patients with chronic diseases, the increasing average life expectancy, and associated healthcare spending. Considerable research is being conducted to tackle these issues and provide practical solutions based on the current Internet architecture and the deployed technologies. However, common challenges related to network and device heterogeneity, complexity, and the lack of full control of the underlying network infrastructure hinder the achievement of desirable end-to-end healthcare-specific requirements. In addition, the massive amount of data generated from the various sensors will make it extremely difficult to use traditional management techniques and tools for fault management, performance analysis, traffic management, and real-time response to any possible malfunction of the software or hardware components. Therefore, we mention two promising technologies that will add value to the architecture and help address the above challenges, namely, software-defined networking (SDN) and deep learning. SDN is an emerging networking paradigm that helps in facilitating network configuration and management [102]. Unlike traditional network configurations used in the current Internet architecture, where network devices implement a control plane and a data-forwarding plane, SDN has two distinct features: SDN separates the control plane from the data plane while enabling programmatic control for network application development. Through this design, SDN aims to achieve more efficient configuration, enhanced performance, and flexible adaptation to emerging network innovations [103].

Four main principles define this SDN architecture:

- Flow-based forwarding rather than traditional destination-based forwarding.
- Externalized control logic, handled by an SDN controller located outside the forwarding devices.
- Programmable network elements, managed by software applications running on the controller.
- Decoupled control and data planes, allowing independent yet coordinated operation.

SDN plays a crucial role in predictively allocating network resources and preventing performance bottlenecks in healthcare applications. The SDN controller dynamically prioritizes critical health data transmissions, such as real-time patient monitoring and emergency response alerts, ensuring preferential network bandwidth for life-saving applications over non-urgent traffic. This proactive approach improves network reliability, reduces congestion, and enhances the overall efficiency of healthcare services [103].

Deep learning (DL) has emerged recently as a powerful tool for machine learning capable of solving complex optimization and predictive analytics problems. It is contributing to solving significant problems that previous attempts by the artificial intelligence community failed to solve for many years. It enables multi-layered computational models to learn abstract representations of healthcare data, providing enhanced fault detection, anomaly prediction, and network optimization. Artificial intelligence techniques have already been explored for fault management in computer networks, where neural networks have been applied for alarm filtering and correlation [104]. DL is also widely used in healthcare analytics and bioinformatics [48,49], demonstrating its potential in predictive network management. DL-based predictive models continuously analyze network performance metrics to anticipate failures, allowing for preemptive intervention. We also envisage the use of deep learning by SDN controller applications. For example, deep learning-based applications can be used to control the switching, intelligent routing, and firewalling behavior of the overall healthcare infrastructure. A DL-enabled SDN architecture for healthcare is presented in Figure 7. Beyond SDN, Network Function Virtualization (NFV) and Management and Orchestration (MANO) play critical roles in modernizing healthcare network infrastructures. NFV virtualizes network functions, decoupling them from dedicated hardware and running them as software instances, thereby reducing dependency on specialized networking equipment, improving scalability, and lowering operational costs [48]. This shift is particularly beneficial in smart healthcare systems, where the rapid deployment and adaptation of network services are essential for real-time patient monitoring and emergency response scenarios.

The MANO framework further enhances NFV by automating the deployment, configuration, and lifecycle management of virtualized network functions (VNFs). This automation reduces manual intervention, minimizes errors, and ensures efficient resource allocation, leading to higher flexibility and reliability in healthcare networks. Moreover, chaining VNFs through MANO enables dynamic service adaptation, allowing healthcare applications to seamlessly scale based on demand, thus enhancing quality of service (QoS) and patient care outcomes [49].

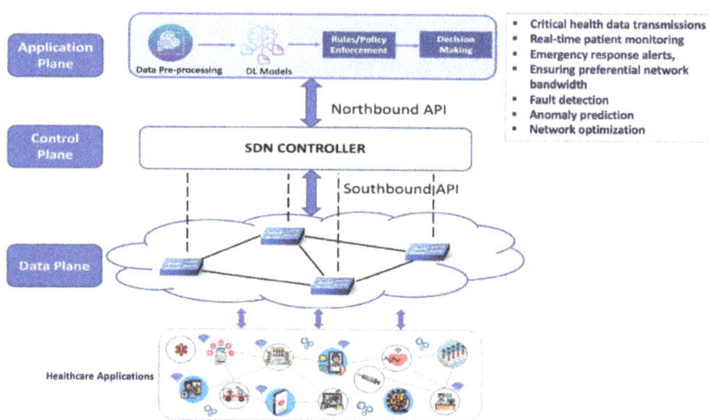

Figure 7. Smart health system architecture implementation plan supported by software-defined networking and deep learning.

6.2. *Low-Power Low-Range Wireless Networks*

WBAN elements are best configured with IEEE 802.15.6, while other IoT devices can be configured with IEEE 801.15.4. IEEE 801.15.4 provides a more extended range compared with IEEE 802.15.6. Implanted sensors can only be configured as relayed devices that should not relay any packets from other sensors [12]. Wearable sensors might be configured as relayed or relaying devices. The PDA (or hub, as defined by IEEE 802.15.6) should be positioned in a place that supports star topology.

Given the anticipated large number of monitored and managed smart healthcare users, the Personal Digital Assistant (PDA) is optimally configured as a software-defined networking (SDN) device, which is programmed by an SDN controller hosted by a third-party provider. Traffic flows within smart healthcare systems can be categorized into three main types: physiological data, environmental data, and sensor health data.

Physiological or vital sign data can be further divided into periodic physiological data and emergency physiological data [105]. Environmental data, on the other hand, may be collected from indoor or outdoor settings. The PDA is responsible for classifying and directing the traffic flows generated by various sensors to their designated destinations.

Figure 8 illustrates the traffic flow from sensors to the PDA and onward to the respective end systems.

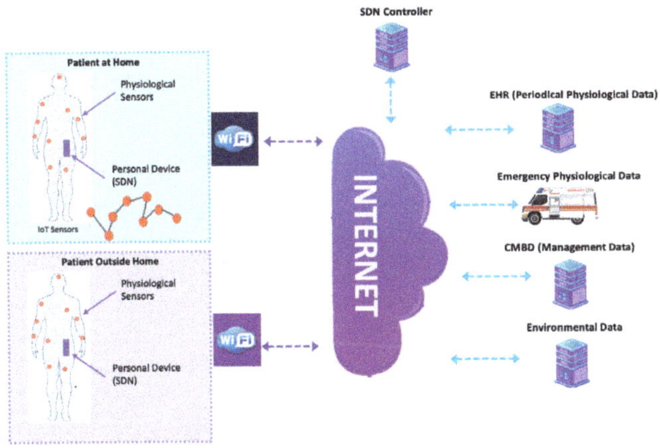

Figure 8. SDN-based traffic classification and management.

6.3. Smart Healthcare Context Management

SDN may also provide substantial help in configuring smart healthcare contexts and switching between contexts by exploiting the capability of running software applications on top of the controller [106]. All elements of the context can be configured and managed directly or indirectly by the controller as long as they are under the administrative authority of the managing entity. The configuration and management of the context elements can be realized by providing a management plane over the control plane. Context management may also include mobility and handover between different network architectures as well as the same network architecture. Mobility is best handled by a mobile IP in the case of all-IP networks.

6.4. Smart Healthcare Service Management

Service management is the focus of the management system to ensure service continuity. Due to the sensitivity of the smart healthcare system, healthcare services should be provisioned without anomalies. Service malfunctions can happen because of hardware or software faults, which cause network elements to produce incorrect outputs. When the smart healthcare system is deployed on a massive scale, it becomes hard to monitor and manage such services, especially with the reception of a considerable number of alarms and sensor-related management data. To provide efficient alarm filtering and real-time decisions, we propose using deep learning techniques for this purpose. To further elaborate on the use of deep learning in fault management, the workflow of the fault management process is outlined as follows:

- Collect monitoring and performance alarms.
- Provide feedback to users to ensure service continuity.
- Filter, eliminate, and correlate the alarms.
- Diagnose faults through analysis and testing to eliminate the cause of the fault.
- Determine a plan for correction, display correction options to patients, and implement the plan.
- Verify the fault is eliminated.
- Record data and determine the effectiveness of the current fault management process.

Deep learning can help in filtering, classifying, and correlating alarms and recognizing basic fault patterns. However, to identify the exact fault location, a further process is necessary. This procedure involves human intervention to make some decisions that might be deemed impossible by deep learning alone. Some work suggests using case-based reasoning to analyze the data further and run tests to identify the exact location of the fault [104]. However, one recent work suggests an architecture for deep learning that explains its reasoning process and learns by observation [107].

7. Advanced Futuristic Network Management Protocols

As networks continue to evolve into self-healing, AI-driven, and hyper-connected ecosystems, traditional network management protocols struggle to meet the demands of real-time automation, security, and predictive analytics. Future networks will be defined by protocols that leverage AI, quantum cryptography, blockchain, and intent-based networking to enable seamless, autonomous operations. These next-generation protocols will be designed to support critical infrastructures such as 6G, healthcare cyber–physical systems [108], smart cities, and autonomous vehicles, ensuring ultra-reliable and intelligent network management. One of the most promising advancements in network management is Intent-Based Networking Protocols (IBNPs). Unlike traditional protocols that require manual configurations, IBNPs allow network administrators to define high-level business objectives, and the network autonomously configures itself to meet these goals. These protocols will rely on AI-driven automation, machine learning, federated learning [109,110], and real-time data analytics to dynamically optimize traffic, predict failures, and enforce security policies. Future versions of IBNPs will also incorporate blockchain for decentralized trust and zero-trust architectures to enhance cybersecurity in highly distributed environments.

Another significant innovation is Predictive Telemetry Protocols (PTPs), which will replace conventional polling-based network monitoring systems with AI-powered proactive anomaly detection. Instead of relying on reactive alerts, PTPs will analyze vast amounts of real-time network data using deep learning models to predict failures before they happen. These protocols will enable self-healing networks by autonomously rerouting traffic, reconfiguring devices, and mitigating security threats before they impact operations. This will be particularly crucial for critical applications such as autonomous transportation, smart factories, and real-time healthcare monitoring. With the rise of quantum computing, current encryption mechanisms will become obsolete, necessitating Quantum-Secure Network Management Protocols [111,112]. These protocols will integrate post-quantum cryptographic algorithms and quantum key distribution (QKD) to protect sensitive network communications. Future networks, especially in financial institutions, government infrastructures, and military communications, will require these quantum-resistant protocols to prevent potential cyber threats posed by quantum decryption capabilities.

In addition, Bio-Inspired Autonomous Network Protocols (BANPs) will revolutionize network resilience by mimicking biological self-organizing systems [113]. Inspired by swarm intelligence and neural networks, BANPs will enable self-adaptive, decentralized, and fault-tolerant network management. These protocols will be vital for managing massive IoT ecosystems, where devices must seamlessly communicate, self-optimize, and self-heal in unpredictable environments. The future of network management will be AI-driven, quantum-secure, predictive, and intent-based, ensuring autonomous, self-optimizing, and ultra-resilient network infrastructures that can support the demands of next-generation technologies.

8. Conclusions

This paper provides a comprehensive review of smart network management protocols for Internet of Things (IoT)-based healthcare systems, addressing critical challenges in ensuring seamless connectivity, secure data exchange, and efficient device management. A reference architecture is proposed to enhance network reliability and service continuity by integrating standards and frameworks into a comprehensive management system that mitigates health risks through robust strategies while proactively predicting and preventing healthcare faults via the collection of data from multiple protocols, which are then forwarded to a central entity for machine learning-based analysis. Additionally, this study explores the integration of advanced technologies, including software-defined networking (SDN) and deep learning (DL), to optimize network performance. Furthermore, it highlights emerging paradigms shaping next-generation network management solutions for future smart healthcare ecosystems.

Author Contributions: Conceptualization, F.M.S.; methodology, F.M.S., A.T. and H.M.K.; writing, F.M.S., A.T. and H.M.K.; review, M.H., K.S., F.M.S., A.T. and E.S.B.; editing, F.M.S., H.M.K. and A.T.; visualization, F.M.S., H.M.K. and A.T.; supervision, F.M.S.; project administration, M.H. All authors have read and agreed to the published version of the manuscript.

Funding: This research received no external funding.

Informed Consent Statement: Not applicable.

Data Availability Statement: No data were used or produced in this study.

Acknowledgments: This work was supported by UAEU.

Conflicts of Interest: The authors declare no conflicts of interest.

References

1. Chen, Y.; Zhong, W.; Da H.; Ruosen, L. 6G is coming: Discussion on key candidate technologies and application scenarios. In Proceedings of the 2020 International Conference on Computer Communication and Network Security (CCNS), Xi'an, China, 21–23 August 2020; IEEE: Piscataway, NJ, USA, 2020; pp. 59–62.
2. Pereira, F.; Correia, R.; Pinho, P.; Lopes, S.I.; Carvalho, N.B. Challenges in resource-constrained IoT devices: Energy and communication as critical success factors for future IoT deployment. *Sensors* **2020**, *20*, 6420. [CrossRef] [PubMed]
3. Yang, X.; Liu, Z.; Tang, K.; Yin, X.; Zhuo, C.; Wei, Q.; Qiao, F. Breaking the energy-efficiency barriers for smart sensing applications with "Sensing with Computing" architectures. *Sci. China Inf. Sci.* **2023**, *66*, 200409. [CrossRef]
4. Lin, J.; Yu, W.; Zhang, N.; Yang, X.; Zhang, H.; Zhao, W. A survey on internet of things: Architecture, enabling technologies, security and privacy, and applications. *IEEE Internet Things J.* **2017**, *4*, 1125–1142. [CrossRef]
5. Tariq, M.U. Advanced wearable medical devices and their role in transformative remote health monitoring. In *Transformative Approaches to Patient Literacy and Healthcare Innovation*; IGI Global: Hershey, PA, USA, 2024; pp. 308–326.
6. Chui, M.; Collins, M.; Patel, M. *The Internet of Things: Catching Up to an Accelerating Opportunity*; McKinsey & Company: New York, NY, USA, 2021.
7. Wehde, M. Healthcare 4.0. *IEEE Eng. Manag. Rev.* **2019**, *47*, 24–28. [CrossRef]
8. Khater, H.M.; Sallabi, F.; Serhani, M.A.; Turaev, S.; Barka, E. Efficient hybrid fault-management clustering algorithm (hfmca) in wbans based on weighted bipartite graph. *IEEE Access* **2023**, *11*, 57977–57990. [CrossRef]
9. Saha, E.; Rathore, P. The impact of healthcare 4.0 technologies on healthcare supply chain performance: Extending the organizational information processing theory. *Technol. Forecast. Soc. Change* **2024**, *201*, 123256. [CrossRef]
10. Sallabi, F.; Shuaib, K. Internet of Things Network Management System Architecture for Smart Healthcare. In Proceedings of the 2016 Sixth International Conference on Digital Information and Communication Technology and its Applications (DICTAP), Konya, Turkey, 21–23 July 2016; pp. 165–170.
11. IEEE. *IEEE Std 802.15.4-2020*; IEEE Standard for Low-Rate Wireless Networks. IEEE: New York, NY, USA, 2020; pp. 1–800.
12. IEEE. *IEEE Std 802.15.6-2012*; IEEE Standard for Local and Metropolitan Area Networks—Part 15.6: Wireless Body Area Networks. IEEE: New York, NY, USA, 2012.

13. Detwal, P.K.; Agrawal, R.; Samadhiya, A.; Kumar, A.; Garza-Reyes, J.A. Revolutionizing healthcare organizations with Operational Excellence and Healthcare 4.0: A systematic review of the state-of-the-art literature. *Int. J. Lean Six Sigma* **2024**, *15*, 80–102. [CrossRef]
14. Gopich, G.; Sarath, T.; Dumka, A.; Goyal, H.R.; Singh, R.; Gehlot, A.; Gupta, L.R.; Thakur, A.K.; Priyadarshi, N.; Twala, B. Use of IoT sensor devices for efficient management of healthcare systems: A review. *Discov. Internet Things* **2024**, *4*, 8. [CrossRef]
15. Whitmore, A.; Agarwal, A.; Da Xu, L. The Internet of Things A survey of topics and trends. *Inf. Syst. Front.* **2015**, *17*, 261–274. [CrossRef]
16. Razzaque, M.A.; Milojevic-Jevric, M.; Palade, A.; Cla, S. Middleware for internet of things: A survey. *IEEE Internet Things J.* **2016**, *3*, 70–95. [CrossRef]
17. Botta, A.; De Donato, W.; Persico, V.; Pescapé, A. Integration of Cloud computing and Internet of Things: A survey. *Future Gener. Comput. Syst.* **2016**, *56*, 684–700. [CrossRef]
18. Ghaleb, S.M.; Subramaniam, S.; Zukarnain, Z.A.; Muhammed, A. Mobility management for IoT: A survey. *EURASIP J. Wirel. Commun. Netw.* **2016**, *2016*, 165. [CrossRef]
19. Shibata, Y.; Sato, G. IoT based mobility information infrastructure in challenged network environment toward aging society. In Proceedings of the 31st IEEE International Conference on Advanced Information Networking and Applications Workshops, WAINA 2017, Taipei, Taiwan, 27–29 March 2017; pp. 645–648.
20. Ulrich. T. *5G: An Introduction to the 5th Generation Mobile Networks*; Walter de Gruyter GmbH & Co KG: Berlin, Germany, 2021.
21. Li, S.; Zhao, S.; Zhao, S. 5G Internet of Things: A Survey. *J. Ind. Inf. Integr.* **2018**, *10*, 1–9. [CrossRef]
22. Stankovic, J.A. Research Directions for the Internet of Things. *IEEE Internet Things J.* **2014**, *1*, 3–9. [CrossRef]
23. Zimmermann, A.; Schmidt, R.; Sandkuhl, K.; Wißotzki, M.; Jugel, D.; Möhring, M. Digital enterprise architecture-transformation for the internet of things. In Proceedings of the 2015 IEEE 19th International Enterprise Distributed Object Computing Workshop EDOCW 2015, Adelaide, SA, Australia, 21–25 September 2015; pp. 130–138.
24. Al-Fuqaha, A.; Khreishah, A.; Guizani, M.; Rayes, A.; Mohammadi, M. Toward better horizontal integration among IoT services. *IEEE Commun. Mag.* **2015**, *53*, 72–79. [CrossRef]
25. Aswale, P.; Shukla, A.; Bharati, P.; Bharambe, S.; Palve, S. An Overview of Internet of Things: Architecture, Protocols and Challenges. In *Information and Communication Technology for Intelligent Systems*; Springer: Singapore, 2019; pp. 299–308.
26. Gigli, M. Internet of Things: Services and Applications Categorization. *Adv. Internet Things* **2011**, *1*, 27–31. [CrossRef]
27. Cook, D.J.; Crandall, A.S.; Thomas, B.L.; Krishnan, N.C. CASAS: A Smart Home in a Box. *Computer* **2013**, *46*, 62–69. [CrossRef]
28. Li, X.; Lu, R.; Liang, X.; Shen, X.; Chen, J.; Lin, X. Smart community: An internet of things application. *IEEE Commun. Mag.* **2011**, *49*, 68–75. [CrossRef]
29. Zanella, A.; Bui, N.; Castellani, A.; Vangelista, L.; Zorzi, M. Internet of things for smart cities. *IEEE Internet Things J.* **2014**, *1*, 22–32. [CrossRef]
30. Dohr, A.; Modre-Opsrian, R.; Drobics, M.; Hayn, D.; Schreier, G. The Internet of Things for Ambient Assisted Living. In Proceedings of the 2010 Seventh International Conference on Information Technology: New Generations, Las Vegas, NV, USA, 12–14 April 2010; pp. 804–809.
31. Marques, G. Ambient Assisted Living and Internet of Things. In *Harnessing the Internet of Everything (IoE) for Accelerated Innovation Opportunities*; IGI Global: Hershey, PA, USA, 2019; pp. 100–115.
32. Bui, N.; Zorzi, M. Health care applications: A solution based on the internet of things. In Proceedings of the 4th International Symposium on Applied Sciences in Biomedical and Communication Technologies, Barcelona, Spain, 26–29 October 2011; pp. 1–5.
33. Jara, A.J.; Zamora, M.A.; Skarmeta, A.F.G. An Internet of things-based personal device for diabetes therapy management in ambient assisted living (AAL). *Pers. Ubiquitous Comput.* **2011**, *15*, 431–440. [CrossRef]
34. Jara, A.J.; Zamora-Izquierdo, M.A.; Skarmeta, A.F. Interconnection Framework for mHealth and Remote Monitoring Based on the Internet of Things. *IEEE J. Sel. Areas Commun.* **2013**, *31*, 47–65. [CrossRef]
35. Jung, Y. Hybrid-aware model for senior wellness service in smart home. *Sensors* **2017**, *17*, 1182. [CrossRef] [PubMed]
36. Bennett, J.; Rokas, O.; Chen, L. Healthcare in the Smart Home: A Study of Past, Present and Future. *Sustainability* **2017**, *9*, 840. [CrossRef]
37. Liao, Z.; Kong, L.; Wang, X.; Zhao, Y.; Zhou, F.; Liao, Z.; Fan, X. A Visual Analytics Approach for Detecting and Understanding Anomalous Resident Behaviors in Smart Healthcare. *Appl. Sci.* **2017**, *7*, 254. [CrossRef]
38. Solanas, A.; Patsakis, C.; Conti, M.; Vlachos, I.S.; Ramos, V.; Falcone, F.; Postolache, O.; Pérez-Martínez, P.A.; Di Pietro, R.; Perrea, D.N.; et al. Smart health: A context-aware health paradigm within smart cities. *IEEE Commun. Mag.* **2014**, *52*, 74–81. [CrossRef]
39. Ghose, A.; Bhaumik, C.; Das, D.; Agrawal, A.K. Mobile healthcare infrastructure for home and small clinic. In Proceedings of the 2nd ACM international Workshop on Pervasive Wireless Healthcare—MobileHealth '12, Hilton Head, SC, USA, 11–14 June 2012; p. 15.

40. Santos, J.; Rodrigues, J.J.P.C.; Silva, B.M.C.; Casal, J.; Saleem, K.; Denisov, V. An IoT-based mobile gateway for intelligent personal assistants on mobile health environments. *J. Netw. Comput. Appl.* **2016**, *71*, 194–204. [CrossRef]
41. Pereira, C.; Pinto, A.; Ferreira, D.; Aguiar, A. Experimental Characterization of Mobile IoT Application Latency. *IEEE Internet Things J.* **2017**, *4*, 1082–1094. [CrossRef]
42. Khattak, H.A.; Ruta, M.; Eugenio, E.; Sciascio, D. CoAP-based healthcare sensor networks: A survey. In Proceedings of the 2014 11th International Bhurban Conference on Applied Sciences & Technology (IBCAST), 2014, Islamabad, Pakistan, 14–18 January 2014; pp. 499–503.
43. Rahmani, A.-M.; Thanigaivelan, N.K.; Gia, T.N.; Granados, J.; Negash, B.; Liljeberg, P.; Tenhunen, H. Smart e-Health Gateway: Bringing intelligence to Internet-of-Things based ubiquitous healthcare systems. In Proceedings of the 2015 12th Annual IEEE Consumer Communications and Networking Conference CCNC 2015, Las Vegas, NV, USA, 9–12 January 2015; pp. 826–834.
44. Islam, S.M.R.; Kwak, D.; Kabir, H.; Hossain, M.; Kwak, K.-S. The Internet of Things for Health Care: A Comprehensive Survey. *IEEE Access* **2015**, *3*, 678–708. [CrossRef]
45. Farahani, B.; Firouzi, F.; Chang, V.; Badaroglu, M.; Constant, N.; Mankodiya, K. Towards fog-driven IoT eHealth: Promises and challenges of IoT in medicine and healthcare. *Future Gener. Comput. Syst.* **2018**, *78*, 659–676. [CrossRef]
46. Pagán, J.; Zapater, M.; Ayala, J.L. Power transmission and workload balancing policies in eHealth mobile cloud computing scenarios. *Future Gener. Comput. Syst.* **2018**, *78*, 587–601. [CrossRef]
47. Adame, T.; Bel, A.; Carreras, A.; Melià-Seguí, J.; Oliver, M.; Pous, R. CUIDATS: An RFID–WSN hybrid monitoring system for smart health care environments. *Future Gener. Comput. Syst.* **2018**, *78*, 602–615. [CrossRef]
48. Miotto, R.; Wang, F.; Wang, S.; Jiang, X.; Dudley, J.T. Deep learning for healthcare: Review, opportunities and challenges. *Briefings Bioinform.* **2017**, *19*, 1236–1246. [CrossRef] [PubMed]
49. Ravì, D.; Wong, C.; Deligianni, F.; Berthelot, M.; Andreu-Perez, J.; Lo, B.; Yang, G.Z. Deep Learning for Health Informatics. *IEEE J. Biomed. Health Inform.* **2017**, *21*, 4–21. [CrossRef] [PubMed]
50. Cavallari, R.; Martelli, F.; Rosini, R.; Buratti, C.; Verdone, R. A Survey on Wireless Body Area Networks: Technologies and Design Challenges. *IEEE Commun. Surv. Tutor.* **2014**, *16*, 1635–1657. [CrossRef]
51. Movassaghi, S.; Member, S.; Abolhasan, M.; Member, S. Wireless Body Area Networks: A Survey. *IEEE Commun. Surv. Tutor.* **2013**, *16*, 1658–1686. [CrossRef]
52. Ghamari, M.; Janko, B.; Sherratt, R.S.; Harwin, W.; Piechockic, R.; Soltanpur, C. A survey on wireless body area networks for ehealthcare systems in residential environments. *Sensors* **2016**, *16*, 831. [CrossRef]
53. Sinche, S.; Silva, J.S.; Raposo, D.; Rodrigues, A.; Pereira, V.; Boavida, F. Towards Effective IoT Management. In Proceedings of the 2018 IEEE SENSORS, New Delhi, India, 28–31 October 2018; pp. 18–21.
54. Sheng, Z.; Wang, H.; Yin, C.; Hu, X.; Yang, S.; Leung, V. Lightweight Management of Resource Constrained Sensor Devices in Internet-of-Things. *IEEE Internet Things J.* **2015**, *2*, 402–411. [CrossRef]
55. Li, X.; Ji, H.; Li, Y. Layered fault management scheme for end-to-end transmission in internet of things. *Mob. Netw. Appl.* **2013**, *18*, 195–205. [CrossRef]
56. Lindholm-Ventola, H.; Bilhanan, S. *CoAP-SNMP Interworking IoT Scenarios*; Report No. 3; Department of Pervasive Computing, Tampere University of Technology: Tampere, Finland, 2014.
57. Huang, H.-P.; Xiao, S.-D.; Meng, X.-Y. Applying SNMP Technology to Manage the Sensors in Internet of Things. *Open Cybern. Syst. J.* **2015**, *15*, 1019–1024.
58. Zhou, S.; Lin, K.J.; Na, J.; Chuang, C.C.; Shih, C.S. Supporting Service Adaptation in Fault Tolerant Internet of Things. In Proceedings of the 2015 IEEE 8th International Conference on Service-Oriented Computing and Applications (SOCA), Rome, Italy, 19–21 October 2015; pp. 65–72.
59. Hassanalieragh, M.; Page, A.; Soyata, T.; Sharma, G.; Aktas, M.; Mateos, G.; Kantarci, B.; Andreescu, S. Health Monitoring and Management Using Internet-of-Things (IoT) Sensing with Cloud-Based Processing: Opportunities and Challenges. In Proceedings of the 2015 IEEE International Conference on Services Computing SCC, New York, NY, USA, 27 June–2 July 2015; pp. 285–292.
60. Gia, T.N.; Rahmani, A.-M.; Westerlund, T.; Liljeberg, P.; Tenhunen, H. Fault tolerant and scalable IoT-based architecture for health monitoring. In Proceedings of the 2015 IEEE Sensors Applications Symposium (SAS), Zadar, Croatia, 13–15 April 2015; pp. 1–6.
61. Salahuddin, M.A.M.A.; Al-Fuqaha, A.; Guizani, M.; Shuaib, K.; Sallabi, F. Softwarization of internet of things infrastructure for secure and smart healthcare. *Computer* **2017**, *50*, 74–79. [CrossRef]
62. Gharaibeh, A.; Salahuddin, M.A.; Hussini, S.J.; Khreishah, A.; Khalil, I.; Guizani, M.; Al-Fuqaha, A. Smart Cities: A Survey on Data Management, Security and Enabling Technologies. *IEEE Commun. Surv. Tutor.* **2017**, *19*, 2456–2501. [CrossRef]
63. Tariq, A.; Lakas, A.; Sallabi, F.; Qayyum, T.; Serhani, M.A.; Barka, E. Empowering Trustworthy Client Selection in Edge Federated Learning Leveraging Reinforcement Learning. In Proceedings of the Eighth ACM/IEEE Symposium on Edge Computing, Wilmington, DE, USA, 7–9 December 2023; pp. 372–377.

64. Lightweight M2M (LWM2M)—OMA SpecWorks. [Online]. Available online: https://www.omaspecworks.org/what-is-oma-specworks/iot/lightweight-m2m-lwm2m/ (accessed on 20 October 2018).
65. Kombate, Y.; Houngue, P.; Ouya, S. Securing MQTT: Unveiling Vulnerabilities and Innovating Cyber Range Solutions. *Procedia Comput. Sci.* **2024**, *241*, 69–76. [CrossRef]
66. Weqar, M.; Mehfuz, S.; Gupta, D.; Urooj, S. Adaptive Switching-Based Data-Communication Model for Internet of Healthcare Things Networks. *IEEE Access* **2024**, *12*, 11530–11540. [CrossRef]
67. Qin, W. Construction of Computer Network Management System Based on NETCONF and Artificial Intelligence. In Proceedings of the 2024 International Conference on Intelligent Algorithms for Computational Intelligence Systems (IACIS), Hassan, India, 23–24 August 2024; pp. 1–8. [CrossRef]
68. Alkwiefi, M.; Khalifeh, A. M2M Protocols: An Overview on LwM2M and XMPP Machine-to-Machine Protocols in IoT Context. In Proceedings of the IEEE International Conference on Computer and Information Science (ICIS), Shanghai, China, 20–22 September 2024; Volume 24, pp. 209–216. [CrossRef]
69. Varshney, U. Pervasive healthcare and wireless health monitoring. *Mob. Netw. Appl.* **2007**, *12*, 113–127. [CrossRef]
70. Navarro, K.F.; Lawrence, E. WSN applications in personal healthcare monitoring systems: A heterogeneous framework. In Proceedings of the 2010 Second International Conference on eHealth, Telemedicine, and Social Medicine BUSMMed 2010, Saint Maarten, Netherlands Antilles, 10–16 February 2010; pp. 77–83.
71. Mota, L.D.C.; Moreno, E.D.; Ribeiro, A.L.; Salgueiro, R.J.P.D.B. A Comparative Analysis of Network Management Protocols in IoT Applications. *J. Comput. Sci.* **2018**, *14*, 1238–1246. [CrossRef]
72. Alshammari, H.H. The internet of things healthcare monitoring system based on MQTT protocol. *Alex. Eng. J.* **2023**, *69*, 275–287. [CrossRef]
73. Soni, D.; Makwana, A. A survey on mqtt: A protocol of internet of things (iot). In Proceedings of the International Conference on Telecommunication, Power Analysis and Computing Techniques (ICTPACT-2017), Chennai, India, 6–8 April 2017; Volume 20, pp. 173–177.
74. Enns, R.; Bjorklund, M.; Schoenwaelder, J.; Bierman, A. *RFC 6241*; Network Configuration Protocol (NETCONF). Internet Engineering Task Force: Fremont, CA, USA, 2011. [CrossRef]
75. Silverajan, B.; Luoma, J.P.; Vajaranta, M.; Itäpuro, R. Collaborative cloud-based management of home networks. In Proceedings of the 2015 IFIP/IEEE International Symposium on Integrated Network Management (IM), Ottawa, ON, Canada, 11–15 May 2015; IEEE: Piscataway, NJ, USA, 2015; pp. 786–789.
76. Buzhin, I.G.; Derevyankin, A.Y.; Antonova, V.M.; Perevalov, A.P. Comparative Analysis of the REST and gRPC Used in the Monitoring System of Communication Network Virtualized Infrastructure. *T-Comm- Телекоммуникации и Транспорт* **2023**, *17*, 50–55. [CrossRef]
77. Perelman, V.; Ersue, M.; Schönwälder, J.; Watsen, K. *Network Configuration Protocol Light (NETCONF Light)*; Internet Engineering Task Force (IETF): Fremont, CA, USA, 2012.
78. Choi, H.; Kim, J.; Choi, S.; Kwon, T. 6LoWPAN-SNMP: Simple Network Management Protocol for 6LoWPAN. In Proceedings of the 2009 11th IEEE International Conference on High Performance Computing and Communications, Seoul, Republic of Korea, 25–27 June 2009; pp. 305–313. Available online: https://ieeexplore.ieee.org/document/5167008/ (accessed on 16 February 2025).
79. Veillette, M.; van der Stok, P.; Pelov, A.; Bierman, A.; Bormann, C. CoAP Management Interface (CORECONF); Internet Engineering Task Force, Internet-Draft Draft-Ietf-Core-Comi-19, 2024. Available online: https://datatracker.ietf.org/doc/draft-ietf-core-comi/ (accessed on 29 January 2025).
80. Lalhriatpuii, R.; Ruchi; Wasson, V. Comprehensive Exploration of IoT Communication Protocol: CoAP, MQTT, HTTP, LoRaWAN and AMQP. In Proceedings of the ICMLA 2024: Machine Learning Algorithms, Miami, FL, USA, 18–20 December 2024; Springer: Cham, Switzerland, 2025; Communications in Computer and Information Science, Volume 2238; pp. 261–274.
81. Rao, S.; Chendanda, D.; Deshpande, C.; Lakkundi, V. Implementing LWM2M in Constrained IoT Devices. In Proceedings of the 2015 IEEE Conference on Wireless Sensors (ICWiSe), Malacca, Malaysia, 24–26 August 2015; IEEE: Piscataway, NJ, USA, 2015; pp. 52–57.
82. Agyemang, J.O.; Kponyo, J.J.; Gadze, J.D.; Nunoo-Mensah, H.; Yu, D. A Lightweight Messaging Protocol for Internet of Things Devices. *Technologies* **2022**, *10*, 21. [CrossRef]
83. Słabicki, M.; Grochla, K. Performance evaluation of snmp, netconf and cwmp management protocols in wireless network. In *Electronics, Communications and Networks IV, Proceedings of the 4th International Conference on Electronics, Communications and Networks, Bhopal, India, 7–9 April 2014*; CRC Press: Boca Raton, FL, USA, 2014; pp. 12–15.
84. Kang, J.M.; Ju, H.T.; Choi, M.J.; Hong, J.W.K.; Kim, J.G. OMA DM-based remote software fault management for mobile devices. *Int. J. Netw. Manag.* **2009**, *19*, 491–511. [CrossRef]

85. Khaled, A.E.; Helal, S. Interoperable communication framework for bridging RESTful and topic-based communication in IoT. *Future Gener. Comput. Syst.* **2019**, *92*, 628–643. [CrossRef]
86. McKeown, N.; Anderson, T.; Balakrishnan, H.; Parulkar, G.; Peterson, L.; Rexford, J.; Shenker, S.; Turner, J. OpenFlow: Enabling Innovation in Campus Networks. *ACM SIGCOMM Comput. Commun. Rev.* **2008**, *38*, 69–74. . [CrossRef]
87. Claise, B.; Trammell, B.; Aitken, P. *RFC 7011*; Specification of the IP Flow Information Export (IPFIX) Protocol for the Exchange of Flow Information. Internet Engineering Task Force: Fremont, CA, USA, 2013. [CrossRef]
88. Khang, A.; Nazila, A.R.; Vugar, A.H.; Abuzarova, V.A. Advanced technologies and data management in the smart healthcare system. In *AI-Centric Smart City Ecosystems*; CRC Press: Boca Raton, FL, USA, 2022; pp. 261–270.
89. Lombardi, M.; Pascale, F.; Santaniello, D. Internet of things: A general overview between architectures, protocols and applications. *Information* **2021**, *12*, 87. [CrossRef]
90. Indrakumari, R.; Poongodi, T.; Sumathi, D.; Suganthi, S.; Suresh, P. IoT-based biomedical sensors for pervasive and personalized healthcare. In *Internet of Things, Artificial Intelligence and Blockchain Technology*; Springer: Cham, Switzerland, 2021; pp. 111–130.
91. Navaz, A.N.; Serhani, M.A.; El Kassabi, H.T.; Al-Qirim, N.; Ismail, H. Trends, technologies, and key challenges in smart and connected healthcare. *IEEE Access* **2021**, *9*, 74044–74067. [CrossRef] [PubMed]
92. Sehgal, A.; Perelman, V.; Kuryla, S.; Schönwälder, J. Management of resource constrained devices in the internet of things. *IEEE Commun. Mag.* **2012**, *50*, 144–149. [CrossRef]
93. MQTT Version 3.1.1. [Online]. Available online: http://docs.oasis-open.org/mqtt/mqtt/v3.1.1/mqtt-v3.1.1.html (accessed on 1 March 2025).
94. oneM2M. *Functional Architecture*, Technical Specification TS-0001, Version 3.11.1; oneM2M: Sophia Antipolis, France, 2022. Available online:https://www.onem2m.org/technical (accessed on 13 March 2025).
95. Savic, M. Bridging the SNMP gap: Simple network monitoring the internet of things. *Facta Univ. Ser. Electron. Energetics* **2016**, *29*, 475–487. [CrossRef]
96. Swetina, J.; Lu, G.; Jacobs, P.; Ennesser, F.; Song, J. Toward a standardized common M2M service layer platform: Introduction to oneM2M. *IEEE Wirel. Commun.* **2014**, *21*, 20–26. [CrossRef]
97. Woo, M.W.; Lee, J.W.; Park, K.H. A reliable IoT system for Personal Healthcare Devices. *Future Gener. Comput. Syst.* **2018**, *78*, 626–640. [CrossRef]
98. ISO/IEEE. *Health Informatics—Medical Device Communication—Part 11073: Service-Oriented Device Connectivity (SDC) Standards Series. ISO/IEEE 11073*; International Organization for Standardization (ISO)/IEEE: Geneva, Switzerland; New York, NY, USA, 2021.
99. Yao, J.; Warren, S. Applying the ISO/IEEE 11073 standards to wearable home health monitoring systems. *J. Clin. Monit. Comput.* **2005**, *19*, 427–436. [CrossRef]
100. Mekrache, A.; Ksentini, A.; Verikoukis, C. Machine Reasoning in FCAPS: Towards Enhanced Beyond 5G Network Management. *IEEE Commun. Surv. Tutor.* **2024**, *26*, 2769–2797. [CrossRef]
101. Amin, A.A.; Hasan, K.M. A review of fault tolerant control systems: Advancements and applications. *Measurement* **2019**, *143*, 58–68. [CrossRef]
102. Tariq, A.; Rehman, R.A.; Kim, B.S. Energy Efficient Priority Aware Forwarding in SDN-Enabled Named Data Internet of Things. In Proceedings of the 2020 International Conference on Electronics, Information, and Communication (ICEIC), Barcelona, Spain, 19–22 January 2020; IEEE: Piscataway, NJ, USA, 2020; pp. 1–4.
103. Tariq, A.; Rehman, R.A.; Kim, B.S. An Intelligent Forwarding Strategy in SDN-Enabled Named-Data IoV. *Comput. Mater. Contin.* **2021**, *69*, 3. [CrossRef]
104. Kreutz, D.; Ramos, F.M.V.; Verissimo, P.; Rothenberg, C.E.; Azodolmolky, S.; Uhlig, S. Software-Defined Networking: A Comprehensive Survey. *Proc. IEEE* **2014**, *103*, 14–76. [CrossRef]
105. Deepak, K.S.; Babu, A.V. Improving Reliability of Emergency Data Frame Transmission in IEEE 802.15.6 Wireless Body Area Networks. *IEEE Syst. J.* **2017**, *12*, 2082–2093. [CrossRef]
106. Ahad, A.; Tahir, M.; Yau, K.L.A. 5G-based smart healthcare network: Architecture, taxonomy, challenges and future research directions. *IEEE Access* **2019**, *7*, 100747–100762. [CrossRef]
107. Li, O.; Liu, H.; Chen, C.; Rudin, C. Deep Learning for Case-based Reasoning through Prototypes: A Neural Network that Explains Its Predictions. In Proceedings of the AAAI Conference on Artificial Intelligence, San Francisco, CA, USA, 4–9 February 2017.
108. Khater, H.M.; Sallabi, F.; Serhani, M.A.; Barka, E.; Shuaib, K.; Tariq, A.; Khayat, M. Empowering Healthcare with Cyber-Physical System–A Systematic Literature Review. *IEEE Access* **2024**, *12*, 83952–83993. [CrossRef]
109. Tariq, A.; Serhani, M.A.; Sallabi, F.M.; Barka, E.S.; Qayyum, T.; Khater, H.M.; Shuaib, K.A. Trustworthy Federated Learning: A Comprehensive Review, Architecture, Key Challenges, and Future Research Prospects. *IEEE Open J. Commun. Soc.* **2024**, *5*, 4920–4998. [CrossRef]

110. Tariq, A.; Sallabi, F.; Serhani, M.A.; Qayyum, T.; Barka, E.S. Leveraging Game Theory and XAI for Data Quality-Driven Sample and Client Selection in Trustworthy Split Federated Learning. *IEEE Trans. Consum. Electron.* **2025**. [CrossRef]
111. Babu, P.R.; Kumar, S.A.; Reddy, A.G.; Das, A.K. Quantum Secure Authentication and Key Agreement Protocols for IoT-Enabled Applications: A Comprehensive Survey and Open Challenges. *Comput. Sci. Rev.* **2024**, *54*, 100676. [CrossRef]
112. Qayyum, T.; Tariq, A.; Haseeb, M.W.; Lakas, A.; Serhani, M.A.; Trabelsi, Z. Quantum-Enhanced Convolutional Neural Networks for Image Classification. In Proceedings of the 2023 IEEE Future Networks World Forum (FNWF), Baltimore, MD, USA, 13–15 November 2023; IEEE: Piscataway, NJ, USA, 2023; pp. 1–7.
113. Conti, V.; Militello, C.; Rundo, L.; Vitabile, S. A Novel Bio-Inspired Approach for High-Performance Management in Service-Oriented Networks. *IEEE Trans. Emerg. Top. Comput.* **2020**, *9*, 1709–1722. [CrossRef]

mathematics

MDPI

Article

Enhancing Uplink Communication in Wireless Powered Communication Networks Through Rate-Splitting Multiple Access and Joint Resource Optimization

Iqra Hameed [1], Mario R. Camana [2], Mohammad Abrar Shakil Sejan [3] and Hyoung Kyu Song [1,*]

1 Department of Information and Communication Engineering and Convergence Engineering for Intelligent Drone, Sejong University, Seoul 05006, Republic of Korea; engriqrahameed056@gmail.com
2 Interdisciplinary Centre for Security, Reliability and Trust, University of Luxembourg, 4365 Luxembourg, Luxembourg; mario.camana@uni.lu
3 Department of Electrical Engineering, Sejong University, Seoul 05006, Republic of Korea; sejan@sejong.ac.kr
* Correspondence: songhk@sejong.ac.kr

Abstract: Wireless powered communication networks (WPCNs) provide a sustainable solution for energy-constrained IoT devices by enabling wireless energy transfer (WET) in the downlink and wireless information transmission (WIT) in the uplink. However, their performance is often limited by interference in uplink communication and inefficient resource allocation. To address these challenges, we propose an RSMA-aided WPCN framework, which optimizes rate-splitting factors, power allocation, and time division to enhance spectral efficiency and user fairness. To solve this non-convex joint optimization problem, we employ the simultaneous perturbation stochastic approximation (SPSA) algorithm, a gradient-free method that efficiently estimates optimal parameters with minimal function evaluations. Compared to conventional optimization techniques, SPSA provides a scalable and computationally efficient approach for real-time resource allocation in RSMA-aided WPCNs. Our simulation results demonstrate that the proposed RSMA-aided framework improves sum throughput by 12.5% and enhances fairness by 15–20% compared to conventional multiple-access schemes. These findings establish RSMA as a key enabler for next-generation WPCNs, offering a scalable, interference-resilient, and energy-efficient solution for future wireless networks.

Keywords: rate-splitting multiple access (RSMA); wireless powered communication networks (WPCNs); wireless energy transfer (WET); wireless information transmission (WIT); sum throughput maximization; fairness optimization; simultaneous perturbation stochastic approximation (SPSA); interference management; energy efficiency; resource allocation

MSC: 94-08; 94-10

1. Introduction

The rapid evolution of 6G networks aims to deliver seamless connectivity, ultra-reliable low-latency communication (URLLC), and enhanced mobile broadband (eMBB), and support massive machine-type communications (mMTC) [1,2]. A crucial aspect of 6G is the integration of low-power Internet of Things (IoT) devices, which will be widely used in applications such as smart cities and industrial automation [3–5]. However, these devices face significant challenges due to limited energy resources. Frequent battery replacement is often impractical, especially for devices deployed in remote or hard-to-reach locations. This highlights the urgent need for efficient energy management strategies to prolong their operational lifetimes [6,7].

1.1. Wireless Powered Communication Networks

Wireless powered communication networks (WPCNs) offer a promising solution to the energy constraints of low-power IoT devices [8–11]. In WPCNs, devices harvest energy from dedicated power sources, such as energy transmitters (ETs), and use this energy for uplink communication. This eliminates the reliance on frequent battery replacements by leveraging wireless energy transfer. As a result, WPCNs present a sustainable energy solution for low-power IoT devices, making them a viable option for 6G applications [12,13]. However, the performance of a WPCN relies on the coordinated design of energy harvesting and data transmission. Achieving the desired quality of service (QoS) requires optimizing the time allocation between wireless energy transfer (WET) and wireless information transmission (WIT) [14,15]. This allocation directly impacts the network's energy efficiency and overall throughput. In addition, advance applications such as 4K video streaming, virtual reality (VR), and augmented reality (AR) are central to 6G's promise, but these applications demand exceptionally high data rates, low latency, and seamless connectivity [16–18]. VR and AR, in particular, involve immersive experiences where even minor delays or insufficient bandwidth can significantly degrade the user experience. Similarly, 4K video streaming requires stable, high-throughput channels to ensure uninterrupted playback and maintain video quality. To meet these demands, efficient resource allocation strategies in multiple-access channels (MAC) become critical. In 6G networks, multiple-access technologies enable multiple users or devices to share limited communication resources such as frequency, time, or space. These channels ensure the delivery of high data rates to multiple users simultaneously while managing interference and optimizing spectral efficiency.

1.2. Rate-Splitting Multiple Access

Rate-splitting multiple access (RSMA) emerges as a promising candidate to address these challenges in multiple-access networks [19,20]. Unlike traditional multiple-access schemes like time-division multiple access (TDMA) or orthogonal multiple access (OMA), RSMA allows users to split their messages into common and private parts. These parts are transmitted over superimposed signals, enabling the network to manage interference more effectively. RSMA leverages successive interference cancellation (SIC) at the receiver to decode signals in a hierarchical manner, achieving a balance between spectral efficiency and user fairness [21,22]. By allocating power dynamically and managing interference flexibly, RSMA offers the capacity to support the high data rates required by 6G applications while ensuring robust performance in multi-user environments. This integrated design of RSMA and multiple-access technologies establishes the foundation for optimizing energy and information transmission, enabling 6G networks to meet the stringent demands of modern applications.

1.3. Related Works

WPCNs have been extensively studied to address energy constraints in modern communication systems. Significant focus has been placed on throughput maximization by optimizing resource allocation strategies. For instance, Ju et al. proposed a time allocation framework to maximize system throughput by balancing WET and WIT [15]. Energy beamforming, another key aspect, was explored in [23–25], where directional energy transmission enhanced the efficiency of power delivery to devices in WPCNs. In the context of multiple-access techniques, non-orthogonal multiple access (NOMA) has been integrated into WPCNs to improve spectral efficiency [26,27]. Other works include incorporating hybrid multiple-access schemes, such as the integration of TDMA and FDMA [28], and of NOMA and TDMA [29], to optimize energy and throughput trade-offs in WPCNs. These studies collectively highlight the advancements in WPCNs, focusing on innovative methods to optimize energy efficiency and communication performance in energy-constrained networks. However, no prior research has investigated the performance of WPCNs integrated with RSMA.

RSMA has been widely explored for its ability to optimize spectral efficiency and manage interference in wireless networks. Clerckx et al. laid the foundation of RSMA, showcasing its superiority over traditional multiple-access techniques like NOMA and SDMA in multi-user MISO systems with imperfect channel state information (CSI) [19]. Mao et al. extended this work, demonstrating how RSMA enables simultaneous wireless information and power transfer (SWIPT) in multi-antenna systems, optimizing both communication and energy harvesting performance [30]. Further, Camana et al. explored RSMA in cognitive radio settings, integrating SWIPT to optimize power allocation and beamforming, achieving improved energy and data trade-offs [31]. The integration of RSMA into intelligent reflecting surface (IRS)-assisted SWIPT networks has also been investigated, where dynamic beamforming and rate splitting significantly enhanced energy efficiency and throughput [32].

While most of the early focus was on downlink RSMA, recent studies have highlighted the importance of RSMA in uplink communications. In uplink scenarios, RSMA enables users to transmit their messages in non-orthogonal modes while addressing inter-user interference effectively through message splitting and SIC. Abbasi et al. introduced cooperative RSMA (C-RSMA), leveraging amplify-and-forward relaying and rate-splitting to improve throughput and fairness compared to NOMA [33]. Khisa et al. developed a joint beamforming and power allocation framework for C-RSMA, demonstrating superior SINR and performance over baseline schemes using successive convex approximation (SCA) [34]. Xiao et al. proposed an RSMA-enabled coordinated direct and relay transmission (CDRT) scheme, optimizing power allocation to enhance the reliability and throughput of far users while maintaining fairness for near users [35]. These works highlight uplink RSMA's potential in managing interference and optimizing performance in multi-user networks. Please refer to Table 1 for a summary of the related works.

Table 1. Comparison of existing works and our proposed approach.

Reference	Technique	Objective	Key Limitation	How Our Work Addresses Limitation
[15]	Time allocation in WPCNs	Maximizes throughput by balancing WET and WIT	Does not address interference and fairness issues	Our work integrates RSMA for interference mitigation and fairness enhancement
[24]	Energy beamforming in WPCNs	Improves power transfer efficiency	Limited by multi-user interference and inefficient resource allocation	Our joint optimization of RSMA-based power allocation overcomes these issues
[26]	NOMA-based WPCN	Enhance spectral efficiency	Suffers from high inter-user interference in uplink	RSMA effectively manages interference using rate splitting
[30]	RSMA for SWIPT	Optimizse energy harvesting and data rate trade-offs	Focuses only on downlink RSMA, no uplink extension	Our work extends RSMA to WPCNs in the uplink

Table 1. *Cont.*

Reference	Technique	Objective	Key Limitation	How Our Work Addresses Limitation
[33]	Cooperative RSMA in uplink	Improve fairness and throughput using amplify-and-forward relaying	No consideration of energy harvesting in WPCNs	We integrate RSMA into WPCNs while optimizing energy efficiency
Our Proposed Work	**RSMA-aided WPCN**	**Optimizes throughput, fairness, and energy efficiency**	**Addresses all limitations above**	**Joint optimization of RSMA parameters in WPCNs using SPSA**

1.4. Motivation and Contributions

Building upon the challenges outlined above regarding the need for efficient resource allocation in WPCNs, this work investigate a key gap in the literature, the integration of advanced multiple-access techniques to optimize both energy transfer and uplink communication. While traditional approaches have been extensively explored, they fall short in terms of flexibly managing interference and ensuring an optimal trade-off between throughput and fairness. RSMA has proven to be an effective approach for interference management and resource allocation in traditional wireless systems. Despite its demonstrated potential in improving throughput and fairness, no prior work has explored the integration of RSMA into WPCNs. The absence of such studies motivated us to investigate how RSMA can enhance WPCN performance by jointly optimizing the energy transfer and uplink communication processes.

The main contributions of this paper are summarized as follows:

- We propose an RSMA-aided WPCN framework for a two-user system, where WET occurs in the downlink and RSMA is applied in the uplink WIT. This design exploits RSMA's interference management capabilities to enhance uplink communication efficiency.
- Two distinct optimization problems are formulated: (a) maximizing the sum throughput to improve overall network efficiency and (b) optimizing fairness to ensure balanced resource allocation among users. These problems are addressed by jointly optimizing the rate-splitting factors and time allocations for the WET and WIT phases. The non-convex nature of these problems is tackled using the simultaneous perturbation stochastic approximation (SPSA) method, which provides an efficient and scalable solution.
- Through comprehensive numerical simulations, we analyze the trade-offs between sum throughput and fairness optimization objectives, providing critical insights into the applicability of RSMA in WPCNs. Our results demonstrate RSMA's potential to achieve balanced performance in energy and communication efficiency, underscoring its relevance for next-generation WPCNs.

2. System Model

In this section, we provide a detailed description of the system model used to analyze the performance of RSMA in a WPCN. As shown in Figure 1, the network comprises a hybrid access point (H-AP) equipped with a single antenna. There are two users, denoted as $\{U_1, U_2\}$, and each is equipped with a single antenna. The system operates in two distinct phases with energy harvesting on the downlink and data transmission on the uplink. The H-AP serves two primary functions by acting as an energy transmitter (ET) in the downlink phase, where it transmits energy signals to the users, allowing them to harvest energy for their uplink data transmission operations, and as a receiver in the uplink phase, where it collects the data transmitted by the users. Users U_1 and U_2 harvest energy from the H-AP during the downlink phase and then use the harvested energy to transmit their data to the H-AP during the uplink phase.

The communication links between the H-AP and the users, denoted as h_1, h_2, are modeled using quasi-static, flat-fading channels. The channel coefficients incorporate both large-scale path loss and small-scale fading effects. Specifically, we define the channel coefficient for user U_i as

$$h_i = \frac{g_i}{\sqrt{d_i^\beta}}, \quad i \in \{1,2\} \tag{1}$$

where g_i represents small-scale fading, d_i is the distance from the H-AP, and β is the path-loss exponent. This formulation explicitly captures the impact of distance on both energy harvesting and uplink transmission. We assume that the channel conditions remain constant during each transmission coherence interval but may vary between different transmission intervals. We assume reciprocal channels, meaning the channel gain remains the same for both downlink and uplink transmissions for each user, which is a reasonable assumption in time-division duplex (TDD) systems. Without loss of generality, we assume that $|h_1|^2 \geq |h_2|^2$. The key system parameters and their descriptions are summarized in Table 2 for clarity.

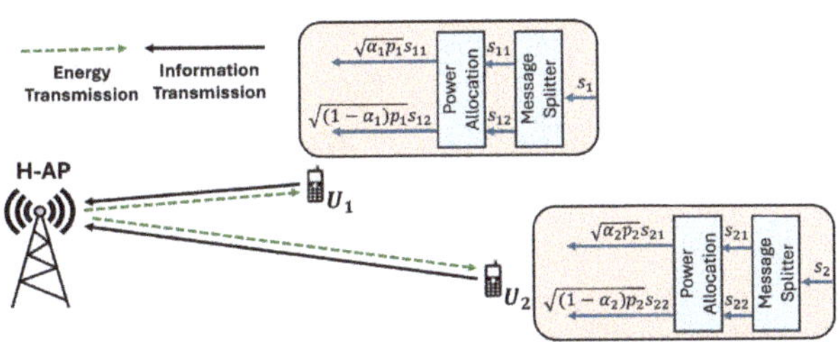

Figure 1. The RSMA-aided wireless powered communication network (WPCN) framework. The hybrid access point (H-AP) first transmits energy to users U_1 and U_2 via wireless energy transfer (WET), followed by wireless information transmission (WIT) in the uplink. Each user employs RSMA, where messages are split into two parts, optimally allocated for power and transmitted to the H-AP.

Table 2. List of parameters and their descriptions.

Symbol	Description	Unit
h_i	Channel coefficient between H-AP and user i	-
P_A	Transmit power at the H-AP	dBm
E_i	Energy harvested by user i	Joules
η_i	Energy conversion efficiency of user i	-
τ_d	Time allocated for downlink energy harvesting	Seconds
τ_u	Time allocated for uplink transmission	Seconds
p_i	Transmit power of user i in uplink	Watts
α_i	Power allocation coefficient for RSMA for user i	-
γ_{ij}	Signal-to-Interference-plus-Noise Ratio (SINR) of stream s_{ij}	-
R_i	Achievable rate of user i	bps/Hz
σ^2	Variance of additive Gaussian noise at H-AP	dBm
$L(x)$	Objective function in optimization problem	-
a_k, c_k	SPSA step size and perturbation constant	-
Δ_k	Perturbation vector in SPSA optimization	-
N	Number of iterations in SPSA algorithm	-

2.1. Phase 1: Downlink Energy Harvesting

During the downlink phase, the H-AP transmits an energy signal with power P_A. The users receive this energy and convert it into usable power through an energy harvesting mechanism. The signal received at U_i is expressed as

$$y_i = \sqrt{P_A} h_i + z_i, \quad i \in \{1,2\} \tag{2}$$

where z_i represents the noise at U_i during the energy harvesting phase, which is typically negligible due to the high power of the energy signal. The energy harvested by U_i is given by

$$E_i = \eta_i \tau_d P_A |h_i|^2, \quad i \in \{1,2\} \tag{3}$$

where η_i denotes the energy conversion efficiency of U_i, representing the efficiency of the user's energy harvesting circuitry, and τ_d is the duration of the energy harvesting phase. This harvested energy E_i is stored in the user's battery and will be used during the uplink transmission phase.

2.2. Phase 2: Uplink Data Transmission Using RSMA

In the uplink phase, U_1 and U_2 use the energy harvested during the downlink phase to transmit their data simultaneously to the H-AP. We apply the RSMA technique to manage the uplink transmissions and mitigate interference. The transmit power available for each user in the uplink is derived from the energy harvested during the downlink phase. This is mathematically expressed as

$$p_i = \frac{\eta_i \tau_d P_A |h_i|^2}{\tau_u}, \quad i \in \{1,2\} \tag{4}$$

where τ_u is the duration of the uplink phase.

- τ_u determines how efficiently the harvested energy is allocated for uplink transmission.
- The ratio $\frac{\tau_d}{\tau_u}$ determines how much energy is allocated for uplink transmission relative to the total harvested energy.
- Increasing τ_d results in more energy harvested, but it also reduces the available time for data transmission, creating a trade-off.

This equation highlights the fundamental trade-off between energy harvesting and data transmission duration, which is a critical aspect of system design in WPCNs. For further details on similar optimization frameworks, refer to [15]. We split the messages of U_1 and U_2 denoted as x_1 and x_2, respectively, into two sub-messages that are independently encoded into streams $\{s_{11}, s_{12}\}$ and $\{s_{21}, s_{22}\}$. These streams are then assigned specific power levels and superposed at U_1 and U_2. The received power at the H-AP is determined by the transmit power of each user and their respective channel gains. Since each user transmits a superposition of sub-messages in the uplink, the received signal at the H-AP is given by

$$y_A = \sqrt{\alpha_1 p_1 h_1} s_{11} + \sqrt{(1-\alpha_1) p_1 h_1} s_{12} + \sqrt{\alpha_2 p_2 h_2} s_{21} + \sqrt{(1-\alpha_2) p_2 h_2} s_{22} + z_A, \tag{5}$$

where the equation elements are defined as follows:

- p_1 and p_2 are the transmit powers of U_1 and U_2, respectively, which depend on the harvested energy.
- h_1 and h_2 are the respective channel coefficients of the users.
- $0 \leq \alpha_1, \alpha_2 \leq 1$ control the sub-messages.

- z_A represents the additive white Gaussian noise at the H-AP with zero mean and variance σ^2.
- $s_j, j \in \{11, 12, 21, 22\}$ satisfies $\mathbb{E}[|s_j|^2] = 1$.

  The received power at the H-AP for user U_i can be expressed as

$$P_{\text{received},i} \propto p_i |h_i|^2, \quad i \in \{1, 2\} \tag{6}$$

From this equation, it is evident that a stronger channel gain h_i leads to a higher received power at the H-AP, improving the signal decoding performance. However, due to interference and noise, optimal allocation of α_1 and α_2 is required to balance the trade-off between throughput and interference mitigation. The parameters α_1 and α_2 control the power allocation between sub-messages in the RSMA-based uplink transmission. Their values significantly impact the SINR at the H-AP, directly affecting the achievable rates of the users. To ensure an optimal trade-off between throughput and fairness, we optimize α_1 and α_2 using the SPSA algorithm. This allows dynamic adjustment of power allocation in response to changing channel conditions. Mathematically, the optimal values of α_1 and α_2 are determined as follows:

$$\{\alpha_1^*, \alpha_2^*\} = \arg\max_{\alpha_1, \alpha_2} \sum_{i=1}^{2} R_i(\alpha_1, \alpha_2) \tag{7}$$

where R_i represents the achievable rate for user U_i, and the constraints $0 \leq \alpha_1, \alpha_2 \leq 1$ ensure feasible power allocation. The SPSA algorithm iteratively updates α_1 and α_2 based on gradient approximations, making it computationally efficient for real-time optimization. This optimization ensures efficient interference management, leading to improved SINR, fair resource distribution between sub-messages, and better adaptability to dynamic channel conditions. By jointly optimizing α_1, α_2, and the time allocation parameters, the RSMA-based system achieves superior performance compared to traditional schemes.

RSMA Decoding Process at the H-AP

The H-AP applies RSMA and executes the successive interference cancellation (SIC) to decode the messages in the order of $s_{11} \to s_{21} \to s_{12} \to s_{22}$. The SINR for s_{11} at the H-AP is expressed as

$$\gamma_{11} = \frac{\alpha_1 p_1 |h_1|^2}{(1-\alpha_1) p_1 |h_1|^2 + \alpha_2 p_2 |h_2|^2 + (1-\alpha_2) p_2 |h_2|^2 + \sigma^2} \tag{8}$$

After successfully decoding and removing s_{11}, the H-AP decodes s_2. The SINR of s_{21} is expressed as

$$\gamma_{21} = \frac{\alpha_2 p_2 |h_2|^2}{(1-\alpha_1) p_1 |h_1|^2 + (1-\alpha_2) p_2 |h_2|^2 + \sigma^2} \tag{9}$$

Similarly, after successfully decoding s_{21} and subtracting it, the SINR of s_{12} is expressed as

$$\gamma_{12} = \frac{(1-\alpha_1) p_1 |h_1|^2}{(1-\alpha_2) p_2 |h_2|^2 + \sigma^2} \tag{10}$$

Similarly, after successfully decoding s_{12} and subtracting it, the SINR of s_{22} is expressed as

$$\gamma_{22} = \frac{(1-\alpha_2) p_2 |h_2|^2}{\sigma^2} \tag{11}$$

The achievable rates of users U_1 and U_2 in RSMA are computed using the Shannon capacity formula:

$$R_1 = \tau_u \log_2(1 + \gamma_{11}) + \tau_u \log_2(1 + \gamma_{12}), \tag{12}$$

$$R_2 = \tau_u \log_2(1 + \gamma_{21}) + \tau_u \log_2(1 + \gamma_{22}), \tag{13}$$

Each SINR term represents the quality of the received signal at the H-AP.

3. Problem Formulation and Network Resource Optimization

In this section, we formulate two distinct optimization problems to address critical challenges in RIS-aided RSMA-enabled WPCNs, the sum throughput optimization problem and the fairness problem. For each problem, we jointly optimize the power allocation coefficients for RSMA and the time allocation between the WET and WIT phases. These formulations aim to investigate the trade-offs between maximizing the system's sum throughput and ensuring fairness by maximizing a minimum throughput among users.

3.1. Sum Throughput Maximization Problem

In this problem, the objective is to maximize the total uplink throughput of the system by jointly optimizing the rate-splitting coefficients for the split messages and the time allocations for the downlink and uplink phases. The optimization is conducted under the constraint that all the harvested energy during the downlink phase is fully utilized for uplink communication, i.e., $p_i = \frac{\eta_i \tau_d P_A |h_i|^2}{\tau_u}; i \in \{1, 2\}$.

$$\mathcal{P}1 \max_{\alpha_1, \alpha_2, \tau_d, \tau_u} \quad R_1 + R_2 \tag{14a}$$

$$\text{subject to} \quad \tau_d + \tau_u \leq 1 \tag{14b}$$

$$0 \leq \alpha_1, \alpha_2 \leq 1 \tag{14c}$$

$$\tau_d \geq 0; \ \tau_u \geq 0 \tag{14d}$$

Constraint (14b) ensures that the total time allocation for the downlink (energy harvesting) and uplink (data transmission) phases does not exceed the overall time frame, normalized to 1. This enforces realistic partitioning of time resources between the two phases. Constraint (14c) ensures that it remains within feasible bounds. This parameter determines how power is divided between the split messages of U_1 in the RSMA-based uplink transmission. Finally, constraint (14d) ensures that all the optimization variables are non-negative, which aligns with practical physical limitations on time allocations.

3.2. Fairness Problem

Here, we address the fairness problem in uplink WIT. The primary objective is to maximize the minimum throughput for U_1 and U_2, thereby ensuring fairness across the network. The optimization problem is formulated as

$$\mathcal{P}2 \max_{\alpha_1, \alpha_2, \tau_d, \tau_u} \quad \min(R_1, R_2) \tag{15a}$$

$$\text{subject to} \quad \text{(14b)–(14d)} \tag{15b}$$

This problem ensures balanced resource allocation while considering the limitations imposed by energy harvesting and channel conditions. The focus on fairness addresses disparities in user throughput caused by variations in channel quality and energy harvesting efficiency, promoting equitable access to network resources.

4. Simultaneous Perturbation Stochastic Approximation (SPSA)-Based Solution

$\mathcal{P}1$ and $\mathcal{P}2$ represent non-convex objective functions and include non-convex constraints, primarily due to the highly coupled optimization variables, i.e., time allocations, and the RSMA-based split power coefficients. Their non-linear inter-dependencies, combined with strict system constraints, make conventional optimization techniques impractical. Therefore, we propose utilizing the simultaneous perturbation stochastic approximation (SPSA) algorithm as a robust and efficient solution framework. SPSA is a stochastic optimization algorithm that is well suited for high-dimensional problems with noisy objective functions. Unlike gradient-based methods, SPSA estimates the gradient of the objective function using a small number of noisy measurements, making it computationally efficient and scalable. This feature is particularly advantageous for non-convex optimization problems like $\mathcal{P}1$ and $\mathcal{P}2$, where the exact gradients are difficult to compute.

SPSA operates by approximating the gradient of the objective function $L(\mathbf{x})$ using stochastic perturbations. Unlike traditional gradient-based methods that require explicit derivative calculations, SPSA uses only two evaluations of the objective function per iteration, irrespective of the number of optimization variables. This makes it computationally efficient and scalable, particularly for non-convex problems in high-dimensional spaces.

The process begins by initializing the optimization variables:

$$\mathbf{x}_0 = [\alpha_1, \alpha_2, \tau_d, \tau_u]^T \tag{16}$$

where these variables are chosen to satisfy the system constraints. For instance, the time allocation variables τ_d and τ_u are subject to the constraint (14b). The RSMA coefficients α_1 and α_2 must satisfy (14c). These constraints ensure that all variables remain physically feasible within the operational limits of the system.

At each iteration k, SPSA estimates the gradient of the objective function using a stochastic perturbation vector:

$$\Delta_k = [\Delta_{k,1}, \Delta_{k,2}, \ldots, \Delta_{k,n}]^T \tag{17}$$

where each component $\Delta_{k,i}$ is randomly drawn from a symmetric Bernoulli distribution (± 1). The objective function is evaluated at two perturbed points:

$$L_+ = L(\mathbf{x}_k + c_k\Delta_k), \quad L_- = L(\mathbf{x}_k - c_k\Delta_k) \tag{18}$$

where c_k is a perturbation constant that scales the perturbations. The gradient is then approximated as

$$\hat{\mathbf{g}}_k = \frac{L_+ - L_-}{2c_k} \cdot \Delta_k^{-1} \tag{19}$$

where Δ_k^{-1} denotes the element-wise inverse of the perturbation vector. This estimated gradient guides the iterative update of the optimization variables as follows:

$$\mathbf{x}_{k+1} = \mathbf{x}_k - a_k\hat{\mathbf{g}}_k \tag{20}$$

where a_k is the step size, defined as

$$a_k = \frac{a}{(A+k)^\gamma} \tag{21}$$

with $a, A, \gamma > 0$ being tunable parameters. This step size decreases over iterations to ensure convergence. To maintain feasibility, the updated variables are projected back onto the constrained region defined by the system's physical and operational limits.

The stochastic perturbations in SPSA introduce randomness that allows the algorithm to explore the solution space effectively, avoiding local optima. As the algorithm progresses, the step size reduces, refining the search and ensuring convergence to a near-optimal solution. The termination criteria are based on either the stability of the objective function,

$$|L(\mathbf{x}_{k+1}) - L(\mathbf{x}_k)| < \epsilon \tag{22}$$

or reaching a maximum number of iterations. These conditions ensure computational efficiency while achieving high-quality solutions.

The application of SPSA to RSMA-enabled WPCNs offers several advantages. The proposed method summarized in Algorithm 1 efficiently handles the coupled, non-linear nature of the optimization variables, balancing system throughput and fairness. Its stochastic nature makes it robust to noise and fluctuations in channel conditions, which are inherent in wireless networks. Additionally, its low computational cost makes it practical for real-time implementation in dynamic environments.

Algorithm 1 Proposed solution based on SPSA for solving $\mathcal{P}1$ and $\mathcal{P}2$

1: **Input:** Objective function $L(\mathbf{x})$, initial values $\mathbf{x}_0 = [\alpha_1, \alpha_2, \tau_d, \tau_u]^T$, step size schedule a_k, perturbation scale c_k, maximum iterations $k_{\max}$, tolerance ϵ.

2: **Initialization:** Set iteration $k = 0$. Initialize $\mathbf{x}_k$ within feasible bounds:

$$0 \le \tau_d + \tau_u \le 1, \quad 0 \le \alpha_1, \alpha_2 \le 1.$$

3: **while** $k < k_{\max}$ and $|L(\mathbf{x}_{k+1}) - L(\mathbf{x}_k)| \ge \epsilon$ **do**

4: Generate a random perturbation vector $\Delta_k = [\Delta_{k,1}, \Delta_{k,2}, \ldots, \Delta_{k,n}]^T$, where $\Delta_{k,i} \sim \{-1, 1\}$.

5: Compute perturbed objective function values:

$$L_+ = L(\mathbf{x}_k + c_k\Delta_k), \quad L_- = L(\mathbf{x}_k - c_k\Delta_k).$$

6: Estimate the gradient:

$$\hat{g}_k = \frac{L_+ - L_-}{2c_k} \cdot \Delta_k^{-1},$$

where Δ_k^{-1} denotes the element-wise inverse of Δ_k.

7: Update the optimization variables:

$$\mathbf{x}_{k+1} = \mathbf{x}_k - a_k\hat{g}_k,$$

with $a_k = \frac{a}{(A+k)^\gamma}$, where $a, A, \gamma > 0$ are tunable parameters.

8: Project $\mathbf{x}_{k+1}$ onto the feasible region:

$$0 \le \tau_d + \tau_u \le 1, \quad 0 \le \alpha_1, \alpha_2 \le 1.$$

9: Increment iteration counter: $k = k + 1$.

10: **end while**

11: **Output:** Optimized variables $\mathbf{x}^* = \mathbf{x}_k$.

5. Simulation Results and Discussion

To validate the proposed RSMA-aided optimization framework for WPCNs, simulations were performed under realistic and practical system settings. The channel model adopted is Rayleigh fading, capturing the random nature of wireless propagation. U_1 and U_2 are deployed at distances of 100 m and 200 m, respectively, from the hybrid access point (H-AP). The channel incorporates path loss effects with a path loss coefficient of 3, simulating the attenuation experienced over a distance. The H-AP transmits with a fixed power of 30 dBm during the downlink phase, while the noise variance is set at -100 dBm to model typical environmental noise in wireless communication systems.

The proposed optimization framework leverages the SPSA algorithm to address the non-convex nature of the resource allocation problem. The key parameters for the SPSA algorithm were configured as perturbation size, $c = 0.1$, ensuring the perturbations remain small and controlled for efficient gradient estimation. The step size decay parameters $\gamma = 0.602$ define the rate of decay for the step size, balancing convergence speed and accuracy. The initial step size was $a = 0.002$, providing a stable starting point for iterative updates in the optimization process. The simulations were designed to evaluate the performance of the RSMA-aided solution in terms of key metrics such as sum throughput and fairness. Comparisons were made against benchmark schemes as follows:

- RSMA (Rate-Splitting Multiple Access):
  RSMA shows the proposed scheme where the power allocation coefficients and time allocation for the users are both optimized using the SPSA algorithm.
- FPRSMA (Fixed Power RSMA):
  FPRSMA is a variation of RSMA where the power allocation coefficients for the users are fixed and equally divided ($\alpha_1 = 0.5$, $\alpha_2 = 0.5$). However, the time allocation is dynamically optimized using the SPSA algorithm to maximize system performance.
- NONRSMA (Non-RSMA, Time-Division Multiple Access):
  NONRSMA represents a scheme that adopts a time-division multiple access (TDMA) approach for uplink data transmission. In this scheme, time allocation between users is optimized using SPSA, but RSMA principles (e.g., power splitting) are not employed.
- OMA (Orthogonal Multiple Access):
  OMA represents a fully orthogonal scheme where time allocation is fixed and predefined. Unlike the other schemes, OMA does not incorporate SPSA-based optimization, serving as a baseline for comparison.

Figure 2 depicts the convergence behavior of the proposed SPSA-based optimization algorithm under two different transmit power levels of the H-AP. The results demonstrate that the algorithm converges efficiently within approximately 50 iterations for both transmit power levels, reflecting the effectiveness of the SPSA method in addressing non-convex optimization problems. For $P_t = 40$ dBm, the objective function stabilizes at approximately 13.5 bps/Hz, while for $P_t = 30$ dBm, it converges to around 12 bps/Hz. This illustrates the significant impact of increased transmit power on achievable throughput, with higher power enabling better network performance. The proposed algorithm's rapid convergence and stable results confirm its robustness in optimizing resource allocation in RSMA-aided wireless powered communication networks.

Figure 3 shows the solution to $\mathcal{P}1$ and demonstrates the relationship between the sum throughput (in bps/Hz) and the transmit power of the H-AP (in dBm) for four different schemes, RSMA-$\mathcal{P}1$, FPRSMA-$\mathcal{P}1$, NONRSMA-$\mathcal{P}1$, and OMA. The results highlight the superior performance of RSMA-$\mathcal{P}1$, which involves optimizing both power allocation coefficients and time allocation using the SPSA algorithm. FPRSMA-$\mathcal{P}1$, using fixed power allocation and time optimization, achieves slightly lower throughput. NONRSMA-$\mathcal{P}1$, a TDMA-based scheme with optimized time allocation, performs moderately but remains significantly below RSMA-based schemes. OMA, with fixed power and time allocations, consistently achieves the lowest throughput. The figure demonstrates that as the H-AP transmit power increases, all schemes experience improved throughput. However, RSMA-$\mathcal{P}1$ outperforms the others significantly, showcasing the effectiveness of RSMA-based optimization in improving system performance under varying power levels.

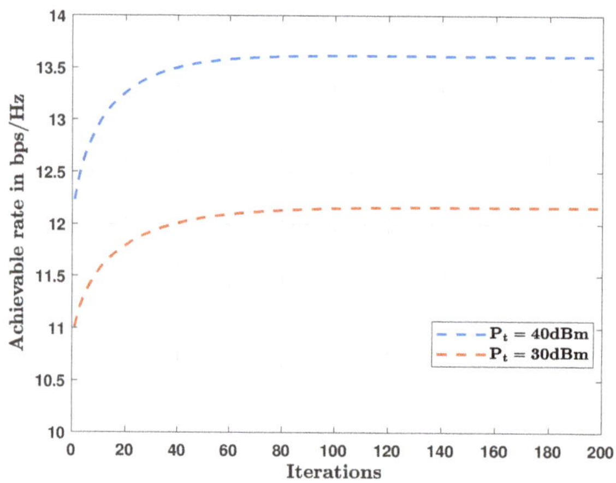

Figure 2. Convergence of proposed SPSA-based algorithm for different transmit power levels.

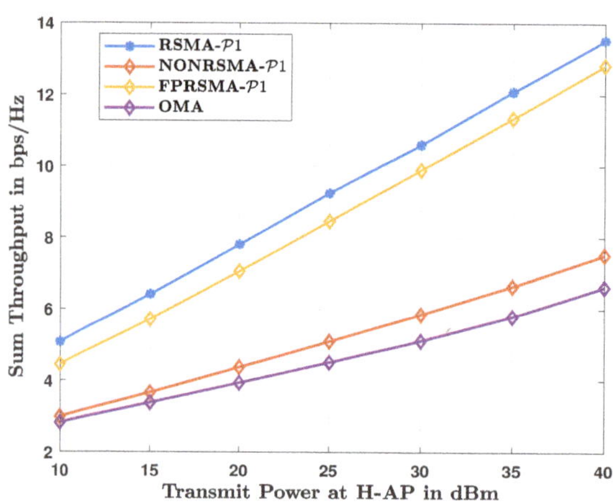

Figure 3. Sum throughput versus transmit power at H-AP for different schemes.

Figure 4 illustrates the throughput of U_2 (in bps/Hz) as a function of the transmit power of the H-AP (in dBm) for $\mathcal{P}1$ and $\mathcal{P}2$. RSMA-$\mathcal{P}1$ optimizes the sum throughput of the system, achieving a steady increase in U_2's throughput as the transmit power grows. RSMA-$\mathcal{P}2$, which optimizes the minimum throughput to enhance fairness between users, significantly improves U_2's throughput, particularly at higher power levels, surpassing RSMA-$\mathcal{P}1$. This demonstrates the effectiveness of RSMA-$\mathcal{P}2$ in ensuring balanced resource allocation between users. In comparison, NONRSMA-$\mathcal{P}1$, a TDMA-based scheme focused on sum throughput optimization, shows limited improvement in U_2's throughput, reflecting the inefficiency of non-RSMA schemes in managing interference and allocating resources effectively. NONRSMA-$\mathcal{P}2$, which optimizes the minimum throughput for fairness in TDMA, provides better throughput for U_2 than NONRSMA-$\mathcal{P}1$ but remains significantly below the RSMA-based schemes, particularly at higher transmit power levels.

The figure highlights the superiority of RSMA-based schemes over NONRSMA schemes in improving U_2's performance. RSMA-$\mathcal{P}2$, in particular, demonstrates remarkable fairness by ensuring higher throughput for U_2, even as the system prioritizes minimum throughput optimization.

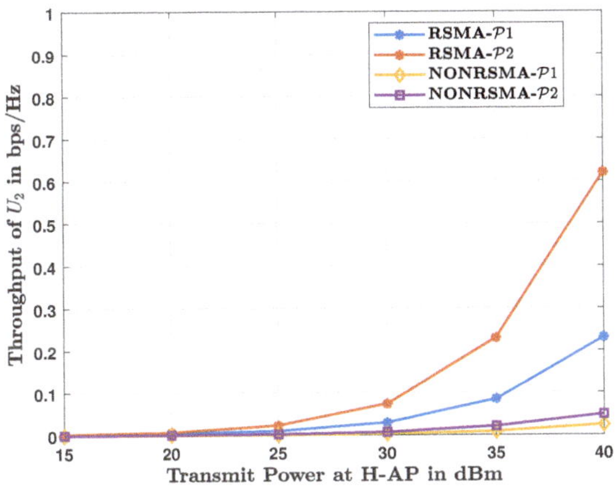

Figure 4. Throughput of U_2 versus transmit power at H-AP for different schemes.

Figure 5 illustrates the sum throughput (in bps/Hz) as a function of the distance between U_1 and U_2 (in meters). RSMA-$\mathcal{P}1$ achieves the highest and most consistent throughput, demonstrating its efficiency in handling interference and resource allocation regardless of user distance. FPRSMA also maintains a nearly constant throughput but performs slightly worse due to fixed power allocation. NONRSMA-$\mathcal{P}1$ shows a decreasing trend as distance increases, reflecting its limited ability to manage interference and adapt resource allocation. OMA performs the worst, with significant throughput degradation, as orthogonal resource allocation and fixed scheduling fail to handle the challenges of increasing user separation. The figure highlights the robust performance of RSMA-$\mathcal{P}1$ compared to other schemes.

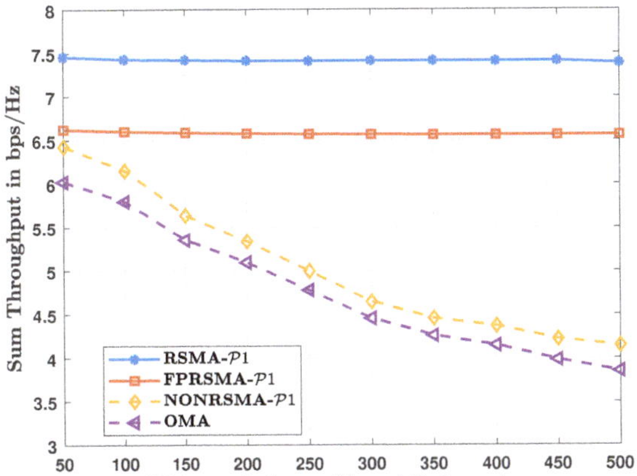

Figure 5. Sum throughput versus distance between U_1 and U_2 for different schemes.

Figure 6 illustrates the throughput of U_2 (in bps/Hz) as a function of the distance between U_1 and U_2 (in meters) for four schemes: RSMA-$\mathcal{P}1$, RSMA-$\mathcal{P}2$, NONRSMA-$\mathcal{P}1$, and NONRSMA-$\mathcal{P}2$. As the distance between users increases, the throughput of U_2 decreases across all schemes due to increased path loss and weakened interference management. RSMA-$\mathcal{P}1$ achieves the highest throughput for U_2 at shorter distances, while RSMA-$\mathcal{P}2$ performs better at larger distances, prioritizing fairness and balancing resource allocation. NONRSMA-$\mathcal{P}1$ and NONRSMA-$\mathcal{P}2$ exhibit significant degradation in throughput with increasing distance, reflecting the limitations of TDMA-based approaches in handling interference and optimizing resource allocation. These results emphasize the superiority of RSMA in maintaining higher throughput and fairness under varying user separations.

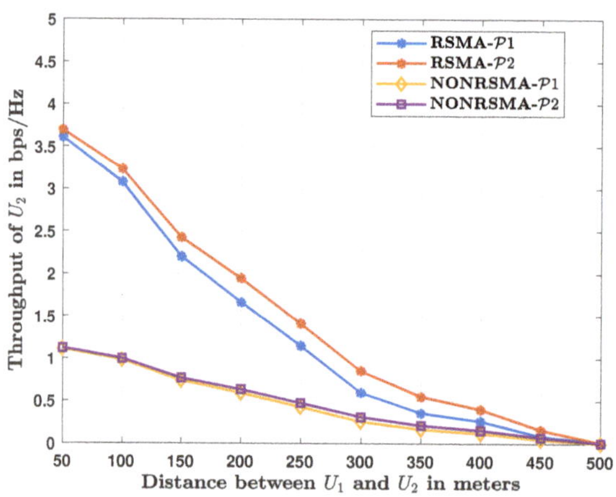

Figure 6. Throughput of U_2 versus distance between U_1 and U_2 for different schemes.

Figure 7 illustrates the sum throughput (in bps/Hz) as a function of the transmit power at the H-AP (in dBm). RSMA-$\mathcal{P}1$, which optimizes the sum throughput, achieves the highest performance across all transmit power levels, highlighting its efficiency in resource allocation and interference management. In contrast, RSMA-$\mathcal{P}2$, which focuses on maximizing the minimum throughput for fairness, provides slightly lower overall throughput but ensures better resource distribution among users, particularly benefiting those with weaker channel conditions. NONRSMA-$\mathcal{P}1$ and NONRSMA-$\mathcal{P}2$, representing TDMA-based schemes, perform significantly worse due to their limited ability to handle interference and allocate resources effectively. The trade-off between $\mathcal{P}1$ and $\mathcal{P}2$ reflects opposing objectives: $\mathcal{P}1$ maximizes efficiency but sacrifices fairness, while $\mathcal{P}2$ prioritizes fairness at the cost of reduced overall throughput. These results emphasize the superiority of RSMA schemes in achieving higher sum throughput and balancing the trade-off between fairness and efficiency.

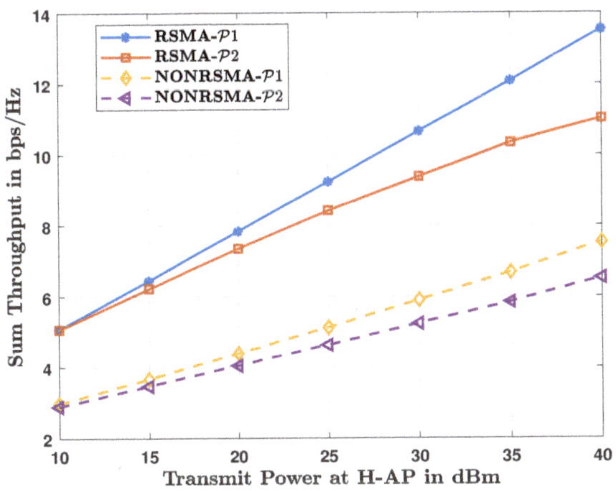

Figure 7. Sum throughput versus transmit power at H-AP.

6. Conclusions

In this paper, we proposed a novel RSMA-aided framework for WPCNs to address the dual challenges of energy efficiency and interference management in uplink communications. By integrating RSMA into the WIT phase and optimizing resource allocation jointly with WET, we investigated two key optimization problems: sum throughput maximization and fairness optimization. These non-convex problems were effectively addressed using the SSPSA method, which proved efficient in handling the complex joint optimization of rate-splitting factors and time allocation. Our numerical results demonstrated the trade-offs between the two objectives, highlighting RSMA's capability to achieve both high spectral efficiency and fairness under practical WPCN settings. Furthermore, the simulations validated RSMA's superior performance over traditional multiple-access schemes in terms of throughput, fairness, and energy efficiency. These findings establish the potential of RSMA as a promising approach for enhancing WPCN performance, paving the way for its adoption in next-generation wireless networks, especially in energy-constrained scenarios. Future work could explore extending this framework to multi-user and multi-antenna WPCNs, as well as investigating adaptive RSMA strategies for real-time resource allocation under dynamic network conditions.

Author Contributions: Conceptualization, I.H. and M.R.C.; Formal Analysis, I.H., M.A.S.S. and H.K.S.; Methodology, I.H., M.A.S.S. and M.R.C.; Supervision, H.K.S.; Validation, I.H., M.R.C. and M.A.S.S.; Writing—Original Draft, I.H. and M.A.S.S.; Writing—Review and Editing, I.H., M.R.C., M.A.S.S. and H.K.S.; Funding Acquisition, H.K.S. All authors have read and agreed to the published version of the manuscript.

Funding: This work was supported in part by an Institute of Information and Communications Technology Planning and Evaluation (IITP) grant funded by the Korean Government (Ministry of Science and ICT (MSIT), South Korea); in part under the Metaverse Support Program to Nurture the Best Talents, Grant IITP-2024-RS-2023-00254529; in part by the Basic Science Research Program through the National Research Foundation of Korea (NRF) funded by the Ministry of Education under Grant 2020R1A6A1A03038540; and in part by the MSIT (Ministry of Science and ICT, Korea) under the ITRC (Information Technology Research Center) support program (IITP-2025-RS-2024-00438007) supervised by the IITP (Institute for Information and Communications Technology Planning and Evaluation).

Data Availability Statement: The data are contained within the article.

Conflicts of Interest: The authors declare that they have no conflicts of interest regarding the publication of this paper.

References

1. Kamath, S.; Anand, S.; Buchke, S.; Agnihotri, K. A Review of Recent Developments in 6G Communications Systems. *Eng. Proc.* **2023**, *59*, 167. [CrossRef]
2. Rojek, I.; Kotlarz, P.; Dorożyński, J.; Mikołajewski, D. Sixth-Generation (6G) Networks for Improved Machine-to-Machine (M2M) Communication in Industry 4.0. *Electronics* **2024**, *13*, 1832. [CrossRef]
3. Zhang, Z.; Xiao, Y.; Ma, Z.; Xiao, M.; Ding, Z.; Lei, X.; Karagiannidis, G.K.; Fan, P. 6G Wireless Networks: Vision, Requirements, Architecture, and Key Technologies. *IEEE Veh. Technol. Mag.* **2019**, *14*, 28–41. [CrossRef]
4. Tataria, H.; Shafi, M.; Molisch, A.F.; Dohler, M.; Sjöland, H.; Tufvesson, F. 6G Wireless Systems: Vision, Requirements, Challenges, Insights, and Opportunities. *Proc. IEEE* **2021**, *109*, 1166–1199. [CrossRef]
5. Guo, F.; Yu, F.R.; Zhang, H.; Li, X.; Ji, H.; Leung, V.C.M. Enabling Massive IoT Toward 6G: A Comprehensive Survey. *IEEE Internet Things J.* **2021**, *8*, 11891–11915. [CrossRef]
6. Zhang, H.; Shlezinger, N.; Guidi, F.; Dardari, D.; Imani, M.F.; Eldar, Y.C. Near-Field Wireless Power Transfer for 6G Internet of Everything Mobile Networks: Opportunities and Challenges. *IEEE Commun. Mag.* **2022**, *60*, 12–18. [CrossRef]
7. Verma, S.; Kaur, S.; Khan, M.A.; Sehdev, P.S. Toward Green Communication in 6G-Enabled Massive Internet of Things. *IEEE Internet Things J.* **2021**, *8*, 5408–5415. [CrossRef]
8. Lu, X.; Wang, P.; Niyato, D.; Kim, D.I.; Han, Z. Wireless Networks with RF Energy Harvesting: A Contemporary Survey. *IEEE Commun. Surv. Tutor.* **2015**, *17*, 757–789. [CrossRef]
9. Bi, S.; Zeng, Y.; Zhang, R. Wireless powered communication networks: An overview. *IEEE Wirel. Commun.* **2016**, *23*, 10–18. [CrossRef]
10. Bi, S.; Ho, C.K.; Zhang, R. Wireless powered communication: Opportunities and challenges. *IEEE Commun. Mag.* **2015**, *53*, 117–125. [CrossRef]
11. Ramezani, P.; Jamalipour, A. Toward the Evolution of Wireless Powered Communication Networks for the Future Internet of Things. *IEEE Netw.* **2017**, *31*, 62–69. [CrossRef]
12. Jamshed, M.A.; Ali, K.; Abbasi, Q.H.; Imran, M.A.; Ur-Rehman, M. Challenges, Applications, and Future of Wireless Sensors in Internet of Things: A Review. *IEEE Sens. J.* **2022**, *22*, 5482–5494. [CrossRef]
13. Moloudian, G.; Hosseinifard, M.; Kumar, S.; Simorangkir, R.B.V.B.; Buckley, J.L.; Song, C.; Fantoni, G.; O'Flynn, B. RF Energy Harvesting Techniques for Battery-Less Wireless Sensing, Industry 4.0, and Internet of Things: A Review. *IEEE Sens. J.* **2024**, *24*, 5732–5745. [CrossRef]
14. Liu, X.; Qin, Z.; Gao, Y.; McCann, J.A. Resource Allocation in Wireless Powered IoT Networks. *IEEE Internet Things J.* **2019**, *6*, 4935–4945. [CrossRef]
15. Ju, H.; Zhang, R. Throughput Maximization in Wireless Powered Communication Networks. *IEEE Trans. Wirel. Commun.* **2014**, *13*, 418–428. [CrossRef]
16. Zhang, L.; Liang, Y.C.; Niyato, D. 6G Visions: Mobile ultra-broadband, super internet-of-things, and artificial intelligence. *China Commun.* **2019**, *16*, 1–14. [CrossRef]
17. Zhang, M.; Shen, L.; Ma, X.; Liu, J. Toward 6G-Enabled Mobile Vision Analytics for Immersive Extended Reality. *IEEE Wirel. Commun.* **2023**, *30*, 132–138. [CrossRef]
18. Zawish, M.; Dharejo, F.A.; Khowaja, S.A.; Raza, S.; Davy, S.; Dev, K.; Bellavista, P. AI and 6G Into the Metaverse: Fundamentals, Challenges and Future Research Trends. *IEEE Open J. Commun. Soc.* **2024**, *5*, 730–778. [CrossRef]
19. Clerckx, B.; Joudeh, H.; Hao, C.; Dai, M.; Rassouli, B. Rate splitting for MIMO wireless networks: A promising PHY-layer strategy for LTE evolution. *IEEE Commun. Mag.* **2016**, *54*, 98–105. [CrossRef]
20. Mao, Y.; Dizdar, O.; Clerckx, B.; Schober, R.; Popovski, P.; Poor, H.V. Rate-Splitting Multiple Access: Fundamentals, Survey, and Future Research Trends. *IEEE Commun. Surv. Tutor.* **2022**, *24*, 2073–2126. [CrossRef]
21. Mishra, A.; Mao, Y.; Dizdar, O.; Clerckx, B. Rate-Splitting Multiple Access for 6G—Part I: Principles, Applications and Future Works. *IEEE Commun. Lett.* **2022**, *26*, 2232–2236. [CrossRef]
22. Clerckx, B.; Mao, Y.; Schober, R.; Poor, H.V. Rate-Splitting Unifying SDMA, OMA, NOMA, and Multicasting in MISO Broadcast Channel: A Simple Two-User Rate Analysis. *IEEE Wirel. Commun. Lett.* **2020**, *9*, 349–353. [CrossRef]
23. Liu, L.; Zhang, R.; Chua, K. Multi-Antenna Wireless Powered Communication with Energy Beamforming. *IEEE Trans. Commun.* **2014**, *62*, 4349–4361. [CrossRef]
24. Hameed, I.; Tuan, P.V.; Camana, M.R.; Koo, I. Optimal Energy Beamforming to Minimize Transmit Power in a Multi-Antenna Wireless Powered Communication Network. *Electronics* **2021**, *10*, 509. [CrossRef]

25. Wang, D.; Tellambura, C. Performance Analysis of Energy Beamforming WPCN Links with Channel Estimation Errors. *IEEE Open J. Commun. Soc.* **2020**, *1*, 1153–1170. [CrossRef]
26. Song, D.; Shin, W.; Lee, J.; Poor, H.V. Sum-Throughput Maximization in NOMA-Based WPCN: A Cluster-Specific Beamforming Approach. *IEEE Internet Things J.* **2021**, *8*, 10543–10556. [CrossRef]
27. Khazali, A.; Tarchi, D.; Shayesteh, M.G.; Kalbkhani, H.; Bozorgchenani, A. Energy Efficient Uplink Transmission in Cooperative mmWave NOMA Networks with Wireless Power Transfer. *IEEE Trans. Veh. Technol.* **2022**, *71*, 391–405. [CrossRef]
28. Liu, Y.; Na, Z.; Liu, A.; Deng, Z. A Hybrid Multiple Access Scheme in Wireless Powered Communication Systems. In *Communications, Signal Processing, and Systems, Proceedings of the 8th International Conference on Communications, Signal Processing, and Systems, Urumqi, China, 20–22 July 2019*; Liang, Q., Wang, W., Liu, X., Na, Z., Jia, M., Zhang, B., Eds.; Springer: Singapore, 2020; pp. 1528–1532.
29. Afridi, A.; Hameed, I.; García, C.E.; Koo, I. Throughput Maximization of Wireless Powered IoT Network with Hybrid NOMA-TDMA Scheme: A Genetic Algorithm Approach. *IEEE Access* **2024**, *12*, 65241–65253. [CrossRef]
30. Mao, Y.; Clerckx, B.; Li, V.O. Rate-Splitting for Multi-User Multi-Antenna Wireless Information and Power Transfer. In Proceedings of the 2019 IEEE 20th International Workshop on Signal Processing Advances in Wireless Communications (SPAWC), Cannes, France, 2–5 July 2019; pp. 1–5. [CrossRef]
31. Camana Acosta, M.R.; Moreta, C.E.G.; Koo, I. Joint Power Allocation and Power Splitting for MISO-RSMA Cognitive Radio Systems with SWIPT and Information Decoder Users. *IEEE Syst. J.* **2021**, *15*, 5289–5300. [CrossRef]
32. Camana, M.R.; Garcia, C.E.; Koo, I. Rate-Splitting Multiple Access in a MISO SWIPT System Assisted by an Intelligent Reflecting Surface. *IEEE Trans. Green Commun. Netw.* **2022**, *6*, 2084–2099. [CrossRef]
33. Abbasi, O.; Yanikomeroglu, H. Transmission Scheme, Detection and Power Allocation for Uplink User Cooperation with NOMA and RSMA. *IEEE Trans. Wirel. Commun.* **2023**, *22*, 471–485. [CrossRef]
34. Khisa, S.; Elhattab, M.; Arfaoui, M.A.; Sharafeddine, S.; Assi, C. Power Allocation and Beamforming Design for Uplink Rate-Splitting Multiple Access with User Cooperation. *IEEE Trans. Veh. Technol.* **2024**, *73*, 10738–10743. [CrossRef]
35. Xiao, F.; Wen, M.; Yang, L.; Tsiftsis, T.A.; Liu, H. Intelligent Rate-Splitting Multiple Access-Enabled Coordinated Direct and Relay Transmission. *IEEE Wirel. Commun. Lett.* **2024**, *13*, 2606–2610. [CrossRef]

mathematics

MDPI

Article

Blind Channel Estimation Method Using CNN-Based Resource Grouping

Gayeon Kim [1], Yumin Kim [2], Daegun Jang [1], Byeong-Gwon Kang [2] and Taehyoung Kim [3,*]

[1] Department of ICT Convergence, Soonchunhayng University, Asan 31538, Republic of Korea; gayeon17@sch.ac.kr (G.K.); wowhensum@sch.ac.kr (D.J.)
[2] Department of Information and Communication Engineering, Soonchunhayng University, Asan 31538, Republic of Korea; kym@sch.ac.kr (Y.K.); bgkang@sch.ac.kr (B.-G.K.)
[3] School of Electrical Engineering, Kookmin University, Seoul 02707, Republic of Korea
* Correspondence: th.kim@kookmin.ac.kr

Abstract: This paper proposes a novel blind channel estimation method using convolutional neural network (CNN)-based resource grouping. The traditional K-means-based blind channel estimation scheme suffers limitations in reflecting fine-grained channel variations in both the time and frequency domains. To address these limitations, we propose dynamic resource grouping based on CNN architecture utilizing a two-step learning process that adapts to various channel conditions. The first step of the proposed method identifies the optimal number of subcarriers for each channel condition, providing a foundation for the second step. The second step adjusts the number of orthogonal frequency division multiplexing (OFDM) symbols, a parameter for determining the proposed pattern in the time domain, to adapt to dynamic channel variations. Simulation results demonstrate that the proposed CNN-based blind channel estimation method achieves high channel estimation accuracy across various signal-to-noise ratio (SNR) levels, attaining the highest accuracy of 82.5% at an SNR of 10 dB. Even when classification accuracy is relatively low, the CNN effectively mitigates signal distortion, delivering superior performance compared to conventional methods in terms of mean squared error (MSE) across diverse channel conditions. Notably, the proposed method maintains robust performance under high-mobility scenarios and severe channel variations.

Keywords: channel estimation; K-means algorithm; clustering; CNN

MSC: 94-08

1. Introduction

The 3rd Generation Partnership Project (3GPP) is currently advancing the standardization of Release-19, which corresponds to 5G-Advanced, while also discussing the roadmap for 6G standardization [1]. To realize 5G-Advanced and beyond, including 6G, active research is being conducted to integrate artificial intelligence (AI)-based machine learning (ML) and deep learning (DL) into communication services. AI has been widely applied to various technologies, such as traffic prediction, beamforming and beam management, cell coverage optimization, and channel estimation [2–6].

Several methods have been proposed to improve channel estimation performance, which is a core aspect of wireless communication systems [7–10]. In [7], a deep neural network (DNN)-based channel prediction framework was introduced to mitigate the degradation of channel estimation accuracy in scenarios with rapidly varying consecutive channel responses. Another approach to enhancing channel estimation performance proposed the sliding bidirectional gated recurrent unit (SBGRU) estimator, which integrates the recurrent neural network (RNN) structure with the sliding window concept [8]. To further enhance channel estimation, a convolutional neural network (CNN)-based method was proposed to tackle performance degradation caused by spreading factor interference in long-range (LoRa) systems [9]. Moreover, for doubly selective channels in high-speed mobile scenarios, channel estimation was enhanced through the compression-and-reconstruction channel estimation network (CRCENet) [10]. Therefore, accurate channel estimation is crucial to ensuring network stability and improving the quality of received signals across various environments.

For channel estimation, the demodulation reference signal (DM-RS), commonly referred to as a pilot signal, has been adopted as a standard technology for channel estimation. DM-RS is allocated to radio resources and transmitted along with uplink and downlink data, enabling it to effectively respond to real-time channel variations. However, due to its inherent characteristic of occupying radio resources, inefficient resource management of DM-RS can lead to resource wastage. Several studies have explored neural network-based approaches to optimize DM-RS resource management efficiently. Beyond optimizing existing DM-RS resources, channel estimation algorithms leveraging the most significant lag (MSL) have been proposed, along with semi-blind channel estimation methods that combine conventional DM-RS with detected symbols [11,12]. Moreover, blind channel estimation methods utilizing the K-means clustering algorithm, a widely used unsupervised learning scheme, have also been introduced [13]. K-means clustering-based channel estimation enhances wireless resource efficiency by relying solely on received signals, eliminating the need for DM-RS consumption. This approach provides an advantage over semi-blind channel estimation methods that require a significant number of DM-RS resources to ensure reliable operation. Therefore, this paper aims to improve the K-means algorithm-based channel estimation method, which is one of the blind channel estimation schemes that do not utilize DM-RS.

K. Jung and H. Wang [13] proposed a K-means-based blind channel estimation method that performs clustering on the received signals without transmitting DM-RS. In this study, the received signals are clustered using the K-means algorithm, and the centroids of the clusters are used to identify the original positions of quadrature phase shift keying (QPSK) symbols of the transmitted signal. By identifying the phases of QPSK symbols significantly altered due to multipath fading back to their original positions, this method achieves channel estimation without pilot signals, effectively canceling noise. However, this method has the limitation of determining the same channel for all received signals processed by the K-means clustering algorithm, making it ineffective in addressing channel variations at the subcarrier and orthogonal frequency division multiplexing (OFDM) symbol levels.

This paper overcomes the limitations of the methods proposed in [13] by introducing resource grouping patterns for K-means-based blind channel estimation and proposing a two-step learning method using CNN to select optimal patterns. Unlike existing methods, this study focuses on improving the efficiency of radio resource usage by effectively adapting to fine-grained and dynamically changing channel characteristics. The proposed two-step learning structure identifies the optimal number of subcarriers under each channel condition in the first step and refines the number of OFDM symbols based on time-domain variations in the second step. By designing dynamic resource grouping patterns adaptable to various channel conditions, such as signal-to-noise ratio (SNR), UE velocity, and delay spread, this study distinguishes itself from existing K-means algorithm-based approaches. Performance analysis evaluates the channel estimation accuracy of the proposed method and compares it to existing methods in terms of mean squared error (MSE) and resource efficiency. This study provides a foundation for robust and resource-efficient channel estimation in dynamic wireless communication systems.

2. System Model

2.1. NR Frame Structure

In New Radio (NR), transmitted data is allocated to time-frequency resources at the physical layer for transmission. Uplink and downlink data can be assigned to a single or multiple orthogonal frequency division multiplexing (OFDM) symbol(s) in the time domain within a single slot, as well as to a single or multiple RB(s) in the frequency domain. A single slot can consist of 14 OFDM symbols, and a single RB consists of 12 subcarriers [14].

2.2. K-Means-Based Blind Channel Estimation

Channel estimation based on the K-means clustering algorithm is performed on received signals. The generated transmission bits undergo channel coding, QPSK modulation, and OFDM modulation to generate the transmitted signal [14]. The transmitted signal then passes through a wireless fading channel, with noise added to produce the received signal. The noise is modeled as additive white Gaussian noise (AWGN). The receiver performs OFDM demodulation, channel estimation, channel decoding, and QPSK demodulation to estimate the transmitted bits [15]. The channel estimation is carried out using the K-means clustering algorithm. The estimated bits are compared with the original transmitted bits to calculate the bit error rate (BER). The received signal is written as:

$$y_{k,l} = h_{k,l}x_{k,l} + n_{k,l}, \tag{1}$$

where $y_{k,l}$ denotes the received signal for the k-th OFDM symbol and the l-th subcarrier, $h_{k,l}$ represents the wireless fading channel, $x_{k,l}$ denotes the transmitted signal, and $n_{k,l}$ denotes AWGN. The UE receiving $y_{k,l}$ preforms channel estimation based on the K-means clustering algorithm on the OFDM demodulated signal to obtain data information. This channel estimation method appropriately considers the delay spread and the velocity of the UE, classifying the signal into multiple groups. For instance, if the UE is moving at high velocity, the signal should be classified into finer units of OFDM symbols along the time domain. Conversely, if the UE exhibits significant delay spread, classification should occur at finer subcarrier units along the frequency domain. Within each appropriately classified group, the wireless resources are mapped to one of the four QPSK constellation points based on the K-means clustering algorithm, written as:

$$x_{k,l} = c^{\frac{j(2m-1)\pi}{4}}. \tag{2}$$

within each cluster, four clusters and their respective centroids are derived, as shown in Figure 1.

The data symbols, represented in four distinct colors, are identified as the centroids of each cluster. This process effectively cancels distortions caused by wireless channel effects and noise. However, there is an ambiguity issue, as it is unclear which of the original QPSK constellation points corresponds to the four QPSK symbols. To resolve this, channel estimation is performed based on Equation (3) to determine each centroid as one of the four QPSK symbols:

$$\hat{h}_{k,l} = \frac{y_{k,l}}{x_{k,l}}, \tag{3}$$

where $\hat{h}_{k,l}$ represents the estimated channel, and channel estimation is performed for all four clusters within each group, resulting in a total of 16 estimated channels. The estimated channel for all symbols within the range where the algorithm is applied is considered as the channel estimated by the K-means algorithm. To determine the best $\hat{h}_{k,l}$ among the 16 estimated channels, transmission bit estimation is performed based on:

$$\hat{x}_{k,l} = \frac{y_{k,l}}{\hat{h}_{k,l}}, \tag{4}$$

where $\hat{x}_{k,l}$ represents the estimated transmission signal, used along with the original transmitted signal $x_{k,l}$ to calculate the BER. The channel that achieves the highest BER is ultimately determined as the final estimated channel, $\tilde{h}_{k,l}$.

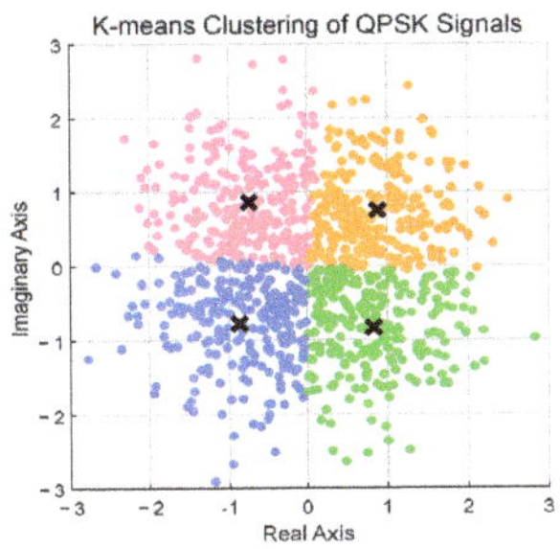

Figure 1. Four classified clusters based on K-means clustering of QPSK signal.

Subsequently, the channel estimation performance is analyzed through the MSE calculation between $\tilde{h}_{k,l}$ and $h_{k,l}$, which is based on:

$$MSE = \frac{1}{n}\sum_{k=1}^{n}\left(\frac{1}{m}\sum_{l=1}^{m}\left|h_{k,l} - \tilde{h}_{k,l}\right|^2\right), \tag{5}$$

where n denotes the number of OFDM symbols in the resource grid and m is the number of subcarriers in the resource grid.

3. Proposed Method

3.1. Resource Grouping Pattern

The channel estimation based on the K-means algorithm identifies the centroids of each cluster, effectively counteracting signal distortion. However, this channel estimation method must adequately account for significant channel variability along the time and frequency domains to perform clustering effectively. For instance, if the channel varies rapidly in the time domain due to high Doppler effects, the groups must be densely configured along the time domain. Similarly, if the channel exhibits rapid variation in the frequency domain due to significant delay spread, the groups must be densely configured with single or multiple subcarriers along the frequency domain. Such clustering-based K-means algorithm channel estimation cannot be optimally performed for all UEs. In this paper, to address this limitation, we propose a method that generates resource group patterns within the resource grid for received signals to enable the K-means-based blind channel estimation to operate effectively by identifying the characteristics of the dynamically changing channel. The resource group pattern consists of t consecutive OFDM symbols in the time domain and f consecutive subcarriers in the frequency domain. The number of consecutive OFDM symbols, t, can range from {1, 2, . . ., 14}. If t is not a divisor of 14, which is the total number of OFDM symbols in a single slot, the remaining OFDM symbols that do not satisfy this condition are grouped into the maximum possible number of consecutive OFDM symbols.

In this paper, such groups are defined as remainder groups. Additionally, the number of consecutive subcarriers, f, can take any value among the divisors of 120 within 10 RBs. The divisors of 120, $D(120)$ are expressed in Equation (6) and are listed as {1, 2, 3, 4, 5, 6, 8, 10, 12, 15, 20, 24, 30, 40, 60, 120}, for a total of 16 values:

$$D(120) = \left\{ 2^a \cdot 3^b \cdot 5^c \,\middle|\, 0 \le a \le 3,\ 0 \le b \le 1,\ 0 \le c \le 1 \right\}. \tag{6}$$

In other words, the resource grouping patterns for the received signal can consider 14 cases in the time domain and 16 cases in the frequency domain, resulting in a total of 224 possible patterns. For instance, in a pattern where $t = 7$ and $f = 60$ within 1 slot in the time domain and 10RBs in the frequency domain, one resource group consists of 7 consecutive OFDM symbols in the time domain and 60 consecutive subcarriers in the frequency domain.

As shown in Figure 2, the clustering pattern consists of a total of four groups, and K-means clustering-based channel estimation is performed for each group within the pattern. In Figure 2, within the resource grid corresponding to one slot in the time domain and 10 RB in the frequency domain, each group is distinguished by different colors, and K-means clustering-based channel estimation is executed for each group. Channel estimation is performed for a single resource group, as shown by the red dashed lines. The estimated channel values are represented as $\tilde{h}_1$, $\tilde{h}_2$, $\tilde{h}_3$, $\tilde{h}_4$, and each resource group is estimated to have the same channel.

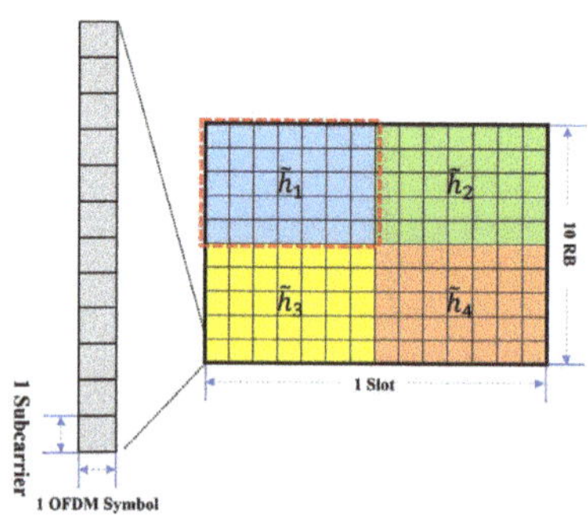

Figure 2. Resource grouping pattern for $t = 7$, $f = 60$.

The generated resource groups must identify QPSK symbols through K-means-based channel estimation, requiring at least four REs within each group. In this paper, this requirement is defined as the minimum RE count condition. Patterns that do not satisfy this condition are excluded. For example, in the pattern where $t = 2$ and $f = 1$, a single group consists of two consecutive OFDM symbols in the time domain and one subcarrier in the frequency domain, resulting in a total of 840 groups. Since each group in this pattern contains only two REs, channel estimation using the K-means algorithm cannot be performed. Therefore, this pattern is excluded for failing to meet the minimum RE count condition.

Another case is the pattern where $t = 1$ and $f = 3$, in which a single group consists of one OFDM symbol in the time domain and three subcarriers in the frequency domain, resulting in a total of 560 groups for the entire pattern. Since the REs within each group of this pattern are four or fewer, it is excluded. Additionally, the minimum RE count condition is equally applied to the remainder groups. For example, in the grouping pattern where $t = 6$ and $f = 1$, a resource group consists of six consecutive OFDM symbols in the time domain and a single subcarrier in the frequency domain. In this case, the 13th and 14th OFDM symbols in the time domain combined with a single subcarrier in the frequency domain form the remainder group, which contains only two REs and, thus, fails to meet the minimum RE count condition. Consequently, this resource grouping pattern is also excluded. Therefore, considering t and f, 212 patterns can be configured out of the 224 resource group patterns, excluding those that do not meet the minimum resource count condition.

3.2. Data Set Generation

This paper evaluates channel estimation performance by supporting optimal resource groups for all UEs. To this end, among all configured resource grouping patterns, the pattern that achieves the highest channel estimation performance in arbitrary channel environments is derived based on Equation (5) and utilized as the correct label for the CNN. The channel environment considers factors such as the UE's velocity and delay spread. For instance, when the UE's velocity is 30 km/h, the channel environment is relatively stable, making a pattern composed of resource groups with many OFDM symbols effective. Conversely, when the velocity is 150 km/h, a pattern composed of resource groups with a smaller number of OFDM symbols will be more effective.

The delay spread is considered from 50 ns to 300 ns in increments of 50 ns, and velocity is considered from 30 km/h to 150 km/h in increments of 30 km/h, resulting in 30 channel conditions. The SNR is assumed to be perfectly calculated. The parameters used for dataset generation, including these, are summarized in Table 1 [16].

Table 1. Simulation parameters.

Parameters	Value
Subcarrier spacing	15 [kHz]
Number of RBs	10
Number of OFDM symbols	14
FFT size	2048
Sample rate	61,440,000 [Hz]
Carrier frequency	4 [GHz]
Channel model	tapped-delay line-D
Velocity	30:30:150 [km/h]
Delay spread	50:50:300 [ns]
SNR	−6:2:10 [dB]
Noise	AWGN
N	100,000
Iteration	100,000
Batch size	256
Epoch	20
Learning rate	0.002
Dropout period	9
Dropout factor	0.1
Optimizer	Adam

The K-means algorithm is employed for channel estimation without relying on pilot signals. In this study, an exhaustive search is performed to identify the optimal pattern among the 212 proposed resource grouping patterns for each of the 30 considered channel conditions using the K-means algorithm. The MSE for each pattern is computed using Equation (5), and the pattern with the best performance is selected as the optimal pattern for the corresponding channel condition. The identified optimal patterns are used to generate correct labels for each step of the two-step training process, which is detailed in Section 3.3. Subsequently, a CNN is employed to effectively group the data according to various channel conditions, including delay spread, velocity, and SNR. The CNN is trained to learn the optimal grouping method for each channel condition and outputs the corresponding grouping pattern for each received signal, thereby enabling more accurate channel estimation. K-means algorithm-based channel estimation requires a sufficient number of data samples to achieve accurate channel estimation, and the minimum number of required data samples for channel estimation may vary with the SNR level. When performing K-means algorithm-based channel estimation under low SNR conditions, a relatively larger number of resource grouping patterns from the 212 proposed in this study have proven to be effective. In contrast, at higher SNR levels, patterns with a relatively smaller number of resources are utilized for channel estimation. The proposed 212 patterns can be effective depending not only on the SNR but also on the delay spread, which affects channel variations in the frequency domain, and the velocity, which impacts the time domain.

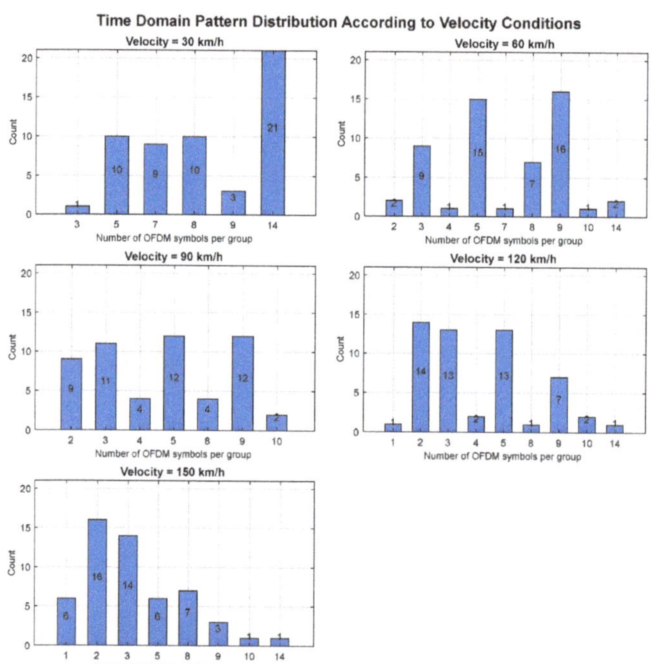

Figure 3. Time domain pattern distribution according to velocity conditions.

Therefore, in this study, the optimal pattern obtained through an exhaustive search of the proposed 212 patterns is analyzed to examine the influence of delay spread, velocity, and SNR on channel estimation performance, as illustrated in Figures 3–6. The 212 patterns proposed in this paper are grouped by each channel parameter and presented in Figures 3–6 to illustrate the distribution of the optimal patterns. Figure 3 shows the distribution of the optimal number of OFDM symbols as t, which is the parameter determining the pattern in the time domain under different velocity conditions. The x-axis in each graph, labeled "Number of OFDM symbols per group", represents the optimal t. In Figure 3, as the velocity of the UE increases, channel state variations in the time domain become more frequent and rapid, resulting in smaller values of t appearing more frequently at higher velocities. This indicates that as velocity increases, the number of OFDM symbols per group decreases, necessitating channel estimation at smaller group sizes. Consequently, for relatively low velocities, such as 30 km/h, patterns with larger values of t per group are determined to be optimal. In contrast, at higher velocities, such as 150 km/h, smaller group sizes are more suitable, leading to patterns with fewer OFDM symbols per group being identified as optimal.

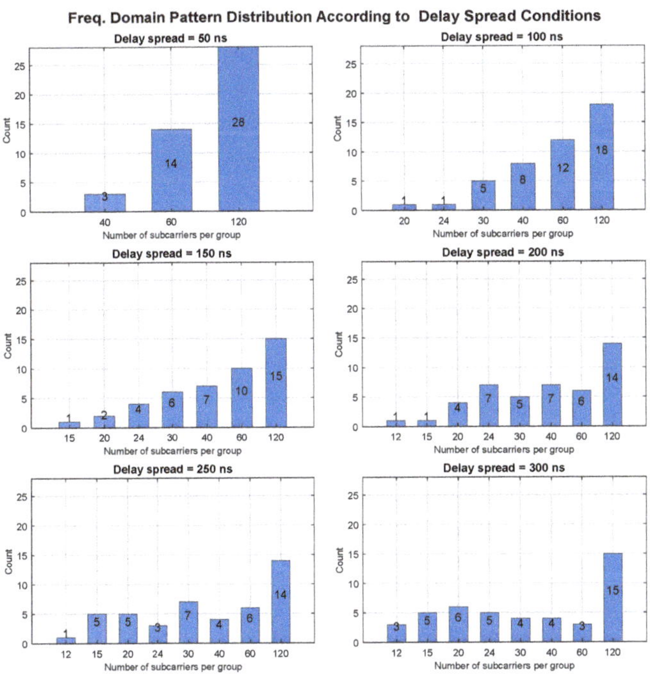

Figure 4. Frequency domain pattern distribution according to delay spread conditions.

Figure 4 illustrates the distribution of the optimal number of subcarriers, f, which is the parameter that determines the resource grouping pattern in the frequency domain under different delay spread conditions. This is represented on the x-axis of each graph.

In Figure 4, as the delay spread increases, channel variations in the frequency domain occur more rapidly and irregularly, requiring smaller values of f per group for channel estimation in the frequency domain. For a low delay spread, such as 50 ns, forming groups is frequently determined as optimal patterns. In contrast, for a high delay spread, such as 300 ns, the group size becomes smaller to respond more finely to channel variations, and smaller values of f are more widely distributed. This indicates that creating groups with smaller units is more effective under conditions of high delay spread, emphasizing the necessity of finer resource grouping in the frequency domain as the delay spread increases. Figures 3 and 4 demonstrate that resource grouping into smaller units is effective under relatively high levels of velocity and delay spread. However, a substantial proportion of cases were identified where larger units of t and f were optimal, which shows that the f with a value of 120 was the most frequently identified optimal value. Because the channel environments considered in this study include both low and high SNR levels, channel estimation based on the K-means algorithm requires an appropriate amount of data to achieve sufficient channel estimation performance based on the SNR level. This requirement increases as the SNR decreases, requiring more data. To meet this requirement, f, which offers more data points, is more effective than t in terms of data quantity. Consequently, when performing K-means-based channel estimation under high SNR conditions, smaller units of t and f are determined while satisfying the relatively small data requirements.

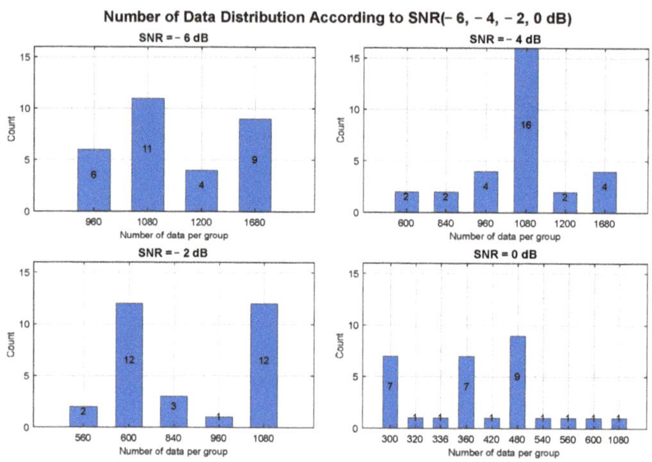

Figure 5. Number of data distribution according to SNR (−6, −4, −2, 0 dB).

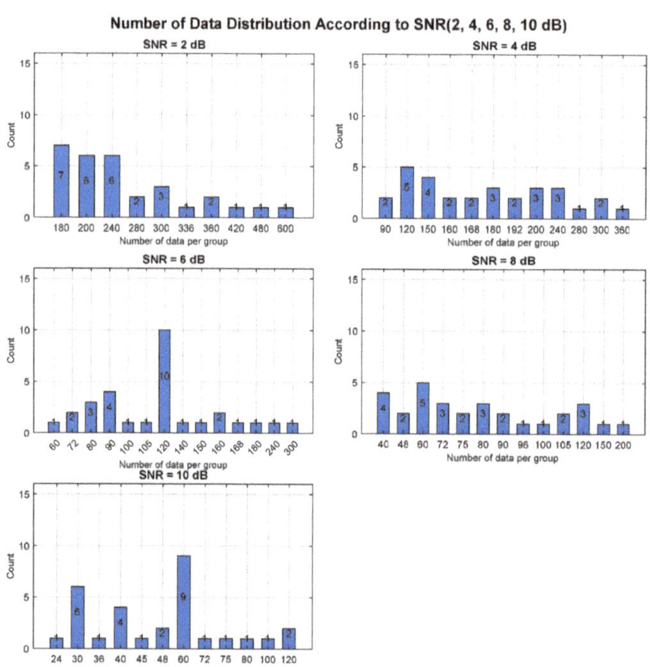

Figure 6. Number of data distribution according to SNR (2, 4, 6, 8, 10 dB).

The optimal data requirements for channel estimation under varying SNR levels are shown in Figures 5 and 6. Figure 5 illustrates the optimal number of data counts per group at SNR levels of −6, −4, −2, and 0 dB, while Figure 6 shows the corresponding values at SNR levels of 2, 4, 6, 8, and 10 dB. In low-SNR conditions, relatively large data quantities, such as 960, 1080, 1200, and 1680, were detected to ensure channel estimation performance. In contrast, under high-SNR conditions, smaller data quantities were more broadly distributed. As shown in Figures 3–6, the analysis of optimal patterns demonstrates that appropriate resource grouping in the time and frequency domains is essential for K-means-based channel estimation methods, considering channel factors such as UE velocity, delay spread, and SNR. Furthermore, it was confirmed that the amount of data required to ensure the performance of K-means-based channel estimation must also be considered.

The input data for the CNN is generated based on simulations. The two-step CNN training process is conducted using the final optimal patterns derived from exhaustive search, characterized by the frequency domain parameter f and the time domain parameter t. Thus, each step involves distinct input data and correct labels. For instance, during training in the frequency domain, the optimal patterns are classified based on the number of subcarriers f within the patterns. The grouped patterns are then defined with f as the correct label. The input data is generated based on the channel conditions associated with the patterns sharing the same correct label. Single or multiple patterns with the defined correct label f have associated channel conditions, such as delay spread, velocity, and SNR. These channel conditions are used to generate a tapped-delay line (TDL)-D channel. The channel generation process is repeated N times, and for each iteration, the channel conditions are randomly selected from the patterns with the same f, following a discrete uniform distribution. The transmitted signal $x_{k,l}$ passes through the generated TDL-D channel, and noise is added to produce $y_{k,l}$, as expressed in Equation (1).

The received N signals are processed as input data for the CNN by separating the real and imaginary components and adjusting the size to $1 \times 1680 \times 2 \times N$. The values are normalized between −1 and 1. In this study, N is set to 100,000. The generated dataset of 100,000 samples is divided into training, validation, and test datasets in proportions of 0.8, 0.1, and 0.1, respectively. The training dataset serves as input data for the CNN.

3.3. Optimal Resource Grouping Pattern Selection

To reduce the complexity of selecting the optimal resource grouping pattern and to efficiently extract features, a two-step learning process is performed by dividing the proposed 212 resource grouping patterns into frequency and time domains. In the first step, learning is focused on selecting the optimal parameter f, which is used to determine the resource grouping pattern in the frequency domain among the proposed resource grouping patterns. In the second step, the learning is focused on the parameter t, which determines the resource grouping pattern in the time domain. After completing the learning process, the trained CNN combines the optimal values of t and f to output the optimal resource grouping pattern within the given channel condition.

In the first step, since the network focuses solely on the frequency domain, learning is performed based exclusively on the number of subcarriers. First, the 212 resource grouping patterns are classified into groups that have the same number of subcarriers, f. Then, for the resource grouping patterns that use the same number of subcarriers, the corresponding subcarrier count, f, is defined as the new correct label for CNN training. The dataset is normalized to values between −1 and 1 under the channel conditions corresponding to this correct value, with a size of $1 \times 1680 \times 2 \times N$, and the network is constructed accordingly. After completing the first-step of training, the CNN outputs the optimal f. In the second step, since the network focuses solely on the time domain, learning focuses exclusively on the number of OFDM symbols in the proposed resource grouping patterns. First, the proposed patterns are grouped according to the same number of OFDM symbols, t. Then, for the patterns with the same number of OFDM symbols, the corresponding number, t, is defined as the new correct label for the second step of CNN training. For the channel conditions corresponding to this correct label, the dataset is normalized to values between -1 and 1, with a size of $1 \times 1680 \times 2 \times N$, and the network is constructed accordingly. Upon completing the second stage of training, the CNN outputs the optimal t, which is combined with the optimal f from the first stage to determine the optimal resource grouping pattern for the given channel environment.

3.4. Proposed CNN Structure

The proposed CNN architecture in this study is illustrated in Figure 7. As shown in Figure 7, the CNN consists of an input layer and 6 convolutional (Conv) layers. The input layer has dimensions of $1 \times 1680 \times 2 \times N$, where 1680 represents the total number of resources in the resource grid considered in this study, 2 corresponds to the real and imaginary parts of the received signal, and N denotes the total number of data samples. The network includes batch normalization (BN), rectified linear unit (ReLU) activation functions, and max pooling (MaxPool) layers, which are sequentially arranged in the structure.

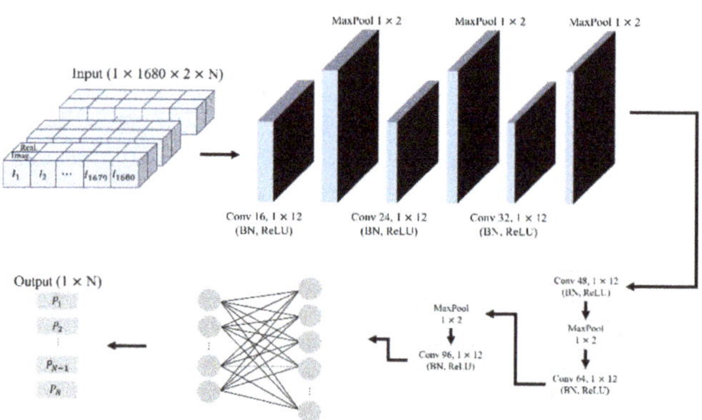

Figure 7. Proposed CNN structure.

The filter size for the convolutional layers is set to 1×12, and the number of filters gradually increases as the layers deepen, specifically to 16, 24, 32, 48, 64, and 96 filters. This incremental increase in the number of filters is designed to enable the CNN to extract progressively complex features, aligning with prior studies [17–19], which have shown that increasing the number of filters enhances data representation capabilities and learning accuracy in large-scale MIMO and mmWave systems. Furthermore, prior examples in OFDM systems demonstrate the effectiveness of this design in enabling the CNN to learn temporal and frequency-domain features effectively [20–22]. The increase in the number of filters in the CNN layer design is a critical criterion, ensuring efficient data processing in bandwidth- and noise-intensive channel environments. This approach maximizes the feature extraction capability, particularly in complex channel conditions [23,24]. The filter size is optimized for effectively learning long time-series data, and previous studies have reported that such designs contribute to improving the SNR [25,26].

Batch normalization and ReLU activation functions are employed as key components to maintain training stability and prevent overfitting. Following each convolutional layer, a

MaxPool layer with a size of 1×2 is used to reduce the data dimensions, thereby decreasing the computational load while retaining essential features [27,28]. This multi-layer structure is designed to enable the CNN to efficiently learn spatial and temporal features.

4. Simulation Results

This section provides a detailed analysis of the accuracy of the proposed two-step learning process and the CNN-based channel estimation performance. The N data samples used in the two-step training process were divided into training, validation, and test datasets in proportions of 0.8, 0.1, and 0.1, respectively. The training results are presented in Table 2. The horizontal axis represents the SNR, and the vertical axis corresponds to the learning accuracy at each step. The accuracy of the first step of learning shows 100% at an SNR of −6 dB, as the optimal pattern consistently selects f = 120. However, as the SNR increases from −4 dB to 2 dB, adjacent f values, such as f = 12, 15 or f = 24, 30 exhibits similar channel estimation performance, resulting in approximately 50% accuracy. When the SNR exceeds 4 dB, the CNN begins to effectively reflect channel states influenced by delay spread and velocity, achieving over 70% accuracy.

Table 2. Classification accuracy for the two-step training process.

Learn Method \ SNR [dB]		−6	−4	−2	0	2	4	6	8	10
[step 1] Learning according to the size of f	Accuracy [%]	100	55.9	51.7	57.1	55.02	72.0	70.6	78.2	87.3
[step 2] Learning based on the size of t when f has the same size	f	120	60	40	24	20	24	15	12	12
	Accuracy [%]	25	100	100	100	100	70.5	100	100	93.4
	f		120	60	30	24	30	20	15	15
	Accuracy [%]		29.8	100	100	100	89.3	84.3	93.3	95.2
	f			120	40	30	40	24	20	20
	Accuracy [%]			50.0	78.8	100	73.5	96.5	93.5	95
	f				60	40	60	30	24	24
	Accuracy [%]				57.9	89.9	73.5	90.6	93.6	98
	f				120	60	120	40	30	30
	Accuracy [%]				57.7	73.8	82.9	86.1	90.6	89.3
	f					120		60	40	40
	Accuracy [%]					74.2		85.3	99.4	92.8
	f							120	40	40
	Accuracy [%]							100	90.2	90.1
Total accuracy [%]		25	36.3	43.1	45.0	49.3	56.2	64.8	73.6	82.5

The accuracy of the two-step learning represents the classification performance of the optimal t for each f selected at the corresponding SNR. At −6 dB, due to the severely poor channel conditions, as shown in Figures 8 and 9, channel estimation performance in this low SNR range is significantly degraded. Consequently, the classification accuracy of the CNN was also found to be considerably low. At −4 dB, the first-stage learning identified f = 60 and f = 120 as optimal. However, due to the same reasons observed at −6 dB, the two-step learning recorded a low accuracy of 36.3%. Similarly, at −2 dB, the accuracy remained low at 43.1%. From 0 dB onward, as the SNR increased, the accuracy of the two-step learning progressively improved. The total accuracy also increased, reaching 82.5% at 10 dB. This improvement indicates that the CNN was able to more effectively learn the optimal t as the stability of the channel conditions improved at higher SNR.

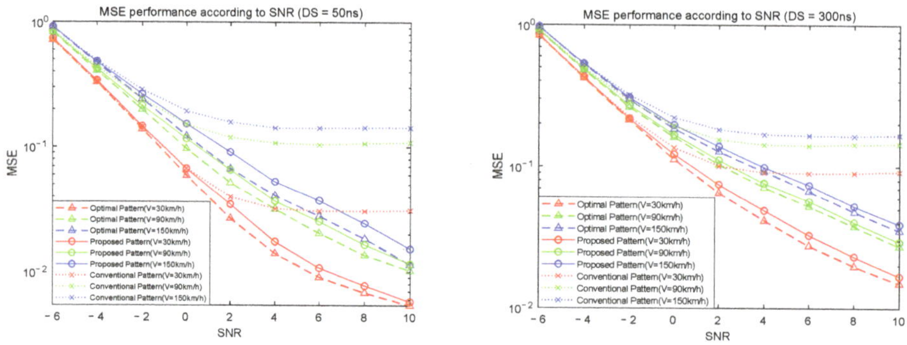

(a) (b)

Figure 8. MSE performance for each pattern with delay spared according to SNR: (**a**) 50 ns delay spread, (**b**) 300 ns delay spread.

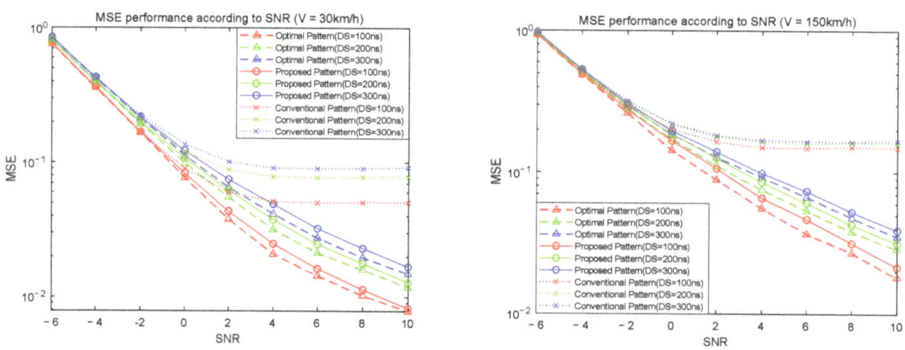

(a) (b)

Figure 9. MSE performance for each pattern with velocity according to SNR: (**a**) 30 km/h velocity, (**b**) 150 km/h velocity.

The proposed two-step learning method demonstrates a complementary structure. The first step of the learning accurately identifies the optimal number of subcarriers per group f, providing a solid foundation for the two-step learning to fine-tune t. This allows the CNN to adapt effectively to dynamic channel conditions. However, at low SNR values such as −6 dB, the significant degradation of channel conditions leads to imbalanced data distribution and excessive noise, which limit the accuracy of the CNN. These limitations highlight the potential for improvement through noise-resilient learning methods or data augmentation strategies.

In conclusion, the proposed two-step learning process adapts effectively to dynamic channel environments, maintaining consistently high accuracy as the SNR increases. Future research could focus on evaluating CNN performance under a wider range of channel conditions or incorporating reinforcement learning methods to enable real-time adaptation and further improve performance.

Figure 8 illustrates the MSE performance of three patterns according to SNR. The left graph represents the case with a delay spread of 50 ns, while the right graph corresponds to 300 ns. The bold dashed lines represent the optimal patterns obtained from the exhaustive search, as shown in Figure 3 through Figure 6. The solid lines represent the Proposed

Pattern obtained using the two-step CNN training introduced in this paper. The thin dashed line representing the Conventional Pattern indicates the channel estimation performance from the most recent study on K-means algorithm-based channel estimation, presented in [13].

In the Conventional Pattern, the K-means algorithm is applied to resources corresponding to one slot in the time domain Iand ten RB in the frequency domain, demonstrating superior performance compared to the traditional least-squares (LS) channel estimation method. Since the method proposed in this paper is designed to address the limitations of [13], its performance is analyzed through a comparison with the Conventional Pattern.

The red, green, and blue curves indicate velocities of 30 km/h, 90 km/h, and 150 km/h, respectively. The optimal pattern exhibited the best performance across all SNR ranges, followed by the proposed pattern and the conventional pattern. The proposed pattern showed slightly inferior performance compared to the optimal pattern, with the performance gap widening in the SNR range of −2 dB to 2 dB. However, beyond 2 dB, the gap gradually decreased, and the proposed pattern achieved performance close to the optimal pattern. In contrast, the conventional pattern exhibited the worst performance across all SNR ranges and tended to converge to a fixed MSE threshold beyond 0 dB. This indicates that the conventional pattern determines the same channel for all received signals processed through clustering, failing to effectively reflect the complexity of the channel environment, which limits performance improvement even in high-SNR conditions. For a delay spread of 50 ns, all patterns demonstrated relatively good MSE performance, and the proposed pattern consistently maintained lower MSE compared to the conventional pattern across all SNR ranges. On the other hand, when the delay spread increased to 300 ns, channel variations became more irregular, causing performance degradation for all patterns. Nonetheless, the proposed pattern continued to outperform the conventional pattern, demonstrating its reliability.

In the velocity analysis, at 30 km/h, the performance gap between the proposed pattern and the optimal pattern was minimal, while the proposed pattern outperformed the conventional pattern by approximately 3 dB. At higher velocities, such as 150 km/h, the performance of the proposed pattern was slightly degraded as a result of increased channel variability but still maintained significantly lower MSE compared to the conventional pattern, demonstrating high reliability even in high-mobility scenarios.

Figure 9 illustrates the MSE performance of three patterns according to SNR. The left graph represents the case where the UE velocity is 30 km/h, and the right graph corresponds to 150 km/h. The red, green, and blue curves represent the delay spread conditions of 100 ns, 200 ns, and 300 ns, respectively. The optimal pattern demonstrated the best performance across all SNR ranges under all conditions, while the proposed pattern showed slightly lower performance but maintained a similar trend. In the low SNR range from −6 dB to −2 dB, the proposed pattern exhibited performance close to the optimal pattern. However, as the SNR increased from −2 dB to 4 dB, the performance gap became more pronounced. Beyond 4 dB, the gap decreased, and the proposed pattern achieved performance close to the optimal pattern. At a velocity of 30 km/h, the performance gap between the proposed pattern and the optimal pattern was relatively small across all SNR ranges. In contrast, at a velocity of 150 km/h, the temporal variability of the channel increased, causing a slight degradation in the performance of the proposed pattern.

Nevertheless, it consistently outperformed the conventional pattern, demonstrating high reliability even in high-mobility scenarios. For a delay spread of 100 ns, all patterns demonstrated stable MSE performance across all SNR ranges, with the proposed pattern maintaining consistently lower MSE compared to the conventional pattern. On the other hand, when the delay spread increased to 300 ns, all patterns experienced performance degradation due to more irregular channel variations. Nonetheless, the proposed pattern still outperformed the conventional pattern, proving its high reliability. The conventional pattern showed a rapid performance degradation as the delay spread increased or the velocity rose, with its performance degradation particularly evident in high-mobility scenarios. Additionally, when the SNR increased beyond 2 dB, the conventional pattern tended to converge to a fixed MSE threshold. This is attributed to the fact that the conventional pattern estimates all received signals processed by the clustering algorithm as the same channel, failing to effectively reflect subtle channel state variations in high-SNR conditions.

5. Conclusions

This paper proposed a novel method to enhance the performance of K-means algorithm-based channel estimation by utilizing a CNN-based two-step learning method. To overcome the limitation of the conventional K-means-based blind channel estimation method, which determines the same channel for all received signals processed through clustering and thus fails to effectively reflect fine-grained channel variations at the subcarrier and OFDM symbol levels, this study designed dynamic clustering grouping patterns and a two-step learning structure adaptable to dynamic channel conditions. The simulation results demonstrated that the proposed method consistently maintained high channel estimation performance under various channel conditions. Compared to the optimal pattern, the proposed scheme showed similar channel estimation performance across all SNR ranges, achieving near-optimal performance in low SNR and progressively improving to closely match the optimal pattern in high SNR. Additionally, the proposed CNN-based learning method outperformed the conventional pattern in all conditions in terms of MSE and maintaining stability, even in high-mobility scenarios with significant channel fluctuations. These results highlighted the proposed scheme's adaptability to dynamic channel conditions.

Author Contributions: Conceptualization and Methodology, G.K. and Y.K.; Data Analysis, G.K.; Data Collection and Experimentation, G.K., Y.K. and D.J.; Experimental Results Analysis—D.J.; Writing—Original Draft, G.K. and D.J.; Writing—Review and Editing, B.-G.K. and T.K.; Supervision and Funding acquisition, T.K. All authors have read and agreed to the published version of the manuscript.

Funding: This work was supported in part by the Soonchunhyang University Research Fund and in part by an Institute of Information Communications Technology Planning Evaluation (IITP) grant, funded by the Korean Government (MSIT) (No. RS-2024-00397789, Development and Standardization of 6G Radio Transmission Standard Technology).

Data Availability Statement: The original contributions presented in this study are included in the article. Further inquiries can be directed to the corresponding author.

Conflicts of Interest: The authors declare no conflicts of interest.

References

1. *Study on Artificial Intelligence (AI)/Machine Learning (ML) for NR Air Interface (Release 18)*; 3GPP TR 38.843 V18.0.0; 3GPP: Sophia Antipolis, France, December 2023.
2. Shubo, S.; Haipeng, H. Network Traffic Prediction Based on the Multi-Time Granularity GRU-BP Neural Network. *IEEE Access* **2024**, *12*, 96997–97003. [CrossRef]
3. Jiang, G.; Qi, C. Near-Field Beam Training Based on Deep Learning for Extremely Large-Scale MIMO. *IEEE Commun. Lett.* **2023**, *27*, 2063–2067. [CrossRef]

4. Sun, Y.; Guo, G.; Zhang, S.; Xu, S.; Wang, T.; Wu, Y. A Cluster-Based Energy-Efficient Resource Management Scheme Wth QoS Requirement for Ultra-Dense Networks. *IEEE Access* **2020**, *8*, 182412–182421. [CrossRef]
5. Kim, E.-H.; Lee, J.-W.; Kim, Y.-M.; Hong, E.-K. Analysis of the Optimal Number of Clusters in UDN Environment. In Proceedings of the 2019 IEEE VTS Asia Pacific Wireless Communications Symposium (APWCS), Singapore, 28–30 September 2019; pp. 1–4.
6. Ke, S.; Li, Y.; Gao, Z.; Huang, L. An Adaptive Clustering Approach for Small Cells in Ultra-Dense Networks. In Proceedings of the 2017 9th International Conference on Advanced Infocomm Technology (ICAIT), Chengdu, China, 25–28 June 2017; pp. 421–425.
7. Ge, L.; Li, J.; Zhang, M.; Zhang, W. Deep Neural Network Based Channel Estimation for Massive MIMO-OFDM Systems with Imperfect Channel State Information. *IEEE Syst. J.* **2022**, *16*, 4675–4685. [CrossRef]
8. Bai, Q.; Zhang, Y.; Song, J. Deep Learning-Based Channel Estimation Algorithm over Time Selective Fading Channels. *IEEE Trans. Cogn. Commun. Netw.* **2020**, *6*, 125–134. [CrossRef]
9. Huang, J.; Cai, G. Joint Channel Estimation and Signal Detection for LoRa Systems Using Convolutional Neural Network. *IEEE Commun. Lett.* **2024**, *28*, 662–666. [CrossRef]
10. Peng, Q.; Li, J.; Shi, H. Deep Learning-Based Channel Estimation for OFDM Systems with Doubly Selective Channel. *IEEE Commun. Lett.* **2022**, *26*, 2067–2071. [CrossRef]
11. Mehrabani, M.R.; Abolhassani, B.; Haddadi, F. Semi-Blind Channel Estimation and Pilot Allocation for Doubly-Selective Massive MIMO-OFDM Systems. *IEEE Access* **2024**, *12*, 111565–111578. [CrossRef]
12. Darya, A.M.; Abdallah, S. Semi-Blind Channel Estimation for Massive MIMO LEO Satellite Communications. *IEEE Commun. Lett.* **2025**, *29*, 75–79. [CrossRef]
13. Jung, K.; Wang, H. Pilotless Channel Estimation Scheme Using Clustering-Based Unsupervised Learning. In Proceedings of the 15th International Symposium on Wireless Communication Systems (ISWCS), Lisbon, Portugal, 28–31 August 2018; pp. 1–5.
14. *Physical Channel and Modulation (Release 17)*; 3GPP TS 38.211 V17.0.0; 3GPP: Sophia Antipolis, France, 2021.
15. Dahlman, E.; Parkvall, S.; Skold, J. *5G NR: The Next Generation Wireless Access Technology*; Academic Press: London, UK, 2020; pp. 141–152.
16. *Study on Channel Model for Frequencies from 0.5 to 100 GHz (Release 17)*; 3GPP TR 38.901 V17.0.0; 3GPP: Sophia Antipolis, France, 2022.
17. Dong, P.; Li, X.; Wang, S.; Wei, H. Deep CNN-Based Channel Estimation for mmWave Massive MIMO Systems. *IEEE J. Sel. Top. Signal Process.* **2019**, *13*, 989–1002. [CrossRef]
18. Gizzini, A.K.; Ali, H.; Hussain, M. CNN-Aided Weighted Interpolation for Channel Estimation in Vehicular Communications. *IEEE Trans. Wirel. Commun.* **2021**, *20*, 4235–4247. [CrossRef]
19. Khichar, S.; Rao, K. Efficient Channel Estimation in OFDM Systems Using a Fast Super-Resolution CNN Model. *J. Sens. Actuator Netw.* **2024**, *8*, 230–245. [CrossRef]
20. Jiang, P.; Wang, Y.; Zhou, H.; Wu, T. Dual CNN-Based Channel Estimation for MIMO-OFDM Systems. *IEEE Trans. Commun.* **2021**, *69*, 2520–2531. [CrossRef]
21. Dong, P.; Wang, X.; Zhou, F. Deep CNN for Wideband mmWave Massive MIMO Channel Estimation Using Frequency Correlation. In Proceedings of the IEEE International Conference on Acoustics, Speech and Signal Processing (ICASSP), Brighton, UK, 12–17 May 2019; pp. 456–460.
22. Kim, J.; Park, H.; Lee, Y.; Choi, S. Feature Optimization in OFDM Systems Using Deep CNNs. *IEEE Trans. Wirel. Commun.* **2024**, *25*, 1021–1032.
23. Patel, A.; Kumar, R.; Singh, H. Impact of Convolutional Layer Depth on MIMO Channel Estimation. *IEEE Trans. Signal Process.* **2023**, *71*, 456–465.
24. Wang, L.; Zhou, X.; Liu, M.; Zhang, W. Multi-Scale Convolutional Neural Networks for mmWave Massive MIMO Systems. *IEEE J. Sel. Areas Commun.* **2023**, *41*, 155–167.
25. Zhou, X.; Lin, H. Advanced Filter Design in Deep Learning for Wireless Communication Systems. *IEEE Access* **2024**, *11*, 7658–7672.
26. Gupta, R.; Lee, T. Temporal Feature Extraction in OFDM Systems Using 1D-CNN Models. *J. Sens. Netw.* **2024**, *19*, 233–245.
27. Zhang, M.; Sun, L.; He, K. Dynamic Filter Allocation for CNN-Based Signal Processing. In Proceedings of the IEEE International Conference on Communication Systems, Paris, France, 18–20 June 2023; pp. 421–429.
28. Park, H.; Kim, K. Improving Feature Extraction in OFDM Systems Using Adaptive CNN Models. In Proceedings of the IEEE Wireless Communications and Networking Conference (WCNC), Shanghai, China, 12–15 April 2024; pp. 21–27.

mathematics

MDPI

Article

Joint Transmit Power and Power-Splitting Optimization for SWIPT in D2D-Enabled Cellular Networks with Energy Cooperation

Dong-Woo Lim [1] and Jae-Mo Kang [2,*]

[1] Department of Information & Communication Engineering, Changwon National University, Changwon 51140, Republic of Korea; dwlim@changwon.ac.kr
[2] Department of Artificial Intelligence, Kyungpook National University, Daegu 41566, Republic of Korea
* Correspondence: jmkang@knu.ac.kr

Abstract: In this paper, we propose a joint optimization scheme for a transmit power and power-splitting ratio in device-to-device (D2D)-enabled simultaneous wireless information and power transfer (SWIPT) cellular networks, considering energy signal transmission. This energy signal facilitates the energy cooperation between the D2D transmitter (DT) and the CU. Under the proposed scheme, the D2D rate is maximized while guaranteeing that the cellular user (CU) achieves the same performance as in scenarios without D2D communications. In order to solve the formulated nonconvex problem, we leverage the monotonically increasing property of logarithmic functions to transform it into an equivalent convex problem. As a result, we obtain the optimal solution in closed form. Also, the optimal D2D performance is analyzed, and useful insights into the performance improvements achievable through the proposed scheme are obtained. Numerical results demonstrate that the proposed scheme significantly outperforms the baseline scheme.

Keywords: device-to-device (D2D) communication; simultaneous wireless information and power transfer (SWIPT); transmit power optimization; power-splitting ratio optimization

MSC: 94A05

1. Introduction

Device-to-device (D2D) communication has become a focal point of research in wireless networks, particularly with its adoption in the 3GPP LTE-A standard (Release 12). D2D technology allows direct communication between devices within close proximity, enabling efficient spectrum utilization, improved throughput, and reduced load on base stations (BSs). In [1–5], resource and transmit power optimization schemes were proposed to improve the D2D performance in cellular networks. In [1], the authors introduced joint resource and transmit power optimization methods under the assumption that each cellular user's (CU's) resource can be shared with at most one D2D pair, and vice versa. Similarly, [2] proposed schemes that jointly optimize mode selection, power control, and resource allocation across three modes: cellular, dedicated-channel D2D, and reuse-channel D2D. In [3], decentralized transmit power optimization strategies were developed for networks with randomly deployed devices. The study in [4] aimed to enhance energy efficiency by jointly optimizing resource allocation and transmit power. The authors in [5] proposed power optimization schemes tailored to relay-assisted D2D communications. Moreover, various resource allocation strategies for D2D communications in future heterogeneous networks were proposed in [6].

In the evolving landscape of 5G and beyond, achieving high spectral efficiency while supporting high energy efficiency is critical, especially in energy-constrained environments, such as Internet of Things (IoT) applications, where devices must operate without frequent battery replacements [7]. To address these challenges, the integration of D2D communication with simultaneous wireless information and power transfer (SWIPT) has been explored in several studies [8–15]. In [8], strategies for optimizing transmit power and resource allocation were proposed, with the assumption that each D2D pair is restricted to reusing only one resource from the CUs. In [9], the authors proposed joint resource and transmit power optimization schemes while accounting for the energy causality constraint. In [10], the authors introduced beamforming design and resource allocation strategies to leverage multiple antennas at the BS effectively. In [11], joint transmit power and subcarrier allocation strategies were proposed for networks using orthogonal frequency-division multiple access (OFDMA). In [12], schemes for power control and resource allocation were suggested to maximize energy efficiency. In [13], an analysis of the outage probability was conducted for cognitive D2D communications. In [14], an analysis was conducted on the ergodic capacity and outage probability for fading channels in networks utilizing non-orthogonal multiple access (NOMA) and full-duplex relaying. Despite the fact that these studies [8–14] ensure a certain level of performance for the CUs, they consistently show that the performance of CUs considerably declines compared to that of CUs without D2D communication, mainly due to the interference from D2D transmitters (DTs).

To overcome these issues, the authors in [15] proposed transmit power optimization schemes for time-switching SWIPT-based D2D communications. However, compared to the time-switching method assumed in [15], the power-splitting approach can achieve a larger energy-rate region [16]. Additionally, [15] did not consider energy signal transmission. With the utilization of energy signal transmission, it is possible to assist energy harvesting at the CU without interfering with its information decoding, which allows for higher transmit power in D2D communications and ultimately improves performance. The absence of energy signal transmission, combined with the use of the time-switching method, may result in significant performance degradation in D2D communications due to the inefficient utilization of transmit power at the DT.

In this paper, we propose a joint transmit power and power-splitting optimization scheme for SWIPT in D2D-enabled cellular networks, addressing the limitations of [15] by incorporating power-splitting SWIPT and energy signal transmission, while ensuring that the CU's performance remains at the same level as in scenarios without D2D communications. Our approach utilizes the concept of 'energy cooperation', where energy signal transmission assists energy harvesting at the CU without disrupting its information decoding. This harvested energy can then support higher transmit power in D2D communications, improving the overall performance and expanding the energy-rate region. By leveraging energy cooperation, the proposed scheme enhances the harvested energy and data rate performance of D2D communications. The main contributions of this study are as follows:

- In considering power-splitting SWIPT and energy signal transmission, we formulate a joint optimization problem for the transmit power and power-splitting ratio, subject to the constraint that the CU's harvested energy and data rate are maintained at the same level as in scenarios without D2D communications. The formulated problem is inherently nonconvex and thus challenging to solve directly. However, by leveraging the monotonically increasing property of logarithmic functions, we transform the original problem into an equivalent convex problem. As a result, the optimal solution is derived in closed form.

- The optimal D2D performance is analyzed to obtain insights into the effectiveness of the proposed scheme. First, we obtain a sufficient condition for achieving optimal D2D performance without energy signal transmission. Second, a sufficient condition is derived under which D2D communication is not feasible with the baseline (BL) scheme but becomes feasible with the proposed scheme. These conditions provide valuable insights into the scenarios where the proposed scheme significantly enhances D2D communication performance.
- Numerical results are provided to demonstrate that the proposed scheme significantly surpasses the performance of the baseline scheme.

2. System Model and Problem Formulation

2.1. System Model

In this subsection, we present two communication cases to describe the considered system model. The first case is a non-D2D (ND) scenario in which the BS transmits signals to the CU, but the DT does not send any signals. Let h_{bc} and P_b denote the complex channel coefficient from the BS to the CU and the transmit power of the BS, respectively. For the SWIPT, with a power-splitting ratio $0 \leq \rho_c^{\mathrm{ND}} \leq 1$, the CU divides the received signals into two parts: one part, with power $\rho_c^{\mathrm{ND}} P_b |h_{bc}|^2$, is allocated to energy harvesting, while the other part, with power $(1-\rho_c^{\mathrm{ND}}) P_b |h_{bc}|^2$, is used for information decoding. Over the block time T, the CU's harvested energy and data rate are expressed as follows:

$$E_c^{\mathrm{ND}} = \zeta \rho_c^{\mathrm{ND}} T P_b |h_{bc}|^2, \tag{1}$$

$$R_c^{\mathrm{ND}} = \log_2\left(1 + \frac{P_b |h_{bc}|^2}{\sigma_{c,\mathrm{ant}}^2 + \sigma_{c,\mathrm{cov}}^2/(1-\rho_c^{\mathrm{ND}})}\right), \tag{2}$$

where $0 < \zeta \leq 1$ denotes the power conversion efficiency, $\sigma_{c,\mathrm{ant}}^2$ represents the antenna noise power at the CU, and $\sigma_{c,\mathrm{cov}}^2$ is the conversion noise power at the CU. In Section 2.2, the harvested energy in (1) and the data rate in (2) are used as thresholds for the CU's performance constraints.

As shown in Figure 1, the second case is a D2D scenario where both the BS and DT simultaneously transmit signals. In a centralized manner, the BS manages both cellular and D2D communications. Let h_{bd}, h_{dc}, and h_{dd} represent the complex channel coefficients from the BS to the D2D receiver (DR), from the DT to the CU, and from the DT to the DR, respectively. The BS maintains a fixed transmit power of P_b. To facilitate energy harvesting at the CU, the DT sends both information and energy signals. Since the energy signals are predetermined, they do not interfere with information decoding at the CU and DR; they can be canceled out prior to information decoding [17]. Under the coordination of the BS, the CU cooperates to cancel the received energy signals. The DT's transmit powers for the information and energy signals are denoted as $P_{d,\mathrm{I}}$ and $P_{d,\mathrm{E}}$, respectively. In this D2D case, the CU must adjust its power-splitting ratio to ensure its previously achieved energy and data rates, as expressed in (1) and (2). This adjusted power-splitting ratio is denoted by $0 \leq \rho_c \leq 1$. Also, for the SWIPT of D2D communications, the DR divides the received signals into two parts with a power-splitting ratio $0 \leq \rho_d \leq 1$. Then, the harvested energies and data rates at the CU and DR are given by

$$E_c = \zeta T \rho_c (P_b |h_{bc}|^2 + P_{d,\mathrm{I}} |h_{dc}|^2 + P_{d,\mathrm{E}} |h_{dc}|^2), \tag{3}$$

$$R_c = \log_2\left(1 + \frac{P_b|h_{bc}|^2}{P_{d,\mathrm{I}}|h_{dc}|^2 + \sigma^2_{c,\mathrm{ant}} + \sigma^2_{c,\mathrm{cov}}/(1-\rho_c)}\right), \tag{4}$$

$$E_d = \zeta T \rho_d (P_b|h_{bd}|^2 + P_{d,\mathrm{I}}|h_{dd}|^2 + P_{d,\mathrm{E}}|h_{dd}|^2), \tag{5}$$

$$R_d = \log_2\left(1 + \frac{P_{d,\mathrm{I}}|h_{dd}|^2}{P_b|h_{bd}|^2 + \sigma^2_{d,\mathrm{ant}} + \sigma^2_{d,\mathrm{ant}}/(1-\rho_d)}\right), \tag{6}$$

where $\sigma^2_{d,\mathrm{ant}}$ is the antenna noise power at the DR, and $\sigma^2_{d,\mathrm{cov}}$ is the conversion noise power at the DR. It is important to note that both the information signals and energy signals transmitted from the DT assist the CU in energy harvesting. On the other hand, the information signals transmitted from the DT interfere with the information decoding at the CU, while the energy signals do not interfere.

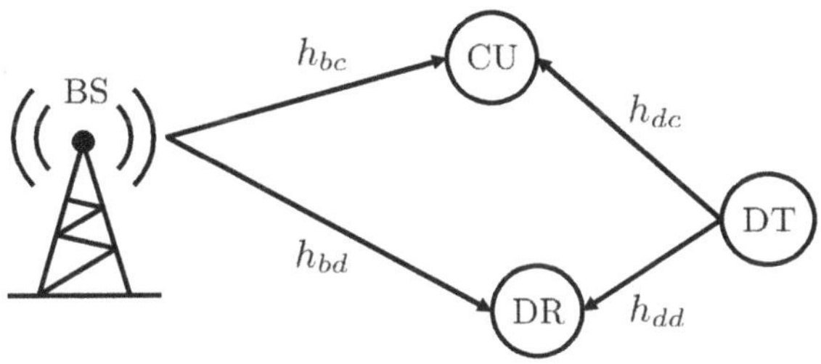

Figure 1. SWIPT in D2D-enabled cellular networks.

2.2. Problem Formulation

In the D2D-enabled cellular networks, to prevent the CU's performance degradation due to the DT's information signals, the transmit powers of the DT, $P_{d,\mathrm{I}}$, $P_{d,\mathrm{E}}$, and the power-splitting ratios, ρ_c and ρ_d, must be optimized to ensure that E_c in (3) and R_c in (4) are not less than E_c^{ND} in (1) and R_c^{ND} in (2), respectively. Given that the BS is assumed to coordinate the entire network and to know ρ_c^{ND} and global channel state information (CSI), the BS is responsible for carrying out this optimization. For this purpose, a D2D rate maximization problem can be formulated under constraints on E_c in (3) and R_c in (4) as follows:

$$[\rho_c^{\mathrm{opt}}, \rho_d^{\mathrm{opt}}, P_{d,\mathrm{I}}^{\mathrm{opt}}, P_{d,\mathrm{E}}^{\mathrm{opt}}] = \arg \max_{\rho_c, \rho_d, P_{d,\mathrm{I}}, P_{d,\mathrm{E}}} R_d \tag{7a}$$

$$\text{s.t.} \quad E_c \geq E_c^{\mathrm{ND}}, \tag{7b}$$

$$R_c \geq R_c^{\mathrm{ND}}, \tag{7c}$$

$$E_d \geq Q_d, \tag{7d}$$

$$P_{d,\mathrm{I}} + P_{d,\mathrm{E}} \leq P_d^{\max}, \tag{7e}$$

where $Q_d \geq 0$ represents the minimum required harvested energy at the DR, and $P_d^{\max}$ denotes the maximum transmit power of the DT.

3. Optimal Solution to Problem (7)

Since E_c in (3), R_c in (4), E_d in (5), and R_d in (6) are not concave functions, problem (7) is not convex. Therefore, it is challenging to solve problem (7). However, in first optimizing for ρ_c and ρ_d, followed by optimization with respect to $P_{d,\mathrm{I}}$ and $P_{d,\mathrm{E}}$, it is possible to obtain the optimal solution. To this end, problem (7) can be rewritten as follows:

$$\max_{P_{d,\mathrm{I}}, P_{d,\mathrm{E}}} \quad f(P_{d,\mathrm{I}}, P_{d,\mathrm{E}}) \tag{8a}$$

$$\text{s.t.} \quad P_{d,\text{I}} + P_{d,\text{E}} \leq P_d^{\max}, \tag{8b}$$

where

$$\begin{aligned} f(P_{d,\text{I}}, P_{d,\text{E}}) = \max_{\rho_c, \rho_d} \quad & R_d & \text{(9a)} \\ \text{s.t.} \quad & E_c \geq E_c^{\text{ND}}, & \text{(9b)} \\ & R_c \geq R_c^{\text{ND}}, & \text{(9c)} \\ & E_d \geq Q_d. & \text{(9d)} \end{aligned}$$

Since $\log_2\left(1 + \frac{P_b|h_{bc}|^2}{\sigma_{c,\text{ant}}^2 + x}\right)$ on both sides of (9c) is a strictly decreasing function, inequality (9c) implies that

$$P_{d,\text{I}}|h_{dc}|^2 + \frac{\sigma_{c,\text{cov}}^2}{1-\rho_c} \leq \frac{\sigma_{c,\text{cov}}^2}{1-\rho_c^{\text{ND}}}. \tag{10}$$

Then, by using this fact along with some additional manipulations, the problem (9) can be rewritten as follows:

$$\begin{aligned} f(P_{d,\text{I}}, P_{d,\text{E}}) = \max_{\rho_c, \rho_d} \quad & R_d & \text{(11a)} \\ \text{s.t.} \quad & \rho_c \geq \frac{\rho_c^{\text{ND}} P_b|h_{bc}|^2}{P_b|h_{bc}|^2 + (P_{d,\text{I}} + P_{d,\text{E}})|h_{dc}|^2}, & \text{(11b)} \\ & \rho_c \leq 1 - \frac{(1-\rho_c^{\text{ND}})\sigma_{c,\text{cov}}^2}{\sigma_{c,\text{cov}}^2 - (1-\rho_c^{\text{ND}})P_{d,\text{I}}|h_{dc}|^2}, & \text{(11c)} \\ & \rho_d \geq \frac{Q_d}{\zeta T(P_b|h_{bd}|^2 + (P_{d,\text{I}} + P_{d,\text{E}})|h_{dd}|^2)}. & \text{(11d)} \end{aligned}$$

Since R_d in (11a) is not a function of ρ_c and is a strictly decreasing function of ρ_d, the optimal solution to problem (11) can be obtained as follows:

$$\rho_c' \in \left[\frac{\rho_c^{\text{ND}} P_b|h_{bc}|^2}{P_b|h_{bc}|^2 + (P_{d,\text{I}} + P_{d,\text{E}})|h_{dc}|^2}, 1 - \frac{(1-\rho_c^{\text{ND}})\sigma_{c,\text{cov}}^2}{\sigma_{c,\text{cov}}^2 - (1-\rho_c^{\text{ND}})P_{d,\text{I}}|h_{dc}|^2}\right], \tag{12}$$

$$\rho_d' = \frac{Q_d}{\zeta T(P_b|h_{bd}|^2 + (P_{d,\text{I}} + P_{d,\text{E}})|h_{dd}|^2)}. \tag{13}$$

Consequently, by substituting $\rho_c' = \frac{\rho_c^{\text{ND}} P_b|h_{bc}|^2}{P_b|h_{bc}|^2 + (P_{d,\text{I}} + P_{d,\text{E}})|h_{dc}|^2}$ in (12) and $\rho_d' = \frac{Q_d}{\zeta T(P_b|h_{bd}|^2 + (P_{d,\text{I}} + P_{d,\text{E}})|h_{dd}|^2)}$ in (13) into (7), we obtain

$$\begin{aligned} \max_{P_{d,\text{I}}, P_{d,\text{E}}} \quad & \log_2\left(1 + \frac{P_{d,\text{I}}|h_{dd}|^2}{P_b|h_{bd}|^2 + \sigma_{d,\text{ant}}^2 + \frac{\sigma_{d,\text{ant}}^2}{1 - \frac{Q_d}{\zeta T(P_b|h_{bd}|^2 + (P_{d,\text{I}} + P_{d,\text{E}})|h_{dd}|^2)}}}\right) & \text{(14a)} \\ \text{s.t.} \quad & \frac{\rho_c^{\text{ND}} P_b|h_{bc}|^2}{P_b|h_{bc}|^2 + (P_{d,\text{I}} + P_{d,\text{E}})|h_{dc}|^2} \leq 1 - \frac{(1-\rho_c^{\text{ND}})\sigma_{c,\text{cov}}^2}{\sigma_{c,\text{cov}}^2 - (1-\rho_c^{\text{ND}})P_{d,\text{I}}|h_{dc}|^2}, & \text{(14b)} \\ & P_{d,\text{I}} + P_{d,\text{E}} \leq P_d^{\max}. & \text{(14c)} \end{aligned}$$

Since the objective function of problem (14a) is an increasing function of $P_{d,\mathrm{I}}$ and $P_{d,\mathrm{E}}$, the solution to problem (14) is given by

$$P_{d,\mathrm{I}}^{\mathrm{opt}} = \min\left(\frac{\rho_c^{\mathrm{ND}}\sigma_{c,\mathrm{cov}}^2 P_d^{\max}}{(1-\rho_c^{\mathrm{ND}})^2 P_b|h_{bc}|^2 + (1-\rho_c^{\mathrm{ND}})P_d^{\max}|h_{dc}|^2}, P_d^{\max}\right), \tag{15}$$

$$P_{d,\mathrm{E}}^{\mathrm{opt}} = P_d^{\max} - P_{d,\mathrm{I}}^{\mathrm{opt}}. \tag{16}$$

Then, through substituting $P_{d,\mathrm{I}}^{\mathrm{opt}}$ in (15) and $P_{d,\mathrm{E}}^{\mathrm{opt}}$ in (16) into (12) and (13), the optimal solutions for ρ_c and ρ_d can be obtained as follows:

$$\rho_c^{\mathrm{opt}} \in \left[\frac{\rho_c^{\mathrm{ND}} P_b|h_{bc}|^2}{P_b|h_{bc}|^2 + P_d^{\max}|h_{dc}|^2}, 1 - \frac{(1-\rho_c^{\mathrm{ND}})\sigma_{c,\mathrm{cov}}^2}{\sigma_{c,\mathrm{cov}}^2 - (1-\rho_c^{\mathrm{ND}})P_{d,\mathrm{I}}^{\mathrm{opt}}|h_{dc}|^2}\right], \tag{17}$$

$$\rho_d^{\mathrm{opt}} = \frac{Q_d}{\zeta T(P_b|h_{bd}|^2 + P_d^{\max}|h_{dd}|^2)}. \tag{18}$$

Since the BS coordinates the entire network, it calculates and forwards the solution in (15)–(18) to the CU and D2D pair.

With the proposed scheme in (15)–(18), the CU's harvested energy and the data rate can be ensured not to be degraded by the D2D signals, i.e., (7b) and (7c) can be satisfied with $P_{d,\mathrm{I}} > 0$. This is achievable because adjusting ρ_c as in (17) enables the CU to offset the rate loss caused by $P_{d,\mathrm{I}}|h_{dc}|^2$ in (4) using the harvested energy obtained from $P_{d,\mathrm{I}}|h_{dc}|^2 + P_{d,\mathrm{E}}|h_{dc}|^2$ in (3). In other words, the CU utilizes the energy harvesting benefits from the received energy signals to compensate for its data rate loss caused by the information signals transmitted by the DT. Accordingly, the performance degradation of the CU can be prevented through this energy cooperation between the DT and the CU, for which the DT transmits energy signals along with information signals.

4. Performance Analysis

4.1. Baseline Scheme Without Energy Signal Transmission

Since the proposed scheme in (15)–(18) considers energy signal transmission, unlike the conventional scheme in [15], its effectiveness can be evaluated by comparing its performance with that of the following BL scheme that does not involve energy signal transmission:

$$P_{d,\mathrm{I}}^{\mathrm{BL}} = \min\left(\left(\frac{\rho_c^{\mathrm{ND}}\sigma_{c,\mathrm{cov}}^2 - (1-\rho_c^{\mathrm{ND}})^2 P_b|h_{bc}|^2}{(1-\rho_c^{\mathrm{ND}})|h_{dc}|^2}\right)^+, P_d^{\max}\right), \tag{19}$$

$$P_{d,\mathrm{E}}^{\mathrm{BL}} = 0, \tag{20}$$

$$\rho_c^{\mathrm{BL}} = \frac{\rho_c^{\mathrm{ND}} P_b|h_{bc}|^2}{P_b|h_{bc}|^2 + P_{d,\mathrm{I}}^{\mathrm{BL}}|h_{dc}|^2}, \tag{21}$$

$$\rho_d^{\mathrm{BL}} = \left[\frac{Q_d}{\zeta T(P_b|h_{bd}|^2 + P_{d,\mathrm{I}}^{\mathrm{BL}}|h_{dd}|^2)}\right]_0^1. \tag{22}$$

The BL scheme in (19)–(22) can be derived through the process of obtaining the proposed scheme in (15)–(18), with the assumption that $P_{d,\mathrm{E}} = 0$. Let R_d^{O} and R_d^{BL} denote the D2D rates achieved with the optimal and BL schemes, respectively, which are obtained by substituting (15)–(18) and (19)–(22) into (6), respectively.

4.2. Performance Comparison

From the optimal solution in (15)–(18), the following theorem can be obtained.

Theorem 1. *If $g(P_d^{max}) \leq \rho_c^{ND} \leq 1$, we have $P_{d,I}^{opt} = P_d^{max}$ and $P_{d,E}^{opt} = 0$, where*

$$g(x) = \frac{2P_b|h_{bc}|^2 + x|h_{dc}|^2 + \sigma_{c,cov}^2 - \sqrt{(x|h_{dc}|^2 + \sigma_{c,cov}^2)^2 + 4P_b|h_{bc}|^2\sigma_{c,cov}^2}}{2P_b|h_{bc}|^2}. \quad (23)$$

Proof. In (15), $\frac{\rho_c^{ND}\sigma_{c,cov}^2 P_d^{max}}{(1-\rho_c^{ND})^2 P_b|h_{bc}|^2 + (1-\rho_c^{ND})P_d^{max}|h_{dc}|^2}$ is an increasing function of ρ_c^{ND} and is equal to P_d^{max} when $\rho_c^{ND} = g(P_d^{max})$. Therefore, when $g(P_d^{max}) \leq \rho_c^{ND} \leq 1$, we have $P_{d,I}^{opt} = P_d^{max}$ and $P_{d,E}^{opt} = 0$. □

By the results in Theorem 1, the performance comparison between the proposed scheme in (15)–(18) and the BL scheme in (19)–(22) can be derived as shown in the following theorem.

Theorem 2. *The optimal and BL schemes perform as shown in Table 1.*

Table 1. Performance comparison between the optimal and BL schemes for SWIPT D2D communications in cellular networks.

Power-Splitting Ratio Condition		Performance Comparison Result
Area 1	$0 \leq \rho_c^{ND} \leq g(0)$	$R_d^{O} \geq R_d^{BL} = 0$
Area 2	$g(0) < \rho_c^{ND} < g(P_d^{max})$	$R_d^{O} > R_d^{BL} > 0$
Area 3	$g(P_d^{max}) \leq \rho_c^{ND} \leq 1$	$R_d^{O} = R_d^{BL} > 0$

Proof. This proof is composed of three parts. (1) When $0 \leq \rho_c^{ND} \leq g_2(0)$, $P_{d,I}^{opt}$ in (15) is larger than 0, and $P_{d,I}^{BL}$ in (19) is equal to 0. Therefore, when $0 \leq \rho_c^{ND} \leq g_2(0)$, we have $R_d^{O} \geq R_d^{BL} = 0$. (2) When $g_2(0) < \rho_c^{ND} < g_2(P_d^{max})$, $P_{d,I}^{BL}$ in (19) is smaller than $P_{d,I}^{opt}$ in (15) and is larger than 0. Therefore, when $g_2(0) < \rho_c^{ND} < g_2(P_d^{max})$, we have $R_d^{O} > R_d^{BL} > 0$. (3) When $g_2(P_d^{max}) \leq \rho_c^{ND} \leq 1$, $P_{d,I}^{opt}$ in (15) is larger than 0 and is equal to $P_{d,I}^{BL}$ in (19). Therefore, when $g_2(P_d^{max}) \leq \rho_c^{ND} \leq 1$, we have $R_d^{O} = R_d^{BL} > 0$. □

The detailed explanation of the conditions in Theorem 2 is as follows. When the CU demands a large amount of energy, specifically when $g(P_d^{max}) \leq \rho_c^{ND} \leq 1$, the transmission of energy signals becomes unnecessary to achieve the optimal D2D data rate, i.e., $R_d^{O} = R_d^{BL} > 0$. This occurs because, under this condition, the energy harvested from the DT's interference power, i.e., $P_{d,I}|h_{dc}|^2$ in (3), is sufficient to offset the data rate reduction caused by the same interference term, i.e., $P_{d,I}|h_{dc}|^2$ in (4). Thus, additional energy signal transmission does not further enhance the D2D data rate in this scenario. On the other hand, when the CU's energy demand is moderate, specifically when $g(0) < \rho_c^{ND} < g(P_d^{max})$, energy signal transmission becomes critical for achieving the optimal D2D data rate, as $R_d^{O} > R_d^{BL} > 0$. This indicates that energy signal transmission is necessary to maximize the D2D performance in situations where the CU's energy requirement falls within this range. Finally, when the CU requires only a small amount of energy, i.e., $0 \leq \rho_c^{ND} \leq g(0)$, the BL scheme fails to enable any data transmission by the DT without the assistance of energy signals. In contrast, the proposed scheme in (15)–(18) enables data transmission even under this condition. This demonstrates the effectiveness of the proposed energy signal transmission approach in such low-energy-demand scenarios.

In summary, as the CU's energy requirement decreases, the role of energy signal transmission becomes increasingly significant in improving the D2D data rate. This trend highlights the adaptability of the proposed scheme in accommodating varying CU energy demands while optimizing D2D performance.

5. Numerical Results

In Figure 2, the D2D data rate R_d versus $\rho_c^{\mathrm{ND}} \in [0,1]$ is plotted to demonstrate the effectiveness of the proposed scheme in maximizing the D2D data rate. Although a related study is presented in [15], making a fair comparison of the performance of our proposed scheme with [15] is challenging, as our study focused on the power-splitting architecture, which offers a larger energy rate region than the time-switching architecture considered in [15]. Therefore, we compared the performance of our proposed scheme with that of the baseline scheme in (19)–(22) unlike in [15]. The channel power gains are modeled as $|h_i|^2 = \frac{10^{-3}}{d_i^3}$ for $i \in \{bc, bd, dc, dd\}$, where d_i is the normalized distance. We set $d_{bc} = d_{bd} = 1$, $d_{dc} = 0.5$, $d_{dd} = 0.2$, $\zeta = 1$, $T = 1$ s, $P_b = 30$ dBm, $P_d^{\max} = 20$ dBm, $\sigma^2 = \sigma_{c,\mathrm{ant}}^2 = \sigma_{c,\mathrm{cov}}^2 = \sigma_{d,\mathrm{ant}}^2 = \sigma_{d,\mathrm{cov}}^2$, $\frac{P_d^{\max}|h_{dd}|^2}{\sigma^2} = 10$ dB, and $Q_d = 1.3$ dBm, 8.3 dBm, and 10.8 dBm. Under these simulation parameters, we have $g_2(0) = 0.4624$ and $g_2(P_d^{\max}) = 0.6482$. As predicted by the results of Area 1 and Area 2 in Theorem 1, Figure 2 shows that the proposed optimal scheme outperforms the baseline scheme when $0 \le \rho_c^{\mathrm{ND}} \le g_2(P_d^{\max})$. When $0 \le \rho_c^{\mathrm{ND}} \le 0.4624$ (Area 1), the optimal scheme can transmit data, whereas the BL scheme cannot. When $0.4624 \le \rho_c^{\mathrm{ND}} \le 0.6482$ (Area 2), both the optimal and BL schemes can transmit data, whereas when $0.6482 \le \rho_c^{\mathrm{ND}} \le 1$ (Area 3), the BL scheme performs the same as the optimal scheme because the optimal scheme allocates all of the DT's transmit power to the information signals in the same way as the baseline scheme. In summary, when a smaller portion of the received energy is required for energy harvesting at the CU, the proposed scheme is more effective in improving D2D performance.

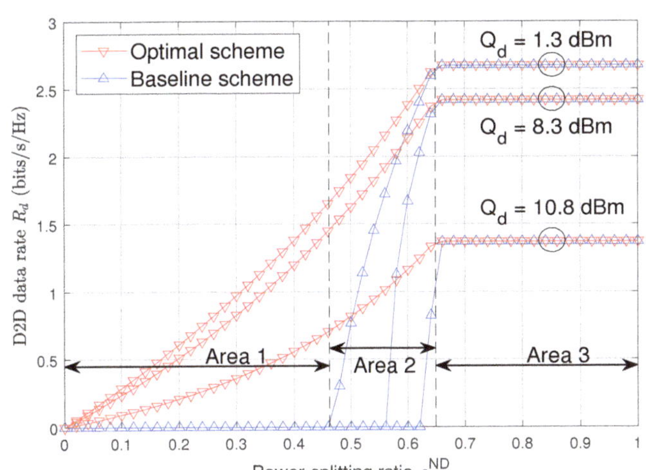

Figure 2. D2D data rate R_d versus ρ_c^{ND}. $d_{bc} = d_{bd} = 1$, $d_{dc} = 0.5$, $d_{dd} = 0.2$, $\zeta = 1$, $T = 1$ s, $P_b = 30$ dBm, $P_d^{\max} = 20$ dBm, $\sigma^2 = \sigma_{c,\mathrm{ant}}^2 = \sigma_{c,\mathrm{cov}}^2 = \sigma_{d,\mathrm{ant}}^2 = \sigma_{d,\mathrm{cov}}^2$, $\frac{P_d^{\max}|h_{dd}|^2}{\sigma^2} = 10$ dB, and $Q_d = 1.3$ dBm, 8.3 dBm, and 10.8 dBm.

To demonstrate the applicability of the proposed scheme to fading channels, which correspond to the scenarios of mobile CU and D2D nodes, Figures 3 and 4 show the D2D performance for Rayleigh fading channels. In Figures 3 and 4, the ergodic rate $\mathbb{E}[R_d]$ and outage probability $\Pr\{R_d < R_d^{\mathrm{t}}\}$ for the fast and slow fading channels are plotted versus $\rho_c^{\mathrm{ND}} \in [0,1]$, respectively. The target rate R_d^{t} is set to $R_d^{\mathrm{t}} = 0.5$ bits/s/Hz, and the channels are modeled as $h_i \sim CN\left(0, \frac{10^{-3}}{d_i^3}\right)$ for $i \in \{bc, bd, dc, dd\}$, where d_i is the normalized distance. The other simulation parameters are the same as those in Figure 2. As predicted by Theorem 2, when the required energy of the CU is small (or large), the optimal scheme outperforms (or performs the same as) the BL scheme. Also, we can observe that the proposed optimal scheme demonstrates robust performance in fading channels.

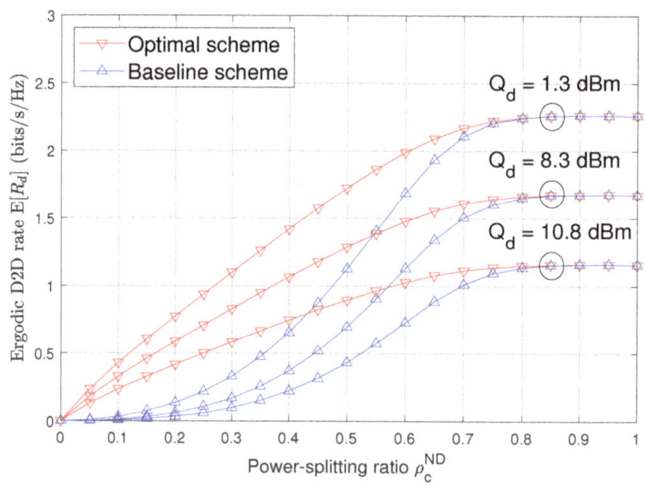

Figure 3. D2D ergodic rate $\mathbb{E}[R_d]$ versus ρ_c^{ND} in Rayleigh fading channel. $d_{bc} = d_{bd} = 1$, $d_{dc} = 0.5$, $d_{dd} = 0.2$, $\zeta = 1$, $T = 1$ s, $P_b = 30$ dBm, $P_d^{\max} = 20$ dBm, $\sigma^2 = \sigma_{c,\mathrm{ant}}^2 = \sigma_{c,\mathrm{cov}}^2 = \sigma_{d,\mathrm{ant}}^2 = \sigma_{d,\mathrm{cov}}^2$, $\frac{P_d^{\max}|h_{dd}|^2}{\sigma^2} = 10$ dB, and Q_d = 1.3 dBm, 8.3 dBm, and 10.8 dBm.

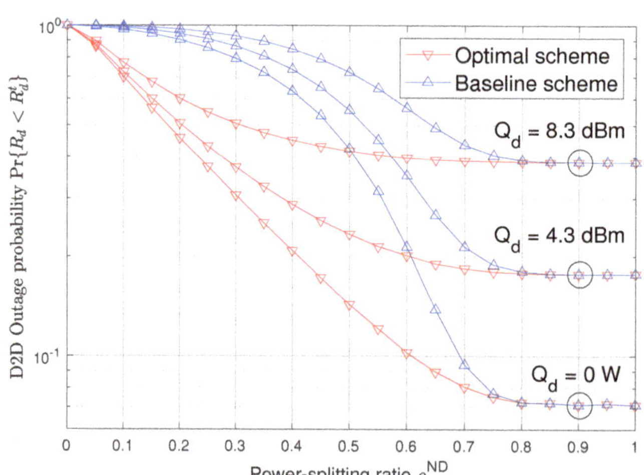

Figure 4. D2D outage probability $\Pr\{R_d < R_d^{\mathrm{t}}\}$ versus ρ_c^{ND} in Rayleigh fading channel. $d_{bc} = d_{bd} = 1$, $d_{dc} = 0.5$, $d_{dd} = 0.2$, $\zeta = 1$, $T = 1$ s, $P_b = 30$ dBm, $P_d^{\max} = 20$ dBm, $\sigma^2 = \sigma_{c,\mathrm{ant}}^2 = \sigma_{c,\mathrm{cov}}^2 = \sigma_{d,\mathrm{ant}}^2 = \sigma_{d,\mathrm{cov}}^2$, $\frac{P_d^{\max}|h_{dd}|^2}{\sigma^2} = 10$ dB, Q_d = 0 W, 4.3 dBm, and 8.3 dBm, and $R_d^{\mathrm{t}} = 0.5$ bits/s/Hz.

6. Conclusions

In this paper, we proposed a transmit power and power-splitting ratio optimization scheme for D2D-enabled SWIPT cellular networks, ensuring that the CU achieves the ND performance. Also, we analyzed the optimal D2D performance and obtained useful insights into the effectiveness of the proposed scheme. Finally, numerical results demonstrated that the proposed scheme outperforms the BL scheme, especially as the CU requires more energy, confirming the effectiveness of the proposed scheme. Furthermore, for the comprehensive performance evaluation, the transmit power and power-splitting optimization problem can be extended to consider channel uncertainty and user mobility in future works.

Author Contributions: Conceptualization, D.-W.L.; supervision, J.-M.K. All authors have read and agreed to the published version of the manuscript.

Funding: This research was supported in part by the National Research Foundation of Korea (NRF) grant funded by the Korea government (MSIT) (No. 2022R1A4A1033830); in part by the "Junior Faculty Research Support Grant" at Changwon National University in 2024; in part by the "Glocal University Project 2024" of the National Research Foundation of Korea (NRF) funded by the Ministry of Education (MOE); and in part by the "Regional Innovation Strategy (RIS)" through the National Research Foundation of Korea (NRF) funded by the Ministry of Education (MOE) (2021RIS-003).

Data Availability Statement: The original contributions presented in this study are included in the article. Further inquiries can be directed to the corresponding author.

Conflicts of Interest: The authors declare no conflicts of interest.

Abbreviations

The following abbreviations are used in this manuscript:

D2D	Device-to-device
BS	Base station
CU	Cellular user
IoT	Internet of Things
SWIPT	Simultaneous wireless information and power transfer
OFDMA	Orthogonal frequency-division multiple access
NOMA	Non-orthogonal multiple access
DT	D2D transmitter
BL	Baseline
ND	Non-D2D
DR	D2D receiver
CSI	Channel state information

References

1. Feng, D.; Lu, L.; Yi, Y.; Li, G.Y. Device-to-device communications underlaying cellular networks. *IEEE Trans. Commun.* **2013**, *61*, 3541–3551. [CrossRef]
2. Yu, G.; Xu, L.; Feng, D.; Yin, R.; Li, G.Y.; Jiang, Y. Joint mode selection and resource allocation for device-to-device communications. *IEEE Trans. Commun.* **2014**, *62*, 3814–3824. [CrossRef]
3. Lee, N.; Lin, X.; Andrews, J.G.; Heath, R.W. Power control for D2D underlaid cellular networks: Modeling, algorithms, and analysis. *IEEE J. Sel. Areas Commun.* **2015**, *33*, 1–13. [CrossRef]
4. Hoang, T.D.; Le, L.B.; Le-Ngoc, T. Energy-efficient resource allocation for D2D communications in cellular networks. *IEEE Trans. Veh. Technol.* **2016**, *65*, 6972–6986. [CrossRef]
5. Han, L.; Zhou, R.; Li, Y.; Zhang, B.; Zhang, X. Power control for two-way AF relay assisted D2D communications underlaying cellular networks. *IEEE Access* **2020**, *8*, 151968–151975. [CrossRef]
6. Xu, Y.; Gui, G.; Gacanin, H.; Adachi, F. A survey on resource allocation for 5G heterogeneous networks: Current research, future trends, and challenges. *IEEE Commun. Surv. Tutor.* **2021**, *23*, 668–695. [CrossRef]

7. Lu, W.; Si, P.; Huang, G.; Han, H.; Qian, L.; Zhao, N.; Gong, Y. SWIPT cooperative spectrum sharing for 6G-enabled cognitive IoT network. *IEEE Internet Things J.* **2021**, *8*, 15070–15080. [CrossRef]
8. Luo, Y.; Hong, P.; Su, R.; Xue, K. Resource allocation for energy harvesting-powered D2D communication underlaying cellular networks. *IEEE Trans. Veh. Technol.* **2017**, *66*, 10486–10498. [CrossRef]
9. Gupta, S.; Zhang, R.; Hanzo, L. Energy harvesting aided device-to-device communication underlaying the cellular downlink. *IEEE Access* **2016**, *5*, 7405–7413. [CrossRef]
10. Ku, M.-L.; Lai, J.-W. Joint beamforming and resource allocation for wireless-powered device-to-device communications in cellular networks. *IEEE Trans. Wirel. Commun.* **2017**, *16*, 7290–7304. [CrossRef]
11. Saleem, U.; Jangsher, S.; Qureshi, H.K.; Hassan, S.A. Joint subcarrier and power allocation in the energy-harvesting-aided D2D communication. *IEEE Trans. Ind. Inform.* **2018**, *14*, 2608–2617. [CrossRef]
12. Kuang, Z.; Liu, G.; Li, G.; Deng, X. Energy efficient resource allocation algorithm in energy harvesting-based D2D heterogeneous networks. *IEEE Internet Things J.* **2019**, *6*, 557–567. [CrossRef]
13. Sakr, A.H.; Hossain, E. Cognitive and energy harvesting-based D2D communication in cellular networks: Stochastic geometry modeling and analysis. *IEEE Trans. Commun.* **2015**, *63*, 1867–1880. [CrossRef]
14. Budhiraja, I.; Kumar, N.; Tyagi, S.; Tanwar, S.; Guizani, M. SWIPT-enabled D2D communication underlaying NOMA-based cellular networks in imperfect CSI. *IEEE Trans. Veh. Technol.* **2021**, *70*, 692–699. [CrossRef]
15. Lim, D.-W.; Kang, J.; Chun, C.-J.; Kim, H.-M. Joint transmit power and time-switching control for device-to-device communications in SWIPT cellular networks. *IEEE Commun. Lett.* **2019**, *23*, 322–325. [CrossRef]
16. Zhou, X.; Zhang, R.; Ho, C.K. Wireless information and power transfer: Architecture design and rate-energy tradeoff. *IEEE Trans. Commun.* **2013**, *61*, 4753–4767. [CrossRef]
17. Xu, J.; Liu, L.; Zhang, R. Multiuser MISO beamforming for simultaneous wireless information and power transfer. *IEEE Trans. Signal Process.* **2014**, *62*, 4798–4810. [CrossRef]

mathematics

Article

Performance Analysis of Artificial-Noise-Based Secure Transmission in Wiretap Channel

Hyukmin Son

Department of Electronic Engineering, Gachon University, Seongnam 13120, Gyeonggi, Republic of Korea; hson102@gachon.ac.kr

Abstract: Artificial noise (AN)-aided techniques have been considered to be promising and practical candidates for enhancing physical layer security. However, there has been a lack of analysis regarding the AN effect on the eavesdropper (EV) from the perspective of the signal-to-interference plus noise ratio (SINR) regarding the existence of the EV's channel state information (CSI) at the legitimate transmitter. In this paper, we analyze the performance of AN-aided secure transmission from the SINR perspective when a legitimate transmitter has and does not have the EV's CSI. Based on the analyzed EV's SINRs for the above two cases, the secrecy gap, which is the difference between the two secrecy capacities, is defined and analyzed. Based on the derived secrecy gap, we have analyzed the asymptotic performances of the secrecy capacity and gap when the number of antennas of the legitimate transmitter and the number of antennas of the EV have large values. Through asymptotic analysis, it is demonstrated that the AN-aided secure transmission under the practical environment (i.e., the case that the EV's CSI is not available at the legitimate transmitter) can nearly achieve an ideal performance (i.e., the performance when the EV's CSI is available at the legitimate transmitter) in a massive antenna system.

Keywords: artificial noise; beamforming; MIMO channels; physical layer security; secrecy capacity; wiretap channel

MSC: 94-10

1. Introduction

Due to the broadcast nature of wireless communication, sensitive information can be easily intercepted by an eavesdropper (EV). Traditionally, cryptographic methods have been employed to prevent eavesdropping, relying on the assumption of limited computational resources [1–3]. However, physical layer security has gained attention in wireless communications, leveraging the inherent randomness of wireless channels for protection.

Several recent studies have explored various system configurations and assumptions in the context of physical layer security [4–7]. In [4], the authors characterize the secrecy capacity of the multiple-input–single-output (MISO) wiretap channel using generalized eigenvalues and demonstrate that transmit beamforming is an optimal strategy. Building on this, the work in [5] extends these techniques to more general setups, such as scenarios with multiple antennas at the destination, developing advanced precoding methods. In [6], a secure unmanned aerial vehicle (UAV) communication system with the assistance of a reconfigurable intelligent surface (RIS) is investigated to prevent eavesdropping from multiple colluding curious users based on the AN. In [7], the authors propose time-domain artificial noise to mitigate the significant impact of multiple intelligent eavesdroppers in uplink multiple-input–multiple-output (MIMO) Internet of Things (IoT) systems. These studies assume that the transmitter has the eavesdropper's (EV) channel state information (CSI). However, obtaining the EV's CSI in multi-antenna wiretap channels with passive eavesdroppers becomes challenging as passive EVs generally conceal their presence from the transmitter.

Artificial noise (AN)-aided techniques have emerged as promising to enhance physical layer security without requiring the EV's CSI. The core idea is that the transmitter introduces additive noise signals into the null space of the channel between the transmitter and the intended receiver [8]. This ensures that the desired information reaches the receiver unaffected while the AN interferes with the EV's reception. In addition to AN, beamforming strategies in multi-antenna systems have been widely explored to improve security further. However, most of these studies focus on mitigating eavesdropping in wiretap channels at the expense of increased resource consumption in generating AN, or higher computational complexity in beamforming [9–15]. In [9], the authors derive an analytical closed-form expression of an achievable secrecy rate and use it to optimize the transmit power allocation between the information signal and the AN. In addition, the impact of imperfect CSI, at both the transmitter and receiver, is investigated. The authors in [10,11] investigate secure downlink transmission in a multi-cell massive MIMO system with matched-filter precoding and AN generation when the EV's CSI is not available at the transmitter.

The effect of imperfect CSI is considered in [16,17]. In [17], a two-phase transmission scheme with AN injection is introduced to achieve zero-secrecy outage probability despite imperfect channel estimation. Furthermore, secure beamforming schemes have been proposed, assuming that the legitimate transmitter has the statistical CSI of the eavesdropper [18–20] to maximize the secrecy rate. More practical scenarios regarding downlink wiretap channels are considered in [12,15]. In [12], the authors propose a framework for AN-assisted secure transmission in order to maximize the lower bound of the achievable effective ergodic secrecy rate when the receiver's CSI is obtained via forward training and analog feedback. In [15], joint training and data transmission schemes are proposed in which the AN in the training phase is used to prevent the EV from obtaining accurate CSI, and that in the data transmission phase can be used to mask the transmission of confidential messages.

Although various AN-aided secure transmission technologies have been studied recently, there has yet to be an analysis of how AN affects EVs depending on whether the legitimate transmitter has the EV's CSI or not. In addition, there needs to be an analysis of the factors necessary from a system design perspective, such as the number of beams used for AN generation depending on the number of antennas of the legitimate transmitter.

This paper analyzes the AN-aided secure transmission in a multiple-input–multiple-output multiple antenna eavesdropper (MIMOME) channel. From the received signal-to-interference plus noise ratio (SINR) perspective, the AN effect on the EV is analyzed for two cases: when a legitimate transmitter has CSI regarding EVs and when it does not. Based on the analyzed AN and SINR, the secrecy capacity for the two cases is analyzed, and, based on this, the secrecy gap, which is the difference between the two secrecy capacities, is analyzed. The simulation results verify the mathematical derivation results for AN, SINR, secrecy capacity, and secrecy gap. In addition, they demonstrate that AN-aided secure transmission under a practical environment (i.e., the case that the EV's CSI is not available at the legitimate transmitter) can nearly achieve an ideal performance (i.e., the performance when the EV's CSI is available at the legitimate transmitter) in a massive MIMOME channel. The main contributions of this paper can be summarized as follows:

- When the legitimate transmitter does not have the CSI of the EV, we propose a system that generates AN by forming a beam in the null space of the effective channel between the legitimate transmitter and the receiver and sending a jamming signal. In this system, we derive the distribution of AN according to the number of beams and obtain their expected values. This shows that the expected values are the same regardless of the number of beams. In addition, we analyze the upper bound for the EV's SINR based on the analyzed AN, prove that the distribution for the upper bound is a beta prime distribution, and analyze its expected value.
- When the legitimate transmitter has the EV's CSI, we propose a method to generate a single beam for AN generation by projecting the singular vector corresponding to the largest singular value of the channel between the legitimate transmitter and the receiver into the null space of the effective channel. Then, we derive the distribution of AN and show that its expected value is $M-1$ times larger than the case when the legitimate transmitter does not have the EV's CSI, where M is the number of antennas of the legitimate transmitter. Based on this, we prove that the EV's SINR has a beta prime distribution and analyze its expected value.
- Based on the EV's SINRs for the above two cases, the secrecy gap, which is the difference between the two secrecy capacities, is defined and analyzed. In particular, the upper bound of the expected secrecy gap is derived. Based on this derivation, the asymptotic performances are analyzed when the number of antennas of the legitimate transmitter and the number of antennas of the EV have large values.

The remainder of this paper is organized as follows. The system model is described in Section 2. The AN design and analysis are discussed in Section 3, and the asymptotic performance of the secrecy capacity and gap for the null-space-based AN-aided secure transmission is analyzed in Section 4. The simulation results are presented in Section 5, and the concluding remarks are presented in Section 6.

Notations: Throughout the paper, boldface upper- and lower-case symbols represent matrices and vectors, respectively, and $\boldsymbol{I}_N$ is a size-N identity matrix. The conjugate, transpose, inverse, and Hermitian transpose operators are $(\cdot)^*$, $(\cdot)^{\mathrm{T}}$, $(\cdot)^{-1}$, and $(\cdot)^{\mathrm{H}}$, respectively. $|a|$ denotes the absolute value of a complex number a and $\|\mathbf{a}\|$ indicates ℓ_2- norms of vector $\mathbf{a}$. $\mathbf{a} \perp \mathbf{b}$ denotes an orthogonality between vectors $\mathbf{a}$ and $\mathbf{b}$. $\beta(a,b)$ is the beta distribution with parameters a and b; $\beta'(a,b,c,d)$ is the generalized beta prime distribution with parameters a, b, c, and d. χ_a^2 is the chi-squared distribution with a degrees of freedom.

2. System Model

As shown in Figure 1, we consider uplink secure communication that consists of one type of user equipment (UE, legitimate transmitter), one desired base station (BS, legitimate receiver), and one passive EV. The UE, BS, and EV equip M, N, and N_E antennas, respectively. In the proposed scheme, the UE transmits the desired data using a beam while generating a jamming signal and transmits it to the EV using multiple beams to enable secure communications against EV attacks. Regarding channel estimation, the reference signals are transmitted from the UE to the BS so that the channel state information (CSI) between the BS and the UE is available at the BS. Since the EV can listen to the reference signals transmitted from the UE, it is assumed that the CSI from the UE to the EV is available at the EV. However, it is assumed that the CSI between the BS and the UE is not available at the EV. It is generally difficult for a transmitter to acquire the CSI between the transmitter and the passive EV because the passive EV hides its existence. Thus, it is assumed that the CSI from the EV is not available at either the BS or the UE.

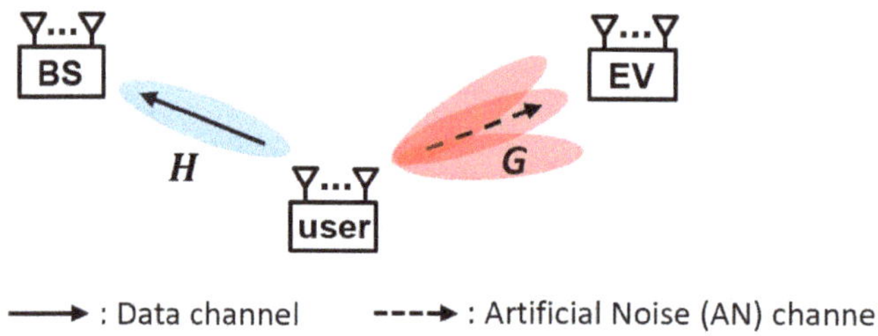

Figure 1. System model for artificial-noise-based uplink cooperative transmission in wiretap channel.

The received signal at the BS is provided as

$$\mathbf{y}_D = \sqrt{P_D}\mathbf{u}_D^H\mathbf{H}\mathbf{w}_D s_D + \sqrt{\frac{P_E}{B}}\sum_{j=1}^{B}\mathbf{u}_D^H\mathbf{H}\mathbf{w}_j s_j + \mathbf{u}_D^H\mathbf{n}_D, \quad (1)$$

$$= \sqrt{P_D}\mathbf{u}_D^H\mathbf{H}\mathbf{w}_D s_D + \mathbf{u}_D^H\mathbf{n}_D, \quad (2)$$

where s_D denotes the transmit symbol, $\mathbb{E}[|s_D|^2] = 1$, P_D is the transmit power allocated for s_D, P_E is the total transmit power allocated for jamming signals, the total transmit power P at the UE is provided as $P_D + P_E$, s_j denotes the jth jamming signal transmitted by UE, $\mathbb{E}[|s_j|^2] = 1$, $\mathbf{H}(\in \mathbb{C}^{N\times M})$ is the channel matrix from the UE to the BS, $\mathbf{u}_D(\in \mathbb{C}^{1\times N})$ is the receive combining vector at the BS, $\mathbf{w}_D(\in \mathbb{C}^{M\times 1})$ is the transmit beamforming vector at the UE, $\mathbf{w}_j(\in \mathbb{C}^{M\times 1})$ is the jth transmit beamforming vector for the jth jamming signal, B is the number of beamforming vectors for generating AN, and $\mathbf{n}_D$ is the additive white Gaussian noise vector. The elements of channel $\mathbf{H}$ and $\mathbf{n}_D$ in (1) are independent and identically distributed (i.i.d.) zero-mean complex Gaussian random variables, with variance 1 and σ^2, respectively. Note that all transmit beamforming vectors and receive combining vectors used in the paper have unit norm values. In (1), it is presumed that the transmit power for each jamming signal is equally allocated because the CSI between the EV and the UE is not available at the UE. To generate AN that cannot degrade the data channel capacity, $\mathbf{w}_j$ is designed in the null space of $\mathbf{u}_D^H\mathbf{H}$, i.e., $\mathbf{w}_j \perp \mathbf{u}_D^H\mathbf{H}$.

Since the interference term $\sqrt{\frac{P_E}{B}}\sum_{j=1}^{B}\mathbf{u}_D^H\mathbf{H}\mathbf{w}_j s_j$ is nullified in (2), the optimal $\mathbf{w}_D$ and $\mathbf{u}_D$ to maximize the SINR of (2) are determined to maximize the term of $||\mathbf{u}_D^H\mathbf{H}\mathbf{w}_D s_D||$, which are, respectively, provided as the right and left singular vectors corresponding to the largest singular value obtained from the singular value decomposition of $\mathbf{H}$. The SINR of (2) is then provided as

$$\mathrm{SINR}_D = \frac{P_D||\mathbf{u}_D^H\mathbf{H}\mathbf{w}_D||^2}{||\mathbf{u}_D^H\mathbf{n}_D||^2} = \rho_D\lambda_{max}^2, \quad (3)$$

where $\rho_D = \frac{P_D}{\sigma^2}$, and λ_{max} is the largest singular value obtained from the singular value decomposition of $\mathbf{H}$.

The received signal at the EV is provided as

$$\mathbf{y}_E = \sqrt{P_D}\mathbf{u}_E^H\mathbf{G}\mathbf{w}_D s_D + \sqrt{\frac{P_E}{B}}\sum_{j=1}^{B}\mathbf{u}_E^H\mathbf{G}\mathbf{w}_j s_j + \mathbf{u}_E^H\mathbf{n}_E, \quad (4)$$

where $\mathbf{G}(\in \mathbb{C}^{N_E\times M})$ is the channel matrix from the UE to the EV, $\mathbf{u}_E(\in \mathbb{C}^{1\times N_E})$ is the receive combining vector at the EV, and $\mathbf{n}_E$ is the additive white Gaussian noise vector.

The elements of channel $\mathbf{G}$ and $\mathbf{n}_E$ in (4) are i.i.d. zero-mean complex Gaussian random variables, with variance 1 and σ^2, respectively. From (4), the SINR at the EV is then provided as

$$\text{SINR}_E = \frac{\rho_D ||\mathbf{u}_E^H \mathbf{G}\mathbf{w}_D||^2}{1 + \frac{\rho_E}{B}\sum_{j=1}^{B} ||\mathbf{u}_E^H \mathbf{G}\mathbf{w}_j||^2}, \tag{5}$$

where $\rho_E = \frac{P_E}{\sigma^2}$. In the proposed scheme, ρ_D is determined as the minimum value that enables SINR_D to satisfy the required SINR, γ. Thus, ρ_D is calculated as $\frac{\gamma}{\lambda_{max}^2}$. ρ_E is then calculated as $\frac{P}{\sigma^2} - \rho_D$.

The secrecy capacity of the UE is bounded by the difference in mutual information between the legitimate transmitter and receiver versus the legitimate transmitter and the EV [8,21]. From (3) and (5), the secrecy capacity is then determined by

$$C = [C_D - C_E]^+, \tag{6}$$

where

$$\begin{aligned} C_D &= \log_2(1 + \text{SINR}_D), \quad (7)\\ C_E &= \log_2(1 + \text{SINR}_E), \quad (8) \end{aligned}$$

and $[x]^+ = \max\{x, 0\}$.

3. Artificial Noise Design and Analysis

3.1. The Case That CSI Between the UE and EV Is Not Available at the UE

B beamforming vectors should be generated in the null space of $\mathbf{u}_D^H \mathbf{H}$ not to generate any interference to the BS. The null space can be provided as

$$\mathbf{H}_n = \mathbf{I}_M - \overline{\mathbf{h}}^H \overline{\mathbf{h}}, \tag{9}$$

where $\mathbf{h} = \mathbf{u}_D^H \mathbf{H}$ and $\overline{\mathbf{h}} = \frac{\mathbf{h}}{||\mathbf{h}||}$. Thus, B beamforming vectors to generate the AN can be selected and used from $M-1$ eigenvectors of $\mathbf{H}_n \mathbf{H}_n^H$, which have non-zero eigenvalues. The reason for using the vectors orthogonal to each other in the given null space is that the orthogonal vectors can transmit the jamming signals to the overall direction, except for the BS, when the CSI between the UE and EV is not available at the UE.

When B equals to $M-1$, the AN term in (5) can be written as

$$\sum_{j=1}^{M-1} \left\| \mathbf{u}_E^H \mathbf{G}\mathbf{w}_j \right\|^2 = \sum_{j=1}^{M-1} ||\mathbf{g}||^2 ||\overline{\mathbf{g}}\mathbf{w}_j||^2 = ||\mathbf{g}||^2 \left(1 - ||\overline{\mathbf{g}}\mathbf{w}_D||^2\right), \tag{10}$$

where $\mathbf{g} = \mathbf{u}_E^H \mathbf{G}$ and $\overline{\mathbf{g}} = \frac{\mathbf{g}}{||\mathbf{g}||}$. Since $\overline{\mathbf{g}}$ and $\mathbf{w}_D$ in (10) are arbitrary unit vectors on the M dimensional hyperplane, the distribution of $||\overline{\mathbf{g}}\mathbf{w}_D||^2$ is $\beta(1, M-1)$. By the properties of beta distribution, $1 - \beta(1, M-1)$ is $\beta(M-1, 1)$. In addition, the product of χ_{2M}^2 and $\beta(M-1, 1)$ is equal to $\chi_{2(M-1)}^2$ [22]. The distribution of (10) is then derived as

$$||\mathbf{g}||^2 \left(1 - ||\overline{\mathbf{g}}\mathbf{w}_D||^2\right) = \chi_{2M}^2 (1 - \beta(1, M-1)) = \chi_{2M}^2 \beta(M-1, 1) = \chi_{2(M-1)}^2. \tag{11}$$

From (5) and (11), the expected AN is then provided as

$$\mathbb{E}\left[\frac{\rho_E}{M-1}\chi_{2(M-1)}^2\right] = \frac{\rho_E}{M-1}\mathbb{E}\left[\chi_{2(M-1)}^2\right] = \rho_E. \tag{12}$$

When B is less than $M-1$, the distribution of the term for the AN in (5) is derived as

$$\sum_{j=1}^{B}\|\mathbf{g}\|^2\|\bar{\mathbf{g}}\mathbf{w}_j\|^2=\sum_{j=1}^{B}\chi^2_{2M}\beta(1,M-1)=\sum_{j=1}^{B}\chi^2_2=\chi^2_{2B}. \quad (13)$$

Since $\bar{\mathbf{g}}$ and $\mathbf{w}_j$ in (13) are arbitrary unit vectors on the M dimensional hyperplane, the distribution of $\|\bar{\mathbf{g}}\mathbf{w}_j\|^2$ is $\beta(1,M-1)$. In addition, the product of χ^2_{2M} and $\beta(1,M-1)$ is χ^2_2 [22]. From the additivity of chi-squared distribution, the sum of independent chi-squared variables, $\sum_{j=1}^{B}\chi^2_2$, is χ^2_{2B}. From (5) and (13), the expected AN is then provided as

$$\mathbb{E}\left[\frac{\rho_E}{B}\chi^2_{2B}\right]=\frac{\rho_E}{B}\mathbb{E}\left[\chi^2_{2B}\right]=\rho_E. \quad (14)$$

From (12) and (14), it is observed that the expected AN is ρ_e regardless of the number of beams for generating the AN. The AN is only increased according to ρ_e when the CSI between the UE and the EV is not available at the UE. Since the AN effect is not related to the number of beams used to generate AN, the computational complexity of generating the AN beams can be reduced by using a single AN beam in the null-space-based AN-aided secure transmission. Hereafter, B is set to $M-1$ for ease of analysis.

From (5), the EV's SINR can be upper-bounded as

$$\mathrm{SINR}_E=\frac{\rho_D||\mathbf{u}_E^H\mathbf{G}\mathbf{w}_D||^2}{1+\frac{\rho_E}{M-1}\sum_{j=1}^{M-1}||\mathbf{u}_E^H\mathbf{G}\mathbf{w}_j||^2}\le\frac{\rho_D||\mathbf{G}\mathbf{w}_D||^2}{\frac{\rho_E}{M-1}\sum_{j=1}^{M-1}||\mathbf{u}_E^H\mathbf{G}\mathbf{w}_j||^2}, \quad (15)$$

where $\mathbf{u}_E$ is assumed as the maximum ratio combiner to maximize the EV's channel gain, i.e., $\mathbf{u}_E=\frac{\mathbf{G}\mathbf{w}_D}{||\mathbf{G}\mathbf{w}_D||}$.

Lemma 1. *The statistical behavior of upper-bound of* SINR_E *in (15) follows the generalized beta prime distribution,* $\beta'(N_E,M-1,1,1/c_1)$, *where* $c_1=\frac{\rho_E}{(M-1)\rho_D}$.

Proof. From (15), the SINR_E can be re-written as

$$\mathrm{SINR}_E=\frac{\rho_D\chi^2_{2N_E}}{\frac{\rho_E}{M-1}\chi^2_{2(M-1)}}=\frac{S_2}{c_1S_1}, \quad (16)$$

where c_1 is $\frac{\rho_E}{(M-1)\rho_D}$, S_1 is the random variable with $\chi^2_{2(M-1)}$, and S_2 is the random variable with $\chi^2_{2N_E}$. Conditioning on S_1, the probability distribution function (PDF) of SINR_E, $f_S(s)$, can be written as

$$\begin{aligned}f_S(s)&=\int_0^\infty f_{S|S_1}(s|s_1)f_{S_1}(s_1)ds_1\\&=\int_0^\infty c_1s_1\frac{(c_1s_1s)^{N_E-1}}{\Gamma(N_E)}e^{-(c_1s_1s)}\frac{s_1^{M-2}}{\Gamma(M-1)}e^{-s_1}ds_1\\&=\frac{s^{N_E-1}}{\Gamma(N_E)\Gamma(M-1)}c_1^{N_E}\int_0^\infty s_1^{M+N_E-2}e^{-(c_1s+1)s_1}ds_1\\&=\frac{s^{N_E-1}}{\Gamma(N_E)\Gamma(M-1)}c_1^{N_E}(c_1s+1)^{-M-N_E+1}\int_0^\infty\tau^{M+N_E-2}e^{-\tau}d\tau\\&=\frac{s^{N_E-1}}{\Gamma(N_E)\Gamma(M-1)}c_1^{N_E}(c_1s+1)^{-M-N_E+1}\Gamma(M+N_E-1),\end{aligned} \quad (17)$$

where $\Gamma(\cdot)$ is the gamma function, $\Gamma(n) = (n-1)!$ for positive integer n, $\Gamma(z) = \int_0^\infty x^{z-1}e^{-x}dx$ for complex number z, and $\tau = (c_1 s + 1)s_1$. The PDF of generalized beta prime distribution $\beta'(\alpha, \beta, p, q)$ is denoted in [22]

$$f_X(x; a, b, p, q) = \frac{p(x/q)^{ap-1}(1 + (x/q)^p)^{-a-b}}{qB(a,b)}, \tag{18}$$

where $B(a,b)\left(= \frac{\Gamma(a)\Gamma(b)}{\Gamma(a+b)}\right)$ is the beta function. Therefore, the PDF of $\beta'(N_E, M-1, 1, 1/c_1)$ is identical to (17). □

Since the mean of the generalized beta prime distribution is denoted as [22]

$$\mathbb{E}[\beta'(a,b,p,q)] = \frac{q\Gamma(a+1/p)\Gamma(b-1/p)}{\Gamma(a)\Gamma(b)}, \tag{19}$$

the expected SINR_E of (16) is upper-bounded as

$$\mathbb{E}[\text{SINR}_E] \le \mathbb{E}\left[\beta'\left(N_E, M-1, 1, \frac{(M-1)\rho_D}{\rho_E}\right)\right] = \frac{\rho_D N_E (M-1)}{\rho_E (M-2)}. \tag{20}$$

3.2. The Case That CSI Between the UE and the EV Is Available at the UE

Suppose the CSI between the UE and the EV is available at the UE. In that case, the UE can determine and use the optimal beamforming vector to intensively transmit the jamming signal to the EV considering the channel $\mathbf{G}$, i.e., $B = 1$. The optimal beamforming vector to transmit the jamming signal from the UE to EV is the right singular vector corresponding to the largest singular value obtained from the singular value decomposition of $\mathbf{G}$. In the proposed scheme, the beamforming vector for the AN should be designed in the null space of (9). In other words, the optimal beamforming vector for the AN can be determined to maximize the AN to the EV and nullify the AN to the BS. Let the right singular vector corresponding to the largest singular value obtained from the singular value decomposition of $\mathbf{G}$ be $\mathbf{w}_g$. For maximizing the AN to the EV in the null space, the optimal beamforming vector for the AN is determined as the projected vector of $\mathbf{w}_g$ onto the null space, $\mathbf{H}_n$, which is provided as

$$\mathbf{w}_E = \frac{\mathbf{Q}_n \mathbf{Q}_n^H \mathbf{w}_g}{||\mathbf{Q}_n \mathbf{Q}_n^H \mathbf{w}_g||}, \tag{21}$$

where $\mathbf{Q}_n$ is $M-1$ eigenvectors of $\mathbf{H}_n$ corresponding to $N-1$ non-zero eigenvalues. The SINR at the EV is then provided as

$$\text{SINR}_E^{CSI} = \frac{\rho_D ||\mathbf{u}_E^H \mathbf{G} \mathbf{w}_D||^2}{1 + \rho_E ||\mathbf{u}_E^H \mathbf{G} \mathbf{w}_E||^2}. \tag{22}$$

The AN term in (22) can be written as

$$\left\|\mathbf{u}_E^H \mathbf{G} \mathbf{w}_E\right\|^2 = ||\mathbf{g}||^2 ||\bar{\mathbf{g}} \mathbf{w}_E||^2. \tag{23}$$

Based on the relation of (9) and (21), and $\mathbf{w}_g$ can be orthogonally decomposed by $\mathbf{w}_D$ and $\mathbf{w}_E$ as shown in Figure 2. The distribution of the AN term in (23) is then derived as

$$\begin{aligned} ||\mathbf{g}||^2 ||\bar{\mathbf{g}} \mathbf{w}_E||^2 &= ||\mathbf{g}||^2 \left(1 - ||\bar{\mathbf{g}} \mathbf{w}_D||^2\right) \\ &= \chi^2_{2M}(1 - \beta(1, M-1)) = \chi^2_{2M}\beta(M-1, 1) = \chi^2_{2(M-1)}. \end{aligned} \tag{24}$$

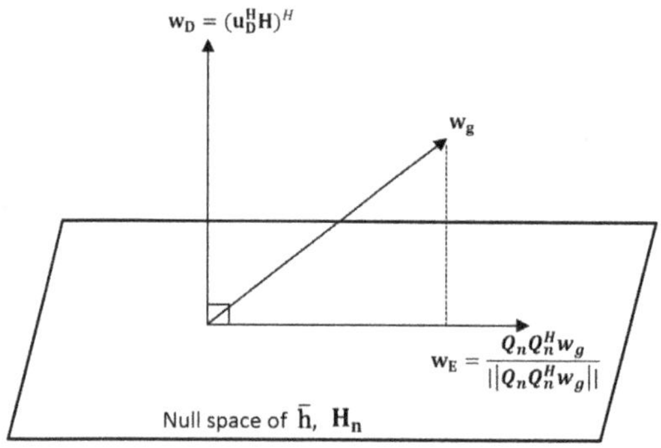

Figure 2. Geometric relation between $\mathbf{w}_D$ and $\mathbf{w}_E$.

From (22) and (24), the expected AN is then provided as

$$\mathbb{E}\left[\rho_E \chi^2_{2(M-1)}\right] = \rho_E \mathbb{E}\left[\chi^2_{2(M-1)}\right] = (M-1)\rho_E. \tag{25}$$

When Equation (25) is compared with Equation (12) and Equation (14), it is observed that the expected AN can increase $M-1$ times depending on the presence or absence of CSI between the UE and the EV. The EV's SINR of (22) is then approximated as

$$\text{SINR}_E^{CSI} \approx \frac{\rho_D ||\mathbf{u}_E^H \mathbf{G} \mathbf{w}_D||^2}{\rho_E ||\mathbf{u}_E^H \mathbf{G} \mathbf{w}_E||^2}, \tag{26}$$

because of $\rho_E ||\mathbf{u}_E^H \mathbf{G} \mathbf{w}_E||^2 \gg 1$. From (24) and (26), the SINR_E can be re-written as

$$\text{SINR}_E^{CSI} = \frac{\rho_D \chi^2_{2N_E}}{\rho_E \chi^2_{2(M-1)}} = \frac{S_2}{c_2 S_1}, \tag{27}$$

where c_2 is $\frac{\rho_E}{\rho_D}$, S_1 is the random variable with $\chi^2_{2(M-1)}$, and S_2 is the random variable with $\chi^2_{2N_E}$. From *Lemma 1*, the statistical behavior of SINR_E^{CSI} in (27) follows the generalized beta prime distribution, $\beta'(N_E, M-1, 1, 1/c_2)$. From (19), the expected SINR_E^{CSI} of (27) is provided as

$$\mathbb{E}\left[\text{SINR}_E^{CSI}\right] = \mathbb{E}\left[\beta'\left(N_E, M-1, 1, \frac{\rho_D}{\rho_E}\right)\right] = \frac{\rho_D N_E}{\rho_E (M-2)}. \tag{28}$$

4. Secrecy Capacity Analysis

For secrecy capacity comparison according to existence of the EV's CSI (i.e., $\mathbf{G}$) at the UE, the secrecy gap is defined as

$$\triangle \text{C} = \text{C}^{CSI} - \text{C} = \text{C}_E - \text{C}_E^{CSI}, \tag{29}$$

where $\text{C}^{CSI} = \text{C}_D - \text{C}_E^{CSI}$, C is defined in (6), and C_E^{CSI} is the capacity obtained from (22). The expected C_E can be provided as

$$\begin{aligned} \mathbb{E}[\text{C}_E] &= \mathbb{E}[\log_2(1+\text{SINR}_E)] \le \log_2(1+\mathbb{E}[\text{SINR}_E]) &(30)\\ &\le \log_2\left(1+\frac{\rho_D N_E (M-1)}{\rho_E (M-2)}\right), &(31) \end{aligned}$$

where Jensen's inequality is used in (30), and the upper bound in (31) is maintained by inserting the upper bound value of $\mathbb{E}[\mathrm{SINR}_E]$, i.e., (20). From (28), the expected C_E^{CSI} is approximated to

$$\begin{aligned}\mathbb{E}\left[\mathrm{C}_E^{CSI}\right] &= \mathbb{E}\left[\log_2\left(1+\mathrm{SINR}_E^{CSI}\right)\right] \approx \log_2\left(1+\mathbb{E}\left[\mathrm{SINR}_E^{CSI}\right]\right) &(32)\\ &= \log_2\left(1+\frac{\rho_D N_E}{\rho_E(M-2)}\right). &(33)\end{aligned}$$

In (32), SINR_E^{CSI} is a very small value owing to the optimal $\mathbf{w}_E$, i.e., $\mathrm{SINR}_E^{CSI} \ll 1$. The log function in (32) can then be interpreted as a linear function of SINR_E^{CSI}. From (31) and (33), the expected secrecy gap is provided as

$$\begin{aligned}\mathbb{E}[\triangle\mathrm{C}] &= \mathbb{E}[\mathrm{C}_E] - \mathbb{E}\left[\mathrm{C}_E^{CSI}\right],\\ &\leq \log_2\frac{\rho_E(M-2)+(M-1)\rho_D N_E}{\rho_E(M-2)+\rho_D N_E}. &(34)\end{aligned}$$

4.1. Asymptotic Performance for Large Value of M

As M increases, λ_{max} increases, and ρ_D becomes a value close to zero. Then, the EV's expected capacity of (31) for a sufficiently large value of M asymptotically approaches

$$\begin{aligned}\mathbb{E}[\mathrm{C}_E]_{M\to\infty} &\leq \lim_{M\to\infty}\log_2\left(1+\frac{\rho_D N_E(M-1)}{\rho_E(M-2)}\right)\\ &\sim \log_2(1)=0. &(35)\end{aligned}$$

Similarly, the EV's expected capacity of (33) for a sufficiently large value of M asymptotically approaches

$$\begin{aligned}\mathbb{E}\left[\mathrm{C}_E^{CSI}\right]_{M\to\infty} &= \lim_{M\to\infty}\log_2\left(1+\frac{\rho_D N_E}{\rho_E(M-2)}\right)\\ &\sim \log_2(1)=0. &(36)\end{aligned}$$

Therefore, the secrecy gap of (34) for a sufficiently large value of M is provided as

$$\begin{aligned}\mathbb{E}[\triangle\mathrm{C}]_{M\to\infty} &\leq \lim_{M\to\infty}\log_2\frac{\rho_E(M-2)+(M-1)\rho_D N_E}{\rho_E(M-2)+\rho_D N_E}\\ &\sim \log_2(1)=0. &(37)\end{aligned}$$

4.2. Asymptotic Performance for Large Value of N_E

As N_E increases, the EV's eavesdropping ability increases. This can be observed from the increase in the numerators' values in (20) and (28) as N_E increases, increasing the EV's SINR. Thus, the EV's expected capacity of (31) for a sufficiently large value of N_E diverges to infinity, as indicated below:

$$\mathbb{E}[\mathrm{C}_E]_{N_E\to\infty} \leq \lim_{N_E\to\infty}\log_2\left(1+\frac{\rho_D N_E(M-1)}{\rho_E(M-2)}\right) \sim \infty. \quad (38)$$

Similarly, the EV's expected capacity of (33) for a sufficiently large value of N_E diverges to infinity, as indicated below:

$$\mathbb{E}\left[\mathrm{C}_E^{CSI}\right]_{N_E\to\infty} = \lim_{N_E\to\infty}\log_2\left(1+\frac{\rho_D N_E}{\rho_E(M-2)}\right) \sim \infty. \quad (39)$$

However, the secrecy gap of (34) for a sufficiently large value of N_E asymptotically approaches

$$\begin{aligned}\mathbb{E}[\triangle C]_{N_E\to\infty} &\leq \lim_{N_E\to\infty}\log_2\frac{\rho_E(M-2)+(M-1)\rho_D N_E}{\rho_E(M-2)+\rho_D N_E}\\ &\sim \log_2(M-1). \end{aligned} \tag{40}$$

4.3. Asymptotic Performance for Large Values of $M = N_E$

The EV's expected capacity of (31) for a sufficiently large value of $M = N_E$ asymptotically approaches

$$\mathbb{E}[C_E]_{M,N_E\to\infty} \leq \lim_{M,N_E\to\infty}\log_2\left(1+\frac{\rho_D N_E(M-1)}{\rho_E(M-2)}\right) \sim \log_2\left(1+\frac{\rho_D N_E}{\rho_E}\right), \tag{41}$$

because ρ_D is a value close to zero and ρ_E is a value to close to $\frac{P}{\sigma^2}$. The simultaneous increase in the values of N_E and M simultaneously increases the eavesdropping capability and AN, thereby increasing both values of the numerator and the denominator in the SINR of the EV. Therefore, it is evident that the SINR value converges to a specific value, as shown in (41).

On the other hand, the EV's expected capacity of (33) for a sufficiently large value of $M = N_E$ asymptotically approaches

$$\mathbb{E}\left[C_E^{CSI}\right]_{M,N_E\to\infty} = \lim_{M,N_E\to\infty}\log_2\left(1+\frac{\rho_D N_E}{\rho_E(M-2)}\right) \sim \log_2(1) = 0. \tag{42}$$

because ρ_D becomes a value close to zero.

As the value of M and N_E approaches infinity, the asymptotical $\mathbb{E}[C_E]$ approaches $\log_2\left(1+\frac{\rho_D N_E}{\rho_E}\right)$, as shown in (41), while the asymptotical $\mathbb{E}\left[C_E^{CSI}\right]$ converges to 0 as shown in (42). Therefore, the secrecy gap of (34) for sufficiently large values of M and N_E approaches

$$\begin{aligned}\mathbb{E}[\triangle C]_{M,N_E\to\infty} &\leq \lim_{M,N_E\to\infty}\log_2\frac{\rho_E(M-2)+(M-1)\rho_D N_E}{\rho_E(M-2)+\rho_D N_E}\\ &\approx \lim_{M,N_E\to\infty}\log_2\frac{\rho_E+\rho_D N_E}{\rho_E+\rho_D} \sim \log_2\left(1+\frac{\rho_D N_E}{\rho_E}\right), \end{aligned} \tag{43}$$

where ρ_D is a value close to zero.

5. Simulation Results and Analysis

This section provides the simulation and numerical results to validate the proposed AN-aided secrecy performance and its numerical analysis. The fading channels between the transmitter (i.e., UE) and receivers (i.e., BS and EV) are randomly generated from i.i.d. Rayleigh fading. In the simulation, $\frac{P}{\sigma^2}$ and the required SINR, γ, are set to 10 dB.

Figure 3 shows the expected AN at the EV with or without the availability of the EV's CSI at the UE when N and N_E are set to 8. In the case when the EV's CSI is not available at the UE, the expected AN values based on (12) and (14), i.e., estimated AN, are identical to the corresponding simulation results, i.e., measured AN, as shown in Figure 3a, and its value is equal to ρ_E. When M is 4, 8, and 16, ρ_E is provided as 9.34, 9.56, and 9.72, respectively, which is the identical value of the estimated AN for each M. As discussed in Section 3.1, the expected AN is ρ_e regardless of the number of beams, B. The expected AN increases as M increases because an increase in M increases λ_{max}, decreasing ρ_D, and, accordingly, ρ_E increases. In the case when the EV's CSI is available at the UE, the expected

AN values based on (25), i.e., estimated AN, are identical to the corresponding simulation results, i.e., measured AN, as shown in Figure 3b, and its value is equal to $(M-1)\rho_E$. When M is 4, 8, and 16, the measured AN values in Figure 3a are 9.34, 9.56, and 9.72, and the measured AN values in Figure 3b are 28.0348, 67.03, and 145.5. As discussed in Section 3.1, when the CSI of the EV is present, we observe that the expected AN is increased by a factor of $M-1$ compared to the expected AN when the CSI of the EV is absent.

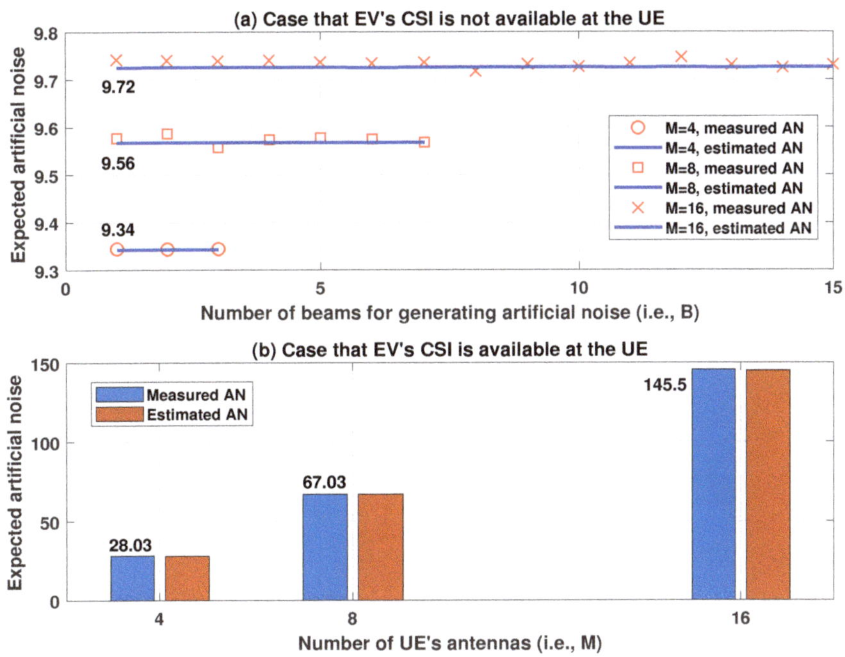

Figure 3. Artificial noise comparison ($N = N_E = 8$).

Figure 4 shows the expected SINR_E when N and N_E are set to 8. In the case when the EV's CSI is not available at the UE, the upper bound of the expected SINR_E based on (20) shows the tight upper bound for the simulated SINR_E, i.e., measured SINR_E. In the case when the EV's CSI is available at the UE, the expected SINR_E^{CSI} based on (28), i.e., estimated SINR_E^{CSI}, is tightly matched to the corresponding simulation results, i.e., measured SINR_E^{CSI}. Furthermore, we can observe that the upper bounds of the expected SINR_E and expected SINR_E^{CSI} differ by a value of $10log10(M-1)$dB, according to (20) and (28). For example, when M is 8 and 16, the SINR gaps between the estimated SINR_E^{CSI} and SINR_E are 8.45 dB $(= 10\log_{10} 7)$ and 11.76 dB $(= 10\log_{10} 15)$, respectively.

Figure 5 shows the secrecy capacity and secrecy gap according to M when N and N_E are set to 8. Increasing M makes ρ_D smaller and increases ρ_E, which in turn increases the AN in both the cases with and without the EV's CSI, as shown in (12), (14), and (25). Thus, increasing M decreases the expected EV's SINR, as shown in (20) and (28). Since the case when the EV's CSI is available at the UE can show the optimal performance in terms of secrecy capacity due to efficient AN generation, it shows ideal performance compared to the case without the EV's CSI at the UE. However, as M increases, the secrecy capacity in a practical environment without the EV's CSI at the UE approaches the ideal performance. As shown in Figure 5a, when M is 8, the secrecy capacity without the EV's

CSI achieves 89.4% of the secrecy capacity with the EV's CSI. However, as M increases to 128, the secrecy capacity without the EV's CSI achieves 98.4% of the secrecy capacity with the EV's CSI. In other words, the secrecy capacity increases with M, and its gap, i.e., secrecy gap, converges to zero, as analyzed in (37). The analyzed asymptotic secrecy gap based on (37), i.e., estimated secrecy gap, shows the tight upper bound for the corresponding simulation results, i.e., measured secrecy gap, as shown in Figure 5b. It is also shown that, as M increases from 8 to 128, the measured secrecy gap effectively decreases from 0.3566 bps/Hz to 0.0629 bps/Hz.

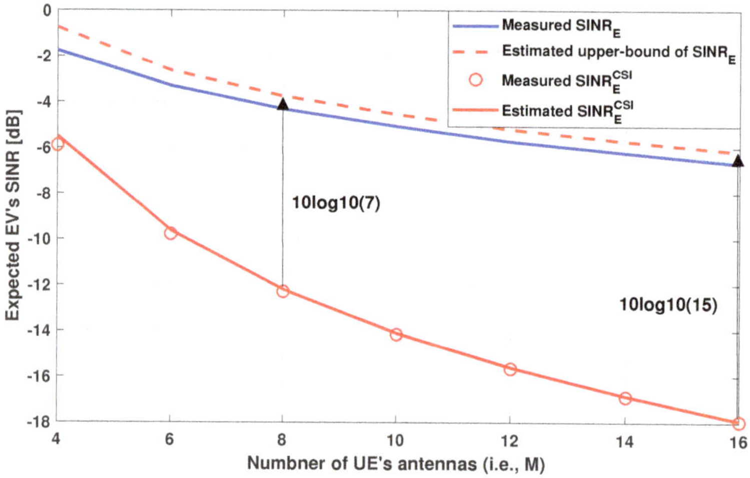

Figure 4. EV's SINR comparison ($N = N_E = 8$).

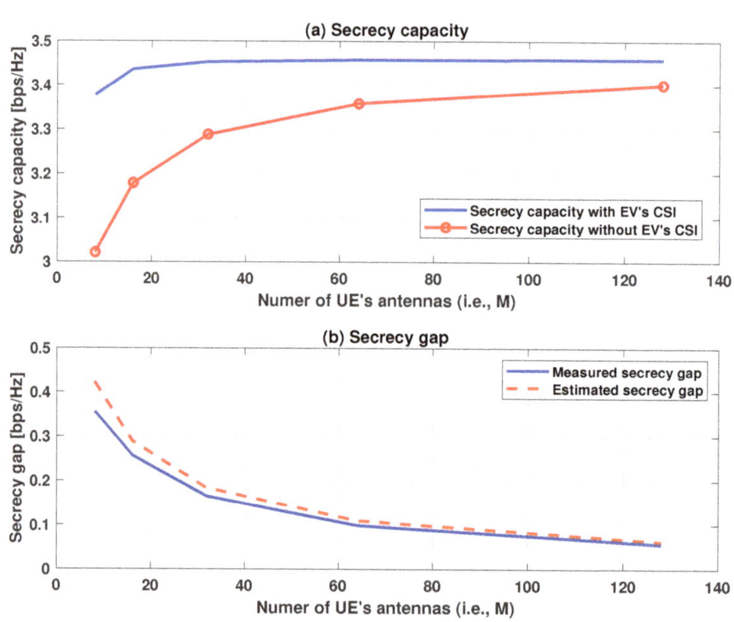

Figure 5. Secrecy capacity and secrecy gap ($N = N_E = 8$).

Figure 6a shows the measured secrecy gap and its tight upper bound as N_E increases when M and N are set to 4 and 8, respectively. The increase in N_E leads to an increase in the EV's eavesdropping ability, and, in turn, both the expected SINR_E and SINR_E^{CSI} are increased. As analyzed in (40), the estimated secrecy gap for a sufficiently large value of N_E converges to 1.5850 bps/Hz $(= \log_2 3)$, as shown in Figure 6a. Figure 6b shows the measured secrecy gap and its tight upper bound as M and N_E simultaneously increase when N is set to 8. As analyzed in (43), the secrecy gap for a sufficiently large value of M and N_E converges to 0.88 bps/Hz $\left(= \log_2\left(1 + \frac{\rho_D}{\rho_E} N_E\right)\right)$.

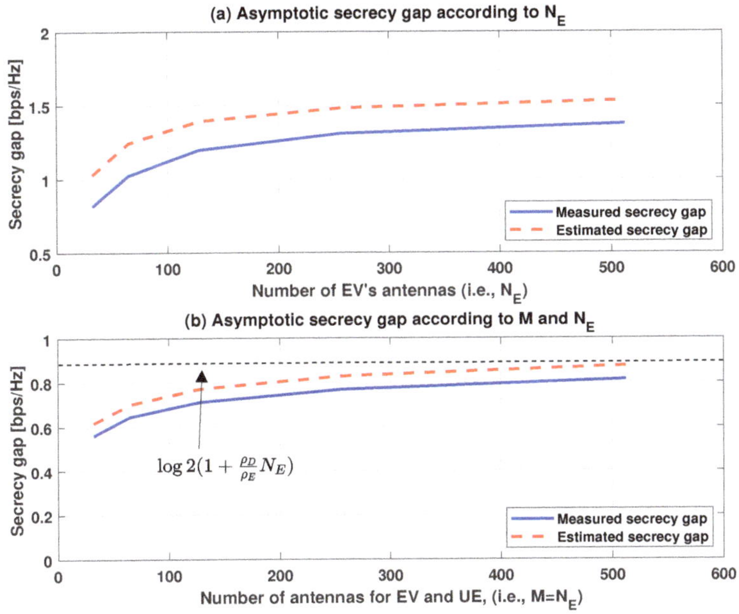

Figure 6. Asymptotic secrecy gap comparison ($N = 8$).

6. Conclusions

This paper analyzes AN-aided secure transmission, focusing on the AN effect on EVs. We derived the distribution of the AN as a closed form. We proved that the expected AN in the case when the EV's CSI is not available at the UE is equal to ρ_E regardless of B, and the expected AN in the case when the EV's CSI is available at the UE is equal to $\rho_E(M-1)$. Consequently, it was shown that the increase in M leads to an increase in the AN for both cases. Furthermore, it was proved that the upper bounds of SINR_E and SINR_E^{CSI} have beta prime distributions. Through this derivation, the expected upper bound of SINR_E and the expected SINR_E^{CSI} were calculated. Similarly to the analysis for the AN, the expected upper bound of SINR_E has $M-1$ times larger than the expected SINR_E^{CSI}. Based on the analyzed AN and SINR, the secrecy capacities for the two cases were analyzed, and, based on these, the secrecy gap was analyzed.

From an asymptotic performance perspective, when M increases from 8 to 128, the asymptotic secrecy capacity without the EV's CSI achieves from 89.4% to 98.4% of the secrecy capacity with the EV's CSI. This means that C_E nearly achieves C_E^{CSI}; i.e., the secrecy gap is close to zero. Through the analyzed asymptotic performance results, it is demonstrated that the AN-aided secure transmission under the practical environment (i.e., the case that the EV's CSI is not available at the legitimate transmitter) can nearly achieve an ideal performance (i.e., the performance when the EV's CSI is available at the legitimate transmitter) in a massive antenna system. Massive MIMO systems can transmit the signal by adopting full digital beamforming or hybrid beamforming techniques. Our research is based on the full digital beamforming system rather than hybrid beamforming systems. Leveraging the derived analyzed results in the full digital beamforming system, we expect our results to extend to the hybrid beamforming system in future works.

Funding: This work was supported by the National Research Foundation of Korea (NRF) grant funded by the Korea government (MSIT) (No. 2021R1C1C1012950).

Data Availability Statement: No new data were created or analyzed in this study.

Conflicts of Interest: The author declares no conflicts of interest.

References

1. Diffie, W.; Hellman, M. New directions in cryptography. *IEEE Trans. Inf. Theory* **1976**, *6*, 644–654. [CrossRef]
2. Shafi, G.; Micali, S. Probabilistic encryption. *J. Comput. Syst. Sci.* **1984**, *2*, 270–299.
3. Hoffstein, J.; Pipher, J.; Silverman, J.H. Ntru: A ring-based public key cryptosystem. In Proceedings of the International Algorithmic Number Theory Symposium, Portland, OR, USA, 21–25 June 1998.
4. Khisti, A.; Wornell, G.W. Secure transmission with multiple antennas I: The MISOME wiretap channel. *IEEE Trans. Inf. Theory* **2010**, *7*, 3088–3104. [CrossRef]
5. Khisti, A.; Wornell, G.W. Secure transmission with multiple antennas II: The MIMOME wiretap channel. *IEEE Trans. Inf. Theory* **2010**, *11*, 5515–5532. [CrossRef]
6. Wen, Y.; Chen, G.; Fang, S.;Wen, M.; Tomasin, S.; Renzo, M.D. RIS-Assisted UAV Secure Communications With Artificial Noise-Aware Trajectory Design Against Multiple Colluding Curious Users. *IEEE Trans. Inform. Forensics Sec.* **2024**, *19*, 3064–3076. [CrossRef]
7. Wu, Y.; Huo, Y.; Gao, Q.; Mao, J.; Yang, Z.; Jing, T. Secure Uplink Transmission Against Multiintelligent Eavesdroppers With Time-Domain Artificial Noise in MIMO IoT Systems. *IEEE Internet Things J.* **2024**, *19*, 31662–31674. [CrossRef]
8. Goel, S.; Negi, R. Guaranteeing secrecy using artificial noise. *IEEE Trans. Wirel. Commun.* **2008**, *7*, 2180–2189. [CrossRef]
9. Zhou, X.; McKay, M.R. Secure transmission with artificial noise over fading channels: Achievable rate and optimal power allocation. *IEEE Trans. Veh. Technol.* **2010**, *8*, 3831–3842. [CrossRef]
10. Zhu, J.; Schober, R.; Bhargava, V.K. Secure transmission in multicell massive MIMO systems. *IEEE Trans. Wirel. Commun.* **2014**, *9*, 4766–4781. [CrossRef]
11. Zhu, J.; Schober, R.; Bhargava, V.K. Linear precoding of data and artificial noise in secure massive MIMO systems. *IEEE Trans. Wirel. Commun.* **2016**, *3*, 2245–2261. [CrossRef]
12. Wang, H.M.; Wang, C.; Ng, D.W.K. Artificial noise assisted secure transmission under training and feedback. *IEEE Trans. Signal Process.* **2015**, *23*, 6285–6298. [CrossRef]
13. Wallace, J.W.; Sharma, R.K. Automatic secret keys from reciprocal MIMO wireless channels: Measurement and analysis. *IEEE Trans. Inform. Forensics Sec.* **2010**, *3*, 381–392.
14. Taha, H.; Alsusa, E. Secret key exchange using private random precoding in MIMO FDD and TDD systems. *IEEE Trans. Veh. Technol.* **2016**, *6*, 4823–4833. [CrossRef]
15. Liu, T.Y.; Lin, S.C.; Hong, Y.W.P. On the role of artificial noise in training and data transmission for secret communications. *IEEE Trans. Inform. Forensics Sec.* **2017**, *3*, 516–531. [CrossRef]
16. Bai, J.; Dong, T.; Zhang, Q.; Wang, S.; Li, N. Coordinated Beamforming and Artificial Noise in the Downlink Secure Multi-Cell MIMO Systems Under Imperfect CSI. *IEEE Wireless Commun. Lett.* **2020**, *9*, 1023–1026. [CrossRef]
17. Ta, H.Q.; Cao, L.; Oh, H. Novel Noise Injection Scheme to Guarantee Zero Secrecy Outage under Imperfect CSI. *Entropy* **2023**, *25*, 1594. [CrossRef]
18. Zappone, A.; Lin, P.H.; Jorswieck, E. Energy efficiency of confidential multi-antenna systems with artificial noise and statistical CSI. *IEEE J. Sel. Top. Signal Process.* **2016**, *10*, 1462–1477. [CrossRef]
19. Wang, W.; Teh, K.C.; Li, K.H. Secrecy throughput maximization for MISO multi-eavesdropper wiretap channels. *IEEE Trans. Inf. Forensics Secur.* **2016**, *12*, 505–515. [CrossRef]
20. Wang, W.; Chen, X.; You, L.; Yi, X.; Gao, X. Artificial noise assisted secure massive MIMO transmission exploiting statistical CSI. *IEEE Commun. Lett.* **2019**, *23*, 2386–2389. [CrossRef]

21. Csiszár, I.; Korner, J. Broadcast channels with confidential messages. *IEEE Trans. Inf. Theory* **1978**, *24*, 339–348. [CrossRef]
22. Gupta, A.K.; Nadarajah, S. *Handbook of Beta Distribution and Its Applications*; CRC Press: Boca Raton, FL, USA, 2004.

mathematics

MDPI

Article

Performance Analysis of Reconfigurable Intelligent Surface (RIS)-Assisted Satellite Communications: Passive Beamforming and Outage Probability

Minchae Jung [1] and Hyukmin Son [2,*]

[1] Department of Electronics and Information Engineering, Sejong University, Seoul 05006, Republic of Korea; mcjung@sejong.ac.kr

[2] Department of Electronic Engineering, Gachon University, Gyeonggi 13120, Republic of Korea

* Correspondence: hson102@gachon.ac.kr

Abstract: Reconfigurable intelligent surfaces (RISs), which consist of numerous passive reflecting elements, have emerged as a prominent technology to enhance energy and spectral efficiency for future wireless networks. RISs have the capability to intelligently reconfigure the incident wave, reflecting it towards the intended target without requiring energy for signal processing. Consequently, they have become a promising solution to support the demand for high-throughput satellite communication (SatCom) and enhanced coverage for areas inaccessible to terrestrial networks. This paper presents an asymptotic analysis of an RIS-assisted SatCom system. In this system, an unmanned aerial vehicle equipped with an RIS operates as a mobile reflector between a satellite and users. In particular, a passive beamformer is designed with the aim of asymptotically attaining optimal performance, considering the limitations imposed by practical SatCom systems. Moreover, the closed-form expressions for the ergodic achievable rate and outage probability are derived considering the proposed passive beamforming technique. Furthermore, we extend the system model to a multicast system and asymptotically analyze the optimality of the proposed scheme, leveraging the derived asymptotic results in the unicast system. The results of the simulations confirm that our analyses can precisely and analytically assess the performance of the RIS-assisted SatCom system, confirming the asymptotic optimality of the proposed scheme.

Keywords: satellite communication; reconfigurable intelligent surface (RIS); passive beamforming; outage probability

MSC: 94A12

1. Introduction

The demand for high-throughput satellite communication (SatCom) and enhanced coverage for areas inaccessible to terrestrial and nonterrestrial networks has been increasing rapidly, driven mainly by the Internet of Things and upcoming 6G applications, such as drones, vehicles, and terrestrial/nonterrestrial sensors [1–3]. Accordingly, reconfigurable intelligent surfaces (RISs) have been investigated to support this growing demand for SatCom systems [4–7]. RISs intelligently manipulate the amplitude, phase, and frequency of the incident signals and reflect them to a target destination using passive reflecting elements [8–10]. In particular, an RIS provides the capability to control the frequency of the incident electromagnetic wave, allowing a controlled frequency shift to the incident RF signal [11]. As RISs are composed of massive reflecting elements, typically constructed from passive components like inductors, capacitors, and resistors, they have the capability to control the phase of the incident signal without requiring additional energy consumption. In this regard, the utilization of an unmanned aerial vehicle (UAV) equipped with RIS has garnered considerable attention in recent years because most UAVs are battery-limited and must be operated energy-efficiently to be power-sustainable in the network [12,13]. A UAV equipped with an RIS can complement existing SatCom systems and enhance the network capacity by covering inaccessible terrestrial/nonterrestrial areas and providing additional capacity to satisfy the steep increase in data requirements [14].

Prior studies [15–17] have investigated the challenges of SatCom systems in which a UAV operates as a mobile relay between the satellite and users. In [15], a relay structure for a UAV between the satellites and the ground is examined. A blind beam-tracking approach for SatCom systems was proposed in [16] when a UAV is equipped with a large-scale antenna array and operates as a relay between the satellite and terrestrial networks. Another study [17] analyzed the performance of SatCom systems in which a UAV is equipped with a multi-antenna array and operates as a moving decode-and-forward relay. However, in [15–17], a UAV was considered as a mobile active relay that consumes additional energy due to the weight of active elements and signal processing, which is an impractical assumption because UAVs are battery-limited and must be operated energy-efficiently. Also, the authors in [7,18] analyzed the performance of SatCom systems with a UAV equipped with an RIS. However, these works focus on a scenario where all users share the same frequency and consider only the unicast SatCom system. In contrast, this paper examines a scenario where the earth station (ES) allocates different frequencies to each user and analyzes the performance of both unicast and multicast systems.

Meanwhile, in [4], the authors proposed a joint optimization problem that considers the beamforming weights at the satellite and the phase shift at the RIS. In this work, the authors proposed an alternative optimization scheme because the optimized beamforming weight vectors at the satellite and BS and phase shifters at the RIS are coupled. In [5], the authors focused on designing an RIS-assisted SatCom system for Internet of Things (IoT) networks. The primary objective of their study was to improve energy efficiency within the RIS-assisted SatCom system. Furthermore, in [6], the authors tackled the challenge of maximizing the channel capacity in an RIS-assisted SatCom system. They formulated a joint optimization problem that involves power allocation and RIS phase shift to achieve this goal and proposed a sub-optimal algorithm because the problem was formulated as nonconvex due to coupled variables. However, those studies [4–6] did not fully address the practical limitations associated with estimating individual channels between the satellite and the RIS, as well as between the RIS and the UE, which remain significant challenges in practical SatCom scenarios. Therefore, it is essential to analyze the performance bounds achievable when utilizing only the practically available information in an RIS-assisted SatCom system.

The main contribution of this work lies in conducting an asymptotic analysis of the data rate within a SatCom system assisted by RIS. Also, the asymptotic analysis takes into consideration the practical RIS environment and its limitations. To address the aforementioned limitations, we present an asymptotic analysis of an RIS-assisted SatCom system that employs a UAV equipped with an RIS. In this system, the UAV acts as a mobile reflector between a satellite and users, while relying solely on the practically available information. In particular, we first derive the asymptotic upper bound of the achievable rate and propose a passive beamformer that can asymptotically achieve the optimality by considering the presence of numerous elements on the RIS. We then derive the closed-form expressions for the ergodic achievable rate and outage probability using the proposed passive beamforming technique. Our analyses can accurately determine the performance of the SatCom system without requiring extensive simulations. Leveraging the derived asymptotic results, we extend our results to cover an additional scenario, such as an RIS-assisted SatCom system that enables a UAV to serve multiple user equipments (UEs) simultaneously. From the simulation results, we can observe that the proposed scheme asymptotically achieves optimal performance and also achieves its upper bound, verifying its asymptotic optimality.

The remainder of this paper is organized as follows. The system model is described in Section 2. The asymptotic performance and outage probability are discussed in Section 3, and the performance of a SatCom system with an RIS serving multiple UEs is analyzed in Section 4. The simulation results are presented in Section 5 and the concluding remarks are presented in Section 6.

Notations: Throughout the paper, matrices are represented by boldface uppercase symbols, while vectors are denoted by boldface lowercase symbols, and $\boldsymbol{I}_N$ and $\mathbf{1}_{M\times N}$ are, respectively, a size-N identity matrix and an $M \times N$ matrix of ones. The transpose, Hermitian transpose, and conjugate operators are $(\cdot)^{\mathrm{T}}$, $(\cdot)^{\mathrm{H}}$, and $(\cdot)^*$, respectively. The operator $\mathrm{diag}(x_1, \cdots, x_n)$ is a diagonal operator that constructs a diagonal matrix from a vector, $[x_1, \cdots, x_n]^{\mathrm{T}}$. The phase and amplitude of a complex number, y, are denoted by $\angle y$ and $|y|$, respectively, and mod is the modulo operation. $\mathcal{O}(\cdot)$ is the big O notation, and $\mathrm{E}[\cdot]$ and $\mathrm{Var}[\cdot]$ denote, respectively, expectation and variance operators. $\mathcal{CN}(m, \sigma^2)$ denotes a complex Gaussian distribution with mean m and variance σ^2, and $\mathcal{LN}(m, \sigma^2)$ and $\mathcal{L}^2\mathcal{N}(m, \sigma^2)$ denote a lognormal distribution and double-lognormal distribution, respectively.

2. System Model

Consider a SatCom system comprising a single ES serving K single-antenna user equipments (UEs), a single multibeam satellite, and K UAVs with individual RISs (i.e., K RISs) equipped with N reflecting elements. In our system model, a UAV equipped with an RIS operates as a mobile reflector between the satellite and the UEs (even though UAV-based communications may have flight-time issues caused by battery constraints, the proposed scheme can also apply to an RIS attached to manned aircraft or building surfaces, which can be relatively free from battery constraints), as shown in Figure 1. The satellite is equipped with a reflective array antenna consisting of K feeds (i.e., a single feed per beam) in which a transparent satellite payload is considered. The data for the K UEs are multiplexed in frequency at the ES and transmitted to the satellite through the feeder link operating in the optical band. Subsequently, at the satellite side, the received signal is demultiplexed, amplified, translated to RF streams, and then transmitted to the UEs via the corresponding UAVs through multiple beams. Each UAV can passively reflect the incident signal from the satellite using an RIS, and each reflecting element induces a phase shift of the incident wave (i.e., passive beamforming). We assume that the ES allocates a suitable RIS to each UE (i.e., one RIS per UE) and sends a control signal, including the reflection phase shift information, to each RIS through a dedicated wireless control link. In Section 3, we consider the unicast system, i.e., each RIS serves each corresponding UE. Subsequently, in Section 4, the unicast system is extended to the multicast system, i.e., each RIS serves multiple UEs in a corresponding multicast group.

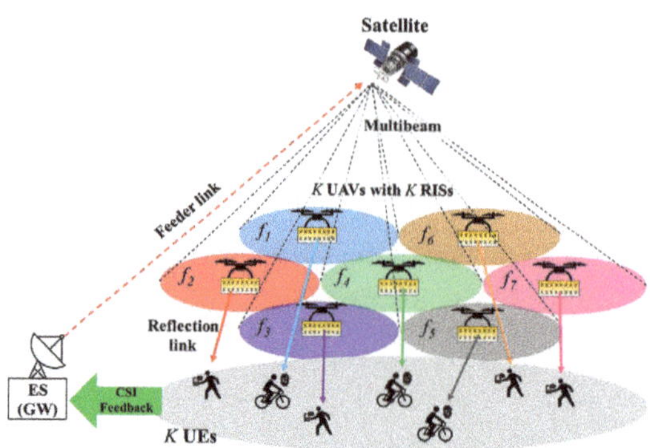

Figure 1. Illustrative system model of an RIS-assisted SatCom system.

Note that multibeam SatCom systems can use the advantages of beamspace isolation and resource allocation for improved spectral efficiency [19]. Depending on the location of the UAV, the UAVs located far away from each other will not suffer from interbeam interference due to the beamspace isolation, whereas the UAVs with similar locations can suffer from interbeam interference, which can degrade the performance. To tackle this challenge, we make the assumption that the ES allocates orthogonal frequencies among UAVs/UEs located in similar positions, thereby supporting dynamic spectrum and power allocation [20], as shown in Figure 1. Consequently, multibeam satellite systems can be an interbeam interference-free environment. The received signal at UE k is then obtained as

$$y_k = \sqrt{\rho_k} \boldsymbol{f}_k^{\mathrm{H}} \boldsymbol{\Phi}_k \boldsymbol{g}_k x_k + n_k, \tag{1}$$

where ρ_k is the signal-to-noise ratio (SNR) for UE k and $\boldsymbol{f}_k \in \mathbb{C}^{N\times 1}$ and $\boldsymbol{g}_k \in \mathbb{C}^{N\times 1}$ are, respectively, the fading channels between UE k and RIS k and between the satellite and RIS k. Moreover, x_k and n_k are the transmit signal and additive complex Gaussian noise with zero mean and unit variance, respectively, and $\boldsymbol{\Phi}_k \in \mathbb{C}^{N\times N}$ is a diagonal reflection matrix functioning as a passive beamformer. The reflection matrix is determined from the control signal transmitted by the ES, and $\boldsymbol{\Phi}_k$ can be obtained as follows:

$$\boldsymbol{\Phi}_k = \mathrm{diag}\Big(\sqrt{\alpha_{k,1}}e^{j\theta_{k,1}}, \sqrt{\alpha_{k,2}}e^{j\theta_{k,2}}, \cdots, \sqrt{\alpha_{k,N}}e^{j\theta_{k,N}}\Big), \tag{2}$$

where $\sqrt{\alpha_{k,n}} \in [0,1]$ and $\theta_{k,n} \in [0, 2\pi)$ are, respectively, the amplitude and phase shifting response of the reflection element n in RIS k. Assuming lossless reflection power at all the RISs (i.e., $\alpha_{k,n} = 1, \forall k, n$), the reflection matrix is given by

$$\boldsymbol{\Phi}_k = \mathrm{diag}\Big(e^{j\theta_{k,1}}, e^{j\theta_{k,2}}, \cdots, e^{j\theta_{k,N}}\Big), \forall k. \tag{3}$$

In (1), the signals directly transmitted from the satellite to the UEs are neglected because of severe path loss and shadowing. For a practical air-to-ground (A2G) channel, $\boldsymbol{f}_k = \sqrt{\zeta_k}\hat{\boldsymbol{f}}_k$ is generated by correlated Rician fading, which consists of both line-of-sight (LoS) and non-line-of-sight (NLoS) paths, where ζ_k is the path loss of the RIS–UE link and $\hat{\boldsymbol{f}}_k$ is Rician fading, as follows:

$$\zeta_{k,\mathrm{dB}} = -10\log_{10}\left(\frac{4\pi d_k^{\mathrm{ru}}}{\lambda}\right)^2 - \eta_{\mathrm{L}} P_k^{\mathrm{L}} - \eta_{\mathrm{NL}} P_k^{\mathrm{NL}}, \tag{4}$$

$$\hat{\boldsymbol{f}}_k = \sqrt{\frac{\kappa_k}{\kappa_k+1}}\bar{\boldsymbol{f}}_k + \sqrt{\frac{1}{\kappa_k+1}}\tilde{\boldsymbol{f}}_k. \tag{5}$$

Here, $\zeta_{k,\mathrm{dB}}$ is the path loss in dB scale, d_k^{ru} is the distance from RIS k to UE k, λ represents the wavelength of a signal, and the average additional losses for LoS and NLoS conditions are denoted as η_{L} and η_{NL}, respectively. Based on [21], the presence of a LoS path is contingent upon the distance between the RIS and the UE. Then, the probability of LoS is expressed as follows:

$$P_k^{\mathrm{L}} = \frac{1}{1 + c_1 \exp\left[-c_2\left(\arctan\left(\frac{d_k^{\mathrm{v}}}{d_k^{\mathrm{h}}}\right)\right)\right]}, \tag{6}$$

where d_k^{v} and d_k^{h} are, respectively, vertical and horizontal distances from UAV k and UE k, and c_1 and c_2 are determined by the particular configuration of the environment [21]. Also, P_k^{NL} is the probability of NLoS as given by $P_k^{\mathrm{NL}} = 1 - P_k^{\mathrm{L}}$. In (5), κ_k is the Rician factor between UE k and RIS k, and $\bar{\boldsymbol{f}}_k \in \mathbb{C}^{N\times 1}$, which is composed of deterministic values whose entries are determined depending on the locations of RIS k and UE k. By defining $\bar{f}_{k,n}$ as the n-th element of $\bar{\boldsymbol{f}}_k$, we have

$$\bar{f}_{k,n} = \exp\Big(-j2\pi d_{k,n}^{\mathrm{ru}}/\lambda\Big), \tag{7}$$

where $d_{k,n}^{\mathrm{ru}}$ is the distance between UE k and RIS element n of RIS k [22]. Considering the spatial correlation in the RIS–UE channel, we define $\tilde{\boldsymbol{f}}_k = \boldsymbol{G}_k^{1/2}\check{\boldsymbol{f}}_k$, where $\boldsymbol{G}_k \in \mathbb{C}^{N\times N}$ is the spatial correlation matrix at RIS k and $\check{\boldsymbol{f}}_k \in \mathbb{C}^{N\times 1}$ is composed of NLoS components, with each entry being an i.i.d. complex Gaussian random variable with a mean of zero and a variance of one. For notational simplicity, we define $\bar{\beta}_k = \sqrt{\zeta_k \kappa_k/(\kappa_k+1)}$ and $\tilde{\beta}_k = \sqrt{\zeta_k/(\kappa_k+1)}$. In (1), the channel gain vector between the satellite and the RIS, $\boldsymbol{g}_k$, includes a multibeam pattern, atmospheric fading, and phase rotations, which can be decomposed as follows [23]:

$$\boldsymbol{g}_k = \boldsymbol{D}_k \boldsymbol{b}_k, \tag{8}$$

where $\boldsymbol{D}_k \in \mathbb{C}^{N\times N}$ is a rain fading matrix and $\boldsymbol{b}_k \in \mathbb{C}^{N\times 1}$ is the coefficient vector that represents the beam radiation pattern gain with path loss component. As the channel gains between the satellite and N elements on RIS k are highly correlated, we have the followings [24]:

$$\boldsymbol{b}_k = \xi_k \sqrt{a_k} \mathbf{1}_{N\times 1}, \tag{9}$$

$$\boldsymbol{D}_k = |c_k| \exp(j\psi_k) \boldsymbol{I}_N, \tag{10}$$

where ξ_k and a_k are the coefficients representing, respectively, the effects of the path loss and beam pattern as follows:

$$\xi_k = \frac{\lambda}{4\pi d_k^{\mathrm{sr}}}, \tag{11}$$

$$a_k = G_{\max}\left(\frac{J_1(u_k)}{2u_k} + 36\frac{J_3(u_k)}{u_k^3}\right)^2. \tag{12}$$

Here, d_k^{sr} is the distance from the satellite to RIS k, $G_{\max}$ is the maximum gain of the satellite beam, $J_n(\cdot)$ is the first kind Bessel function of order n, and $u_k = 2.07123\frac{\sin\phi_k}{\sin\phi_{3\mathrm{dB}}}$. Also, ϕ_k and $\phi_{3\mathrm{dB}}$ are the angle between the beam center and RIS k and the 3 dB angle of the beam, respectively. ψ_k is the phase of the fading channel at RIS k that is assumed to be distributed uniformly over $[0, 2\pi)$ as in [23]. $|c_k|$ is the rain fading channel gain that follows a double-lognormal distribution, such as $|c_k| \sim \mathcal{L}^2\mathcal{N}(\eta_k, \sigma_k^2)$, whose probability density function (PDF) and cumulative distribution function (CDF), respectively, follow [25]

$$f_{|c_k|}(x) = -\frac{\exp\left(-\frac{(\ln(-\ln x)-\eta_k)^2}{2\sigma_k^2}\right)}{\sqrt{2\pi}\sigma_k x \ln x}, \tag{13}$$

$$F_{|c_k|}(x) = Q\left(\frac{\ln(-\ln x) - \eta_k}{\sigma_k}\right), \tag{14}$$

where $0 < x < 1$ and $Q(\cdot)$ is a Gaussian Q-function. Rain fading and LoS components in satellite communications are the main channel factors distinguishing them from terrestrial communication systems.

The cascaded channel from the satellite to UE k via RIS k is given by

$$h_k = \boldsymbol{f}_k^{\mathrm{H}} \boldsymbol{\Phi}_k \boldsymbol{g}_k \tag{15}$$

$$= \xi_k \sqrt{a_k} |c_k| \exp(j\psi_k) \boldsymbol{f}_k^{\mathrm{H}} \boldsymbol{\Phi}_k \mathbf{1}_{N\times 1}. \tag{16}$$

Then, the SNR at UE k and the achievable sum-rate can be obtained, respectively, as follows:

$$\gamma_k = \rho_k \xi_k^2 a_k |c_k|^2 \left|\sum_{n=1}^{N} f_{k,n}^* e^{j\theta_{k,n}}\right|^2, \tag{17}$$

$$R = \sum_{k=1}^{K} \log(1+\gamma_k), \tag{18}$$

where $f_{k,n}$ represents the n-th element of a vector $\boldsymbol{f}_k$. We can observe from (17) that, when $\theta_{k,n} = -\angle f^*_{k,n} \forall n$, γ_k attains its maximum value, $\tilde{\gamma}_k$, which is given as follows:

$$\tilde{\gamma}_k = \rho_k \xi_k^2 a_k |c_k|^2 \Big(\sum_{n=1}^{N} |f_{k,n}| \Big)^2. \tag{19}$$

In this case, the achievable rate of UE k is given by $\tilde{R}_k = \log(1 + \tilde{\gamma}_k)$. Note that RISs are generally passive; moreover, they are typically not equipped with RF chains and are installed on UAVs with limited batteries. Thus, it is hard to accurately estimate RIS-related channel information, such as $\boldsymbol{f}_k$, in the considered SatCom system, and UE k cannot practically achieve the data rate of $\tilde{R}_k$. Given that it is practically feasible to estimate the partial channel state information (CSI) of RIS-related channels [23], we propose a practical passive beamformer that is solely based on the partial CSI while achieving optimal performance, i.e., $\tilde{R}_k$, asymptotically.

3. Asymptotic Performance Analysis in Unicast System

We analyze the asymptotic data rate considering the practical SatCom system according to the following steps. Considering the achievable rate $\tilde{R}_k$, we first analyze the upper bound of $\tilde{R}_k$ as N increases to infinity. We then propose a practical passive beamformer and verify the asymptotic optimality of the proposed scheme. We finally derive the mean and variance of the achievable data rate of the proposed scheme.

From (19), the achievable rate of UE k, $\tilde{R}_k$, and its upper bound, $\hat{R}_k$, are obtained as

$$\tilde{R}_k = \log\Big(1 + \rho_k \xi_k^2 a_k |c_k|^2 \Big(\sum_{n=1}^{N} |f_{k,n}| \Big)^2 \Big) \tag{20}$$

$$\underset{(a)}{\leq} \log\left(1 + N^2 \rho_k \xi_k^2 a_k |c_k|^2 \left(\bar{\beta}_k + \tilde{\beta}_k \frac{\sum_n |\tilde{f}_{k,n}|}{N} \right)^2 \right)$$

$$\triangleq \hat{R}_k, \tag{21}$$

where $\tilde{f}_{k,n}$ is the n-th element of a vector, $\tilde{\boldsymbol{f}}_k$, and (a) is obtained from the following inequality: $\sum_n |f_{k,n}| \leq \sum_n \bar{\beta}_k |\bar{f}_{k,n}| + \sum_n \tilde{\beta}_k |\tilde{f}_{k,n}|$, where $|\bar{f}_{k,n}| = 1$ from (7). As N increases to infinity, we derive the asymptotic upper bound of $\hat{R}_k$ as follows:

$$\hat{R}_k \underset{(b)}{\leq} \log\Big(1 + N^2 \rho_k \xi_k^2 a_k |c_k|^2 (\bar{\beta}_k + \tilde{\beta}_k)^2 \Big)$$

$$\approx \log\Big(N^2 \rho_k \xi_k^2 a_k |c_k|^2 \bar{\beta}_k^2 \Big) + 2\log\left(1 + \sqrt{\frac{1}{\kappa_k}}\right)$$

$$\underset{(c)}{\approx} \log\Big(N^2 \rho_k \xi_k^2 a_k |c_k|^2 \bar{\beta}_k^2 \Big), \tag{22}$$

where (b) results from Jensen's inequality: $\left(\sum_n |\tilde{f}_{k,n}|/N\right)^2 \leq \sum_n |\tilde{f}_{k,n}|^2/N$. Here, the sample mean of $|\tilde{f}_{k,n}|^2$ can be obtained from the asymptotic channel hardening effect [26], under the assumption that the spectral norm of $\boldsymbol{G}_k$ is bounded as

$$\frac{\sum_{n=1}^{N} |\tilde{f}_{k,n}|^2}{N} \to 1, \tag{23}$$

almost surely as $N \to \infty$. Also, (c) results from a large N and a small $\sqrt{1/\kappa_k}$ because the Rician factor, κ_k, will be a large value (in the A2G channel model, the maximum value and the average value of the Rician factor introduced in [27] are 20 dB and 10 dB, respectively), considering a practical A2G channel model [27,28]. We can observe from (22) that, as N increases, the upper bound of the SNR increases with $\mathcal{O}(N^2)$, and the upper bound of the achievable rate increases with $\mathcal{O}(\log N)$.

Note that an ES can leverage global navigation satellite systems (GNSSs) to accurately estimate the three-dimensional positions of UAVs and UEs [23]. Thus, the ES can estimate the deterministic LoS components, denoted as $\bar{\boldsymbol{f}}_k$, by using the distance between RIS k and UE k, which is partial CSI; however, this is the only information available at the ES side. Hence, we design a suboptimal passive beamformer exclusively based on $\bar{\boldsymbol{f}}_k$, as follows:

$$\bar{\theta}_{k,n} = -\angle \bar{f}^*_{k,n}. \tag{24}$$

Using (24), the achievable rate is given as follows:

$$\bar{R}_k = \log\left(1 + N^2 \rho_k \xi_k^2 a_k |c_k|^2 \left| \bar{\beta}_k + \frac{\sum_n \tilde{\beta}_k \tilde{f}_{k,n} e^{j\bar{\theta}_{k,n}}}{N} \right|^2 \right). \tag{25}$$

Then, we derive the following lemma from the asymptotic analyses.

Lemma 1. *On the basis of the scaling law for N, the rate ratio between the achievable rate of the proposed passive beamformer and its performance bound becomes one (i.e.,* $\frac{\hat{R}_k}{\bar{R}_k} = 1$*) as N increases to infinity.*

Proof. In (25), the term $\sum_n \tilde{f}_{k,n} e^{j\bar{\theta}_{k,n}}$ represents the sum of N samples, where each sample is an individual random variable whose mean value is equal to zero because $\mathrm{E}\left[\tilde{f}_{k,n}\right] = 0, \forall n$. Hence, as N increases, $\sum_n \tilde{f}_{k,n} e^{j\bar{\theta}_{k,n}} / N$ approaches zero, and we can then approximate $\bar{R}_k$ as follows:

$$\begin{aligned} \bar{R}_k &\approx \log\left(N^2 \rho_k \xi_k^2 a_k |c_k|^2 \left| \bar{\beta}_k + \tilde{\beta}_k \frac{\sum_n \tilde{f}_{k,n} e^{j\bar{\theta}_{k,n}}}{N} \right|^2 \right) \\ &\approx \log\left(N^2 \rho_k \xi_k^2 a_k |c_k|^2 \bar{\beta}_k^2\right) \triangleq \dot{R}_k, \end{aligned} \tag{26}$$

where $\dot{R}_k$ is defined as the asymptotic achievable rate for a large N. From (22) and (26), the rate ratio between $\bar{R}_k$ and $\hat{R}_k$ becomes one as N increases to infinity, which completes the proof. □

Lemma 1 shows that the channel fading of an RIS-assisted SatCom system exclusively depends on the rain fading channel gain, $|c_k|$, and the impact of the Rician fading on the achievable rate becomes negligible as N increases. Lemma 1 also proves that the proposed passive beamformer achieves the upper bound of the achievable rate and asymptotically achieves the SNR order of $\mathcal{O}(N^2)$. Moreover, since $\tilde{R}_k \leq \hat{R}_k$ from (21), it is proven that the proposed passive beamformer achieves the optimal performance, $\tilde{R}_k$, which is only achievable using the full CSI of $\boldsymbol{f}_k$ when $\theta_{k,n} = -\angle f^*_{k,n}$. Note that the proposed $\bar{\theta}_{k,n}$ can be readily obtained from the estimated value of $d^{\mathrm{ru}}_{k,n}$, leading to a low computational complexity (i.e., constant complexity) to calculate all phase shift values of the RIS. Additionally, the proposed scheme exclusively relies on deterministic LoS channel information (i.e., large-scale information), which exhibits slower variation compared to NLoS channels. Consequently, the proposed scheme does not require frequent feedback for CSI and location information, effectively reducing feedback/signaling overhead and rendering the proposed scheme practically feasible.

Next, we analyze the mean (i.e., ergodic achievable rate) and variance of the asymptotic achievable rate for the proposed passive beamformer. From (26), the asymptotic achievable rate, $\dot{R}_k$, can be obtained using (13) as follows:

$$\begin{aligned} \dot{R}_k &= \log\left(N^2 \rho_k \xi_k^2 a_k \bar{\beta}_k^2\right) + 2\log|c_k| & (27) \\ &= \log\left(N^2 \rho_k \xi_k^2 a_k \bar{\beta}_k^2\right) - \frac{\log_2 10}{10} z_1, & (28) \end{aligned}$$

as N increases, where $z_1 \sim \mathcal{LN}(\mu_k, \sigma_k^2)$ and $\mu_k = \eta_k - \log(\log 10) + \log 20$ [25]. As N increases, the mean and variance of the asymptotic achievable rate are obtained, respectively, as follows:

$$\mathrm{E}\big[\dot{R}_k\big] = \log\Big(N^2 \rho_k \bar{\xi}_k^2 a_k \bar{\beta}_k^2\Big) - \frac{\log_2 10}{10} e^{\mu_k + \frac{\sigma_k^2}{2}}, \tag{29}$$

$$\mathrm{Var}\big[\dot{R}_k\big] = \left(\frac{\log_2 10}{10}\right)^2 \Big(e^{\sigma_k^2} - 1\Big) e^{2\mu_k + \sigma_k^2}. \tag{30}$$

Then, the ergodic sum-rate is asymptotically derived as

$$\mathrm{E}\big[\dot{R}\big] = \frac{2K}{\ln 2}\left(\ln N - \frac{\ln 10}{20} e^{\mu_0 + \frac{\sigma_0^2}{2}}\right) + \sum_k \log\Big(\rho_k \bar{\xi}_k^2 a_k \bar{\beta}_k^2\Big), \tag{31}$$

when $\mu_k = \mu_0$ and $\sigma_k = \sigma_0, \forall k$. For notational simplicity, we define $\dot{m}_k$ and $\dot{s}_k^2$ as the asymptotic mean and variance, respectively, of $\dot{R}_k$ (i.e., $\dot{m}_k \triangleq \mathrm{E}\big[\dot{R}_k\big]$ and $\dot{s}_k^2 \triangleq \mathrm{Var}\big[\dot{R}_k\big]$). On the basis of the scaling law for N, we can observe from (29) that the SNR increases with $\mathcal{O}(N^2)$ and the ergodic achievable rate increases with $\mathcal{O}(\log N)$. Moreover, (30) shows that the variance of $\bar{R}_k$ remains constant regardless of N, indicating the reduced channel hardening effect [29]. Using (29) and (30), we can characterize the outage probability, which is defined as the probability that the instantaneous data rate falls below a target rate, as follows.

Lemma 2. *As N increases, the probability of the outage event $\{\dot{R}_k < R_0\}$ tends to zero, where R_0 represents the target desired rate given by $R_0 = \delta \cdot \dot{m}_k$ for $0 < \delta \leq 1$.*

Proof. From (28), the outage probability can be obtained as follows:

$$\begin{aligned}
\Pr\{\dot{R}_k < R_0\} &= \Pr\left\{\ddot{m}_k - \frac{\log_2 10}{10} z_1 < R_0\right\} \\
&= \Pr\left\{z_1 > \frac{10(\ddot{m}_k - R_0)}{\log_2 10}\right\} \\
&= Q\left(\frac{\ln\left(\frac{10(\ddot{m}_k - R_0)}{\log_2 10}\right) - \mu_k}{\sigma_k}\right),
\end{aligned} \tag{32}$$

where $\ddot{m}_k = \log(N^2 \rho_k \bar{\xi}_k^2 a_k \bar{\beta}_k^2)$. Consider that $R_0 = \delta \dot{m}_k$ for $0 < \delta \leq 1$, we then have

$$\ddot{m}_k - R_0 = (1 - \delta) \log\Big(N^2 \rho_k \bar{\xi}_k^2 a_k \bar{\beta}_k^2\Big) + \frac{\delta \log_2 10}{10} e^{\mu_k + \frac{\sigma_k^2}{2}}. \tag{33}$$

It is observed in (33) that $\ddot{m}_k - R_0$ increases as N increases. Therefore, the outage probability $\Pr\{\dot{R}_k < R_0\}$ eventually approaches zero according to the increase in N, which completes the proof. □

Lemma 2 demonstrates that, as N increases, the outage probability decreases, verifying that an RIS can offer reliable SatCom for a large N. Moreover, we can observe from (32) that the outage probability of an RIS-assisted SatCom system can be readily estimated using the deterministic values. Thus, we can assess the outage probability of a SatCom system without requiring extensive simulations.

4. Asymptotic Performance Analysis in a Multicast System

In this section, we extend the system model to a multicast system and analyze the asymptotic optimality of the proposed scheme, leveraging the derived asymptotic results in the unicast system. In particular, we will demonstrate that achieving asymptotic optimality in the multicast system is readily attainable through a simple yet practical passive beamforming approach. Consider a SatCom system consisting of a single multibeam satellite, with K RISs, each capable of simultaneously serving M_k UEs. As the satellite is equipped with a reflective array antenna comprising K feeds, we examine K multicast groups (M_k UEs per group), with each multicast group k being served by RIS k. On the basis of the multicast nature, the achievable rate of multicast group k is obtained as follows [30]:

$$R_{\mathrm{MC},k} = \min_{m \in \mathcal{M}_k} R_{k,m}, \tag{34}$$

where

$$R_{k,m} = \log\left(1 + \rho_k \bar{\xi}_k^2 a_k |c_k|^2 \left|\sum_{n=1}^{N} f_{k,n,m}^* e^{j\theta_{k,n}}\right|^2\right). \tag{35}$$

Here, $\mathcal{M}_k$ is a set of UEs in multicast group k where $|\mathcal{M}_k| = M_k$, and $f_{k,n,m}$ is the fading channel between reflecting element n of RIS k and UE m. As discussed in (19), optimal reflection performance can be achieved by aligning all phase shift values of RIS with the channel phases from RIS to UE. However, in the multicast system, it is impossible to design the optimal passive beamforming to simultaneously match the phases of channels for multiple UEs.

Based on (34), we can observe that maximizing the achievable rate in the multicast system is synonymous with maximizing the individual rate of each UE. Hence, we introduce a practical passive beamforming approach for the multicast system, wherein the reflecting elements of the RIS are evenly divided and allocated to each UE within a multicast group. In particular, N reflecting elements of each RIS are divided into M_k reflecting groups, and the reflection phases of reflecting group m are obtained as follows:

$$\bar{\theta}_{k,n} = -\angle \bar{f}_{k,n,m}^*, \tag{36}$$

for $n \in \mathcal{N}_{k,m}$. Here, $\bar{f}_{k,n,m}$ is the LoS component of $f_{k,n,m}$ and $\mathcal{N}_{k,m}$ is a set of reflecting elements in reflecting group m of RIS k, where $\sum_{m=1}^{M_k} |\mathcal{N}_{k,m}| = N, \forall k$. For notational simplicity, we assume that $N \bmod M_k = 0, \forall k$. In order to achieve the SNR order of $\mathcal{O}(N^2)$ for a large N, N reflecting elements are equally divided into M_k reflecting groups (i.e., $\frac{N}{M_k}$ elements per each reflecting group) and, similarly in (26), we have:

$$\begin{aligned} \bar{R}_{k,m} &\approx \log\left(\rho_k \bar{\xi}_k^2 a_k |c_k|^2 \left|\bar{\beta}_{k,m}\left(\frac{N}{M_k} + \sum_{n \notin \mathcal{N}_{k,m}} \bar{f}_{k,n,m}^* e^{j\bar{\theta}_{k,n}}\right)\right|^2\right) \\ &\approx \log\left(N^2 \rho_k \bar{\xi}_k^2 a_k |c_k|^2 \bar{\beta}_{k,m}^2 / M_k^2\right), \end{aligned} \tag{37}$$

where $\bar{\beta}_{k,m} = \sqrt{\zeta_{k,m}\kappa_{k,m}/(\kappa_{k,m}+1)}$ and $\kappa_{k,m}$ and $\zeta_{k,m}$ are, respectively, the Rician factor and the path loss between RIS k and UE m. From (34) and (37), we can observe that $\bar{R}_{\mathrm{MC},k}$ is determined exclusively depending on $\bar{\beta}_{k,m}$. Define the minimum $\bar{\beta}_{k,m}$ for all $m \in \mathcal{M}_k$ as $\bar{\beta}_{k,m'}$, then we have:

$$\bar{R}_{\mathrm{MC},k} = \log\left(N^2 \rho_k \bar{\xi}_k^2 a_k |c_k|^2 \bar{\beta}_{k,m'}^2 / M_k^2\right). \tag{38}$$

Next, we derive a theoretical ideal performance for the multicast system under consideration, with the goal of capturing the asymptotic optimality of the multicast system. In this regard, we define an ideal RIS capable of reflecting signals to multiple UEs simultaneously with perfect phase alignment for all UEs. Then, similarly to (20) and (21), the theoretical ideal performance, denoted as $\tilde{R}_{k,m}$, and its upper bound, denoted as $\hat{R}_{k,m}$, can be obtained as follows:

$$\begin{aligned}\tilde{R}_{k,m} &= \log\left(1+\rho_k\bar{\xi}_k^2 a_k|c_k|^2\left(\sum_{n=1}^{N}|f_{k,n,m}|\right)^2\right)\\ &\leq \log\left(1+N^2\rho_k\bar{\xi}_k^2 a_k|c_k|^2\left(\bar{\beta}_{k,m}+\tilde{\beta}_{k,m}\frac{\sum_n|\tilde{f}_{k,n,m}|}{N}\right)^2\right)\\ &\triangleq \hat{R}_{k,m},\end{aligned} \tag{39}$$

where $\tilde{\beta}_{k,m} = \sqrt{\zeta_{k,m}/(\kappa_{k,m}+1)}$ and $\tilde{f}_{k,n,m}$ is the NLoS component of $f_{k,n,m}$. As N increases to infinity, we derive the asymptotic upper bound of $\hat{R}_{k,m}$ following a similar approach as in (22), as follows:

$$\begin{aligned}\hat{R}_{k,m} &\leq \log\left(1+N^2\rho_k\bar{\xi}_k^2 a_k|c_k|^2(\bar{\beta}_{k,m}+\tilde{\beta}_{k,m})^2\right)\\ &\approx \log\left(N^2\rho_k\bar{\xi}_k^2 a_k|c_k|^2\bar{\beta}_{k,m}^2\right).\end{aligned} \tag{40}$$

From (34) and (40), we can observe that the minimum value of (40) for all $m \in \mathcal{M}_k$ will be:

$$\hat{R}_{\mathrm{MC},k} = \log\left(N^2\rho_k\bar{\xi}_k^2 a_k|c_k|^2\bar{\beta}_{k,m'}^2\right). \tag{41}$$

From (38) and (41), it is evident that the proposed passive beamformer achieves a theoretical ideal performance with an SNR order of $\mathcal{O}(N^2)$, establishing the optimality of the proposed scheme. Note that the proposed passive beamformer in the multicast system can achieve near-ideal performance if the value of N is sufficient relative to M_k. This holds true even when the number of passive elements is not infinite, as demonstrated in the simulation results presented in Section 5.

Similarly to (29) and (32), the ergodic achievable rate and its outage probability can be obtained, respectively, as follows:

$$\mathrm{E}[\dot{R}_{\mathrm{MC},k}] = \log\left((N/M_k)^2\rho_k\bar{\xi}_k^2 a_k\bar{\beta}_{k,m'}^2\right) - \frac{\log_2 10}{10}e^{\mu_k+\frac{\sigma_k^2}{2}}, \tag{42}$$

$$\Pr\{\dot{R}_{\mathrm{MC},k} < R_0\} = Q\left(\frac{\ln\left(\frac{10(\ddot{m}_{\mathrm{MC},k}-R_0)}{\log_2 10}\right)-\mu_k}{\sigma_k}\right), \tag{43}$$

where $\dot{R}_{\mathrm{MC},k}$ is the asymptotic achievable rate of multicast UE k and

$$\ddot{m}_{\mathrm{MC},k} = \log\left((N/M_k)^2\rho_k\bar{\xi}_k^2 a_k\bar{\beta}_{k,m'}^2\right). \tag{44}$$

Similarly to the unicast system, (43) shows that the outage probability in the multicast system decreases as N increases, verifying that an RIS can offer reliable multicast SatCom for a large N.

5. Simulation Results

In this section, we conduct comprehensive simulations in terms of the achievable rate ratio, ergodic rate, and outage probability. We consider a practical RIS-assisted SatCom system in which a low earth orbit (LEO) satellite serves a cluster of seen UEs via seven RISs (i.e., one RIS per UE and $K = 7$). The parameters used for the simulations are detailed in Table 1. Statistical averaging is performed over a considerable number of independent runs for all simulations. Here, the label "Estimation" represents the estimated result obtained from our analyses, whereas the label "Simulation" represents the simulated results within the practical RIS-assisted SatCom system. In Table 1, the transmission power for UE k is defined as P_k; thus, we have $\rho_k = P_k/N_0$, where N_0 is the noise power and equal power allocation is assumed for all UEs (i.e., $P_1 = P_2 = \cdots = P_K$).

Table 1. Simulation parameters.

Parameter	Value
Orbit	LEO (2520 km)
Carrier frequency	2 GHz (L-band)
Bandwidth per beam, P_k	20 MHz, 50 dBW
K, $G_{\max}$, $\phi_{3\text{dB}}$	$7, 49.5$ dB, $3.2°$
UE antenna gain	38.16 dB
Boltzmann's constant	1.38×10^{-23} J/K
Noise temperature at UE	207 K
RIS element layout	2D square lattice (spacing 0.01 m)
Average rain attenuation	3 dB ($\eta_k = -1.5617, \sigma_k = 1, \forall k$)
κ_k, d_k^{v}, d_k^{h}, $\forall k$	10, 1 km, 0.1 km
η_{L}, η_{NL}, c_1, c_2	$1, 20, 15, 0.4$

In Figure 2, Lemma 1 is verified when the optimal achievable rate and its upper bound, respectively, follow (20) and (21). As shown in Figure 2, as N increases, the achievable rate ratio obtained using the proposed scheme approaches 1, confirming the optimality of the proposed scheme, as proven in Lemma 1. Also, the rate ratio of the optimal performance is larger than that of the proposed scheme. In particular, the proposed scheme achieves approximately 89% of the upper bound, while the optimal scheme achieves approximately 90% of the upper bound when $N = 64$. However, the optimal performance is only achievable using the full CSI (i.e., f_k), which is an impractical assumption in the considered RIS-assisted SatCom system. By contrast, the proposed scheme only requires partial CSI (i.e., $\bar{f}_k$), which is deterministic LoS channel information and is achievable at the ES using the GNSSs.

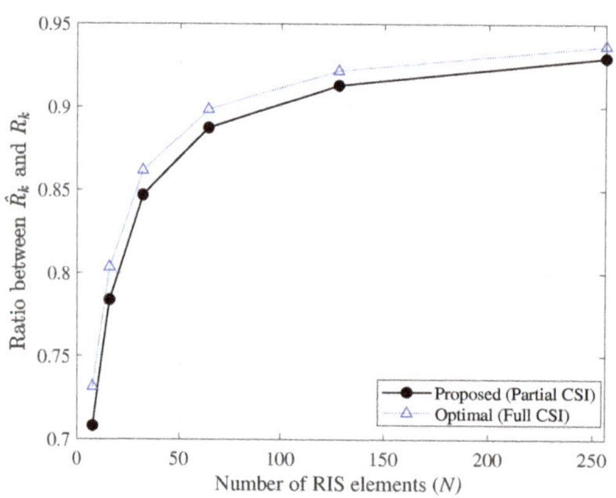

Figure 2. Performance comparison of the achievable rate ratios as a function of N.

In Figures 3 and 4, the asymptotic ergodic sum-rate of the proposed scheme analyzed in (31) is verified. In these figures, the label "Random" represents the case of the random passive beamforming, in which $\theta_{k,n}$ is determined randomly. Both figures show that the proposed scheme utilizes partial CSI, but it can achieve a performance of optimal one, which is operated based on full CSI. Evidently, the asymptotic ergodic sum-rates derived from (31) align with the simulation results across the entire range of ρ_k and N. In Figures 3 and 4, the random passive beamforming technique requires no CSI; however, its ergodic sum-rate is much lower than that of the proposed scheme and the optimal case. As shown in (31), the ergodic sum-rate increases with $\mathcal{O}(\log N)$ and $\mathcal{O}(\log \rho_k)$ as N and ρ_k increase, respectively. Accordingly, the ergodic sum-rates linearly increase with increasing ρ_k in the dBW scale, as shown in Figure 3, and they increase following the logarithm shape as N increases, as shown in Figure 4, verifying (31).

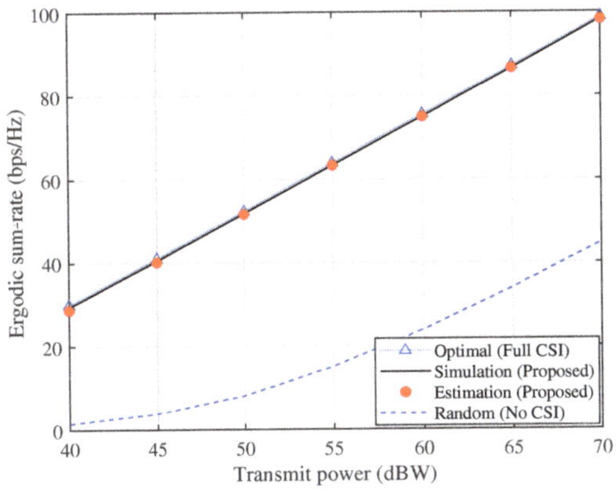

Figure 3. Comparative performance result of ergodic sum-rates based on variations in transmission power when $N = 128$.

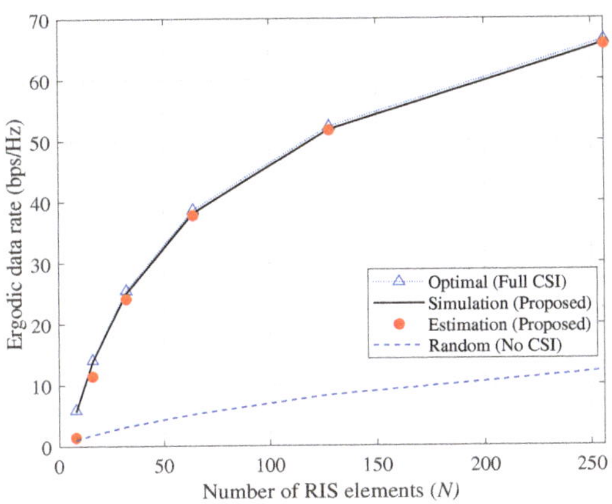

Figure 4. Comparative performance result of ergodic sum-rates based on variations in N.

In Figure 5, the optimality of the proposed scheme in the multicast system is verified. This figure shows the achievable rate ratios between the ergodic rate derived from the proposed scheme and the theoretical ideal performance for various values of M_k. In this simulation, M_k UEs are distributed randomly and uniformly within a circle with a radius of 10 m, centered around the beam's midpoint. As shown in Figure 5, the achievable rate ratios using the proposed scheme approaches 1 as N increases, regardless of M_k. In particular, when $M_k = 16$ and $N = 256$, the proposed scheme achieves approximately 91% of the theoretical ideal performance. Therefore, it is confirmed that the considered RIS-assisted SatCom system enables a UAV to simultaneously serve multiple UEs with optimal performance.

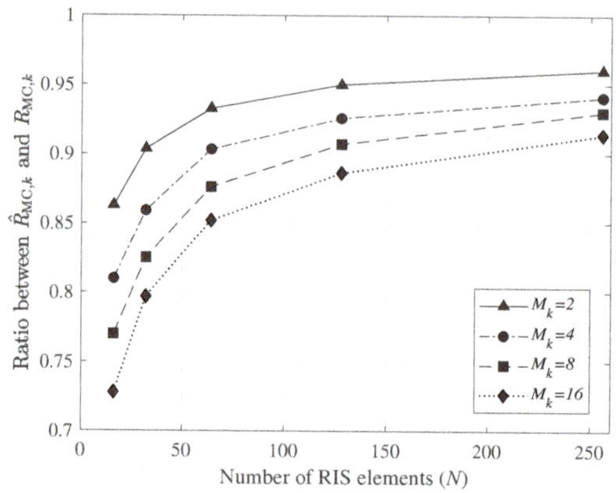

Figure 5. Performance comparison of the achievable rate ratios for a multicast system.

In Figure 6, (32) and (43) are verified. Figure 6 presents a comparison between the outage probabilities derived from simulations and our analytical estimates from (32) and (43). In this figure, we assume $\delta = 0.8$ to analyze the outage event when $R_0 = 0.8\dot{m}_k$ or $0.8\dot{m}_{MC,k}$. As shown in Figure 6, the outage probabilities derived in (32) and (43) are consistent with the simulation results for a large N. Furthermore, the outage probabilities in both unicast and multicast systems approach zero with increasing N, confirming that an RIS can provide reliable SatCom.

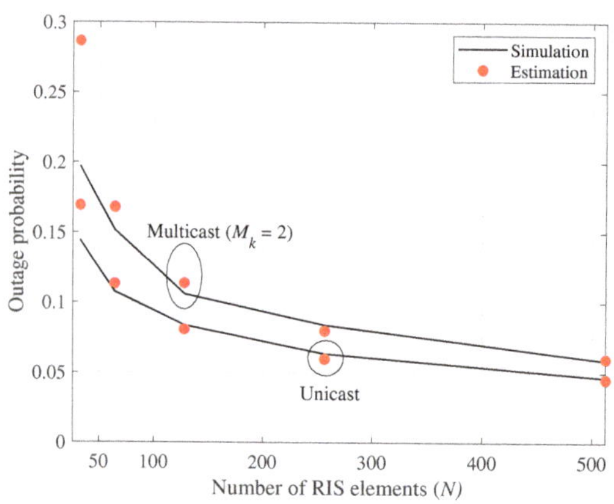

Figure 6. Performance comparison of the outage probabilities as a function of N when $\delta = 0.8$.

6. Conclusions

In this study, we analyzed the achievable data rate of an RIS-assisted SatCom system asymptotically, considering limitations such as practical CSI acquisition. In particular, we derived the asymptotic upper bound of the achievable rate and designed a passive beamformer that can asymptotically achieve the upper bound using numerous reflecting elements on the RIS. Moreover, we asymptotically derived closed-form expressions for the ergodic achievable rate and outage probability of the proposed scheme. We proved that the derived asymptotic analyses can accurately determine the performance of the RIS-assisted SatCom system concerning the ergodic achievable rate and outage probability without requiring extensive simulations. Leveraging the derived asymptotic results in the unicast system, we extended our results to cover the multicast system, such as enabling a UAV to serve multiple UEs simultaneously. From the simulation results, we can observe that the estimated ergodic rate and outage probability were in agreement with the simulation results. Moreover, the proposed scheme asymptotically achieved the optimal performance, verifying the asymptotic optimality of the proposed scheme. We expect that the proposed scheme can significantly reduce signal overheads and the amount of feedback, rendering it a promising technique in practical SatCom systems.

Author Contributions: Conceptualization, M.J.; methodology, H.S.; software, M.J.; validation, M.J. and H.S.; formal analysis, M.J.; investigation, M.J.; resources, M.J.; data curation, M.J.; writing—original draft preparation, M.J.; writing—review and editing, H.S.; visualization, M.J.; supervision, H.S.; project administration, M.J. and H.S.; funding acquisition, M.J. and H.S. All authors have read and agreed to the published version of the manuscript.

Funding: This work was supported in part by the Institute of Information & Communications Technology Planning & Evaluation (IITP) under the metaverse support program to nurture the best talents (IITP-2023-RS-2023-00254529) grant funded by the Korea government (MSIT) and by the National Research Foundation of Korea (NRF) grants funded by the Korea government (MSIT) (No. 2021R1C1C1012950 and No. 2021R1F1A1051075).

Data Availability Statement: No new data were created or analyzed in this study.

Conflicts of Interest: The authors declare no conflicts of interest.

References

1. Wu, Q.; Zhang, R. Beamforming Optimization for Wireless Network Aided by Intelligent Reflecting Surface with Discrete Phase Shifts. *IEEE Trans. Commun.* **2020**, *68*, 1838–1851. [CrossRef]
2. Saad, W.; Bennis, M.; Chen, M. A vision of 6G wireless systems: Applications, trends, technologies, and open research problems. *IEEE Netw.* **2020**, *34*, 134–142. [CrossRef]
3. Mozaffari, M.; Kasgari, A.T.Z.; Saad, W.; Bennis, M.; Debbah, M. Beyond 5G with UAVs: Foundations of a 3D Wireless Cellular Network. *IEEE Trans. Wirel. Commun.* **2019**, *18*, 357–372. [CrossRef]
4. Lin, Z.; Niu, H.; An, K.; Wang, Y.; Zheng, G.; Chatzinotas, S.; Hu, Y. Refracting RIS-Aided Hybrid Satellite-Terrestrial Relay Networks: Joint Beamforming Design and Optimization. *IEEE Trans. Aerosp. Electron. Syst.* **2022**, *58*, 3717–3724. [CrossRef]
5. Tekbıyık, K.; Kurt, G.K.; Yanikomeroglu, H. Energy-Efficient RIS-Assisted Satellites for IoT Networks. *IEEE Internet Things J.* **2022**, *9*, 14891–14899. [CrossRef]
6. Khan, W.U.; Lagunas, E.; Mahmood, A.; ElHalawany, B.M.; Chatzinotas, S.; Ottersten, B. When RIS Meets GEO Satellite Communications: A New Sustainable Optimization Framework in 6G. In Proceedings of the IEEE Vehicular Technology Conference, Helsinki, Finland, 19–22 June 2022; pp. 1–6. [CrossRef]
7. Jung, M.; Kim, T.; Son, H. Performance Analysis of RIS-Assisted SatComs Based on a ZFBF and Co-Phasing Scheme. *Mathematics* **2024**, *12*, 1257. [CrossRef]
8. Wu, Q.; Zhang, R. Intelligent Reflecting Surface Enhanced Wireless Network via Joint Active and Passive Beamforming. *IEEE Trans. Wirel. Commun.* **2019**, *18*, 5394–5409. [CrossRef]
9. Jung, M.; Saad, W.; Debbah, M.; Hong, C.S. On the Optimality of Reconfigurable Intelligent Surfaces (RISs): Passive Beamforming, Modulation, and Resource Allocation. *IEEE Trans. Wirel. Commun.* **2021**, *20*, 4347–4363. [CrossRef]
10. Jung, M.; Saad, W.; Jang, Y.; Kong, G.; Choi, S. Reliability analysis of large intelligent surfaces (LISs): Rate distribution and outage probability. *IEEE Wirel. Commun. Lett.* **2019**, *8*, 1662–1666. [CrossRef]
11. Ramaccia, D.; Sounas, D.L.; Alù, A.; Toscano, A.; Bilotti, F. Phase-Induced Frequency Conversion and Doppler Effect with Time-Modulated Metasurfaces. *IEEE Trans. Antennas Propag.* **2020**, *68*, 1607–1617. [CrossRef]
12. He, J.; Wymeersch, H.; Juntti, M. Leveraging Location Information for RIS-Aided mmWave MIMO Communications. *IEEE Wirel. Commun. Lett.* **2021**, *10*, 1380–1384. [CrossRef]
13. Long, H.; Chen, M.; Yang, Z.; Wang, B.; Li, Z.; Yun, X.; Shikh-Bahaei, M. Reflections in the Sky: Joint Trajectory and Passive Beamforming Design for Secure UAV Networks with Reconfigurable Intelligent Surface. *arXiv* **2020**, arXiv:2005.10559.
14. Zhang, Q.; Saad, W.; Bennis, M.; Lu, X.; Debbah, M.; Zuo, W. Predictive Deployment of UAV Base Stations in Wireless Networks: Machine Learning Meets Contract Theory. *IEEE Trans. Wirel. Commun.* **2021**, *20*, 637–652. [CrossRef]
15. Joo, C.; Choi, J. Low-delay broadband satellite communications with high-altitude unmanned aerial vehicles. *J. Commun. Netw.* **2018**, *20*, 102–108. [CrossRef]
16. Zhao, J.; Gao, F.; Wu, Q.; Jin, S.; Wu, Y.; Jia, W. Beam Tracking for UAV Mounted SatCom on-the-Move with Massive Antenna Array. *IEEE J. Sel. Areas Commun.* **2018**, *36*, 363–375. [CrossRef]
17. Singya, P.K.; Alouini, M.S. Performance of UAV-Assisted Multiuser Terrestrial-Satellite Communication System Over Mixed FSO/RF Channels. *IEEE Trans. Aerosp. Electron. Syst.* **2022**, *58*, 781–796. [CrossRef]
18. Kim, C.; Saad, W.; Jung, M. Performance Analysis of RIS-aided Multibeam Satellite Communication Systems. *IEEE Trans. Aerosp. Electron. Syst.* **2024**, 1–11. [CrossRef]

19. Jia, M.; Zhang, X.; Gu, X.; Guo, Q.; Li, Y.; Lin, P. Interbeam Interference Constrained Resource Allocation for Shared Spectrum Multibeam Satellite Communication Systems. *IEEE Internet Things J.* **2019**, *6*, 6052–6059. [CrossRef]
20. Jia, M.; Gu, X.; Guo, Q.; Xiang, W.; Zhang, N. Broadband Hybrid Satellite-Terrestrial Communication Systems Based on Cognitive Radio toward 5G. *IEEE Wirel. Commun.* **2016**, *23*, 96–106. [CrossRef]
21. Li, Y.; Yin, C.; Do-Duy, T.; Masaracchia, A.; Duong, T.Q. Aerial Reconfigurable Intelligent Surface-Enabled URLLC UAV Systems. *IEEE Access* **2021**, *9*, 140248–140257. [CrossRef]
22. Tse, D.; Viswanath, P. *Fundamentals of Wireless Communication*; Cambridge Univ. Press: Cambridge, UK, 2005.
23. Ahmad, I.; Nguyen, K.D.; Letzepis, N.; Lechner, G.; Joroughi, V. Zero-Forcing Precoding with Partial CSI in Multibeam High Throughput Satellite Systems. *IEEE Trans. Veh. Technol.* **2021**, *70*, 1410–1420. [CrossRef]
24. Ahmad, I.; Nguyen, K.D.; Letzepis, N. Performance Analysis of High Throughput Satellite Systems with Optical Feeder Links. In Proceedings of the GLOBECOM 2017—2017 IEEE Global Communications Conference, Singapore, 4–8 December 2017; pp. 1–7. [CrossRef]
25. Arnau, J.; Christopoulos, D.; Chatzinotas, S.; Mosquera, C.; Ottersten, B. Performance of the Multibeam Satellite Return Link with Correlated Rain Attenuation. *IEEE Trans. Wirel. Commun.* **2014**, *13*, 6286–6299. [CrossRef]
26. Björnson, E.; Hoydis, J.; Sanguinetti, L. Massive MIMO networks: Spectral, energy, and hardware efficiency. *Found. Trends® Signal Process.* **2017**, *11*, 154–655. [CrossRef]
27. Cai, X.; Song, J.; Rodríguez-Piñeiro, J.; Mogensen, E.P.; Tufvesson, F. Characterizing the Small-Scale Fading for Low Altitude UAV Channels. In Proceedings of the International Conference on Wireless and Mobile Communications, Tumkur, India, 3–4 December 2021; pp. 16–19.
28. Lohan, P.; Mishra, D. Utility-Aware Optimal Resource Allocation Protocol for UAV-Assisted Small Cells with Heterogeneous Coverage Demands. *IEEE Trans. Wirel. Commun.* **2020**, *19*, 1221–1236. [CrossRef]
29. Hochwald, B.M.; Marzetta, T.L.; Tarokh, V. Multiple-antenna channel hardening and its implications for rate feedback and scheduling. *IEEE Trans. Inf. Theory* **2004**, *50*, 1893–1909. [CrossRef]
30. Li, Q.; Feng, Z.; Li, W.; Gulliver, T.A. A New Multicast Scheduling Scheme for Cellular Networks. *IEEE Wirel. Commun. Lett.* **2014**, *3*, 26–29. [CrossRef]

mathematics

Article

Deep Learning-Enhanced Autoencoder for Multi-Carrier Wireless Systems

Md Abdul Aziz [1,2], Md Habibur Rahman [1,2], Rana Tabassum [1,2], Mohammad Abrar Shakil Sejan [3], Myung-Sun Baek [3] and Hyoung-Kyu Song [1,2,*]

1 Department of Information and Communication Engineering, Sejong University, Seoul 05006, Republic of Korea; aziz@sju.ac.kr (M.A.A.); habibur@sju.ac.kr (M.H.R.); tanvi@sju.ac.kr (R.T.)
2 Department of Convergence Engineering for Intelligent Drone, Sejong University, Seoul 05006, Republic of Korea
3 Department of Electrical Engineering, Sejong University, Seoul 05006, Republic of Korea; sejan@sejong.ac.kr (M.A.S.S.); msbaek@sejong.ac.kr (M.-S.B.)
* Correspondence: songhk@sejong.ac.kr

Abstract: In a multi-carrier (MC) system, the transmitted data are split across several sub-carriers as a crucial approach for achieving high data rates, reliability, and spectral efficiency. Deep learning (DL) enhances MC systems by improving signal representation, leading to more efficient data transmission and reduced bit error rates. In this paper, we propose an MC system supported by DL for operation on fading channels. Deep neural networks are utilized to model the modulation block, while a gated recurrent unit (GRU) network is used to model the demodulation blocks, acting as the encoder and decoder within an autoencoder (AE) architecture. The proposed scheme, known as MC-AE, differs from existing AE-based systems by directly processing channel state information and the received signal in a fully data-driven way, unlike traditional methods that rely on channel equalizers. This approach enables MC-AE to improve diversity and coding gains in fading channels by simultaneously optimizing the encoder and decoder. In this experiment, we evaluated the performance of the proposed model under both perfect and imperfect channel conditions and compared it with other models. Additionally, we assessed the performance of the MC-AE system against index modulation-based MC systems. The results demonstrate that the GRU-based MC-AE system outperforms the others.

Keywords: multicarrier; autoencoder; deep learning; wireless communication; block error rate; gated recurrent unit (GRU); modulation; demodulation

MSC: 94A14

1. Introduction

Multicarrier (MC) transmission [1] has been recognized and adopted by numerous standards bodies for both wireline and wireless systems [2,3] due to its effectiveness and reliability. This approach offers several benefits, including resistance to single-frequency interference and enhanced capability to reduce inter-symbol interference (ISI). MC's strength in addressing various channel impairments, such as frequency selectivity and impulse noise, makes it a popular choice in emerging applications like wireless local area networks (WLANs) [4] and power line communications (PLCs) [5].

The core concept of the MC protocol is the decomposition of the spectrum into a series of orthogonal narrowband subchannels, using complex exponentials as carriers of information [6]. Two primary MC techniques are commonly employed: discrete multitone (DMT) [7] for wireline systems and orthogonal frequency-division multiplexing (OFDM) [8], which is predominantly used in wireless applications. Both OFDM and DMT rely on fast Fourier transform (FFT) for spectrum decomposition, and data are transmitted in blocks as a result. Many wireless standards, such as IEEE 802.11, IEEE 802.16, 3GPP LTE, and LTE-Advanced, utilize an OFDM-based multicarrier system [9].

A variety of sophisticated MC methods built on OFDM have been investigated recently with the goal of improving spectral or energy efficiency. For example, spread OFDM (S-OFDM), using a Walsh–Hadamard spreading matrix, was proposed in [10], where each subcarrier in the OFDM system is modulated using a spread spectrum technique. In [11], the author introduced OFDM with index modulation (OFDM-IM), employing a maximum likelihood (ML) detector to identify the received signal. Compared to traditional OFDM, OFDM-IM offers enhanced reliability and energy efficiency by activating only a subset of subcarriers. This method transmits data bits through active indices in addition to M-ary symbols. The study in [12] revealed that, like conventional OFDM, OFDM-IM has a diversity order of one and examined its error performance. In particular, spread OFDM-IM (S-OFDM-IM) was developed in [13] by applying a spreading code to OFDM-IM to enhance its diversity advantage. This experiment demonstrates that when low-complexity detection techniques, like minimum mean squared error (MMSE)-based detectors, are used, S-OFDM-IM performs better than S-OFDM. Similarly, the dual model OFDM (DM-OFDM) was introduced in [14] and used several distinct signal constellations.

Recently, deep learning (DL) has been applied to various areas within the communications field, particularly to address physical layer challenges [15,16]. DL has also been widely utilized to tackle issues in MC systems [17]. For instance, deep neural networks (DNNs) have been effectively employed to detect OFDM and OFDM-IM signals [18,19], particularly in the presence of channel impairments. In [20], a convolutional neural network (CNN)-based receiver was designed for joint detection and modulation classification of the received signal. The work in [21] introduced a dual CNN-based decoder for channel estimation in a multiple input multiple output OFDM (MIMO-OFDM) system, aiming to reduce complexity by incorporating an additional neural network, referred to as HyperNet. A long short-term memory (LSTM)-based detector for the OFDM-IM system was proposed in [22]. A Y-shaped net-based bi-diretional LSTM (Bi-LSTM) model was proposed for the OFDM-IM system in [23].

The concept of an end-to-end communication system modeled as an autoencoder (AE) in DNNs was first introduced in [24]. In this approach, the transmitter encodes the input data so that the receiver's output is optimized to match the original input. Additionally, ref. [25] demonstrated the ease of implementing these techniques using open-source DNN frameworks and commercially available software-defined radios. Building on these ideas, ref. [26] proposed an AE-based end-to-end communication system capable of handling unknown channel models. In [27], the authors proposed a novel deep energy EA for noncoherent multicarrier multiuser SIMO (MU-SIMO) systems in fading channels. The EA-based approach, using neural networks for both transmitter and receiver, leverages energy-only input at the decoder and optimizes subcarrier power levels. In [28], the authors proposed a novel peak-to-average power ratio (PAPR) reduction scheme, PRNet, for OFDM systems based on deep AE architecture. The scheme adaptively determines constellation mapping and demapping using DL, minimizing both bit error rate (BER) and PAPR. For fading channels, ref. [29] introduced a radio transformer network (RTN) to counteract fading, but it did not leverage multipath diversity. Similarly, the AE method applying individual OFDM sub-carriers in [29] did not provide diversity advantages. The paper in [30] compared an AE-based OFDM system with IEEE 802.11a Wi-Fi, demonstrating that the AE system adapts better to channel conditions and effectively manages multi-path effects. A CNN-based universal filtered MC system was proposed in [31], where the DL models leverage the typically discarded odd-indexed samples from the received signal. Consequently, the model relies heavily on extensive and diverse training data to perform effectively. A DNN-based MC with AE (MC-AE) for fading channels was proposed in [32], where both modulation and demodulation are handled by neural networks without relying on channel equalizers. The approach improves diversity and coding gains, achieving superior BLER performance over traditional methods, and is extended to MU-MC-AE.

Another DNN-based turbo-style MC-AE was proposed in [33] for the OFDM system, incorporating the exchange of extrinsic information between the MC-AE decoder and the soft-decision channel decoder, which adds computational complexity. The gated recurrent unit (GRU) model captures long-term dependencies in sequential data through gating mechanisms, effectively addressing the vanishing gradient problem. GRUs have a simpler structure with fewer parameters than LSTMs, enabling faster training efficiency. Additionally, GRUs require less memory than LSTMs and generalize well with smaller datasets, making them suitable for real-time and resource-constrained applications [34,35]. Meanwhile, AEs are advantageous for feature extraction and dimensionality reduction, allowing data compression and efficient representation learning, which often enhances performance in downstream tasks [36]. In this paper, we propose a GRU-based MC-AE (GRU-MC-AE) system that leverages the advantages of the GRU model to tackle the previously mentioned challenges. In this paper, we evaluate the BER and block error rate (BLER) performance of the proposed model across various signal-to-noise ratios (SNRs) under both perfect and imperfect channel conditions. Additionally, we calculate the spectral efficiency (SE) and energy efficiency (EE) across various SNR levels. We compare its performance with traditional decoders and other DL-based decoders. The simulation results demonstrate that our proposed GRU-MC-AE model outperforms both conventional and DL-based models. The key contributions of this paper are summarized below:

- This study proposes a single-user MC-AE system, wherein a DNN model is used for the modulation block and the demodulation block is implemented using a GRU, acting as the encoder and decoder in an AE framework. The encoder employs a linear activation function, which passes the weighted sum of inputs directly to produce continuous outputs and the GRU-based decoder effectively manages the temporal dependencies between subcarriers, enhancing the system's ability to adapt to changing channel conditions.
- The suggested method does not require an equalization or domain knowledge because it feeds the decoder both the received signal and the channel state information (CSI) directly. The encoder and decoder are efficiently learned by this totally data-driven method to maximize diversity and coding gains in fading channels.
- We evaluate the performance of the proposed model in terms of BER and BLER under both perfect and imperfect channel conditions. We compare it with traditional and DL-based decoders, and the results show that the GRU-MC-AE model exceeds the performance of both conventional and DL-based models. To further assess the efficiency of the proposed model, we evaluate its SE and EE.

The remainder of this paper is organized as follows: The system modeling is introduced in Section 2. A thorough discussion of the suggested model, including details on offline training and online testing protocols, is given in Section 3. The simulation results are presented in Section 4, while the complexity analysis and detailed discussion are provided in Sections 5 and 6, respectively. Finally, the conclusions are drawn in Section 7.

2. System Model

In an MC-AE system, the incoming message is passed through the modulation section, which consists of a DL-based encoder, generating a transmitted data vector. This data vector is then processed through a time-domain OFDM (t-OFDM) operation. Specifically, an inverse fast Fourier transform (IFFT) is applied to convert the data into the time domain, and a cyclic prefix is added to combat inter-symbol interference before the signal is transmitted through a fading channel with noise. At the receiver, the cyclic prefix is removed, and an inverse time-domain OFDM operation is performed using a fast Fourier transform (FFT) to convert the signal back into the frequency domain. The M-ary QAM/PSK modulation or several newly developed instant messaging systems [37] are examples of modulation schemes. The resulting received signal is then passed through a demodulation block, which consists of a DL-based decoder that generates the estimated output.

The general structure of the proposed MC-AE system is shown in Figure 1. In our proposed system, the incoming message s is fed as input to the encoder. In the MC-AE encoder, each incoming message $s\epsilon S = \{s_1, \ldots, s_M\}$, where $M = 2^m$ and consists of a bit-stream of m bits, is mapped into a one-hot vector s of size $M \times 1$. This vector has all zero entries except for a single one. Afterward, the DL-based encoder produces the transmitted data vector $\bar{\mathbf{x}} = [x_1, x_2, \ldots, x_{N_c}]^T$, where N_c denotes the number of sub-carriers. In a manner similar to the IM-based scheme [38], the proposed MC-AE system splits the total N_c subcarriers into G groups, each containing N subcarriers, such that $N_c = NG$. The AE structure is then applied independently to each group. Following processing by the t-OFDM block, $\mathbf{x}$ is transmitted to the receiver where it is impeded by additive noise $\mathbf{w}$ after first going via the fading channel $\mathbf{H}$. After performing the inverse t-OFDM operation at the receiver, the received signal $\mathbf{y}$ is obtained. The received received frequency domain signal $\mathbf{y}$ is expressed as follows:

$$\mathbf{y} = \mathbf{H} \odot \mathbf{x} + \mathbf{w}, \tag{1}$$

Figure 1. A general configuration of the MC-AE system.

In this equation, w represents the AWGN, where $\mathbf{w}_i \sim \mathcal{CN}(0, \sigma^2)$. The $\mathbf{H} = [H_1, \ldots, H_N]$ denotes the Rayleigh fading channel, where $\mathcal{H}_i \sim \mathcal{CN}(0, 1)$ and the operator $\odot$ signifies element-wise multiplication. It is assumed that the average energy of the transmitted M-ary symbol is E_a, resulting in an average SNR at the receiver of $\overline{\gamma} = E_a/\sigma^2$.

In order to decode the signal, we assume that the received signal $\mathbf{y}$ and the CSI are both fully known at the receiver and are fed into the decoder as inputs. The real vector $\hat{\mathbf{x}} = [\mathbf{y}_R, \mathbf{y}_I, \mathbf{H}_R, \mathbf{H}_I]^T$ is specifically created from the complex vectors $\mathbf{y}$ and $\mathbf{H}$ where the real and imaginary components of $\mathbf{y}$ and $\mathbf{H}$ are denoted, respectively, by $\mathbf{y}_R, \mathbf{y}_I$ and $\mathbf{H}_R, \mathbf{H}_I$. The proposed MC-AE significantly differs from traditional RTN-based AE schemes [29]. RTN is designed for block fading channels where sub-carrier channels remain constant over multiple uses. This allows RTN to operate without CSI, relying instead on the domain knowledge of a channel equalizer. These properties make it a model-driven, noncoherent approach. On the other hand, the proposed MC-AE is designed for time-varying channel conditions, in which the channel coefficients fluctuate arbitrarily with each usage. This flexibility enables the proposed scheme to utilize perfect CSI at the receiver in a fully data-driven manner to exploit frequency diversity across different sub-carriers, moving away from the reliance on domain knowledge characteristic of RTN.

For the imperfect CSI settings, we consider an actual system that faces difficulties because of the erroneous CSI assessment of the receiver. The imprecise calculation of CSI at the receiver leads to system performance degradation, especially in scenarios where accurate channel estimation is crucial for effective signal detection and decoding. The estimated channel $h(\alpha)$ is modeled as follows:

$$h(\alpha) = \hat{h}(\alpha) + E(\alpha), \tag{2}$$

where $E(\alpha) \sim \mathcal{CN}(0, \epsilon^2)$ represents the channel estimation error, and $\hat{h}(\alpha) \sim \mathcal{CN}\left(0, \frac{1}{\epsilon^2}\right)$ is the imperfect estimate of the channel. Here, ϵ^2 denotes the error variance in the CSI estimation. In this model, the channel estimation error variance ϵ^2 is dependent on the average SNR ($\overline{\gamma}$) and is defined as $\epsilon^2 = \frac{1}{1+\overline{\gamma}}$. Thus, as the SNR increases, the channel estimation error variance decreases, reflecting improved CSI accuracy at higher SNR values [12].

3. Proposed GRU-MC-AE Network Architecture

The structure of the proposed GRU-MC-AE network is shown in Figure 2. Our proposed model consists of three main parts: a DL-based encoder, a channel, and a DL-based decoder. The encoder section includes an input layer, a dense layer, a normalization layer, and a reshape layer. In the encoder, one of the $M = 2^m$ possible incoming messages s is mapped to a one-hot vector $\mathbf{s}$ of size $M \times 1$, where all entries are zero except for one, which is set to one. The next layer is a dense layer with a linear activation function. This layer produces a $2N$-dimensional vector $\mathbf{u}$. The operation of this layer is expressed as follows:

$$\mathbf{u} = \mathbf{W}\mathbf{s} + \mathbf{b}, \tag{3}$$

where $\mathbf{w}$ is the weight and $\mathbf{b}$ is the bias vector. The following layer is a normalization layer implemented as a lambda function, which regulates the average transmission power for each subcarrier. The output of this layer is a $2N$-dimensional vector represented as follows:

$$\mathbf{v} = \sqrt{NE_a}\mathbf{u} / \parallel \mathbf{u} \parallel, \tag{4}$$

where E_a is the average energy of the transmitted symbol. The next layer is the reshape layer, which generates the real and imaginary parts by reshaping $\mathbf{v}$ into an $N \times 1$ complex-valued vector. The subsequent section is the channel implemented with a lambda layer, where noise is added to the signal. After passing through the channel, the output consists of both the received signal and the estimated channel coefficients.

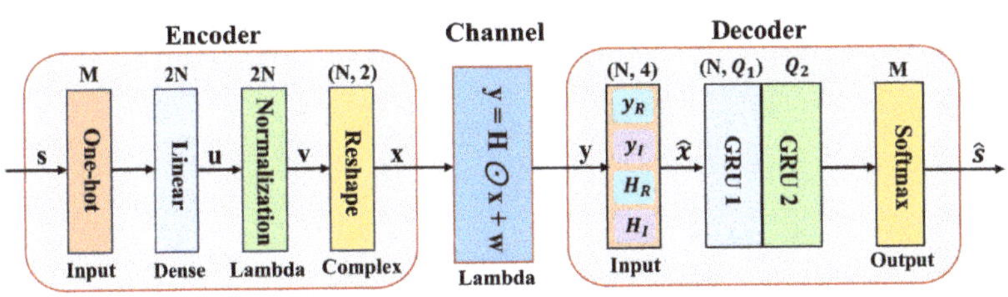

Figure 2. The proposed GRU-MC-AE network model.

The final section is the decoder, which consists of an input layer, two GRU layers, and an output layer. In the input layer, a $4N$-dimensional real vector, denoted as $\hat{\mathbf{x}}$, is constructed from the complex vectors $\mathbf{y}$ and $\mathbf{H}$. This vector is then fed into the GRU layers, which have Q_1 and Q_2 hidden nodes, respectively. The first GRU layer processes the channel output and generates a hidden state with Q_1 units for each subcarrier. The second GRU layer further reduces the sequence to produce a single output vector of Q_2 units. The internal structure of the GRU model is illustrated in Figure 3. The GRU streamlines the traditional RNN by incorporating two key gates: the update gate and the reset gate. The update gate manages the portion of the previous hidden state that is passed to the next time step, enabling the model to retain information across longer sequences. Conversely, the reset gate controls how much of the previous hidden state is discarded, allowing the model to clear its memory when it is necessary. The candidate hidden state is calculated using the current input and the reset hidden state, and the final hidden state is determined by combining the previous hidden state with the candidate state, based on the update gate's influence [39,40].

The update gate, represented as d_t, controls the extent to which the previous hidden state h_{t-1} is retained and passed on to the next time step. The reset gate, denoted as R_t, regulates how much of the previous hidden state h_{t-1} should be discarded or reset before incorporating the new input. The candidate hidden state, $\hat{h}_t$, represents the newly computed memory content, which is generated based on the current input and the modified hidden state. The final hidden state h_t at time step t is a combination of the previous hidden state h_{t-1} and the candidate hidden state $\hat{h}_t$, controlled by the update gate d_t. The operation of the gates and hidden states in the GRU can be expressed as follows:

$$d_t = \sigma(W_d \cdot \hat{x}_t + U_d \cdot h_{t-1} + b_d), \tag{5}$$

$$R_t = \sigma(W_R \cdot \hat{x}_t + U_R \cdot h_{t-1} + b_R), \tag{6}$$

$$\hat{h}_t = \tanh(W_h \cdot \hat{x}_t + U_h \cdot (R_t \odot h_{t-1}) + b_h), \tag{7}$$

$$h_t = (1 - d_t) \odot h_{t-1} + d_t \odot \hat{h}_t, \tag{8}$$

where σ represents the sigmoid activation function, *tanh* denotes the hyperbolic tangent function, $\odot$ represents element-wise (Hadamard) multiplication, b_d, b_R, b_h are the bias vectors, W_d, W_R, W_h are the weight matrices, and U_d, U_R, U_h are the recurrent weight matrices for the update, reset, and candidate hidden state, respectively.

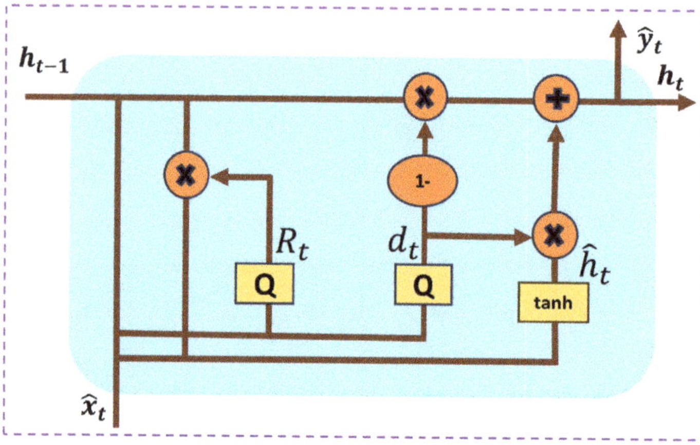

Figure 3. The internal structure of a GRU model.

The final layer of the decoding section is a dense layer with M nodes, utilizing a softmax activation function. By utilizing the softmax, the decoder produces a probability vector $\hat{s} = [\hat{s}_1, \ldots, \hat{s}_M]^T$, where the i-th entry represents the likelihood that the corresponding message is transmitted. The largest entry of $\hat{s}$ is used to determine the estimated message $\hat{s}$. Applying softmax enables the model to produce a clear and interpretable output with values that sum is 1. This allows the output to be evaluated based on the correct predicted symbol, similar to a classification setup in communication systems [41]. If $\theta_{\text{dec}} = \{W_i, b_i\}_{i=1,2}$ denote the parameters and z denotes the input of the output layer, the output is expressed as follows:

$$\hat{s} = \sigma_{\text{Softmax}}(Wz + b), \tag{9}$$

where σ_{Softmax} represent the softmax activation functions. The detailed signal detection process for the proposed GRU-based decoder in the MC-AE system is systematically presented in Algorithm 1. This algorithm provides a structured overview of each stage, including data generation, AE model design, training, and testing. The step-by-step organization offers a clear perspective on the entire workflow.

Algorithm 1 MC-AE Model Setup, Training, and Evaluations

Initialize Parameters: N, M, m, R (bit rate), SNR, batch size, norm type, epochs, learning rate, hidden layers, loss function, train/test sizes, activation functions
Define noise_std based on training SNR
Define Channel Layers
Function `channel(Z)`:
  Initialize channel estimate H_{est} with noise
  Calculate real and imaginary parts with noise added
  Return received signal y
Function `channel_test(Z, noise_std, test_size, imperfect_channel=False)`:
if imperfect_channel **then**
    Calculate imperfect CSI error
end if
  Add noise to channel estimates
  Calculate received signal y
  Return y
Build MC-AE Model
Create input layer for encoder
Add dense encoding layer
Normalize encoder output based on norm type
Reshape encoder output
Define decoder GRU layers and Dense output layer
Construct complete Autoencoder model
Train or Load Model
if pre-trained **then**
  Load encoder and decoder weights
else
  Generate training data
  Compile and train AE model
  Save encoder and decoder weights
end if
Generate test data
Encode test data
BLER and BER Calculation
Function `calculate_BER_BLER`(EbNodB_range, test_data, test_label, imperfect_channel)
Initialize BER and BLER arrays
for each SNR value in EbNodB_range **do**
  Calculate noise_std for given SNR
  Encode test data
  Pass encoded data through `channel_test`
  Decode received data and calculate BER and BLER
end for
Return BER and BLER
Perfect CSI BLER Calculation
Print "Perfect CSI BLER"
Set EbNodB_range = range(0, 15, 4)
BER, BLER ← `calculate_BER_BLER`(EbNodB_range, test_data, test_label, False)
Imperfect CSI BLER Calculation
Print "Imperfect CSI BLER"
Set EbNodB_range = range(0, 15, 4)
BER, BLER ← `calculate_BER_BLER`(EbNodB_range, test_data, test_label, True)

Training and Testing Procedure

A collection of randomly generated incoming messages s or their corresponding one-hot vectors is used to train the MC-AE model offline. During training, randomly generated noise $\mathbf{w}$ and channel $\mathbf{H}$ are added to the encoder's output. The model is trained with 200,000 date samples. The training and testing parameters used in this experiment are shown in Table 1. In DL, the selection of an appropriate loss function is crucial as it directly influences model performance and guides the optimization process. By quantifying errors, the loss function ensures that the model learns relevant patterns for specific tasks, such as regression or classification. For GRU-MC-AE training, we use the mean squared error (MSE) loss function [42], which can be expressed as follows:

$$\mathcal{L}(\mathbf{s}_i, \hat{\mathbf{s}}_i; \theta) = \frac{1}{n} \sum_{i=1}^{n} (s_i - \hat{s}_i)^2, \tag{10}$$

where θ denotes the weight and bias of the model. n is the training batch size and $\hat{\mathbf{s}}_\mathrm{i}$ is the prediction of $\mathbf{s}_\mathrm{i}$. For randomly selected batches from the data sample, the stochastic gradient descent (SGD) algorithm can update the model parameter θ in the following manner:

$$\theta^+ := \theta - \eta \nabla \mathcal{L}(\mathbf{s}_i, \hat{\mathbf{s}}_i; \theta), \tag{11}$$

where η is the learning rate and step size of SGD. Optimizers play a crucial role in DL by enabling efficient weight updates, leading to quicker convergence and improved model performance [43]. The adaptive moment estimation (Adam) optimizer is widely used in DL due to its adaptive learning rates, which combine the strengths of momentum and RMSProp for faster and more stable convergence. It is efficient, robust to noisy gradients, and performs effectively across various tasks, especially with large models and datasets. Leveraging these advantages, we train our proposed model using the Adam optimizer [44] which is extensively accessible in several commercial DL libraries, including TensorFlow [45]. During the training, data labeling is conducted using one-hot encoding. Each data sample represents one of the possible $M = 16$ modulation symbols. The label for each sample is a one-hot encoded vector of length M, where the position corresponding to the true modulation symbol is set to 1, and all other positions are set to 0. This allows the neural network to learn the mapping between the transmitted symbol and its corresponding one-hot label during training.

Table 1. The simulation parameters for the proposed system.

Parameters	Value
Training samples	200,000
Testing samples	30,000
Fading channel	Rayleigh fading channel
Noise type	AWGN
Number of hidden layer	256, 128
Learning rate	0.01, 0.005, 0.001, 0.0005
Batch size	64, 128, 256, 512
Number of epoch	250, 500, 1000
Training SNR (dB)	5, 7, 10, 15
(N,M)	(4, 16), (4, 32), (4, 64)
Optimizer	Adam

Figure 4 illustrates the classification performance metrics of the proposed GRU-based MC-AE model for the (4, 16) data combination trained at 7 dB SNR. In Figure 4a, the precision-recall curve for each modulation class shows a gradual precision decrease as recall increases, reflecting the model's accuracy across different modulation classes. The inset zooms in on variations in the middle precision-recall range, revealing slight class differences in performance. Figure 4b depicts the training loss over epochs, with an initial sharp decline indicating rapid adaptation to the data. The loss stabilizes around the 200th epoch, suggesting the model has reached a stable, low-error state and effectively minimized prediction errors.

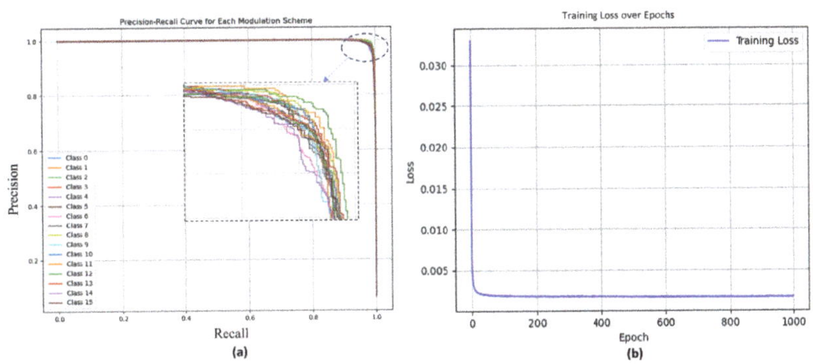

Figure 4. Classification performance metrics of the proposed GRU-based MC-AE model for the (4, 16) data combination trained at 7 dB SNR. (**a**) Precision-recall curve and (**b**) loss performance over epochs.

Training the GRU-MC-AE model requires the careful selection of the SNR level, referred to as $\overline{\gamma}_t$, due to its critical impact on the model's performance. Specifically, it is essential to choose the optimal $\overline{\gamma}_t$ to ensure that the model trained at this SNR level maintains effective performance across a range of relevant SNRs. If $\overline{\gamma}_t$ is set too low, the effects of noise may not be adequately represented during training, leading to a poorly generalized model [46]. The subsequent simulation results will detail the specific training SNR levels used for each experimental setup. All other training parameters, such as learning rate, batch size, and the number of epochs, which have a significant impact on model training, are chosen very carefully.

The testing procedure for the MC-AE system evaluates its performance by transmitting encoded test data through a simulated wireless channel and measuring the error rates. The testing procedure is performed using 30,000 samples. BER and BLER are calculated by comparing the decoded output with the original messages. The performance is tested over a range of SNR values, and the results are plotted to compare BLER and BER under perfect and imperfect CSI conditions. At high SNRs, the diversity gain g_d and coding gain g_c can be used to approximate the BLER as follows [47]:

$$BLER = (g_c\bar{\gamma})^{-g_d} + o\left(\bar{\gamma}^{-g_d}\right). \tag{12}$$

4. Simulation Results

We utilize a variety of advanced MC schemes listed in Table 1 as baseline schemes for the GRU-MC-AE model. In this study, simulations are conducted using the Python environment, with an Intel Core i7-8700 CPU at 3.20 GHz and an NVIDIA GeForce GTX 1660 Ti GPU. The configuration of IM-based schemes, such as OFDM-IM [11], is denoted by (N, K, M), where N represents the number of sub-carriers per block, K is the number of active sub-carriers, and M is the size of the conventional M-ary modulation. In particular, the configuration for OFDM [9] and S-OFDM [10] is represented by (N, M). All of these systems, including OFDM, S-OFDM, and OFDM-IM utilized ML-based detectors, with the exception of the MMSE-based S-OFDM system in [13]. The configurations for both the DNN-based MC-AE [32] and the proposed MC-AE are represented as (N, M). It is essential to note that the M in our approach is distinct from that in the baseline schemes, as it signifies the size of the transmitted message for each block of N sub-carriers, where M is defined as 2^m, with m representing the number of data bits in each message.

The speed at which a model updates its weights during training is determined by the learning rate, a crucial parameter in deep learning. Finding the ideal learning rate is crucial to striking a balance between convergence speed and stability. A low learning rate may result in sluggish or less-than-ideal convergence, whereas an excessively high learning rate may create instability [48]. The BLER of the proposed model under perfect CSI conditions, with a (4, 16) setup, is shown in Figure 5. From the results, we observe that the model demonstrates superior performance at a 0.001 learning rate. While the performance at lower SNR levels is nearly similar for all learning rates, the performance gap widens at higher SNR levels. Specifically, at 15 dB SNR, the 0.001 learning rate outperforms the 0.01, 0.005, and 0.0005 learning rates by approximately 3 dB, 1.2 dB, and 0.9 dB, respectively. Hence, the 0.001 learning rate is used for all setups in this experiment.

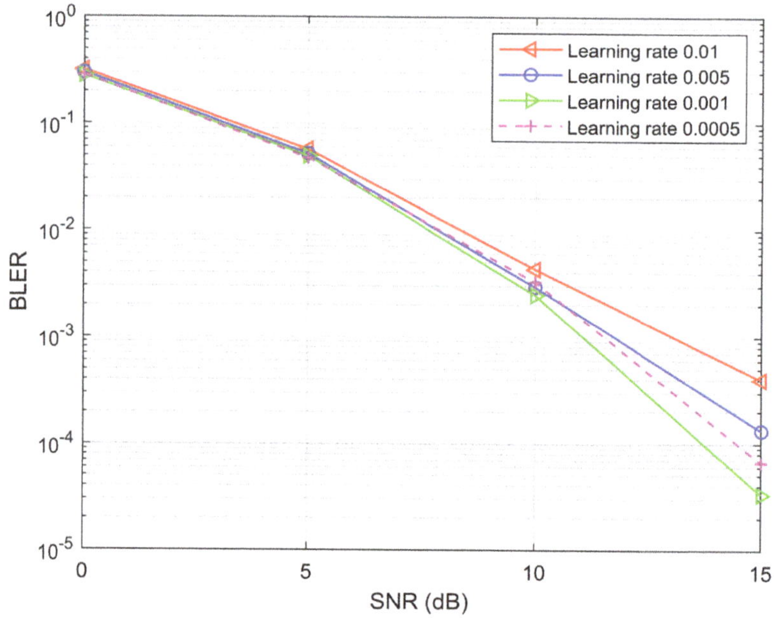

Figure 5. Comparative BLER performance of the proposed model for different learning rates under perfect CSI condition with (4, 16) setup.

Figure 6 shows the BLER performance of the proposed model across different batch sizes and epochs for the (4, 16) data setup. While the model demonstrates strong performance across all batch sizes and epoch configurations, it achieves the best results with a batch size of 512 and 1000 epochs. When the batch size decreases while keeping the epoch count at 1000, performance declines. Similarly, reducing the number of epochs while maintaining a batch size of 512 also results in decreased performance. Moreover, reducing both batch size and epochs further degrades performance. These results confirm that our proposed model performs optimally with a batch size of 512 and 1000 epochs.

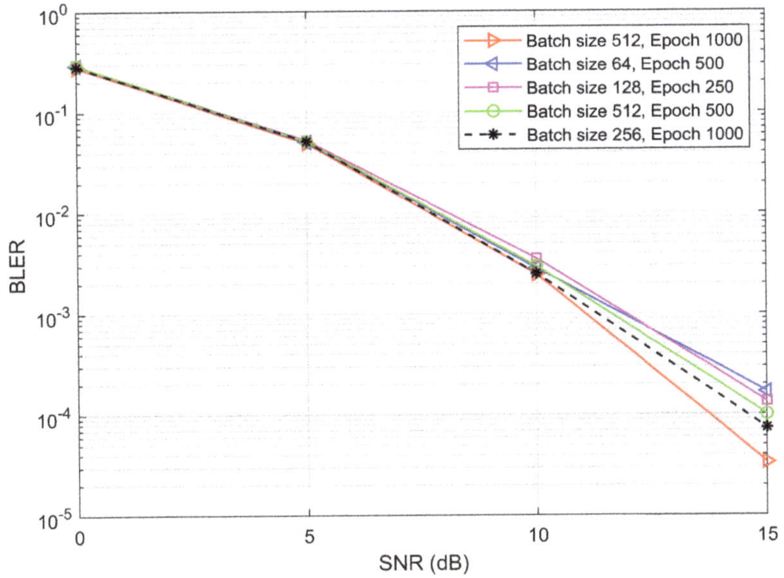

Figure 6. The BLER performance of the proposed GRU-based MC-AE system for varying batch sizes and epochs.

Training SNR significantly affects a DL model's performance. Low training SNR can hinder the model's ability to generalize in high-SNR conditions, while high SNR may reduce its robustness in noisy environments. A balanced training SNR ensures better overall performance across varying SNRs [46]. Figure 7 presents a comparison of the proposed model's BLER performance across different training SNRs. The BLER is measured with a batch size of 512 and a configuration of (4, 16). From the results, we observe that the proposed model performs well for all positive training SNR values. Although the GRU-MC-AE model shows slightly poorer performance at 15 dB SNR, it still demonstrates good and almost similar performance at 5 dB and 10 dB training SNRs. When it is trained at 7 dB SNR, the model performs better compared to other training SNRs. Performance decreases as the training SNR is either increased or decreased from this value. Specifically, the GRU-MC-AE model exhibits around a 1.25 dB improvement when it is trained at 7 dB compared to 5 dB and 10 dB training SNRs at 15 dB SNR. Thus, all the experiments for this model are conducted using a 7 dB training SNR.

The BLER for different values of M is shown in Figure 8. We compare the performance for the configurations (N, M) = (4, 16), (4, 32), and (4, 64). From the results, it can be observed that while our proposed model demonstrates good performance across all setups, the performance decreases as the size of M increases. Specifically, for $M = 32$ and 64, the GRU-MC-AE model shows a degradation of approximately 1.125 dB and 1.7 dB, respectively, compared to $M = 16$.

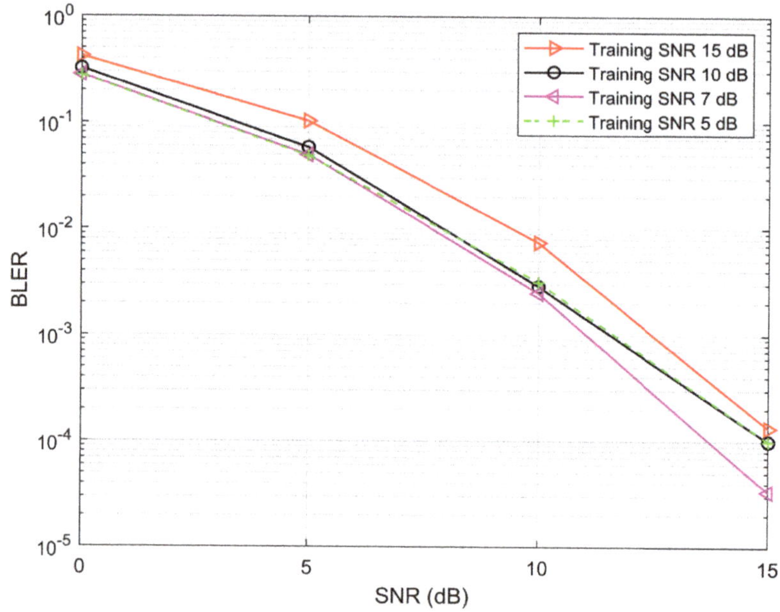

Figure 7. BLER performance of the proposed model for different training SNRs under perfect channel condition with (4,16) setup.

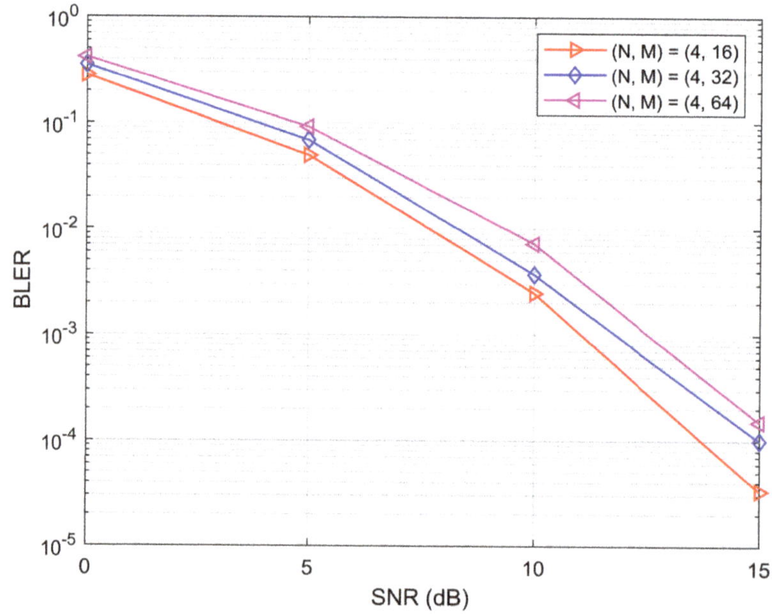

Figure 8. BLER performance of the proposed model for different values of M under perfect channel condition.

In Figure 9, the performance of the proposed model configured as (4, 16), is evaluated against ML-based OFDM and OFDM-IM, MMSE-based S-OFDM, and the DNN-based MC-AE [32] system under ideal channel conditions at a spectral efficiency (SE) of 1 bps/Hz.

The results indicate that the proposed GRU-MC-AE model consistently outperforms the alternative models. While its performance closely aligns with that of the DNN-based MC-AE system at lower SNR levels, the proposed model demonstrates a notable gain of approximately 1.15 dB in performance at higher SNR levels.

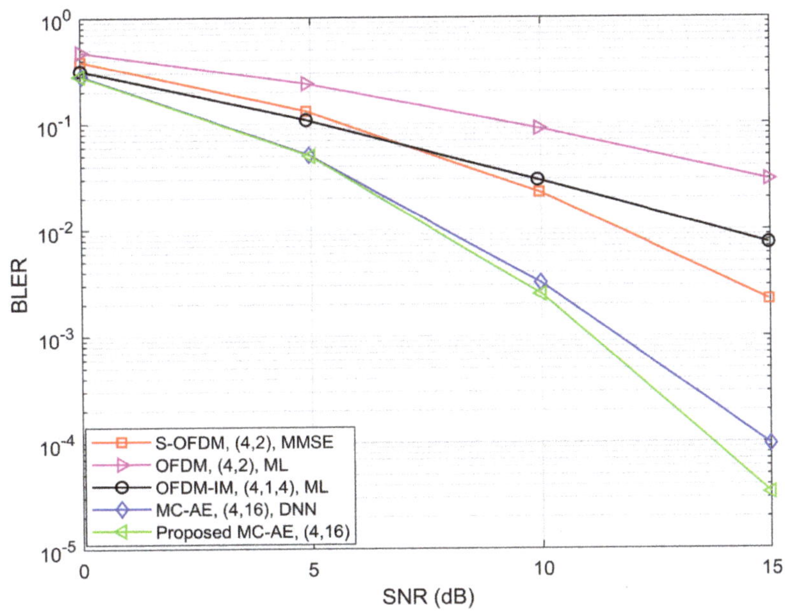

Figure 9. The BLER performance comparison between the proposed model and other models under perfect channel conditions at a SE of 1 bps/Hz.

A comparative BER performance of the proposed model with the configuration $(4, 16)$ is illustrated in Figure 10. This figure considers the results under perfect channel conditions at a SE of 1 bps/Hz. The performance of our proposed system is compared against several benchmarks, including a turbo-style DNN-based MC-AE system [33], an LSTM-based OFDM-IM system [22], a Y-BLSTM-based OFDM-IM system [23], and a greedy detector (GD)-based OFDM-IM system [49]. It is important to note that all OFDM systems are configured with a $(4, 1, 4)$ setup, while the turbo-style MC-AE uses a $(4, 16)$ configuration. From the results, it is evident that the proposed model outperforms all other system models. Although the GRU-MC-AE model demonstrates slightly inferior performance at lower SNRs, particularly around 5 dB SNR, it exhibits approximately a 1 dB performance gain compared to the turbo MC-AE model at higher SNR levels.

The BLER performance of the GRU-MC-AE model under uncertain channel conditions is presented in Figure 11, using a configuration of $(4, 16)$ and a training SNR of 7 dB. The figure compares the performance of the GRU-MC-AE system with a DNN-based MC-AE system, as well as ML-based OFDM, OFDM-IM, and S-OFDM systems, at a SE of 1 bps/Hz. It is important to highlight that the proposed approach is evaluated using MMSE-based imperfect CSI, despite being trained under perfect CSI conditions. As shown, the GRU-MC-AE model surpasses all other systems. While it effectively outperforms S-OFDM and the DNN-based MC-AE, it exhibits a more significant advantage over OFDM and OFDM-IM systems. Notably, at 15 dB SNR, the proposed model offers approximately a 0.7 dB improvement over the DNN-based MC-AE system and 0.9 dB improvement over the ML-based S-OFDM system.

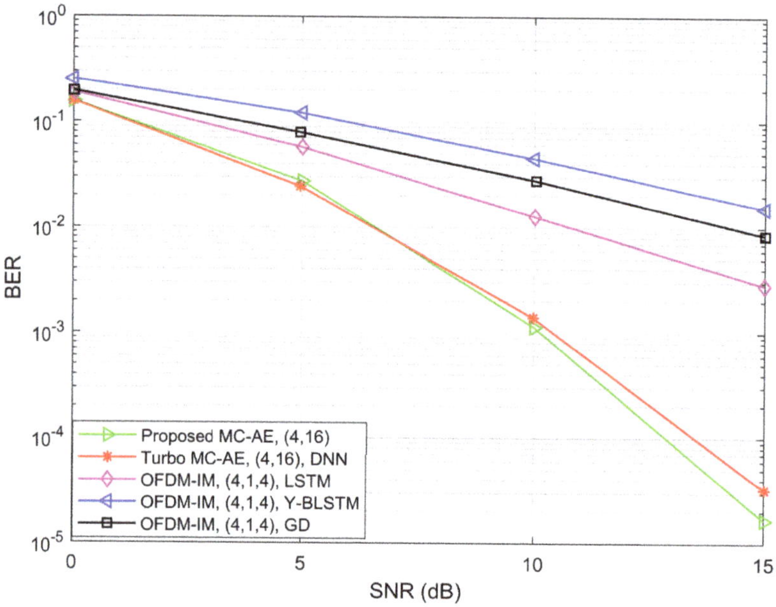

Figure 10. BER performance comparison of the proposed model with other models under perfect channel conditions at a SE of 1 bps/Hz.

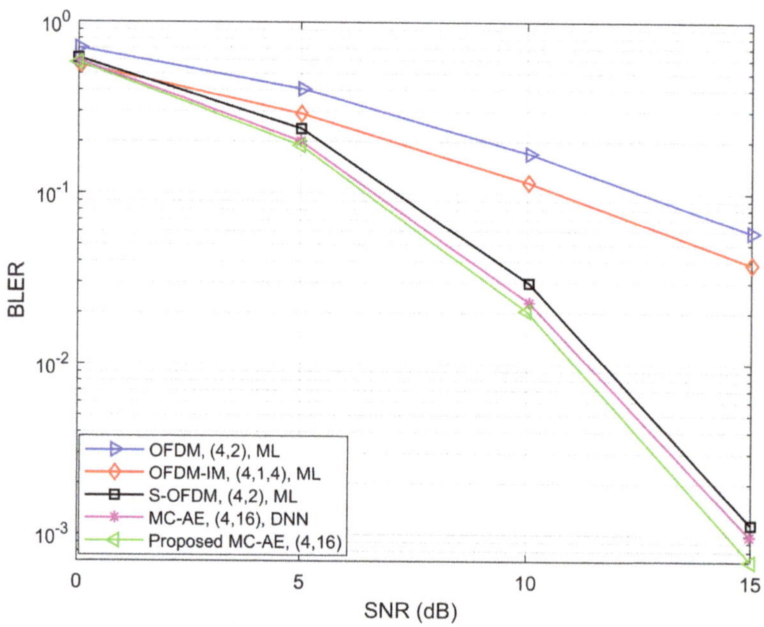

Figure 11. Comparison of BLER performance between the proposed model and other models in imperfect channel conditions, with a SE of 1 bps/Hz.

Figure 12 shows two plots that analyze the performance of a system in terms of SE and EE as a function of SNR under perfect channel conditions for different data settings. This performance is calculated with a training SNR of 7 dB, assuming a fixed circuit power of 0.25 W [50]. In Figure 12a on the left, SE is plotted against SNR for three different configurations of (N, M): (4, 16), (4, 32), and (4, 64). As SNR increases, SE improves for all configurations, indicating that higher SNR values support higher data rates. Among the configurations, (N, M) = (4, 64) achieves the highest SE, followed by (4, 32) and (4, 16), suggesting that the increase of M enhances SE. In Figure 12b on the right, EE is plotted against SNR for the same configurations. EE also improves with SNR, showing that the system becomes more energy-efficient at higher SNR levels. Here, the configuration (N, M) = (4, 64) shows superior EE performance across most SNR values, particularly at higher SNRs. This suggests that larger values of M contribute to higher EE as well. The (4, 32) configuration follows in EE performance, and (4, 16) has the lowest EE across the SNR range. These plots provide insights into how adjusting the (N, M) parameters and SNR affects the SE and EE, helping to optimize the system's performance based on specific requirements.

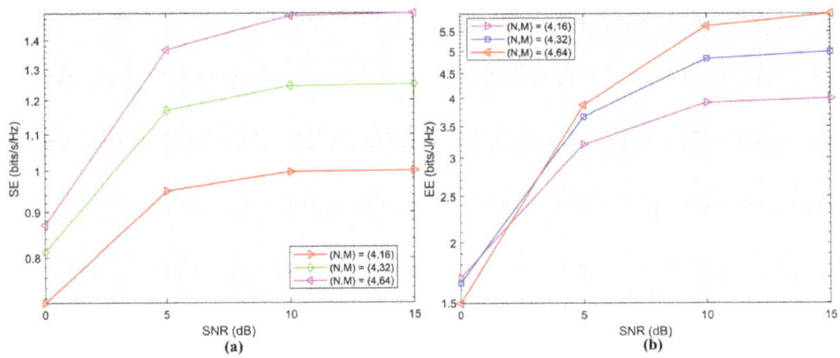

Figure 12. The SE and EE performance of the proposed GRU-based MC-AE system for configurations (N, M) = (4, 16), (4, 32), (4, 64) across SNR values. (**a**) SE performance. (**b**) EE performance.

5. Computational Complexity

The high computational cost of data decoding is a major challenge in today's advanced MC systems, often viewed as a trade-off for performance enhancement. To address this, we analyze the decoding complexity of the proposed DL-based MC schemes by measuring the decoding runtime per sample and comparing it with baseline models included in Figure 9. The transmitter complexity of our schemes is minimal compared to the receiver since the proposed encoders require only a single linear fully connected layer. The decoding complexity of the proposed MC-AE and the baselines, as shown in Figure 9, is compared in Table 2, where the runtime is expressed in milliseconds (ms). Our proposed model achieves a runtime of 0.049 ms per sample. In comparison, the runtimes for S-OFDM and OFDM-IM are 0.121 ms and 0.107 ms per sample, respectively, which are significantly higher than our proposed model. Additionally, we have calculated the training time per sample for different training SNRs to further evaluate the system's efficiency, as presented in Table 3. From the table, it is evident that the runtime shows slight variations with changes in the training SNR value. While our model requires slightly more runtime than the OFDM and DNN-based MC-AE systems, the results clearly indicate that our scheme not only offers higher reliability but also benefits from lower computational complexity.

During the inference (decoding) phase of the MC-AE model, computational costs are minimized through several optimizations. The channel layer, which simulates noise and fading, leverages batch operations in TensorFlow to efficiently compute the effects of the noisy channel with parallel processing. The decoder, which uses GRU layers, is less resource-intensive, helping reduce latency. Additionally, GPU acceleration further improves performance. For class prediction, the use of *argmax* efficiently maps probabilities to the predicted modulation classes, minimizing post-processing time.

Table 2. Computational complexity comparison of the proposed GRU-based MC-AE system with other systems.

Model Name	Complexity per Sample (ms)
OFDM [9]	0.032
S-OFDM [10]	0.121
OFDM-IM [11]	0.107
DNN-based MC-AE [32]	0.028
Proposed MC-AE	0.049

Table 3. Computational complexity of the proposed GRU-based MC-AE system evaluated for various training SNRs.

Training SNR (dB)	Complexity per Sample (ms)
5	0.04995
7	0.049
10	0.0517
15	0.05281

6. Discussion

To optimize diversity and coding gains over fading channels in a fully data-driven manner, we proposed a robust DL-based MC-AE system capable of learning both the encoder and decoder. The encoder uses a linear activation function for continuous outputs, while the decoder employs a GRU neural network to effectively manage temporal dependencies across subcarriers, enhancing the system's adaptability to varying channel conditions. In our investigation, we focused on evaluating the performance of the model, calculating key metrics such as BER, BLER, SE, and EE for different data configurations and DL model parameters. The results were compared with several conventional and DL-based models. While our system demonstrated superior performance in terms of these metrics, it exhibits higher complexity compared to some of the other models. Some techniques like model quantization, static graph execution, and simplifying the channel model through approximations for noise and fading can be employed to optimize performance while reducing complexity. These combined approaches enable faster inference with minimal performance loss. In a prior study [32], the authors designed an AE model with a DNN network and analyzed the BER and BLER for different scenarios, such as uplink-downlink and multiuser systems. Although this approach had lower complexity than ours, our model outperformed it in terms of performance. We intentionally avoid training with negative SNR levels, as excessive noise would overpower the signal, leading to overfitting and reduced information for learning. Our current work primarily focuses on static channel conditions, although we have explored varying training SNR values to account for dynamic channel conditions. Further work could extend this model to fully dynamic channel scenarios.

7. Conclusions

We proposed a robust GRU-based MC-AE system that optimizes diversity and coding gains over fading channels in a data-driven manner. We evaluated the BLER under both perfect and imperfect CSI conditions, comparing the performance of our proposed model with traditional and DL-based MC systems. Additionally, we assessed the BER performance under perfect CSI conditions. For further analysis, we calculated the SE and EE across varying SNR levels. Our findings show that the model performs optimally with a learning rate of 0.001, but it is capable of learning efficiently at different learning rates. Training at moderate SNR values helps capture more relevant features, improving generalization. Our proposed model exhibited superior performance with 1000 epochs and a batch size of 512, achieving optimal results when configured with (4,16) and trained at 7 dB SNR. The results demonstrate that MC-AE outperforms conventional baselines, including recent IM-based and S-OFDM-based schemes, in terms of performance improvements. It also achieves superior performance compared to other DL-based OFDM-IM and MC-AE systems under both perfect and imperfect CSI conditions. Moreover, the SE and EE performance highlights the system's robustness. Our study is primarily simulation-based, and we have not pursued practical implementation at this stage. However, in the future, the research community may consider the design of this experiment for practical implementation.

Author Contributions: Conceptualization, M.A.A. and M.H.R.; methodology, M.A.A., R.T. and M.A.S.S.; software, M.A.A., M.H.R. and M.-S.B.; validation, M.A.A. and M.A.S.S.; formal analysis, M.A.A., M.H.R., R.T. and M.-S.B.; investigation, M.A.A., M.H.R. and M.A.S.S.; resources, H.-K.S.; data curation, M.A.A. and R.T.; writing—original draft preparation, M.A.A.; writing—review and editing, M.A.A. and H.-K.S.; visualization, M.A.A. and M.-S.B.; supervision, H.-K.S.; project administration, H.-K.S.; funding acquisition, H.-K.S. All authors have read and agreed to the published version of the manuscript.

Funding: This work was supported in part by the Institute of Information and Communications Technology Planning and Evaluation (IITP) grant funded by the Korea Government [Ministry of Science and ICT (MSIT)], Republic of Korea, under the Metaverse Support Program to Nurture the Best Talents under Grant IITP-2024-RS-2023-00254529; in part by the Basic Science Research Program through the National Research Foundation of Korea (NRF) funded by the Ministry of Education under Grant 2020R1A6A1A03038540; and in part by the MSIT (Ministry of Science and ICT), Republic of Korea, under the ITRC (Information Technology Research Center) support program (IITP-2024-RS-2024-00437191) supervised by the IITP (Institute for Information and Communications Technology Planning and Evaluation).

Data Availability Statement: The data will be made available by the authors upon request.

Conflicts of Interest: The authors declare no conflicts of interest.

References

1. Come, H. Multicarrier Modulation for Data Transmission: An Idea Whose Time. *IEEE Commun. Mag.* **1990**, *28*, 5–14.
2. Taura, K.; Tsujishita, M.; Takeda, M.; Kato, H.; Ishida, M.; Ishida, Y. A digital audio broadcasting (DAB) receiver. *IEEE Trans. Consum. Electron.* **1996**, *42*, 322–327. [CrossRef]
3. *G.992.1*; Asymmetric Digital Subscriber Line (ADSL) Transceivers. Telecommunication Standardization Sector of ITU: Geneva, Switzerland, 1999.
4. Nee, R.V.; Prasad, R. *OFDM for Wireless Multimedia Communications*; Artech House, Inc.: Norwood, MA, USA, 2000.
5. Biglieri, E. Coding and modulation for a horrible channel. *IEEE Commun. Mag.* **2003**, *41*, 92–98. [CrossRef]
6. Garg, V.K. Fourth generation systems and new wireless technologies. In *Wireless Communications & Networking*; Elsevier: Amsterdam, The Netherlands, 2007; pp. 1–22.
7. Cioffi, J.M. A multicarrier primer. *ANSI T1E1* **1991**, *4*, 91–157.
8. Terry, J.; Heiskala, J. *OFDM Wireless LANs: A Theoretical and Practical Guide*; Sams Publishing: Carmel, IN, USA, 2002.
9. Hwang, T.; Yang, C.; Wu, G.; Li, S.; Li, G.Y. OFDM and its wireless applications: A survey. *IEEE Trans. Veh. Technol.* **2008**, *58*, 1673–1694. [CrossRef]
10. Bury, A.; Egle, J.; Lindner, J. Diversity comparison of spreading transforms for multicarrier spread spectrum transmission. *IEEE Trans. Commun.* **2003**, *51*, 774–781. [CrossRef]
11. Başar, E.; Aygölü, Ü.; Panayırcı, E.; Poor, H.V. Orthogonal frequency division multiplexing with index modulation. *IEEE Trans. Signal Process.* **2013**, *61*, 5536–5549. [CrossRef]
12. Van Luong, T.; Ko, Y. Impact of CSI uncertainty on MCIK-OFDM: Tight closed-form symbol error probability analysis. *IEEE Trans. Veh. Technol.* **2017**, *67*, 1272–1279. [CrossRef]
13. Van Luong, T.; Ko, Y. Spread OFDM-IM with precoding matrix and low-complexity detection designs. *IEEE Trans. Veh. Technol.* **2018**, *67*, 11619–11626. [CrossRef]
14. Mao, T.; Wang, Z.; Wang, Q.; Chen, S.; Hanzo, L. Dual-mode index modulation aided OFDM. *IEEE Access* **2016**, *5*, 50–60. [CrossRef]
15. Rahman, M.H.; Sejan, M.A.S.; Aziz, M.A.; Baik, J.I.; Kim, D.S.; Song, H.K. Deep learning-based improved cascaded channel estimation and signal detection for reconfigurable intelligent surfaces-assisted MU-MISO systems. *IEEE Trans. Green Commun. Netw.* **2023**, *7*, 1515–1527. [CrossRef]

16. Aziz, M.A.; Rahman, M.H.; Sejan, M.A.S.; Baik, J.I.; Kim, D.S.; Song, H.K. Spectral Efficiency Improvement Using Bi-Deep Learning Model for IRS-Assisted MU-MISO Communication System. *Sensors* **2023**, *23*, 7793. [CrossRef] [PubMed]
17. Li, A. Deep Learning for Multi-Carrier Signal Reception. Ph.D. Thesis, University of Surrey, Guildford, UK, 2022.
18. Ye, H.; Li, G.Y.; Juang, B.H. Power of deep learning for channel estimation and signal detection in OFDM systems. *IEEE Wirel. Commun. Lett.* **2017**, *7*, 114–117. [CrossRef]
19. Van Luong, T.; Ko, Y.; Vien, N.A.; Nguyen, D.H.; Matthaiou, M. Deep learning-based detector for OFDM-IM. *IEEE Wirel. Commun. Lett.* **2019**, *8*, 1159–1162. [CrossRef]
20. Yıldırım, Y.; Özer, S.; Cırpan, H.A. Deep receiver design for multi-carrier waveforms using cnns. In Proceedings of the 2020 43rd International Conference on Telecommunications and Signal Processing (TSP), Milan, Italy, 7–9 July 2020; pp. 31–36.
21. Jiang, P.; Wen, C.K.; Jin, S.; Li, G.Y. Dual CNN-based channel estimation for MIMO-OFDM systems. *IEEE Trans. Commun.* **2021**, *69*, 5859–5872. [CrossRef]
22. Aziz, M.A.; Rahman, M.H.; Sejan, M.A.S.; Tabassum, R.; Hwang, D.D.; Song, H.K. Deep Recurrent Neural Network Based Detector for OFDM With Index Modulation. *IEEE Access* **2024**, *12*, 89538–89547. [CrossRef]
23. Zhu, Y.; Wang, B.; Li, J.; Zhang, Y.; Xie, F. Y-shaped net-based signal detection for OFDM-IM systems. *IEEE Commun. Lett.* **2022**, *26*, 2661–2664. [CrossRef]
24. O'shea, T.; Hoydis, J. An introduction to deep learning for the physical layer. *IEEE Trans. Cogn. Commun. Netw.* **2017**, *3*, 563–575. [CrossRef]
25. Dörner, S.; Cammerer, S.; Hoydis, J.; Ten Brink, S. Deep learning based communication over the air. *IEEE J. Sel. Top. Signal Process.* **2017**, *12*, 132–143. [CrossRef]
26. Aoudia, F.A.; Hoydis, J. End-to-end learning of communications systems without a channel model. In Proceedings of the 2018 52nd Asilomar Conference on Signals, Systems, and Computers, Pacific Grove, CA, USA, 28–31 October 2018; pp. 298–303.
27. Van Luong, T.; Ko, Y.; Vien, N.A.; Matthaiou, M.; Ngo, H.Q. Deep energy autoencoder for noncoherent multicarrier MU-SIMO systems. *IEEE Trans. Wirel. Commun.* **2020**, *19*, 3952–3962. [CrossRef]
28. Kim, M.; Lee, W.; Cho, D.H. A novel PAPR reduction scheme for OFDM system based on deep learning. *IEEE Commun. Lett.* **2017**, *22*, 510–513. [CrossRef]
29. Felix, A.; Cammerer, S.; Dörner, S.; Hoydis, J.; Ten Brink, S. OFDM-autoencoder for end-to-end learning of communications systems. In Proceedings of the 2018 IEEE 19th International Workshop on Signal Processing Advances in Wireless Communications (SPAWC), Kalamata, Greece, 25–28 June 2018; pp. 1–5.
30. Oshiro, S.; Toma, T.; Wada, T. Performance Comparison of Autoencoder based OFDM Communication System with Wi-Fi. *Int. J. Comput. Sci. Netw. Secur. IJCSNS* **2023**, *23*, 172–178.
31. Youssef, M.M.; Ibrahim, M.; Abdelhamid, B. Deep Learning-aided Channel Estimation For Universal Filtered Multi-carrier Systems. In Proceedings of the 2023 40th National Radio Science Conference (NRSC), Giza, Egypt, 30 May–1 June 2023; Volume 1, pp. 159–166.
32. Van Luong, T.; Ko, Y.; Matthaiou, M.; Vien, N.A.; Le, M.T.; Ngo, V.D. Deep learning-aided multicarrier systems. *IEEE Trans. Wirel. Commun.* **2020**, *20*, 2109–2119. [CrossRef]
33. Xu, C.; Van Luong, T.; Xiang, L.; Sugiura, S.; Maunder, R.G.; Yang, L.L.; Hanzo, L. Turbo detection aided autoencoder for multicarrier wireless systems: Integrating deep learning into channel coded systems. *IEEE Trans. Cogn. Commun. Netw.* **2022**, *8*, 600–614. [CrossRef]
34. Ali, M.H.E.; Rabeh, M.L.; Hekal, S.; Abbas, A.N. Deep learning gated recurrent neural network-based channel state estimator for OFDM wireless communication systems. *IEEE Access* **2022**, *10*, 69312–69322.
35. Rahman, M.H.; Sejan, M.A.S.; Aziz, M.A.; Kim, D.S.; You, Y.H.; Song, H.K. Deep Convolutional and Recurrent Neural-Network-Based Optimal Decoding for RIS-Assisted MIMO Communication. *Mathematics* **2023**, *11*, 3397. [CrossRef]
36. Chen, S.; Guo, W. Auto-encoders in deep learning—A review with new perspectives. *Mathematics* **2023**, *11*, 1777. [CrossRef]
37. Basar, E.; Wen, M.; Mesleh, R.; Di Renzo, M.; Xiao, Y.; Haas, H. Index modulation techniques for next-generation wireless networks. *IEEE Access* **2017**, *5*, 16693–16746. [CrossRef]
38. Mao, T.; Wang, Q.; Wang, Z.; Chen, S. Novel index modulation techniques: A survey. *IEEE Commun. Surv. Tutorials* **2018**, *21*, 315–348. [CrossRef]
39. Nosouhian, S.; Nosouhian, F.; Khoshouei, A.K. A Review of Recurrent Neural Network Architecture for Sequence Learning: Comparison Between LSTM and GRU. 2021. Available online: https://www.preprints.org/manuscript/202107.0252/v1 (accessed on 18 November 2024).
40. Dey, R.; Salem, F.M. Gate-variants of gated recurrent unit (GRU) neural networks. In Proceedings of the 2017 IEEE 60th International Midwest Symposium on Circuits and Systems (MWSCAS), Boston, MA, USA, 6–9 August 2017; pp. 1597–1600.
41. Schmidhuber, J. Deep learning in neural networks: An overview. *Neural Netw.* **2015**, *61*, 85–117. [CrossRef]
42. Zhou, B.; Chen, Q. A tutorial on minimum mean square error estimation. *Southwest Jiaotong Univ. Sichuan China Tech. Rep.* **2015**. [CrossRef]
43. Hassan, E.; Shams, M.Y.; Hikal, N.A.; Elmougy, S. The effect of choosing optimizer algorithms to improve computer vision tasks: A comparative study. *Multimed. Tools Appl.* **2023**, *82*, 16591–16633. [CrossRef] [PubMed]
44. Diederik, P.K. Adam: A method for stochastic optimization. *arXiv* **2014**, arXiv:1412.6980.

45. Abadi, M.; Agarwal, A.; Barham, P.; Brevdo, E.; Chen, Z.; Citro, C.; Corrado, G.S.; Davis, A.; Dean, J.; Devin, M.; et al. TensorFlow: Large-Scale Machine Learning on Heterogeneous Systems. 2015. Available online: https://arxiv.org/abs/1603.04467 (accessed on 18 November 2024).
46. Zhang, X.; Seyfi, T.; Ju, S.; Ramjee, S.; El Gamal, A.; Eldar, Y.C. Deep learning for interference identification: Band, training SNR, and sample selection. In Proceedings of the 2019 IEEE 20th International Workshop on Signal Processing Advances in Wireless Communications (SPAWC), Cannes, France, 2–5 July 2019; pp. 1–5.
47. Simon, M.K.; Alouini, M.S. Digital communications over fading channels (mk simon and ms alouini; 2005) [book review]. *IEEE Trans. Inf. Theory* **2008**, *54*, 3369–3370. [CrossRef]
48. Wu, Y.; Liu, L.; Bae, J.; Chow, K.H.; Iyengar, A.; Pu, C.; Wei, W.; Yu, L.; Zhang, Q. Demystifying learning rate policies for high accuracy training of deep neural networks. In Proceedings of the 2019 IEEE International Conference on Big Data (Big Data), Los Angeles, CA, USA, 9–12 December 2019; pp. 1971–1980.
49. Crawford, J.; Ko, Y. Low complexity greedy detection method with generalized multicarrier index keying OFDM. In Proceedings of the 2015 IEEE 26th Annual International Symposium on Personal, Indoor, and Mobile Radio Communications (PIMRC), Hong Kong, China, 30 August–2 September 2015; pp. 688–693.
50. Sboui, L.; Rezki, Z.; Sultan, A.; Alouini, M.S. A new relation between energy efficiency and spectral efficiency in wireless communications systems. *IEEE Wirel. Commun.* **2019**, *26*, 168–174. [CrossRef]

mathematics

MDPI

Article

A Hybrid Deep Learning Framework for OFDM with Index Modulation Under Uncertain Channel Conditions †

Md Abdul Aziz [1,2], Md Habibur Rahman [1,2], Rana Tabassum [1,2], Mohammad Abrar Shakil Sejan [3], Myung-Sun Baek [3] and Hyoung-Kyu Song [1,2,*]

1 Department of Information and Communication Engineering, Sejong University, Seoul 05006, Republic of Korea; aziz@sju.ac.kr (M.A.A.); habibur@sju.ac.kr (M.H.R.); tanvi@sju.ac.kr (R.T.)
2 Department of Convergence Engineering for Intelligent Drone, Sejong University, Seoul 05006, Republic of Korea
3 Department of Electrical Engineering, Sejong University, Seoul 05006, Republic of Korea; sejan@sejong.ac.kr (M.A.S.S.); msbeak@sejong.ac.kr (M.-S.B.)
* Correspondence: songhk@sejong.ac.kr

† This paper is an extended version of our paper published in ICTC 2024: The 15th International Conference on ICT Convergence, Jeju Island, Republic of Korea, 16–18 October 2024.

Abstract: Index modulation (IM) is considered a promising approach for fifth-generation wireless systems due to its spectral efficiency and reduced complexity compared to conventional modulation techniques. However, IM faces difficulties in environments with unpredictable channel conditions, particularly in accurately detecting index values and dynamically adjusting index assignments. Deep learning (DL) offers a potential solution by improving detection performance and resilience through the learning of intricate patterns in varying channel conditions. In this paper, we introduce a robust detection method based on a hybrid DL (HDL) model designed specifically for orthogonal frequency-division multiplexing with IM (OFDM-IM) in challenging channel environments. Our proposed HDL detector leverages a one-dimensional convolutional neural network (1D-CNN) for feature extraction, followed by a bidirectional long short-term memory (Bi-LSTM) network to capture temporal dependencies. Before feeding data into the network, the channel matrix and received signals are preprocessed using domain-specific knowledge. We evaluate the bit error rate (BER) performance of the proposed model using different optimizers and equalizers, then compare it with other models. Moreover, we evaluate the throughput and spectral efficiency across varying SNR levels. Simulation results demonstrate that the proposed hybrid detector surpasses traditional and other DL-based detectors in terms of performance, underscoring its effectiveness for OFDM-IM under uncertain channel conditions.

Keywords: orthogonal frequency-division multiplexing (OFDM); wireless communication; index modulation (IM); deep learning (DL); imperfect channel conditions; bidirectional long short-term memory (Bi-LSTM); convolutional neural network (CNN); hybrid deep learning (HDL)

MSC: 94A14

1. Introduction

Because of its dependability and efficiency, many standards organizations have acknowledged and embraced multi-carrier (MC) transmission [1] for wireline and wireless systems [2,3]. Key advantages of this method include its enhanced capacity to reduce inter-symbol interference (ISI) and its resistance to single-frequency interference. MC is a popular choice for emerging applications such as wireless local area networks (WLANs) [4] and power line communications (PLC) [5] thanks to its capacity to withstand a range of channel impairments, including impulse noise and frequency selectivity. The fundamental idea behind the MC protocol is to use complex exponentials as information carriers to break the spectrum down into a number of orthogonal narrowband subchannels [6]. The two most frequently used MC techniques are discrete multi-tone (DMT) [7] for wireline systems and orthogonal frequency-division multiplexing (OFDM) [8], which is mostly utilized in wireless applications.

Recently, numerous advanced multi-carrier (MC) techniques based on OFDM have been explored to provide enhanced energy and spectrum [9,10]. Among these, OFDM employing index modulation (OFDM-IM) stands out as a promising approach. OFDM-IM is a viable alternative for next-generation wireless communication that transmits data more efficiently than conventional OFDM. Index modulation (IM) encodes information not only in data symbols but also in the indices of active elements, enabling additional data transmission without requiring extra bandwidth [11]. This technique enhances data throughput as well as resilience against frequency-selective fading and interference by leveraging diverse subcarriers and indices. Moreover, the selective activation of subcarriers supports energy-efficient operation, making it suitable for energy-conscious applications [12]. The ability of OFDM-IM to mitigate both inter-symbol and inter-carrier interference while balancing spectral efficiency, energy efficiency, and system complexity makes it a versatile choice for modern wireless networks [13].

Several works have been conducted on OFDM-IM [14,15]. In [16], a noncoherent OFDM-IM system based on maximum likelihood (ML) was proposed where the information bits were transmitted solely through the indices of active subcarriers. Low-complexity near-optimal detectors for OFDM-IM were introduced in [11] to address the computational challenges of ML detection, which is intensive due to the presence of index bits. The log-likelihood ratio (LLR) detector was introduced in [17]. Although it achieves near-ML performance, knowledge of the incoming signal's noise power spectral density is required, which represents a drawback. A low-complexity OFDM-IM detector that encodes every potential subcarrier activation pattern (SAP) was presented in [18]. This approach overlooks the possibility of incorrect SAP detection by using every potential SAP to transmit data. In [19], a greedy detector (GD) based on energy detection was proposed to estimate OFDM-IM performance with significantly lower complexity, although it did not achieve near-optimal performance.

Reliable data transmission in OFDM-IM relies on accurate detection, as it ensures proper decoding of both index and data symbols, an enhanced bit error rate (BER), resilience to channel impairments, and preservation of communication quality [20]. Deep learning (DL) excels in handling high-dimensional complex data, adapting to varying conditions, and performing end-to-end optimization, making it ideal for efficient signal detection and data processing in communication systems [21–23]. Thus, DL may serve as a promising approach for OFDM-IM systems. Various DL-based strategies have been explored to simplify OFDM-IM implementations while achieving near-optimal performance [24–26]. A deep neural network (DNN)-based OFDM-IM system called DeepIM was proposed in [27]. The authors investigated the performance of the proposed model by utilizing both rectified linear units (ReLU) and hyperbolic tangent (Tanh) activation functions. A low-complexity dual-mode OFDM-IM system was proposed in [28], which employed a DL model known as DeepDM constructed through a combined DNN and convolutional neural network (CNN) to identify carrier bits and index bits independently. The authors of [29] proposed an index-bit detection neural network for OFDM-IM that incorporated fully connected layers and an inception module to effectively manage different modulation settings. In [30], the authors presented a DL-based index bit estimator (IE-DNN) for active LED index detection in visible-light communication systems. By employing fully connected or convolutional layers, IE-DNN significantly reduces the index error rate (IER) and BER with acceptable complexity. In [31], the authors presented a CNN-based detection framework for OFDM-IM systems. This framework uses convolutional layers to effectively identify indices by converting incoming symbols to polar coordinates, creating 2D matrices. In [32], the authors proposed a multiple-input–multiple-output OFDM-IM system using IMNet, a non-iterative detector. The IMNet design combines two subnets with the conventional least squares method to recover the transmitted signal. In [33], a transformer-based detector called TransIM was proposed. The structure of the transformer adds a great deal of complexity to TransIM. In an earlier work [34], we presented a long short-term memory (LSTM)-based detector for OFDM-IM systems. Although we mostly concentrated on perfect channel conditions in that study, we also assessed BER performance under uncertain channel state information (CSI) settings. Interestingly, under imperfect CSI conditions, our previous experiment outperformed the ML, GD, and DeepIM detectors at higher SNR levels while showing comparable performance at lower SNR levels.

CNN-based models can effectively learn and extract features from received signals [35]. Their ability to capture spatial hierarchies enhances the detection of active subcarriers, enabling more accurate index identification in OFDM-IM systems. The translation invariance of CNNs ensures consistent performance across varying signal positions, while their parameter efficiency reduces the risk of overfitting [31]. Furthermore, by incorporating dependencies from both past and future contexts, bidirectional LSTM (Bi-LSTM) models enhance sequential data processing. They address the vanishing gradient issue, enabling effective learning of long-range relationships, and provide parallel processing, which increases training efficiency [36,37]. In this paper, we propose a hybrid DL (HDL)-based OFDM-IM system designed to operate under uncertain CSI conditions. The proposed HDL detector integrates a one-dimensional CNN (1D-CNN) with a Bi-LSTM model. The 1D-CNN model extracts spatial information from the received signal, focusing on patterns associated with the activated subcarriers by analyzing the received symbols in the OFDM-IM system. This step enhances the model's capacity to differentiate between various indices in the input data. The Bi-LSTM then processes these features, capturing temporal dependencies in the signal by considering both past and future information. This combined approach leverages spatial and temporal insights, significantly improving the OFDM-IM system's BER performance as well as its adaptability to signal variations, resulting in more accurate index detection. We evaluate the BER performance of the proposed model across different signal-to-noise ratios (SNRs) for the OFDM-IM system under imperfect CSI conditions. Additionally, we assess the BER for various optimizers and equalizers. We determine the throughput and spectral efficiency (SE) to further illustrate the resilience of the suggested HDL-based OFDM-IM system. Our simulation results demonstrate that the proposed model outperforms traditional detectors as well as other DL-based detectors. The key contributions of this paper are summarized as follows:

- We introduce an HDL-based OFDM-IM system that operates under uncertain CSI conditions. This HDL-based detector integrates a 1D-CNN model with a Bi-LSTM model; the 1D-CNN model captures spatial information from received signal, while the Bi-LSTM model processes these features. This combined approach improves detection performance even in the presence of CSI uncertainty.
- The channel data and received signal are preprocessed using OFDM-IM domain knowledge before being fed into the HDL model. This preprocessing step enhances the HDL-based detector's accuracy in detecting active subcarrier indices. Additionally, we apply several different equalizers to the received signal and evaluate their comparative performance.
- We assess the BER performance of the proposed HDL-based detector at various SNR levels and explore its performance with different optimizers. Additionally, we calculate the SE and throughput of the model to further demonstrate its effectiveness. The results indicate that the proposed HDL detector significantly enhances detection performance in OFDM-IM systems.

The remainder of this paper is organized as follows: the system modeling is introduced in Section 2; a thorough discussion of the suggested model, including details on offline training and online testing protocols, is provided in Section 3; the simulation results are presented in Section 4 and the complexity in Section 5; finally, the conclusions are presented in Section 6.

2. System Model

OFDM-IM is a sophisticated modulation technique that extends the traditional OFDM architecture. Unlike conventional OFDM, which utilizes all subcarriers to carry data, OFDM-IM conveys additional information by using both data symbols and the indices of active subcarriers. By selectively activating a subset of subcarriers, OFDM-IM enhances spectral efficiency and improves the system's resilience against interference and fading, leading to better BER performance. The signal transmission process of an OFDM-IM system is illustrated in Figure 1.

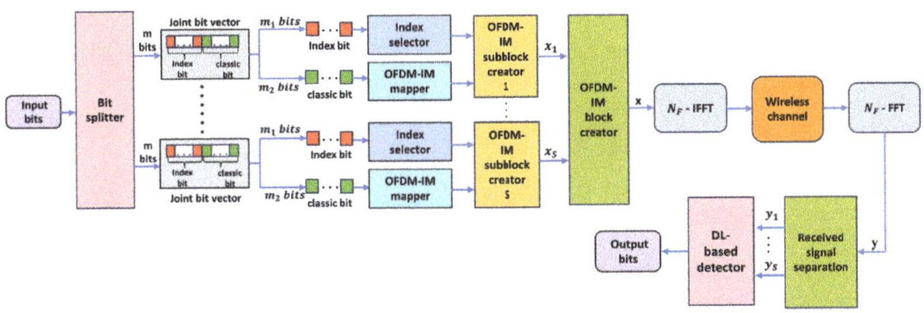

Figure 1. Generalized data transmission process for an OFDM-IM system.

There are N subcarriers in each of the sections S that make up the entire transmitted signal bandwidth. Therefore, the total number of transmitted subcarriers in this OFDM-IM system is N_T, where $N_T = SN$. Each section's signal processing system is independent from the others and operates separately. Of the N subcarriers, only A are active in each section, while the remaining $(N - A)$ subcarriers are deactivated (zero-padded). A total of m bits is transmitted per section in each transmission. These m bits consist of m_1 bits sent by the modulated symbols and m_2 bits sent by the active subcarriers' indices. In this case, m_1 bits is represented as $m_1 = Alog_2M$ and m_2 bits is represented as $m_2 = \lfloor log_2C(N, A) \rfloor$; hence, the total bits m can be expressed as follows:

$$m = m_1 + m_2 = Alog_2M + \lfloor log_2C(N, A) \rfloor \tag{1}$$

where M is the *M-ary* modulation size. The m_1 bits are mapped to a set of A active subcarrier indices using combinatorial methods. As a result, the transmit vector $\mathbf{x} = [x_1, \ldots, x_N]$ is formed by assigning A nonzero data symbols to the A active subcarriers based on the incoming m bits. The mapping from bits to symbols is denoted as $\mathbf{x} = f_{OFDM-IM}(\mathbf{D_i})$, where $\mathbf{D_i}$ represents the input sequence of m bits for each group.

The received signal in the frequency domain is represented as follows:

$$\mathbf{y} = \mathbf{h} \odot \mathbf{x} + \mathbf{w} \tag{2}$$

where $\mathbf{h} = [h_1, \ldots, h_N]$ denotes the Rayleigh fading channel and $\boldsymbol{h}_i \sim \mathcal{CN}(0, 1)$. The term $\mathbf{w}_i \sim \mathcal{CN}(0, \sigma)$ represents the additive white Gaussian noise (AWGN).

For the imperfect CSI settings, we consider a system in which the receiver experiences channel estimation errors. Let $\hat{h}(\alpha)$ denote the estimated channel, while the actual channel is modeled as follows:

$$h(\alpha) = \hat{h}(\alpha) + e(\alpha) \tag{3}$$

where the channel estimation error $e(\alpha)$ follows a complex Gaussian distribution $e(\alpha) \sim \mathcal{CN}(0, \epsilon^2)$, where ϵ^2 represents the channel estimation error variance. The estimated channel $\hat{h}(\alpha)$ is modeled as $\hat{h}(\alpha) \sim \mathcal{CN}(0, 1 - \epsilon^2)$. The CSI error variance ϵ^2 is inversely related to the average SNR by $\bar{\gamma} = \frac{E_a{}^2}{\sigma^2}$, where $E_a{}^2$ is the average energy of a transmitted *M-ary* symbol and σ^2 is the noise variance.

In this experiment, ϵ is computed as $\epsilon = \frac{1}{1+\text{SNR}}$, with $\text{SNR} = 10^{\frac{SNR_{dB}}{10}}$. This relationship adjusts the error variance based on the average SNR, which can more accurately reflect the real-world conditions of imperfect CSI under Rayleigh fading.

3. Architecture of the Proposed HDL Detector

Figure 2 illustrates the structure of the proposed HDL model. Similar to existing OFDM-IM detection techniques, the receiver is assumed to have knowledge of the channel information. As a result, both the channel $\mathbf{h}$ and received signal $\mathbf{y}$ are considered as inputs to the HDL model. To enhance detection performance, $\mathbf{y}$ and $\mathbf{h}$ undergo preprocessing using domain knowledge from OFDM-IM before being fed into the HDL detector. To obtain the equalized received signal vector, the widely used zero-forcing (ZF) equalizer is applied [38]. This is represented as follows:

$$\bar{\mathbf{y}} = \mathbf{y} \odot \mathbf{h}^{-1} \tag{4}$$

where the received signal $\mathbf{y}$ is multiplied element-wise by the inverse of the channel $\mathbf{h}$. Along with this equalizer, we also utilized the minimum mean square error (MMSE) and decision feedback equalizer (DFE) to compare their performance. By minimizing the mean square error (MSE) between the transmitted and received signals, the MMSE equalizer optimizes the tradeoff between inter-symbol interference (ISI) suppression and noise enhancement [39]. The equalization process of the MMSE equalizer is expressed as follows:

$$\bar{\mathbf{y}} = \frac{y \cdot h}{|h|^2 + \sigma^2} \tag{5}$$

where $\bar{\mathbf{y}}$ represents the received signal equalized by MMSE, $|h|^2$ represents the squared magnitude of the channel coefficient, and σ^2 is the noise variance.

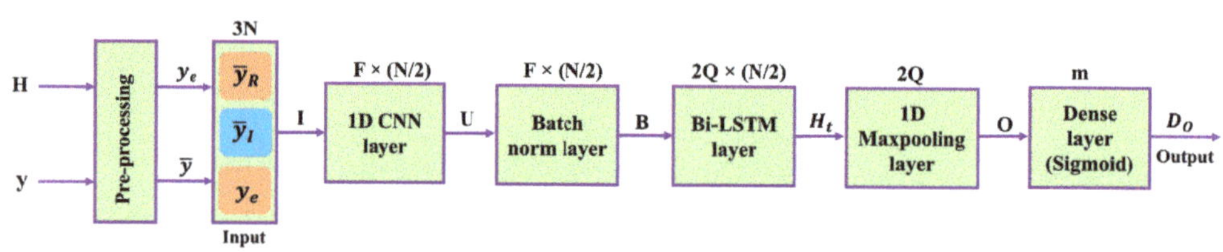

Figure 2. Structure of the proposed HDL detector for OFDM-IM systems.

The DFE equalizer achieves equalization by using the received signal along with feedback from previously detected symbols to reduce the ISI [40]. The equalization process of the DFE equalizer can be represented as follows:

$$\bar{\mathbf{y}} = \mathbf{y} - \mathbf{d} \cdot \hat{\mathbf{y}} \tag{6}$$

where $\bar{\mathbf{y}}$ is the equalized output signal, $\mathbf{d}$ represents the feedback filter applied to the previously detected symbols, and $\hat{\mathbf{y}}$ denotes the decision (or detected) output from the previous symbol decisions. To generate the HDL's input, the energy of the received signal $\mathbf{y}_e = \left[|y_1|^2, |y_2|^2, \ldots, |y_N|^2\right]$ is calculated and then added to $\bar{\mathbf{y}}$. The $3N$-dimensional input vector $\mathbf{I} = [\bar{\mathbf{y}}_R, \bar{\mathbf{y}}_I, \mathbf{y}_e]$ is created by concatenating the real and imaginary parts of $\bar{\mathbf{y}}$, respectively denoted as $\mathbf{y}_R$ and $\mathbf{y}_I$, with the received energy vector $\mathbf{y}_e$, as illustrated in Figure 2.

Our proposed HDL detector employs an organized methodology consisting of several essential elements: a 1D-CNN layer, a batch normalization layer, a Bi-LSTM layer, a one-dimensional max-pooling layer, and a dense layer with a sigmoid activation function in the output layer. The processed data I are fed into the 1D-CNN layer. A total of $F = 64$ filters of various 3×3 sizes are used in the convolutional layer to extract features [41]. The function of the 1D-CNN model with the data I can be explained as follows:

$$\mathbf{U} = f(\mathbf{I}_{-1} \otimes \mathbf{w}_i + \mathbf{b}_i) \tag{7}$$

where the weight and bias vectors of the i-th layer are indicated by $\mathbf{w}_i$ and $\mathbf{b}_i$, respectively. The batch normalization (BN) layer follows next. The BN technique is used to normalize the input of each layer in order to minimize internal covariate shifts and stabilize the learning process. This enables neural networks to achieve faster training, greater stability, and improved performance [42]. The Bi-LSTM layer takes its input from the output of the BN layer, denoted as B. The proposed Bi-LSTM layer is configured with $Q = 128$ hidden units.

Two LSTM layers (forward and backward layers) make up a Bi-LSTM hidden layer. Figure 3 shows that each LSTM cell contains three gates, namely, the input, forget, and output gates [43]. Additionally, the LSTM has two states, the hidden state h_{t-1} and cell state C_{t-1}. The cell state functions as memory to retain information extrapolated from previous inputs, while the hidden state is used to compute the output. Here, t represents the time instant and B_t is the current input. The gates function updates the cell state, enabling the Bi-LSTM cell to add or remove data at each time step. The forget gate and input gate control the amount of cell state that needs to be updated and reset, respectively [44].

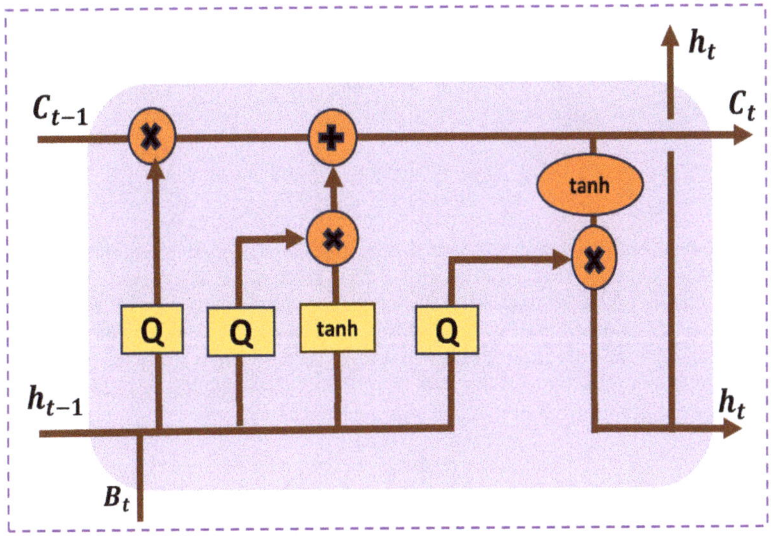

Figure 3. The internal configuration of an LSTM cell.

The functions of the forget, input, output, and candidate gates in the Bi-LSTM model are expressed as follows:

$$Fo_t = \sigma_\psi(W_{Fo}B_t + T_{Fo}h_{t-1} + b_{Fo}), \tag{8}$$

$$In_t = \sigma_\psi(W_{In}B_t + T_{In}h_{t-1} + b_{In}), \tag{9}$$

$$Ou_t = \sigma_\psi(W_{Ou}B_t + T_{Ou}h_{t-1} + b_{Ou}), \tag{10}$$

$$Ca_t = \sigma_{tanh}(W_{Ca}B_t + T_{Ca}h_{t-1} + b_{Ca}), \tag{11}$$

where B_t denotes the input vector of the Bi-LSTM at iteration t, the sigmoid activation function is represented by σ_ψ, the terms Fo_t, In_t, Ou_t, and Ca_t correspond to the respective activation vectors of the forget, input, output, and candidate gates at iteration t, the vector h_{t-1} represents the hidden state from the previous iteration $t-1$, and the weight matrices W_{Fo}, W_{In}, W_{Ou}, and W_{Ca} are associated with the forget, input, output, and candidate gates, respectively, with regard to the input B_t. Similarly, the weight matrices T_{Fo}, T_{In}, T_{Ou}, and T_{Ca} are tied to the previous hidden state h_{t-1}. Finally, the bias vectors for the forget, input, output, and candidate gates are represented by b_{Fo}, b_{In}, b_{Ou}, and b_{Ca}, respectively. The Bi-LSTM model updates the previous cell state c_{t-1} to the current cell state according to the following equation:

$$Up_t = +(In_t \bigodot Ca_t). \tag{12}$$

The hidden state function at any time step t can be represented as follows:

$$h_t = Ou_t \bigodot \sigma_{tanh}(Up_t). \tag{13}$$

The operation of the hidden state output in the Bi-LSTM layer is expressed as follows:

$$H_t = \sigma_f(W_{H\overrightarrow{T}} \overrightarrow{T}_t + W_{H\overleftarrow{T}} \overleftarrow{T}_t + b_v) \tag{14}$$

where $\overrightarrow{T}_t$ and $\overleftarrow{T}_t$ respectively represent the forward and backward sequence of the Bi-LSTM network.

The max-pooling layer now uses the Bi-LSTM layer's output as its input. By choosing the maximum value within predetermined window sizes, the max-pooling layer shortens the input sequences, reducing dimensionality while preserving important temporal features for later layers [45]. The fundamental function of the max-pooling layer can be illustrated as follows:

$$\mathbf{O}(i,j) = \max_{q,r \in K} H_t(i+q, j+r) \tag{15}$$

where $\mathbf{O}(i,j)$ is the output obtained after max-pooling, (q,r) denotes the indices within the pooling window, K is the pooling window, and $H_t(i,j)$ is the input feature map.

The output layer, which consists of a dense layer with a sigmoid activation function, receives the input O from the max-pooling layer. The weighted sum of the inputs is converted by the sigmoid activation function into a value between 0 and 1 indicating the output neuron's probability or activity level [46]. The output $\mathbf{D}_o$ of the HDL detector through the sigmoid activation function can be expressed as follows:

$$\mathbf{D_o} = f_{Sigmoid}(W_s \times O + b_s) \tag{16}$$

where W_s is the weight vector of the output layer and b_s is its bias.

Offline Training and Online Testing Procedure

The DL model must be trained offline using simulation data before applying the proposed HDL detector. Specifically, a set of transmitted vectors is generated by randomly producing different sequences of m bits, denoted as $\mathbf{D}_i$, such that $x = f_{\text{OFDM-IM}}\mathbf{D}_i$. These vectors are subsequently transmitted to the receiver, where they are impacted by AWGN noise and the Rayleigh fading channel. According to their established statistical models, the noise vectors and the channel are also created at random, and vary from one bit-sequence to another in this case. The input dataset $\mathbf{I}$, for which the labels are corresponding bit sequences $\mathbf{D}_i$, is obtained by preprocessing the received signal and channel vectors, i.e., $\mathbf{y}$ and $\mathbf{h}$, as explained in the preceding section. The signal detection process of the proposed HDL-based decoder in the OFDM-IM system is comprehensively outlined in Algorithm 1, featuring each stage from data generation to HDL model design, training, and testing. This step-by-step breakdown offers a clear understanding of the entire workflow.

Algorithm 1 OFDM-IM symbol detection with HDL model

Initialize Parameters:
Set subcarriers N, active subcarriers A, modulation order M, SNR SNR_{dB}, batch size, training epochs, and learning rate.
Calculate derived parameters: bits per symbol m, QAM symbols, power allocation, and noise level.
Define Subcarrier Patterns:
Generate index patterns based on N and A (e.g., index set idx).
Generate OFDM-IM Signal:
Define `OFDM_IM_received(bits, SNRdb)` for training and testing data generation:

- Convert bits into subcarrier indices and QAM symbols.
- Apply channel and noise effects.
- Normalize and return the received signals.

Design Neural Network (HDL model):
Construct `HDL_model(x)`:

- Use `Conv1D` for feature extraction.
- Apply `BatchNormalization`.
- Add bidirectional `LSTM` layers.
- Add `GlobalMaxPool1D`.
- Use a dense output layer with `sigmoid` activation for bit prediction.

Training Phase:
for each epoch from 1 to training epochs **do**
  Initialize average cost, avg_cost = 0
  for each batch from 1 to total batches **do**
    Initialize empty batches `batch_x` and `batch_y`
    for each sample from 1 to batch size **do**
      Generate random bits
      Compute received signal using `OFDM_IM_received`
      Append signal to `batch_x` and bits to `batch_y`
    end for
    Run optimizer, calculate training cost, and update avg_cost
  end for
  Append avg_cost to training loss list
end for
Testing Phase:
for each SNR in range of SNR values **do**
  Set test number, `test_number`, based on SNR
  Initialize empty test batches `batch_x` and `batch_y`
  for each test sample from 1 to `test_number` **do**
    Generate random bits
    Compute received signal using `OFDM_IM_received`
    Append signal to `batch_x` and bits to `batch_y`
  end for
  Calculate BER using model predictions and store result
end for
Display Results:
Plot BER vs. SNR

The HDL model is trained using the gathered data in order to reduce the BER, or more specifically the discrepancy between $\mathbf{D}_i$ and its prediction $\mathbf{D}_o$. Therefore, we use the mean-squared error (MSE) loss function for the training process. The MSE effectively penalizes larger errors more heavily than smaller ones due to squaring of the differences, making it highly sensitive to outliers. Mathematically, it is expressed as follows [47]:

$$\mathcal{L}(\mathbf{D}_i, \mathbf{D}_o; \theta) = \frac{1}{n}\sum_{i=1}^{n}(D_i - D_o)^2 \tag{17}$$

where $\mathbf{D}_i$ represents the true values, $\mathbf{D}_o$ represents the predicted values, and n is the number of samples. Adaptive moment estimation (Adam) is a DL optimization technique that combines the strengths of both RMSprop and momentum. By adjusting the learning rate for each parameter, Adam efficiently handles sparse gradients and noisy data, making it well suited for complex models and large datasets [48]. Adam is readily implementable in a number of commercial DL platforms, including Keras 2.10.0 and TensorFlow 2.10.0. Additionally, we evaluate the performance of the RMSProp, Adagrad, and Gradient Descent optimizers for comparison.

Figures 4 and 5 illustrate the training loss of the proposed model for the different equalizers and modulation schemes, respectively. We evaluate the training loss across each epoch to monitor the model's performance over time. As observed in Figure 4, the proposed model demonstrates strong training loss performance across all equalizers. While the DFE equalizer exhibits slightly higher loss, both the ZF and MMSE equalizers achieve very low and nearly identical training loss levels. Figure 5 illustrates the training loss for different modulation schemes under various data setups, specifically for $(4, 1, 4)$, $(8, 2, 8)$, and $(8, 4, 16)$ with the ZF equalizer. Notably, this training loss is estimated at 15 dB training SNR with a batch size of 1000 and learning rate of 0.005. As can be observed, the training loss tends to increase with higher modulation orders and data setup.

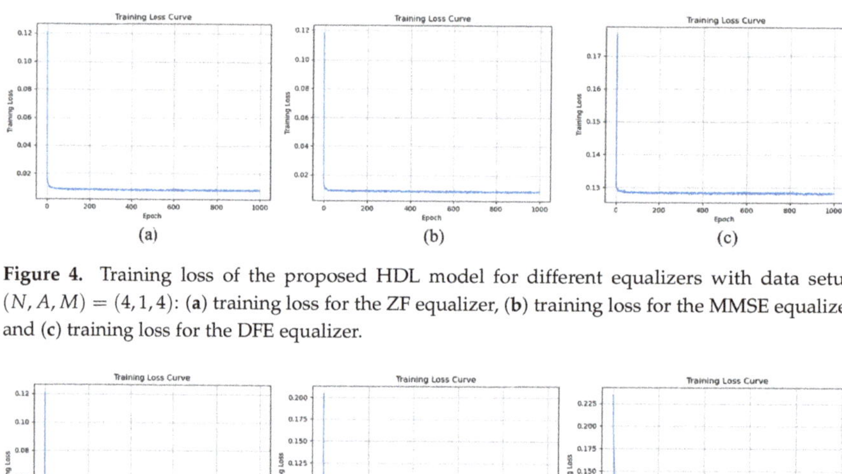

Figure 4. Training loss of the proposed HDL model for different equalizers with data setup $(N, A, M) = (4, 1, 4)$: (**a**) training loss for the ZF equalizer, (**b**) training loss for the MMSE equalizer, and (**c**) training loss for the DFE equalizer.

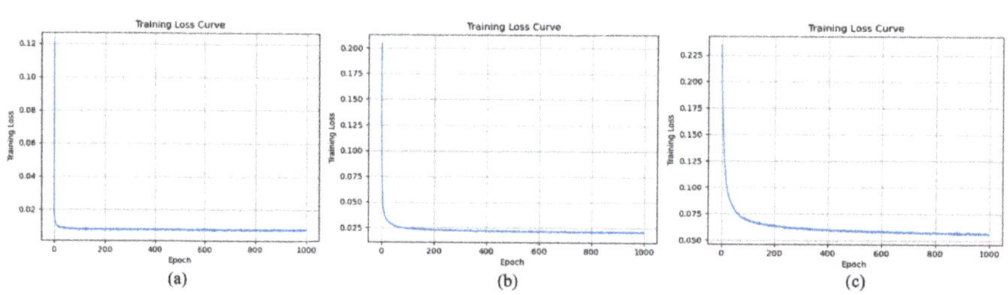

Figure 5. Training loss of the proposed HDL model for different modulation orders and data combinations with the ZF equalizer: (**a**) training loss for the $(N, A, M) = (4, 1, 4)$ setup, (**b**) training loss for the $(N, A, M) = (8, 2, 8)$ setup, and (**c**) training loss for the $(N, A, M) = (8, 4, 16)$ setup.

As the model's performance depends significantly on the SNR level chosen for training, denoted by $\bar{\gamma}_{\text{train}}$; thus, careful selection is crucial in order to optimize HDL's effectiveness. Specifically, the model trained at this SNR level must perform reliably across other relevant SNR values, making it essential to select an optimal $\bar{\gamma}_{\text{train}}$. For instance, if $\bar{\gamma}_{\text{train}}$ is set too low, the model may fail to generalize adequately, as it will not account for the noise impact during training. The simulation results presented in the next section detail the process of selecting an appropriate training SNR for each experimental setup.

We trained our proposed HDL model over 1000 epochs across all experimental settings, with each epoch comprising 20 batches and each batch containing 1000 data samples (often referred to as the batch size). This setup resulted in a total of 2×10^4 batches, yielding approximately 2×10^7 unique training samples. The samples for each batch were generated randomly during the training process. The HDL model was trained for various configurations of (N, A, M). Specifically, we trained the proposed model with modulation orders of QPSK, 8QAM, and 16QAM. All training configurations for the proposed model are detailed in Table 1. Our proposed model was trained using a batch size of 1000 over 1000 epochs.

The proposed system can operate in real time to estimate data bits across different channel fading scenarios without the need for further modifications to θ after it has been trained offline. Upon receiving the signal and channel characteristics, the HDL model efficiently generates an estimated bit and performs comparably to the ML detector in the presence of channel estimation errors. We applied 100,000 samples to test the proposed model.

Table 1. Simulation parameters for the proposed system.

Parameters	Value
Training samples	10,000,000
Testing samples	100,000
Modulation scheme	QPSK, 8QAM, 16QAM
Noise type	AWGN
Fading channel	Rayleigh fading channel
Number of hidden layers	128
Learning rate	0.01, 0.005, 0.002
Number of epoch	1000
Batch size	1000
Training SNR (dB)	0, 5, 10, 15, 20, 25
N	4, 8
A	1, 2, 3, 4
Optimizer	Adam, RMSProp, Adagrad, Gradient Descent

Figure 6 presents the confusion matrix of the proposed model under the $(4, 1, 4)$ data setup and with a training SNR of 15 dB, providing a bit-wise performance evaluation on the test data. The confusion matrix shows the model's accuracy in predicting "0" and "1" bits at the specified SNR level. Test samples with true bit labels were passed through the trained model to obtain predicted values, which were thresholded at 0.5 to produce binary predictions. The confusion matrix indicates that the model correctly classified 197,769 "0" bits and 197,822 "1" bits, with 2214 "0" bits misclassified as "1" bits and 2195 "1" bits misclassified as "0" bits. These results highlight the model's strong bit-wise accuracy with a low misclassification rate.

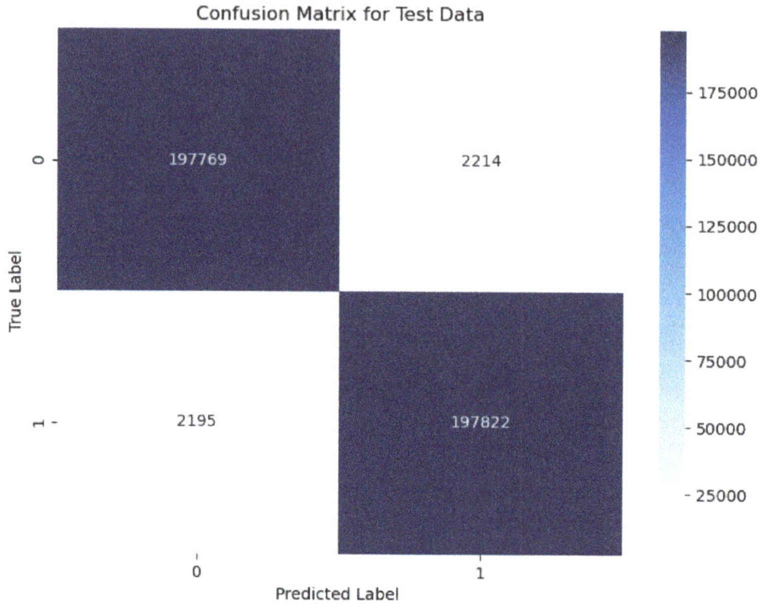

Figure 6. The confusion matrix of the proposed HDL-based model.

4. Results

The BER performance of the proposed HDL detector was evaluated with respect to SNR under imperfect CSI conditions. A model's learning rate balances speed and stability, determining how quickly it learns; an optimal rate accelerates convergence, prevents instability, and ensures efficient and accurate outputs [49]. Our proposed model demonstrates efficient performance across different learning rates, as shown in Figure 7a. The performance of the HDL-based detector at various learning rates was calculated with a training SNR of 15 dB and a $(4, 1, 4)$ data setup. At lower SNR levels, it performs similarly for learning rates of 0.01, 0.005, and 0.002; at higher SNR levels it shows a slight advantage, with a learning rate of 0.005 compared to the others. Consequently, all experimental setups were designed with a learning rate of 0.005.

The BER performance of the proposed HDL-based model for different batch sizes is shown in Figure 7b. The results demonstrate that the model exhibits robust performance across all batch sizes. At lower SNR levels, the performance remains nearly consistent for batch sizes of 500, 1000, and 2000. However, at higher SNR levels, the model achieves slightly superior performance with a batch size of 1000.

The training SNR is critical for DL models in communication systems, as it directly impacts the model's ability to generalize across various channel conditions. The performance of the proposed detector across various training SNR levels is illustrated in Figure 8. Notably, the model demonstrates the most stable performance compared to the other levels at a training SNR of 15 dB. Increasing or decreasing the training SNR away from this optimal point results in diminished performance. For training SNR levels of 0 dB and 10 dB, the HDL model performs well at lower SNRs but exhibits slightly weaker performance at higher SNRs. In contrast, at training SNRs of 20 dB and 25 dB the proposed model demonstrates good performance at higher SNR levels while showing comparatively weaker results at lower SNRs. Thus, the experimental results indicate that the proposed HDL-based detector achieves superior performance at a training SNR of 15 dB. Consequently, all experimental setups were designed with a training SNR of 15 dB.

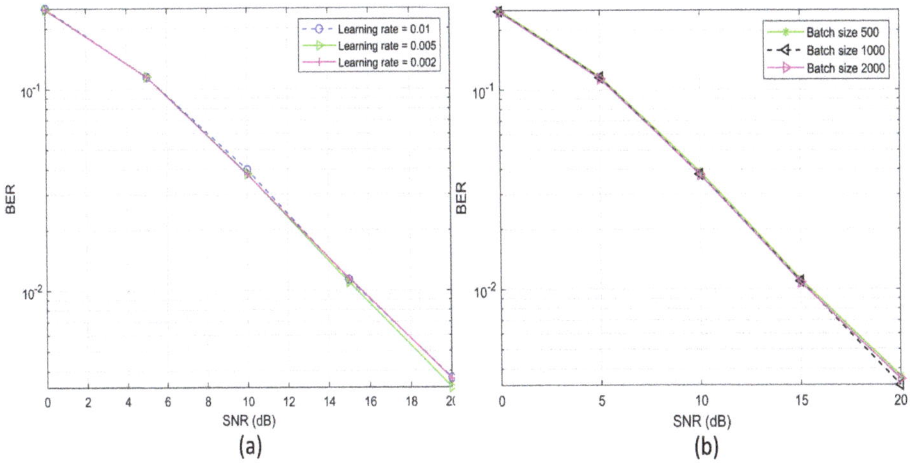

Figure 7. Performance of the HDL-based detector with (**a**) different learning rates and (**b**) different batch sizes in for the $(4, 1, 4)$ data combination.

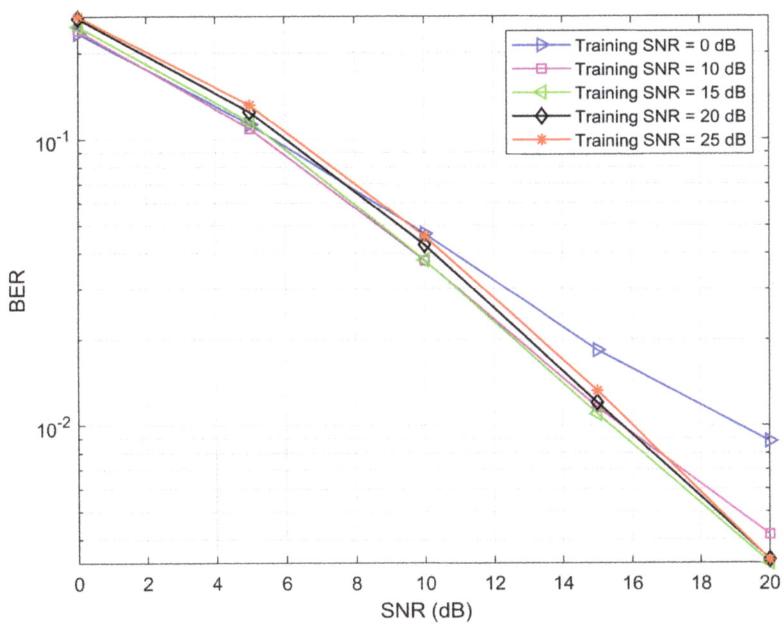

Figure 8. Performance of the HDL-based detector at various training SNRs for the $(4, 1, 4)$ data configuration.

In all experimental settings, the results were calculated using the ZF equalizer and Adam optimizer, except in Figures 9 and 10, where BER performance is compared across different equalizers and optimizers. Additionally, all experiments in this setup utilized $(N, A, M) = (4, 1, 4)$ except in Figure 11, which presents a comparison of model performance across different modulation orders and data setups.

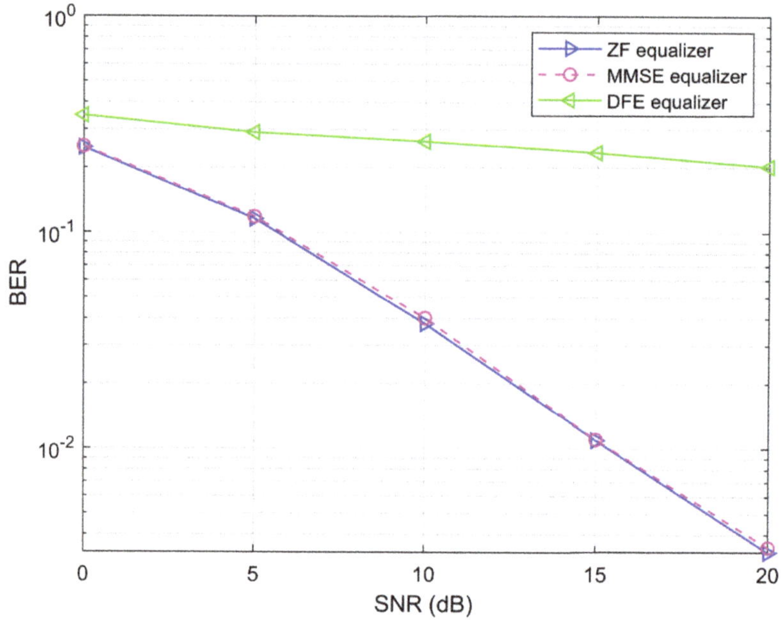

Figure 9. Performance of the proposed HDL-based detector with various equalizers for the (4, 1, 4) data configuration.

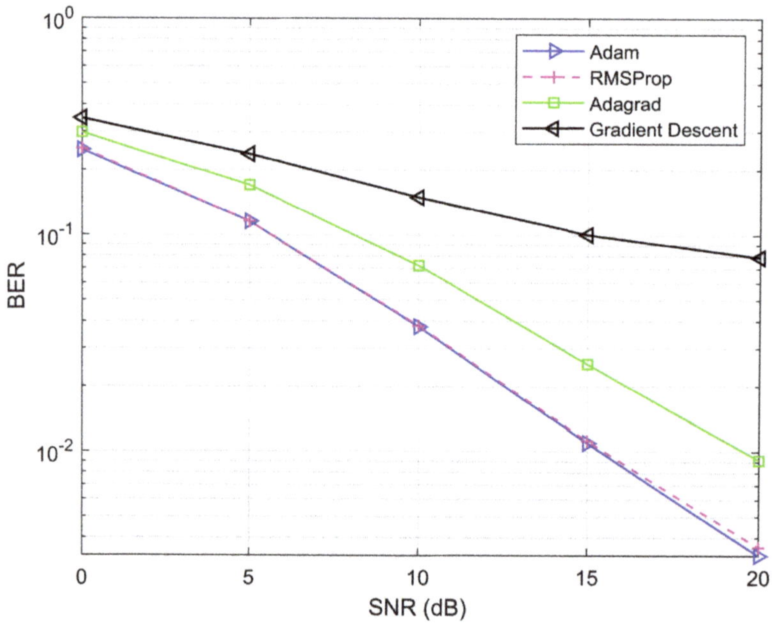

Figure 10. BER performance of the proposed HDL-based detector utilizing different optimizers for the (4, 1, 4) data configuration.

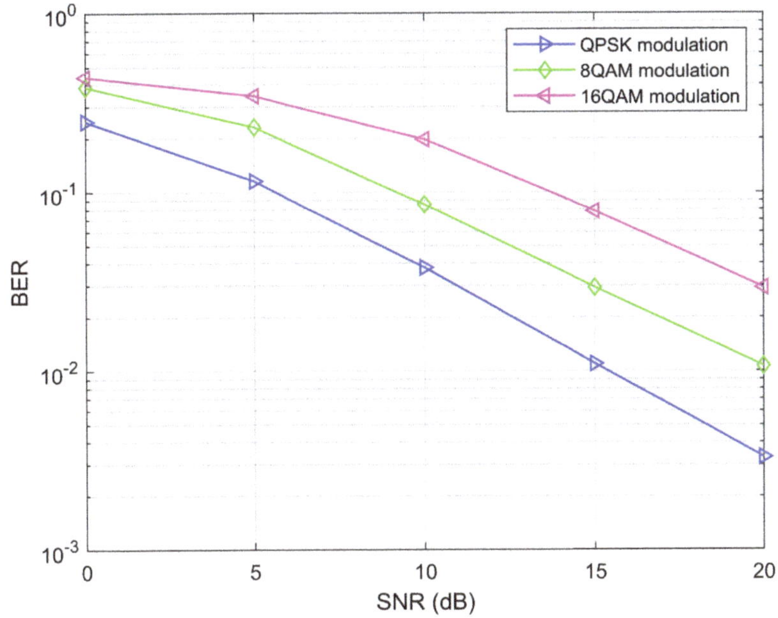

Figure 11. BER performance of the proposed HDL-based detector for various modulation orders and data setup.

Equalizers play a crucial role in communication systems by mitigating channel distortions such as ISI and multipath fading. Equalizers ensure reliable data transmission in noisy environments by improving BER and overall system performance by providing improved signal quality and dependability [50]. The BER performance of the proposed model with various equalizers is depicted in Figure 9. The results reveal that the proposed detector achieves impressive BER performance with both the ZF and MMSE equalizers, while its performance with the DFE equalizer is notably less effective. Although the model delivers nearly equivalent performance with the ZF and MMSE equalizers, it demonstrates slightly enhanced performance when employing the ZF equalizer, highlighting its efficiency in this configuration.

The BER performance of the proposed model with different optimizers is illustrated in Figure 10. Specifically, we examined the performance of the HDL model using the Adam, RMSProp, Adagrad, and Gradient Descent optimizers. The figure clearly indicates that the proposed model demonstrates acceptable performance across all optimizers. However, its performance with the Gradient Descent optimizer is somewhat less favorable compared to the others. Notably, while the HDL detector exhibits nearly comparable performance with both Adam and RMSProp, the Adam optimizer achieves a slightly superior performance in comparison to RMSProp.

Figure 11 illustrates the BER performance of the HDL detector in the OFDM-IM system with different modulation orders under imperfect channel conditions. We evaluated the performance with the QPSK, 8QAM, and 16QAM modulation schemes, where the configurations were set as follows: $(N, A, M) = (4, 1, 4)$ for QPSK, $(8, 2, 8)$ for 8QAM, and $(8, 4, 16)$ for 16QAM. We observed that increasing the modulation order and data setup leads to a reduction in performance; specifically, the model achieves 5 dB and 9 dB superior performance for the $(4, 1, 4)$ configuration compared to the $(8, 2, 8)$ and $(8, 4, 16)$ configurations, respectively.

Figure 12 compares the BER performance of the proposed HDL-based detector against ML, GD, DeepIM [27], and LSTM-IM [34] detectors. This evaluation used a 15 dB training

SNR and a 0.005 learning rate along with the ZF equalizer and Adam optimizer. The results clearly demonstrate that the HDL detector consistently outperforms GD, ML, DeepIM, and LSTM-IM models with the $(4,1,4)$ data setup. Specifically, the proposed detector shows approximately 2.5 dB improved performance over GD and achieves nearly 0.9 dB better BER performance compared to both the ML and DeepIM detectors under imperfect CSI conditions. Additionally, we compared the performance of the proposed model with the $(4,3,4)$ data setting against a CNN-based model [31]. Table 2 presents the BER performance of all comparable models at an SNR of 20 dB, with the results illustrated in Figure 12. The simulation results confirm that the HDL model outperforms the CNN model at higher data rates.

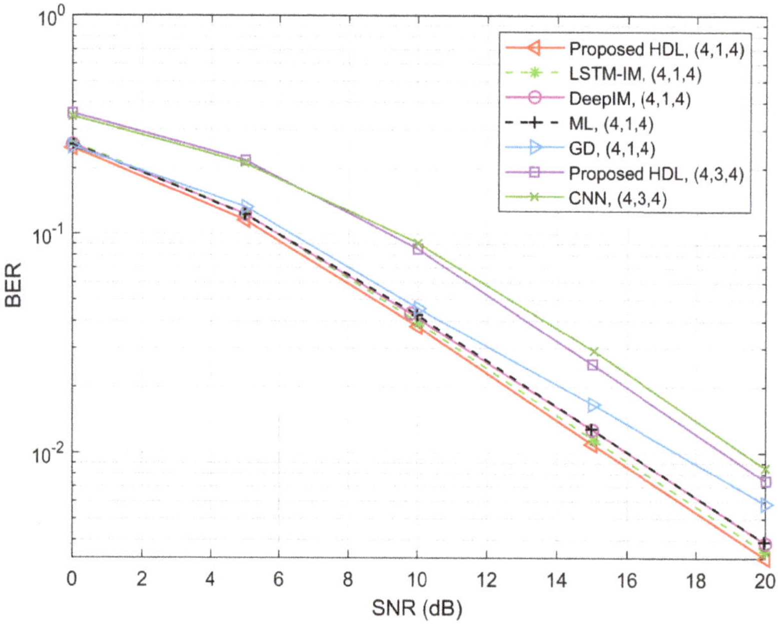

Figure 12. BER performance comparison of the proposed HDL-based detector with other detectors under imperfect CSI conditions for the $(4,1,4)$ data combinations.

Table 2. Comparative BER performance at 20 dB SNR for all models.

SNR	Model Name	BER	Model Name	BER
20 dB	HDL	0.0033074783	DeeIM	0.00387089132682052
	LSTM-IM	0.00355340813397947	ML	0.00392567922218776
	HDL	0.0074599785 (A = 3)	CNN	0.00854427859235968 (A = 3)
	GD	0.00585573211245060		

To further demonstrate the robustness of the proposed HDL-based OFDM-IM system, we calculated the throughput and spectral efficiency (SE), as shown in Figure 13. These metrics were evaluated for the (4,1,4), (4,2,4), and (4,3,4) data configuration, with each setup representing a distinct combination of active indices. The results indicate that both throughput and SE increase with higher active indices. The model demonstrated the highest performance with (4,3,4) setup. These results confirm that our proposed HDL model effectively achieves high throughput and SE.

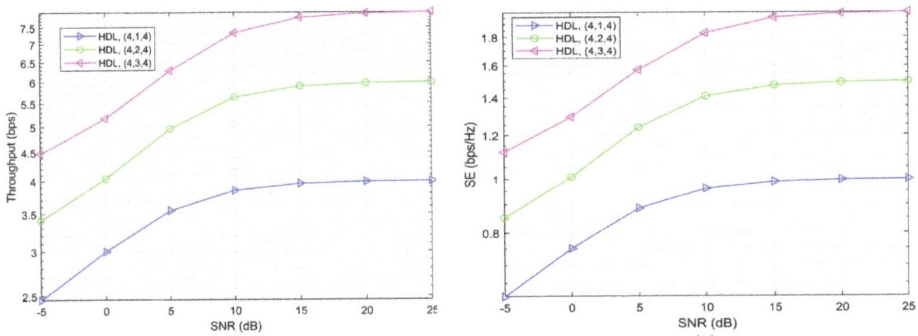

(a) **(b)**

Figure 13. Throughput and SE of the proposed HDL-based OFDM-IM system: (**a**) throughput performance and (**b**) SE performance.

5. Computational Complexity

The computational complexity of a DL model can be estimated by calculating the number of floating point operations (FLOPs) required for each layer [51–53]. In the given network, the computational cost of the 1D-CNN layer with ReLU activation function is $O(2 \times \text{input channels} \times \text{kernel size} \times \frac{N}{2} \times F) = O(576N)$. The subsequent BN layer normalizes the feature maps, with a complexity of $O(2 \times \frac{N}{2} \times F)$ FLOPs. Moving on to the Bi-LSTM layer, the FLOP count is provided by $O(8 \times N \times Q \times (64 + Q))$. The global max-pooling performs a simple pooling operation with a computational cost of $O(\frac{N}{2} \times 2Q)$. Finally, the output dense layer with sigmoid activation function incurs a FLOP count of $O((2 \times Q \times m) + m)$. Summing these operations provides the total FLOPs for the HDL model as $O(197376N + 1028)$, providing a complete picture of the computational demands. This measure is crucial for model optimization and deployment, particularly in environments with limited resources.

6. Conclusions

In this study, we investigate the BER performance of our hybrid DL (HDL) model combining a 1D-CNN and Bi-LSTM under imperfect channel conditions. By preprocessing the channel vector and received signal based on domain-specific knowledge, the HDL model effectively recovers transmitted data bits. Our evaluation included a thorough analysis of various batch sizes, learning rates, optimizers, and equalizers, finding that the HDL model achieves optimal results with the Adam optimizer and ZF equalizer. Although our proposed model demonstrates effective performance at positive training SNR values, its performance is comparatively lower at negative training SNR values, which led us to exclude such values during training. Our comparative results reveal that the HDL model outperforms GD, ML, DeepIM, CNN, and LSTM-IM detectors, demonstrating its robustness compared to both traditional and conventional models. Additionally, we explored the performance of our HDL detector across different modulation orders and data configurations, confirming that the best performance is achieved with QPSK modulation and the $(4, 1, 4)$ data setup. These promising results highlight the advantages of HDL, suggesting that it can provide valuable contributions towards further integration of DL within advanced IM systems. Furthermore, to substantiate the resilience of the proposed HDL-based OFDM-IM system, we evaluated it in terms of both throughput and SE. Our findings should inspire continued research on ways to leverage DL techniques in next-generation communication systems.

Author Contributions: Conceptualization, M.A.A. and M.H.R.; methodology, M.A.A., R.T. and M.A.S.S.; software, M.A.A., M.H.R. and M.-S.B.; validation, M.A.A. and M.A.S.S.; formal analysis, M.A.A., M.H.R., R.T. and M.-S.B.; investigation, M.A.A., M.H.R. and M.A.S.S.; resources, H.-K.S.; data curation, M.A.A. and R.T.; writing—original draft preparation, M.A.A.; writing—review and editing, M.A.A. and H.-K.S.; visualization, M.A.A. and M.-S.B.; supervision, H.-K.S.; project administration, H.-K.S.; funding acquisition, H.-K.S. All authors have read and agreed to the published version of the manuscript.

Funding: This work was supported in part by an Institute of Information and Communications Technology Planning and Evaluation (IITP) grant funded by the Korean Government (Ministry of Science and ICT (MSIT), South Korea); in part under the Metaverse Support Program to Nurture the Best Talents, Grant IITP-2024-RS-2023-00254529; in part by the Basic Science Research Program through the National Research Foundation of Korea (NRF) funded by the Ministry of Education under Grant 2020R1A6A1A03038540; and in part by the MSIT (Ministry of Science and ICT, Korea) under the ITRC (Information Technology Research Center) support program (IITP-2024-RS-2024-00437494) supervised by the IITP (Institute for Information and Communications Technology Planning and Evaluation).

Data Availability Statement: The data will be made available by the authors upon request.

Conflicts of Interest: The authors declare no conflicts of interest.

Abbreviations

The following abbreviations are used in this manuscript:

AWGN	Additive White Gaussian Noise
BER	Bit Error Rate
Bi-LSTM	Bidirectional Long Short-Term Memory
BN	Batch Normalization
CNN	Convolutional Neural Network
CSI	Channel State Information
DFE	Decision Feedback Equalizer
DNN	Deep Neural Network
DL	Deep Learning
GD	Greedy Detector
HDL	Hybrid Deep Learning
ISI	Inter-Symbol Interference
MC	Multi-Carrier
MMSE	Minimum Mean Square Error
ML	Maximum Likelihood
OFDM-IM	Orthogonal Frequency-Division Multiplexing with Index Modulation
16QAM	16-Quadrature Amplitude Modulation
QPSK	Quadrature Phase-Shift Keying
ReLU	Rectified Linear Units
SNR	Signal-to-Noise Ratio
SE	Spectral Efficiency
Tanh	Hyperbolic Tangent
ZF	Zero-Forcing

References

1. Bingham, J.A.C. Multicarrier Modulation for Data Transmission: An Idea Whose Time Has Come. *IEEE Commun. Mag.* **1990**, *28*, 5–14. [CrossRef]
2. Taura, K.; Tsujishita, M.; Takeda, M.; Kato, H.; Ishida, M.; Ishida, Y. A digital audio broadcasting (DAB) receiver. *IEEE Trans. Consum. Electron.* **1996**, *42*, 322–327. [CrossRef]
3. Telecommunication Standardization Sector of ITU. *Asymmetric Digital Subscriber Line (ADSL) Transceivers*; Telecommunication Standardization Sector of ITU: Geneva, Switzerland, 1999.
4. Nee, R.v.; Prasad, R. *OFDM for Wireless Multimedia Communications*; Artech House, Inc.: London, UK, 2000.
5. Biglieri, E. Coding and modulation for a horrible channel. *IEEE Commun. Mag.* **2003**, *41*, 92–98. [CrossRef]
6. Garg, V.K. Fourth generation systems and new wireless technologies. In *Wireless Communications & Networking*; Elsevier: Amsterdam, The Netherlands, 2007; pp. 1–22.

7. Cioffi, J.M. A multicarrier primer. *ANSI T1E1* **1991**, *4*, 91–157.
8. Terry, J.; Heiskala, J. *OFDM Wireless LANs: A Theoretical and Practical Guide*; Sams Publishing: Carmel, IN, USA, 2002.
9. Bury, A.; Egle, J.; Lindner, J. Diversity comparison of spreading transforms for multicarrier spread spectrum transmission. *IEEE Trans. Commun.* **2003**, *51*, 774–781. [CrossRef]
10. Mao, T.; Wang, Z.; Wang, Q.; Chen, S.; Hanzo, L. Dual-mode index modulation aided OFDM. *IEEE Access* **2016**, *5*, 50–60. [CrossRef]
11. Başar, E.; Aygölü, Ü.; Panayırcı, E.; Poor, H.V. Orthogonal frequency division multiplexing with index modulation. *IEEE Trans. Signal Process.* **2013**, *61*, 5536–5549. [CrossRef]
12. Wen, M.; Cheng, X.; Ma, M.; Jiao, B.; Poor, H.V. On the achievable rate of OFDM with index modulation. *IEEE Trans. Signal Process.* **2015**, *64*, 1919–1932. [CrossRef]
13. Mao, T.; Wang, Q.; Wang, Z.; Chen, S. Novel index modulation techniques: A survey. *IEEE Commun. Surv. Tutor.* **2018**, *21*, 315–348. [CrossRef]
14. Abu-Alhiga, R.; Haas, H. Subcarrier-index modulation OFDM. In Proceedings of the 2009 IEEE 20th International Symposium on Personal, Indoor and Mobile Radio Communications, Tokyo, Japan, 13–16 September 2009; pp. 177–181.
15. Fan, R.; Yu, Y.J.; Guan, Y.L. Generalization of orthogonal frequency division multiplexing with index modulation. *IEEE Trans. Wirel. Commun.* **2015**, *14*, 5350–5359. [CrossRef]
16. Choi, J. Noncoherent OFDM-IM and its performance analysis. *IEEE Trans. Wirel. Commun.* **2017**, *17*, 352–360. [CrossRef]
17. Başar, E.; Aygölü, Ü.; Panaylrcl, E. Orthogonal frequency division multiplexing with index modulation in the presence of high mobility. In Proceedings of the 2013 IEEE First International Black Sea Conference on Communications and Networking (BlackSeaCom), Batumi, Georgia, 3–5 July 2013; pp. 147–151.
18. Siddiq, A.I. Low complexity OFDM-IM detector by encoding all possible subcarrier activation patterns. *IEEE Commun. Lett.* **2016**, *20*, 446–449. [CrossRef]
19. Crawford, J.; Ko, Y. Low complexity greedy detection method with generalized multicarrier index keying OFDM. In Proceedings of the 2015 IEEE 26th Annual International Symposium on Personal, Indoor, and Mobile Radio Communications (PIMRC), Hong Kong, China, 30 August–2 September 2015; pp. 688–693.
20. Basar, E. Index modulation techniques for 5G wireless networks. *IEEE Commun. Mag.* **2016**, *54*, 168–175. [CrossRef]
21. Zappone, A.; Di Renzo, M.; Debbah, M. Wireless networks design in the era of deep learning: Model-based, AI-based, or both? *IEEE Trans. Commun.* **2019**, *67*, 7331–7376. [CrossRef]
22. Erpek, T.; O'Shea, T.J.; Sagduyu, Y.E.; Shi, Y.; Clancy, T.C. Deep learning for wireless communications. In *Development and Analysis of Deep Learning Architectures*; Springer: Berlin/Heidelberg, Germany, 2020; pp. 223–266.
23. Sejan, M.A.S.; Rahman, M.H.; Aziz, M.A.; Kim, D.S.; You, Y.H.; Song, H.K. A comprehensive survey on MIMO visible light communication: Current research, machine learning and future trends. *Sensors* **2023**, *23*, 739. [CrossRef]
24. Dhanke, S.S.; Sharma, S.; Kumar, A.; Mandloi, M. Index Modulation Multiple Access via Deep Learning based Detection. In Proceedings of the 2020 IEEE 3rd 5G World Forum (5GWF), Bangalore, India, 10–12 September 2020; pp. 312–317.
25. Esmaiel, H.; Leftah, H.A.; Junejo, N.U.R.; Sun, H. Deep Learning-Based Index Modulation for Underground Communications. *IEEE Open J. Commun. Soc.* **2023**, *4*, 2122–2132. [CrossRef]
26. Liu, H.; Zhang, Y.; Zhang, X.; El-Hajjar, M.; Yang, L.L. Deep learning assisted adaptive index modulation for mmWave communications with channel estimation. *IEEE Trans. Veh. Technol.* **2022**, *71*, 9186–9201. [CrossRef]
27. Van Luong, T.; Ko, Y.; Vien, N.A.; Nguyen, D.H.; Matthaiou, M. Deep learning-based detector for OFDM-IM. *IEEE Wirel. Commun. Lett.* **2019**, *8*, 1159–1162. [CrossRef]
28. Kim, J.; Ro, H.; Park, H. Deep learning-based detector for dual mode OFDM with index modulation. *IEEE Wirel. Commun. Lett.* **2021**, *10*, 1562–1566. [CrossRef]
29. Kim, J.; Park, H. Deep Learning-Based Detector With Modulation Parameter-Independent Structure for OFDM With Index Modulation. *IEEE Access* **2023**, *11*, 130358–130367. [CrossRef]
30. Le-Tran, M.; Kim, S. Deep Learning-Assisted Index estimator for generalized LED index modulation OFDM in visible light communication. *Photonics* **2021**, *8*, 168. [CrossRef]
31. Wang, T.; Yang, F.; Song, J.; Han, Z. Deep convolutional neural network-based detector for index modulation. *IEEE Wirel. Commun. Lett.* **2020**, *9*, 1705–1709. [CrossRef]
32. Liu, J.; Lu, H. IMNet: A learning based detector for index modulation aided MIMO-OFDM systems. In Proceedings of the 2020 IEEE Wireless Communications and Networking Conference (WCNC), Seoul, Republic of Korea, 25–28 May 2020; pp. 1–6.
33. Zhang, D.; Wang, S.; Niu, K.; Dai, J.; Wang, S.; Yuan, Y. Transformer-based detector for OFDM with index modulation. *IEEE Commun. Lett.* **2022**, *26*, 1313–1317. [CrossRef]
34. Aziz, M.A.; Rahman, M.H.; Sejan, M.A.S.; Tabassum, R.; Hwang, D.d.; Song, H.K. Deep Recurrent Neural Network Based Detector for OFDM with Index Modulation. *IEEE Access* **2024**, *12*, 89538–89547. [CrossRef]
35. Alzubaidi, L.; Zhang, J.; Humaidi, A.J.; Al-Dujaili, A.; Duan, Y.; Al-Shamma, O.; Santamaría, J.; Fadhel, M.A.; Al-Amidie, M.; Farhan, L. Review of deep learning: Concepts, CNN architectures, challenges, applications, future directions. *J. Big Data* **2021**, *8*, 53. [CrossRef]
36. Ratnam, D.V.; Rao, K.N. Bi-LSTM based deep learning method for 5G signal detection and channel estimation. *AIMS Electron. Electr. Eng.* **2021**, *5*, 334–341. [CrossRef]

37. Aziz, M.A.; Rahman, M.H.; Sejan, M.A.S.; Tabassum, R.; Hwang, D.d.; Song, H.K. Bidirectional Deep Learning Decoder for Polar Codes in Flat Fading Channels. *IEEE Access* **2024**, *12*, 149580–149592. [CrossRef]
38. Jiang, Y.; Varanasi, M.K.; Li, J. Performance analysis of ZF and MMSE equalizers for MIMO systems: An in-depth study of the high SNR regime. *IEEE Trans. Inf. Theory* **2011**, *57*, 2008–2026. [CrossRef]
39. Hanoon, S.S.; Khazal, H.F.; Jamel, T.M. A Deep Learning Approach to Evaluating SISO-OFDM Channel Equalization. *Wasit J. Eng. Sci.* **2024**, *12*, 110–124. [CrossRef]
40. Belfiore, C.A.; Park, J.H. Decision feedback equalization. *Proc. IEEE* **1979**, *67*, 1143–1156. [CrossRef]
41. Li, H.; Li, Y.; Zhou, S.; Wang, J. Wireless channel feature extraction via GMM and CNN in the tomographic channel model. *J. Commun. Inf. Netw.* **2017**, *2*, 41–51. [CrossRef]
42. Bjorck, N.; Gomes, C.P.; Selman, B.; Weinberger, K.Q. Understanding batch normalization. *arXiv* **2018**, arXiv:1806.02375.
43. Yu, Y.; Si, X.; Hu, C.; Zhang, J. A review of recurrent neural networks: LSTM cells and network architectures. *Neural Comput.* **2019**, *31*, 1235–1270. [CrossRef]
44. Aziz, M.A.; Rahman, M.H.; Sejan, M.A.S.; Baik, J.I.; Kim, D.S.; Song, H.K. Spectral Efficiency Improvement Using Bi-Deep Learning Model for IRS-Assisted MU-MISO Communication System. *Sensors* **2023**, *23*, 7793. [CrossRef]
45. Yu, D.; Wang, H.; Chen, P.; Wei, Z. Mixed pooling for convolutional neural networks. In *Rough Sets and Knowledge Technology, Proceedings of the 9th International Conference, RSKT 2014, Shanghai, China, 24–26 October 2014, Proceedings 9*; Springer: Berlin/Heidelber, Germany, 2014; pp. 364–375.
46. Narayan, S. The generalized sigmoid activation function: Competitive supervised learning. *Inf. Sci.* **1997**, *99*, 69–82. [CrossRef]
47. Hodson, T.O.; Over, T.M.; Foks, S.S. Mean squared error, deconstructed. *J. Adv. Model. Earth Syst.* **2021**, *13*, e2021MS002681. [CrossRef]
48. Zhang, Z. Improved adam optimizer for deep neural networks. In Proceedings of the 2018 IEEE/ACM 26th International Symposium on Quality of Service (IWQoS), Banff, AB, Canada, 4–6 June 2018; pp. 1–2.
49. You, K.; Long, M.; Wang, J.; Jordan, M.I. How does learning rate decay help modern neural networks? *arXiv* **2019**, arXiv:1908.01878.
50. Ahmed, S.; Khan, Y.; Wahab, A. A review on training and blind equalization algorithms for wireless communications. *Wirel. Pers. Commun.* **2019**, *108*, 1759–1783. [CrossRef]
51. Brownlee, J. A gentle introduction to pooling layers for convolutional neural networks. *Mach. Learn. Mastery* **2019**, *22*, 1–16.
52. Zhang, M.; Wang, W.; Liu, X.; Gao, J.; He, Y. Navigating with graph representations for fast and scalable decoding of neural language models. *arXiv* **2018**, arXiv:1806.04189.
53. Chaudhary, N.; Misra, S.; Kalamkar, D.; Heinecke, A.; Georganas, E.; Ziv, B.; Adelman, M.; Kaul, B. Efficient and generic 1d dilated convolution layer for deep learning. *arXiv* **2021**, arXiv:2104.08002.

mathematics

MDPI

Article

Closed-Form Performance Analysis of the Inverse Power Lomax Fading Channel Model

Aleksey S. Gvozdarev

Department of Intelligent Radiophysical Information Systems (IRIS), Physics Faculty, P.G. Demidov Yaroslavl State University, Yaroslavl 150003, Russia; asg.rus@gmail.com

Abstract: This research presents a closed-form mathematical framework for assessing the performance of a wireless communication system in the presence of multipath fading channels with an instantaneous signal-to-noise ratio (SNR) subjected to the inverse power Lomax (IPL) distribution. It is demonstrated that depending on the channel parameters, such a model can describe both severe and light fading covering most cases of the well-renowned simplified models (i.e., Rayleigh, Rice, Nakagami-m, Hoyt, $\alpha-\mu$, Lomax, etc.). This study provides the exact results for a basic statistical description of an IPL channel, including the PDF, CDF, MGF, and raw moments. The derived representation was further used to assess the performance of a communication link. For this purpose, the exact expression and their high signal-to-noise ratio (SNR) asymptotics were derived for the amount of fading (AoF), outage probability (OP), average bit error rate (ABER), and ergodic capacity (EC). The closed-form and numerical hyper-Rayleigh analysis of the IPL channel is performed, identifying the boundaries of weak, strong, and full hyper-Rayleigh regimes (HRRs). An in-depth analysis of the system performance was carried out for all possible fading channel parameters' values. The practical applicability of the channel model was supported by comparing it with real-world experimental results. The derived expressions were tested against a numerical analysis and statistical simulation and demonstrated a high correspondence.

Keywords: fading channel; statistical description; inverse power Lomax; error rate; outage; capacity

PACS: 89.70.+c; 84.40.Ua

MSC: 94A15; 94A40

1. Introduction

Wireless communications channels' statistical description is a mature field of science and engineering that has significantly evolved during the last two decades [1–3]. This is mainly due to the rapid evolution of communication networks and systems and the necessity of high-quality predictions of their performance [4–6]. It is particularly relevant for tasks related to the analysis of ad hoc communication systems, including 5G NR [7], 6G [8], V2X [9], Wi-Fi 7 [10], Wi-Fi 8 and beyond [11], and unmanned aerial vehicle (UAV) systems [12,13], which greatly suffer from wireless link degradation. Although a wide range of measurement campaigns has been conducted [13–19], the thickening of propagation conditions prevents the possible unification and generalization of its statistical description.

An extensive growth in the number of proposed models started with the outstanding works by M.D. Yacoub and colleagues [20–24] and continues to the present day with a wide range of various types, including the most intricate ones [25] and the most simplistic [26,27]. Although these models cover a wide range of practical channels, much effort has been put into modeling heavy fading, which corresponds to poor propagation conditions. Such scenarios are generally covered by the so-called heavy-tailed distributions, with one of the most recent examples being the Lomax channel [26]. Notably, the authors in [26] specifically focused on the most simplistic model, but the Lomax distribution has many modifications (suitable for wireless communications) that combine both flexibility and analytical tractability. Among such modifications, one can mention the inverse power Lomax.

The inverse power Lomax distribution (first proposed in [28]) is a three-parametric distribution family and was addressed for a wide range of applications related to econometrics (e.g., GDP analysis [29]), survey sampling, biological sciences (e.g., biological survival data [30]), engineering sciences (e.g., failure times for repairable items [31]), medical research (e.g., relief times analysis [28]), testing problems [32], and social and environmental sciences (e.g., school education-related problems [33] and flood discharges [34]). Moreover, the specific focus of [28] was on applications where censored data arise (with Type-I and Type-II censoring). Furthermore, it was numerously generalized, yielding more complex models, including the four-parametric model, the so-called modified alpha power transformation-inverse power Lomax [33], and the Type-II Topp–Leone modification for the IPL [35]. In spite of being widely used in various applications, its usage in communication systems design, and specifically in wireless communications, is absent.

Recently, a simplified subcase of the IPL, i.e., the classical Lomax model [36], was proposed to describe the statistical effects of a fading wireless communication channel [26]. In their research, Sanchez and López-Martínez [26] aimed to derive a simplified model capable of handling practically valuable scenarios while at the same time being analytically tractable in closed form without resorting to complex functions. It was demonstrated that all the basic statistical characteristics required for a system performance analysis (i.e., the probability density function (PDF), the cumulative distribution function (CDF) and its inverse (ICDF), its raw moments, the moment-generating function (MGF), and its generalization (GMGF)) can be expressed in closed form in contrast to some classical fading distributions (e.g., Rician, Nakagami-m, and Hoyt distributions) [26]. While defining the term "closed-form", the authors state that "We take the usual convention in the literature, which considers a closed-form as that including a finite number of well-known functions [26]. This classically includes special functions like Bessel or Hypergeometric ones, but not more involved ones like Meijer-G or Fox-H functions". Although the rigor and clarity of the presented results are praiseworthy, the practical aspects of the chosen approach can be argued. For example, the results in [37] (i.e., [26]) are delivered in terms of various bivariate confluent hypergeometric functions, which are nonstandard for most modern computer algebra systems (CASs) (although they can be easily implemented via series summation or numerical integration). In contrast, the classical univariate Meijer-G or Fox-H functions are implemented in most CASs and have the same computational complexity as the ones mentioned above. Nevertheless, [26] presented not only the statistical description but also most of the Lomax channel-performance metrics (the average bit error rate (ABER), ergodic capacity (EC), and outage probability (OP)). The model was further used for a spectrum sensing analysis [38,39]. One of its key properties, as mentioned in [26,39], is that it is always in the hyper-Rayleigh regime (HRR) (at least the weak HRR) and thus can describe heavy fading conditions, which are of paramount importance for modern urban wireless communications. This can be linked to the fact that the Lomax family (with its modifications) belongs to the heavy-tailed distributions [28], which tend to describe poor propagation conditions.

Additionally, in the last few years, the performance of modern wireless communications has been successfully described with the help of the inverse distributions approach (e.g., inverse gamma shadowing models [40] or inverse Gaussian and inverse generalized Gaussian models [41]) and their power-transformed modifications [42–45]. Hence, the inverse power Lomax model incorporates all these approaches.

Although much effort has been applied to the description of the IPL model in various applications (from economics to social sciences), it has not been studied in communication theory.

Thus, motivated by the problem stated above, this research studies the performance of a wireless communications system functioning in the presence of a multipath fading channel with the instantaneous signal-to-noise ratio (SNR) subjected to the inverse power Lomax distribution.

The major contributions of this research can be summarized as follows:

- An easy-to-use statistical formulation (in terms of the PDF and CDF) of the inverse power Lomax fading channel model is introduced.
- A specific positioning of the IPL model among the ones most widely used in wireless communications is performed. The results help to identify its capability to handle both severe and light fading, depending on the channel parameters.
- The closed-form expressions are derived for (a) the moment-generating function of the instantaneous SNR; (b) amount of fading (AoF); (c) outage probability (OP); (d) average bit error rate (ABER); and (e) ergodic channel capacity (EC).
- Closed-form high-SNR asymptotic expressions for all the derived channel-performance metrics are derived.
- The closed-form and numerical hyper-Rayleigh analysis of the IPL channel is performed, identifying the boundaries of weak, strong, and full hyper-Rayleigh regimes.
- A thorough joint analysis of the dependence of all the assumed channel-performance metrics on the channel parameters is performed for different fading scenarios—heavy fading and light fading.
- The numerical and statistical simulation results are presented, proving the correctness of the performed theoretical analysis.
- The practical applicability of the proposed model was verified using a variety of datasets, including both synthetic computer-generated data and real-world experimental measurements.

The remainder of this paper is organized as follows: Section 2 provides some preliminary information about the underlying IPL model and derives its representation (in terms of the PDF and CDF), suitable for communications systems' descriptions, as well as its MGF with its high-SNR asymptotics; Section 3 derives the results for the specific application of the IPL channel in wireless communications (i.e., the exact expressions for the AoF, ABER, OP, and EC); Section 4 delivers an in-depth analysis (theoretical, numerical, and simulation) of the derived expressions for all specific channel and system parameters' values; and conclusions are drawn in Section 5.

2. IPL Channel Statistical Description

2.1. PDF and CDF of the Instantaneous SNR

This research considers that the instantaneous SNR γ follows the inverse power Lomax distribution with the PDF $w_\gamma(\gamma)$ and CDF $F_\gamma(\gamma)$ defined as follows [28]:

$$w_\gamma(\gamma) = \frac{\alpha\beta}{\lambda}\gamma^{-(\beta+1)}\left(1+\frac{\gamma^{-\beta}}{\lambda}\right)^{-(\alpha+1)}, \tag{1}$$

$$F_\gamma(\gamma) = \left(1+\frac{\gamma^{-\beta}}{\lambda}\right)^{-\alpha}, \tag{2}$$

which are valid for arbitrary positive-distribution parameters (i.e., the scale parameter λ and the shape parameters α and β).

Although such a model can be used for further elaborations, it is conventional to perform a reparametrization in terms of the average SNR $\bar{\gamma}$ as given by the following Lemma:

Lemma 1. *The random variable γ is assumed to follow the two-parametric inverse power Lomax model (i.e., $\gamma \sim \mathcal{IPL}(\alpha, \beta)$) with the average value $\bar{\gamma}$ and channel parameters α, β if its PDF and CDF admit the following representation:*

$$\begin{cases} w_\gamma(\gamma) = \dfrac{\alpha\beta}{\mho_{\alpha,\beta}} \dfrac{\bar{\gamma}^\beta}{\gamma^{\beta+1}} \left(1 + \dfrac{1}{\mho_{\alpha,\beta}} \left(\dfrac{\bar{\gamma}}{\gamma}\right)^\beta\right)^{-(\alpha+1)}, \\ F_\gamma(\gamma) = \left(1 + \dfrac{1}{\mho_{\alpha,\beta}} \left(\dfrac{\bar{\gamma}}{\gamma}\right)^\beta\right)^{-\alpha}. \end{cases} \tag{3}$$

Proof of Lemma 1. To perform this step, one needs the expression for the moments of the instantaneous SNR, i.e., $\mathbb{E}\{\gamma^r\}$ (where r is the order of the moment and $\mathbb{E}\{\cdot\}$ is the expectation operator).

Starting from the definition and performing the change in variable $\gamma^{-\beta} = t$, after some straightforward manipulations, yields

$$\mathbb{E}\{\gamma^r\} = \frac{\alpha\beta}{\lambda} \int_0^\infty \gamma^{r-(\beta+1)} \left(1 + \frac{\gamma^{-\beta}}{\lambda}\right)^{-(\alpha+1)} \mathrm{d}\gamma = \tag{4a}$$

$$= \alpha\lambda^\alpha \int_0^\infty \frac{t^{-\frac{r}{\beta}}}{(t+\lambda)^{\alpha+1}} \mathrm{d}t = \tag{4b}$$

$$= \lambda^{-\frac{r}{\beta}} \frac{\Gamma\left(1 - \frac{r}{\beta}\right)\Gamma\left(\alpha + \frac{r}{\beta}\right)}{\Gamma(\alpha)}, \tag{4c}$$

where $\Gamma(\cdot)$ is the Euler gamma function [46].

It must be pointed out that (4c) is valid for any positive α and λ but for $\beta > r$, which implies a new restriction on β, when evaluating the moments of γ.

Calculating the average SNR, i.e., $\bar{\gamma} = \mathbb{E}\{\gamma\}$, and evaluating λ leads to

$$\lambda = \mho_{\alpha,\beta} \frac{1}{\bar{\gamma}^\beta}, \tag{5}$$

where $\mho_{\alpha,\beta}$ is introduced for notation clarity in the following way:

$$\mho_{\alpha,\beta} = \left(\frac{\Gamma\left(1 - \frac{1}{\beta}\right)\Gamma\left(\alpha + \frac{1}{\beta}\right)}{\Gamma(\alpha)}\right)^\beta. \tag{6}$$

This allows one to rewrite (1) and (2) as follows:

$$w_\gamma(\gamma) = \frac{\alpha\beta}{\mho_{\alpha,\beta}} \frac{\bar{\gamma}^\beta}{\gamma^{\beta+1}} \left(1 + \frac{1}{\mho_{\alpha,\beta}} \left(\frac{\bar{\gamma}}{\gamma}\right)^\beta\right)^{-(\alpha+1)}, \tag{7}$$

$$F_\gamma(\gamma) = \left(1 + \frac{1}{\mho_{\alpha,\beta}} \left(\frac{\bar{\gamma}}{\gamma}\right)^\beta\right)^{-\alpha}, \tag{8}$$

which finalizes the proof of Lemma 1 and will be used further as a starting point for a performance analysis. □

For further analysis, one resorts to α as a "power-transformation parameter" and β as a "shape parameter" of the distribution.

It must be pointed out that the form (3) is quite similar to the corresponding expression for the $\alpha - \mathcal{F}$ model (see [42]); thus, some explanations are needed. At its core, the difference between the two lies in the way they are derived: to derive the instantaneous SNR statistics, [42] assumes the α-transformed signal's envelope, which experiences inverse Nakagami-m shadowing. In contrast, here, the α-transformed inverse of the SNR is assumed. Moreover, it can be demonstrated that the models cannot be derived from one another by reparameterization. For example, the closest match can be achieved when m_s in [42] is set to one, but this is not allowed for the shadowing model assumed in [42] (i.e., $m_s > 1$).

Generally, the basic statistical representation of a fading model requires not only the PDF and CDF but also an expression for the moment-generating function, since many practical issues are resolved via the so-called MGF approach [47–50].

2.2. MGF of the Instantaneous SNR

Theorem 1. *The moment-generating function $\mathcal{M}_\gamma(s)$ of the IPL channel and its high-SNR asymptotics are given by*

$$\begin{cases} \mathcal{M}_\gamma(s) = \dfrac{1}{\Gamma(\alpha)} \mathrm{H}_{2,1}^{1,2}\left(\dfrac{\mho_{\alpha,\beta}}{(s\bar{\gamma})^\beta} \middle| \begin{matrix} (1,1),(0,\beta) \\ (\alpha,1) \end{matrix} \right), \\ \mathcal{M}_\gamma(s)\Big|_{\bar{\gamma}\to\infty} \backsimeq \dfrac{\Gamma(\alpha\beta+1)\mho_{\alpha,\beta}^\alpha}{(s\bar{\gamma})^{\alpha\beta}}. \end{cases} \tag{9}$$

Proof of Theorem 1.

$$\mathcal{M}_\gamma(s) \triangleq = \mathbb{E}\{e^{-\gamma s}\} = s\int_0^\infty F_\gamma(\gamma) e^{-\gamma s} \mathrm{d}\gamma \tag{10a}$$

$$= s\int_0^\infty \left(1 + \frac{1}{\mho_{\alpha,\beta}}\left(\frac{\bar{\gamma}}{\gamma}\right)^\beta\right)^{-\alpha} e^{-\gamma s}\mathrm{d}\gamma \tag{10b}$$

$$= \frac{s\bar{\gamma}}{\beta}\int_0^\infty \left(1 + \mho_{\alpha,\beta}^{-1} t\right)^{-\alpha} e^{-s\bar{\gamma}t^{-\frac{1}{\beta}}} t^{-1-\frac{1}{\beta}}\mathrm{d}t, \tag{10c}$$

where the equality (10c) is obtained via the transformation of variable $t = \left(\frac{\bar{\gamma}}{\gamma}\right)^\beta$:

$$e^{-xt^{-\frac{1}{\beta}}} = \mathrm{H}_{0,1}^{1,0}\left(xt^{-\frac{1}{\beta}} \middle| \begin{matrix} \text{—} \\ (0,1) \end{matrix} \right) \tag{11a}$$

$$= \mathrm{H}_{1,0}^{0,1}\left(\left(x^{-\beta}t\right)^{\frac{1}{\beta}} \middle| \begin{matrix} (1,1) \\ \text{—} \end{matrix} \right) \tag{11b}$$

$$= \beta\, \mathrm{H}_{1,0}^{0,1}\left(x^{-\beta}t \middle| \begin{matrix} (1,\beta) \\ \text{—} \end{matrix} \right), \tag{11c}$$

where $\mathrm{H}_{p,q}^{m,n}\left(z \middle| \begin{matrix} (a_p, A_p) \\ (b_q, B_q) \end{matrix} \right)$ is the Fox H-function [46], (11a) is due to (1.39) from [51], and (11b) and (11c) are derived with the help of (1.58)–(1.59) from [51]:

$$\left(1 + \mho_{\alpha,\beta}^{-1}t\right)^{-\alpha} = \mathrm{H}_{1,1}^{1,1}\left(\mho_{\alpha,\beta}^{-1}t \middle| \begin{matrix} (1-\alpha,1) \\ (0,1) \end{matrix} \right), \tag{12}$$

Combining (10c) with (11c) and (12) and using the power-transformation property of the Fox H-function (see (1.59) in [51]), the expression for the moment-generating function can be expressed as

$$\mathcal{M}_\gamma(s) = \frac{(s\bar{\gamma})^{-\beta}}{\Gamma(\alpha)}\int_0^\infty \mathrm{H}_{1,1}^{1,1}\left(\mho_{\alpha,\beta}^{-1}t \middle| \begin{matrix} (1-\alpha,1) \\ (0,1) \end{matrix} \right) \mathrm{H}_{1,0}^{0,1}\left(t(s\bar{\gamma})^{-\beta} \middle| \begin{matrix} (-\beta,\beta) \\ \text{—} \end{matrix} \right)\mathrm{d}t. \tag{13}$$

The integral can be evaluated with the help of (2.25.1.1) from [52], yielding

$$\mathcal{M}_\gamma(s) = \frac{(s\bar{\gamma})^{-\beta}}{\Gamma(\alpha)}\mho_{\alpha,\beta} \mathrm{H}_{2,1}^{1,2}\left(\frac{\mho_{\alpha,\beta}}{(s\bar{\gamma})^\beta} \middle| \begin{matrix} (0,1),(-\beta,\beta) \\ (\alpha-1,1) \end{matrix} \right) \tag{14a}$$

$$= \frac{1}{\Gamma(\alpha)} \mathrm{H}_{2,1}^{1,2}\left(\frac{\mho_{\alpha,\beta}}{(s\bar{\gamma})^\beta} \middle| \begin{matrix} (1,1),(0,\beta) \\ (\alpha,1) \end{matrix} \right). \tag{14b}$$

To obtain a high-SNR asymptotic expression for (14b), one can resort to [53] (see expression (3) and (4) in [53]). In this case, one obtains

$$\mathrm{H}_{2,1}^{1,2}\left(\frac{\mho_{\alpha,\beta}}{(s\bar{\gamma})^{\beta}} \middle| \begin{matrix} (1,1),(0,\beta) \\ (\alpha,1) \end{matrix} \right) \Bigg|_{\bar{\gamma}\to\infty} \backsimeq \left(\frac{\mho_{\alpha,\beta}}{(s\bar{\gamma})^{\beta}} \right)^{\omega} h_j, \tag{15}$$

where ω and h_j are defined in [53] (see expression (3) and (4), respectively). For the IPL model, $\omega = \alpha$ and $h_j = \Gamma(\alpha)\Gamma(\alpha\beta + 1)$, which finally helps to derive the high-SNR asymptotics for the MGF of the IPL model as

$$\mathcal{M}_{\gamma}(s)\Big|_{\bar{\gamma}\to\infty} \backsimeq \frac{\Gamma(\alpha\beta+1)\mho_{\alpha,\beta}^{\alpha}}{(s\bar{\gamma})^{\alpha\beta}}. \tag{16}$$

□

It must be noted that the univariate Fox H-functions are implemented in most modern computer algebra systems (e.g., Wolfram Mathematica, MATLAB, Maple, etc.). Thus, (14b) can be easily computed for arbitrary channel parameters, although the computational time can vary, and for some values of α and β, evaluations can be time-consuming. To speed up the computations, the integral representation of the Fox H-function can be implemented, and efficient numerical integration strategies can be employed. In this case, the MGF (14b) can be rewritten in the following form:

$$\mathcal{M}\gamma(s) = \frac{1}{\Gamma(\alpha)} \frac{1}{2\pi i} \int \mathcal{L}\Gamma(-\sigma)\Gamma(1-\beta\sigma)\Gamma(\alpha+\sigma) \left(\frac{\mho_{\alpha,\beta}}{(s\bar{\gamma})^{\beta}} \right)^{-\sigma} d\sigma, \tag{17}$$

where $\mathcal{L}$ is the contour chosen in such a way as to separate the poles of the gamma function with opposite signs for the integration variable. For instance, it can be chosen as $\mathcal{L} = (-\frac{\alpha}{2} - i\infty; -\frac{\alpha}{2} + i\infty)$.

3. IPL Fading Channel Model Application

The derived expressions for the PDF, CDF, and MGF provide a complete first-order statistical representation for assessing the performance of a wireless communication system operating under fading conditions. According to the general framework, communication link reliability is typically quantified in terms of the outage probability, while link quality is measured by the average error rate and ergodic capacity. The impact of fading is conventionally described by the amount of fading and the hyper-Rayleigh regimes.

3.1. *Amount of Fading*

The amount of fading is defined as the variance $\mathbb{V}\mathrm{ar}\{\gamma\}$ of the instantaneous SNR γ normalized by its squared average value:

$$\mathrm{AoF} = \frac{\mathbb{V}\mathrm{ar}\{\gamma\}}{(\mathbb{E}\{\gamma\})^2} = \frac{\mathbb{E}\{\gamma^2\}}{(\mathbb{E}\{\gamma\})^2} - 1, \tag{18}$$

Theorem 2. *The amount of fading* (AoF) *for the IPL channel model is given by*

$$\mathrm{AoF} = \Gamma(\alpha) \frac{\Gamma\left(1-\frac{2}{\beta}\right)\Gamma\left(\alpha+\frac{2}{\beta}\right)}{\left(\Gamma\left(1-\frac{1}{\beta}\right)\Gamma\left(\alpha+\frac{1}{\beta}\right)\right)^2} - 1. \tag{19}$$

Proof of Theorem 2. The proof of Theorem 2 is trivially obtained by combining the definition of the amount of fading (18) with the expression for the SNR r-th order moment (4c) obtained in Lemma 1. □

Treating the AoF as a function of channel parameters (i.e., AoF(α, β)), one can see that

$$\lim_{\alpha \to \infty} \text{AoF}(\alpha, \beta) = \frac{\Gamma\left(1 - \frac{2}{\beta}\right)}{\Gamma\left(1 - \frac{1}{\beta}\right)^2} - 1 \gtrless 1, \tag{20a}$$

$$\lim_{\beta \to \infty} \text{AoF}(\alpha, \beta) = \lim_{\alpha, \beta \to \infty} \text{AoF}(\alpha, \beta) = 0, \tag{20b}$$

which means that the IPL channel model can exhibit a hyper-Rayleigh regime (according to [54]) depending on the parameters.

3.2. Outage Probability

The outage probability is defined as the probability that the instantaneous SNR γ falls below the specified threshold value γ_{th}:

$$\text{P}_{\text{out}}(\gamma_{th}) = \mathbb{P}\{\gamma \leq \gamma_{th}\}. \tag{21}$$

Theorem 3. *The outage probability P_{out} for the IPL channel model and its high-SNR asymptotics are given by*

$$\begin{cases} \text{P}_{\text{out}}(\gamma_{th}) = \left(1 + \dfrac{1}{\mho_{\alpha,\beta}}\left(\dfrac{\bar{\gamma}}{\gamma_{th}}\right)^{\beta}\right)^{-\alpha}, \\ \text{P}_{\text{out}}(\gamma_{th})\,|_{\bar{\gamma}\to\infty} \backsimeq \left(\mho_{\alpha,\beta}\left(\dfrac{\gamma_{th}}{\bar{\gamma}}\right)^{\beta}\right)^{\alpha}. \end{cases} \tag{22}$$

Proof of Theorem 3. The proof of Theorem 3 is trivially obtained by noting that $\text{P}_{\text{out}}(\gamma_{th}) = \mathbb{P}\{\gamma \leq \gamma_{th}\} = F_\gamma(\gamma_{th})$ with the help of Lemma 1. □

It is well known (see [55]) that under certain conditions, the outage probability can asymptotically be expressed as $\text{P}_{\text{out}}|_{\bar{\gamma}\to\infty} = (\mathcal{G}_c\bar{\gamma})^{-\mathcal{G}_d}$, where $\mathcal{G}_d$ and $\mathcal{G}_c$ are termed as the diversity and coding gains.

Corollary 1. *The diversity and coding gains of the IPL channel model are expressed in terms of the channel parameters as follows: $\mathcal{G}_d = \alpha\beta, \mathcal{G}_c = \mho_{\alpha,\beta}^{-\frac{1}{\beta}}$.*

3.3. Error Rate Analysis

The average bit error rate is defined as the instantaneous error rate averaged with the respect to the PDF of the instantaneous SNR:

$$\bar{\text{P}}_e = \int_0^\infty \text{P}_e(\gamma) f_\gamma(\gamma) \text{d}\gamma = \delta_1 \sum_{j=1}^{\delta_3} \int_0^\infty Q\left(\sqrt{2\delta_{2,j}\gamma}\right) f_\gamma(\gamma) \text{d}\gamma, \tag{23}$$

where $Q(x)$ is the Gaussian Q-function [46] and the coefficients $\{\delta_1, \delta_{2,j}, \delta_3\}$ depend on the choice of the modulation format. For M-QAM or M-PSK, they are given in Table 1.

Table 1. Modulation-specific coefficients for M-QAM and M-PSK.

	δ_1	$\delta_{2,j}$	δ_3
M-QAM	$\frac{4}{\log_2 M}\left(1 - \frac{1}{\sqrt{M}}\right)$	$\frac{3(2j-1)^2}{2(M-1)}$	$\frac{\sqrt{M}}{2}$
M-PSK	$\frac{1}{\max(2, \log_2 M)}$	$2\sin^2\left(\frac{(2j-1)\pi}{M}\right)$	$\max\left(1, \frac{M}{4}\right)$

Theorem 4. *The average error rate for the IPL channel model* $\overline{\mathrm{P}}_{er}$ *and its high-SNR asymptotics are given by*

$$\begin{cases} \overline{\mathrm{P}}_{er} = \dfrac{\delta_1}{2\sqrt{\pi}\Gamma(\alpha)} \sum\limits_{j=1}^{\delta_3} \mathrm{H}_{2,1}^{1,2}\left(\dfrac{\mho_{\alpha,\beta}}{(\delta_{2,j}\bar{\gamma})^{\beta}} \middle| \begin{matrix} (1,1), \left(\frac{1}{2}, \beta\right) \\ (\alpha, 1) \end{matrix} \right), \\ \overline{\mathrm{P}}_{er}\,\Big|_{\bar{\gamma}\to\infty} \backsimeq \dfrac{\delta_1 \mho_{\alpha,\beta}^{\alpha}}{2\sqrt{\pi}\bar{\gamma}^{\alpha\beta}} \Gamma\left(\alpha\beta + \dfrac{1}{2}\right) \sum\limits_{j=1}^{\delta_3} \delta_{2,j}^{-\alpha\beta}. \end{cases} \tag{24}$$

Proof of Theorem 4. To prove Theorem 4, one can make use of the connection between the Gaussian Q-function and the complementary error function $\mathrm{erfc}(\cdot)$ [46]:

$$\overline{\mathrm{P}}_{er} = \frac{\delta_1}{2} \sum_{j=1}^{\delta_3} \int_0^{\infty} \mathrm{erfc}\left(\sqrt{\delta_{2,j}\gamma}\right) f_{\gamma}(\gamma) \mathrm{d}\gamma. \tag{25}$$

Denoting the integral over γ as $\mathrm{J_Q}$, it can be further expressed as

$$\mathrm{J_Q} = \int_0^{\infty} \mathrm{erfc}\left(\sqrt{\delta_{2,j}\gamma}\right) f_{\gamma}(\gamma) \mathrm{d}\gamma = \sqrt{\frac{\delta_{2,j}}{\pi}} \int_0^{\infty} \frac{F_{\gamma}(\gamma)}{\sqrt{\gamma}} e^{-\delta_{2,j}\gamma} \mathrm{d}\gamma, \tag{26a}$$

$$= \sqrt{\frac{\delta_{2,j}}{\pi}} \int_0^{\infty} \frac{\left(1 + \frac{1}{\mho_{\alpha,\beta}}\left(\frac{\bar{\gamma}}{\gamma}\right)^{\beta}\right)^{-\alpha}}{\sqrt{\gamma}} e^{-\delta_{2,j}\gamma} \mathrm{d}\gamma, \tag{26b}$$

$$= \sqrt{\frac{\delta_{2,j}\bar{\gamma}}{\pi\beta^2}} \int_0^{\infty} \left(1 + \frac{t}{\mho_{\alpha,\beta}}\right)^{-\alpha} t^{-\frac{1}{2\beta}-1} e^{-\delta_{2,j}\bar{\gamma} t^{-\frac{1}{\beta}}} \mathrm{d}t\,, \tag{26c}$$

where (26a) is obtained with the help of integration by parts and (26c) via the change in the variable $t = \left(\frac{\bar{\gamma}}{\gamma}\right)^{\beta}$.

Making use of (11c) and (12), $\mathrm{J_Q}$ can be represented as

$$\mathrm{J_Q} = \frac{1}{\sqrt{\pi}(\delta_{2,j}\bar{\gamma})^{\beta}\Gamma(\alpha)} \int_0^{\infty} \mathrm{H}_{1,1}^{1,1}\left(\frac{t}{\mho_{\alpha,\beta}} \middle| \begin{matrix} (1-\alpha, 1) \\ (0,1) \end{matrix} \right) \mathrm{H}_{1,0}^{0,1}\left(\frac{t}{(\delta_{2,j}\bar{\gamma})^{\beta}} \middle| \begin{matrix} \left(\frac{1}{2}-\beta, \beta\right) \\ — \end{matrix} \right) \mathrm{d}t \tag{27a}$$

$$= \frac{\mho_{\alpha,\beta}}{\sqrt{\pi}(\delta_{2,j}\bar{\gamma})^{\beta}\Gamma(\alpha)} \mathrm{H}_{2,1}^{1,2}\left(\frac{\mho_{\alpha,\beta}}{(\delta_{2,j}\bar{\gamma})^{\beta}} \middle| \begin{matrix} (0,1), \left(\frac{1}{2}-\beta, \beta\right) \\ (\alpha-1, 1) \end{matrix} \right), \tag{27b}$$

where (27b) is evaluated with equality (2.25.1.1) from [52].

Finally, combining (27b) with (25) and making use of the power-transformation property of the Fox H-function (see (1.59) in [51]), the closed-form expression for the average error rate can be equated as

$$\overline{\mathrm{P}}_{er} = \frac{\delta_1}{2\sqrt{\pi}\Gamma(\alpha)} \sum_{j=1}^{\delta_3} \mathrm{H}_{2,1}^{1,2}\left(\frac{\mho_{\alpha,\beta}}{(\delta_{2,j}\bar{\gamma})^{\beta}} \middle| \begin{matrix} (1,1), \left(\frac{1}{2}, \beta\right) \\ (\alpha, 1) \end{matrix} \right). \tag{28}$$

To derive the high-SNR asymptotics of (28), one can make use of Fox H-function expansion in the case of an infinitely small argument (see expression (3) and (4) in [53]):

$$\mathrm{H}_{2,1}^{1,2}\left(\frac{\mho_{\alpha,\beta}}{(\delta_{2,j}\bar{\gamma})^{\beta}} \middle| \begin{matrix} (1,1), \left(\frac{1}{2}, \beta\right) \\ (\alpha, 1) \end{matrix} \right)\Bigg|_{\bar{\gamma}\to\infty} \backsimeq \left(\frac{\mho_{\alpha,\beta}}{(\delta_{2,j}\bar{\gamma})^{\beta}} \right)^{\omega} h_j, \tag{29}$$

where ω and h_j are defined in [53] (see expression (3) and (4), respectively). For the IPL model, $\omega = \alpha$ and $h_j = \Gamma(\alpha)\Gamma\left(\alpha\beta + \frac{1}{2}\right)$, which finally helps to derive the high-SNR asymptotics for the average error rate of the IPL model as

$$\overline{\mathrm{P}}_{\mathrm{er}}\big|_{\bar{\gamma}\to\infty} \simeq \frac{\delta_1 \mho_{\alpha,\beta}^{\alpha}}{2\sqrt{\pi}\bar{\gamma}^{\alpha\beta}}\Gamma\left(\alpha\beta+\frac{1}{2}\right)\sum_{j=1}^{\delta_3}\delta_{2,j}^{-\alpha\beta}. \tag{30}$$

□

Similar to the approach used for the MGF expression, calculations for (28) can be efficiently performed through direct numerical integration. This method can significantly speed up calculations for certain parameter values by utilizing the integral representation of the Fox H-function as follows:

$$\overline{\mathrm{P}}_{er} = \frac{\delta_1}{2\sqrt{\pi}\Gamma(\alpha)}\sum_{j=1}^{\delta_3}\frac{1}{2\pi i}\int_{-\frac{\alpha}{2}-i\infty}^{-\frac{\alpha}{2}+i\infty}\Gamma(-\sigma)\Gamma\left(\frac{1}{2}-\beta\sigma\right)\Gamma(\alpha+\sigma)\left(\frac{\mho_{\alpha,\beta}}{(\delta_{2,j}\bar{\gamma})^{\beta}}\right)^{-\sigma}\mathrm{d}\sigma. \tag{31}$$

Analogous to the outage probability asymptotic representation, the average bit error rate can be represented as (see [55]) $\overline{\mathrm{P}}_{er}\big|_{\bar{\gamma}\to\infty} = (\mathcal{O}_c\bar{\gamma})^{-\mathcal{O}_d}$, where $\mathcal{O}_d$ and $\mathcal{O}_c$ are the diversity and coding gains derived from the ABER.

Corollary 2. *The ABER diversity and coding gains of the IPL channel model are expressed in terms of the channel parameters as follows:* $\mathcal{O}_d = \alpha\beta, \mathcal{O}_c = \frac{\delta_1\mho_{\alpha,\beta}^{\alpha}}{2\sqrt{\pi}}\Gamma\left(\alpha\beta+\frac{1}{2}\right)\sum_{j=1}^{\delta_3}\delta_{2,j}^{-\alpha\beta}$.

It can be noted that the diversity gains derived from the OP and ABER are equal, and the ABER coding gain accounts for modulation-specific parameters.

3.4. Ergodic Channel Capacity Analysis

The ergodic channel capacity is defined as the instantaneous capacity averaged over the stochastic channel realizations:

$$\overline{\mathrm{C}} = \int_0^{\infty}\log_2(1+\gamma)f_{\gamma}(\gamma)\mathrm{d}\gamma, \tag{32}$$

Theorem 5. *The ergodic capacity of the IPL channel* $\overline{\mathrm{C}}$ *and its high-SNR asymptotics are given by*

$$\begin{cases}\overline{\mathrm{C}} = \dfrac{1}{\Gamma(\alpha)\ln 2}\mathrm{H}_{2,2}^{2,2}\left(\dfrac{\mho_{\alpha,\beta}}{\bar{\gamma}^{\beta}}\middle|\begin{matrix}(1,1),(1,\beta)\\(\alpha,1),(1,\beta)\end{matrix}\right),\\[2ex] \overline{\mathrm{C}}\big|_{\bar{\gamma}\to\infty} \simeq \log_2\bar{\gamma} + \dfrac{1}{\beta\ln 2}\left(G_e + \psi^{(0)}(\alpha) - \ln\mho_{\alpha,\beta}\right),\end{cases} \tag{33}$$

where $G_e \approx 0.577$ *is the Euler–Mascheroni constant and* $\psi^{(0)}(\cdot)$ *is the digamma function [46].*

Proof of Theorem 5 .

$$\overline{\mathrm{C}} = \frac{\alpha\beta\bar{\gamma}^{\beta}}{\mho_{\alpha,\beta}\ln 2}\int_0^{\infty}\ln(1+\gamma)\left(1+\frac{1}{\mho_{\alpha,\beta}}\left(\frac{\bar{\gamma}}{\gamma}\right)^{\beta}\right)^{-(\alpha+1)}\gamma^{-(\beta+1)}\mathrm{d}\gamma \tag{34a}$$

$$= \frac{\alpha\beta\bar{\gamma}^{\beta}}{\Gamma(\alpha+1)\mho_{\alpha,\beta}\ln 2}\int_0^{\infty}\gamma^{-\beta-1}\mathrm{H}_{1,1}^{1,1}\left(\frac{\mho_{\alpha,\beta}}{\bar{\gamma}^{\beta}}\gamma^{\beta}\middle|\begin{matrix}(1,1)\\(\alpha+1,1)\end{matrix}\right)\mathrm{H}_{2,2}^{1,2}\left(\gamma\middle|\begin{matrix}(1,1),(1,1)\\(1,1),(0,1)\end{matrix}\right)\mathrm{d}\gamma, \tag{34b}$$

where (12) was considered for the power multiplier, and it was noted that

$$\ln(1+\gamma) = \mathrm{H}_{2,2}^{1,2}\left(\gamma\middle|\begin{matrix}(1,1),(1,1)\\(1,1),(0,1)\end{matrix}\right). \tag{35}$$

Applying (1.59) from [51] and (2.25.1.1) from [52] after some simplifications, the expression for the IPL channel capacity can be obtained in the following form:

$$\overline{C} = \frac{\alpha\beta}{\Gamma(\alpha+1)\ln 2}\int_0^{\infty} H_{1,1}^{1,1}\left(\frac{\mho_{\alpha,\beta}}{\bar{\gamma}^{\beta}}\gamma^{\beta}\,\middle|\,\begin{matrix}(0,1)\\(\alpha,1)\end{matrix}\right) H_{2,2}^{1,2}\left(\gamma\,\middle|\,\begin{matrix}(0,1),(0,1)\\(0,1),(-1,1)\end{matrix}\right)\mathrm{d}\gamma \tag{36a}$$

$$= \frac{\alpha\beta}{\Gamma(\alpha+1)\ln 2} H_{3,3}^{3,2}\left(\frac{\mho_{\alpha,\beta}}{\bar{\gamma}^{\beta}}\,\middle|\,\begin{matrix}(0,\beta),(0,1),(1,\beta)\\(\alpha,1),(0,\beta),(0,\beta)\end{matrix}\right). \tag{36b}$$

Although (36b) delivers the closed-form result, it can be further simplified. For this, one can recall the integral representation of $H_{3,3}^{3,2}(\cdot)$ in (36b):

$$H_{3,3}^{3,2}\left(\frac{\mho_{\alpha,\beta}}{\bar{\gamma}^{\beta}}\,\middle|\,\begin{matrix}(0,\beta),(0,1),(1,\beta)\\(\alpha,1),(0,\beta),(0,\beta)\end{matrix}\right)=$$

$$= \frac{1}{2\pi i}\int_{\frac{1}{2\beta}-i\infty}^{\frac{1}{2\beta}+i\infty} \frac{\Gamma(1-\beta\sigma)\Gamma(1-\sigma)\Gamma(\alpha+\sigma)\Gamma(\beta\sigma)\Gamma(\beta\sigma)}{\Gamma(1+\beta\sigma)}\left(\frac{\mho_{\alpha,\beta}}{\bar{\gamma}^{\beta}}\right)^{-\sigma}\mathrm{d}\sigma \tag{37a}$$

$$= \frac{1}{\beta}\frac{1}{2\pi i}\int_{\frac{1}{2\beta}-i\infty}^{\frac{1}{2\beta}+i\infty} \frac{(-\pi)}{\sin(\pi\sigma\beta)}\Gamma(-\sigma)\Gamma(\alpha+\sigma)\left(\frac{\mho_{\alpha,\beta}}{\bar{\gamma}^{\beta}}\right)^{-\sigma}\mathrm{d}\sigma \tag{37b}$$

$$= \frac{1}{\beta}\frac{1}{2\pi i}\int_{\frac{1}{2\beta}-i\infty}^{\frac{1}{2\beta}+i\infty} \Gamma(-\beta\sigma)\Gamma(-\sigma)\Gamma(\alpha+\sigma)\Gamma(1+\beta\sigma)\left(\frac{\mho_{\alpha,\beta}}{\bar{\gamma}^{\beta}}\right)^{-\sigma}\mathrm{d}\sigma \tag{37c}$$

$$= \frac{1}{\beta} H_{2,2}^{2,2}\left(\frac{\mho_{\alpha,\beta}}{\bar{\gamma}^{\beta}}\,\middle|\,\begin{matrix}(1,1),(1,\beta)\\(\alpha,1),(1,\beta)\end{matrix}\right), \tag{37d}$$

where it was noted that

$$\frac{\Gamma(1-\beta\sigma)\Gamma(1-\sigma)\Gamma(\beta\sigma)\Gamma(\beta\sigma)}{\Gamma(1+\beta\sigma)} = \frac{1}{\beta}\frac{(-\pi)}{\sin(\pi\sigma\beta)}\Gamma(-\sigma), \tag{38}$$

$$\frac{(-\pi)}{\sin(\pi\sigma\beta)} = \Gamma(-\beta\sigma)\Gamma(1+\beta\sigma). \tag{39}$$

After some simplifications, the final closed-form expression for the ergodic capacity of the IPL channel model thus will be given by

$$\overline{C} = \frac{1}{\Gamma(\alpha)\ln 2} H_{2,2}^{2,2}\left(\frac{\mho_{\alpha,\beta}}{\bar{\gamma}^{\beta}}\,\middle|\,\begin{matrix}(1,1),(1,\beta)\\(\alpha,1),(1,\beta)\end{matrix}\right). \tag{40}$$

To derive the high-SNR asymptotics for the IPL channel capacity, one must first note that the asymptotic representation of the Fox H-function in (40) will not be useful because

$$H_{2,2}^{2,2}\left(\frac{\mho_{\alpha,\beta}}{\bar{\gamma}^{\beta}}\,\middle|\,\begin{matrix}(1,1),(1,\beta)\\(\alpha,1),(1,\beta)\end{matrix}\right)\Bigg|_{\bar{\gamma}\to\infty} \sim O(1), \tag{41}$$

which means that the obtained result does not depend on $\bar{\gamma}$.

To overcome that issue, one can resort to the well-known relation (see, for example, [54]):

$$\overline{C}\,|_{\bar{\gamma}\to\infty} \simeq \log_2\bar{\gamma} + \frac{1}{\ln 2}\frac{\mathrm{d}}{\mathrm{d}r}\left(\frac{\mathbb{E}\{\gamma^r\}}{\bar{\gamma}^r}\right)\Bigg|_{r=0}. \tag{42}$$

Since the raw moments were derived in the proof of Lemma 1,

$$\frac{\mathbb{E}\{\gamma^r\}}{\bar{\gamma}^r} = \frac{\Gamma\left(1-\frac{r}{\beta}\right)\Gamma\left(\alpha+\frac{r}{\beta}\right)}{\Gamma(\alpha)\mho_{\alpha,\beta}^{\frac{r}{\beta}}}. \tag{43}$$

Finally, differentiating (43) and taking a limit as $r \to 0$ yields

$$\overline{C}\,|_{\bar{\gamma}\to\infty} \backsimeq \log_2 \bar{\gamma} + \frac{1}{\beta \ln 2}\Big(G_e + \psi^{(0)}(\alpha) - \ln \mho_{\alpha,\beta}\Big). \tag{44}$$

□

3.5. Hyper-Rayleigh Analysis

Hyper-Rayleigh fading is identified in terms of the amount of fading, outage probability (OP), and ergodic capacity (EC) [54]. If one of these characteristics is worse than in the Rayleigh case, the channel is said to be in the weak hyper-Rayleigh regime (HRR); if two, in the strong HRR; and if all factors are below the corresponding levels for the Rayleigh case, in the full HRR.

It is worth mentioning that these regimes are identified asymptotically. It is well known that for the classical Rayleigh channel, the assumed metrics are equal to

$$\begin{cases} \mathrm{AoF}^{Ray} = 1, \\ \mathrm{P}_{\mathrm{out}}^{Ray}\big|_{\bar{\gamma}\to\infty} \backsimeq \dfrac{\gamma_{th}}{\bar{\gamma}}, \\ \overline{C}^{Ray}\,\big|_{\bar{\gamma}\to\infty} \backsimeq \log_2 \bar{\gamma} + \dfrac{G_e}{\ln 2}. \end{cases} \tag{45}$$

Thus, one can state the following:

Lemma 2. *The IPL channel model exhibits hyper-Rayleigh behavior in the following ways:*

- *In the OP sense if $\alpha\beta < 1$ or $\alpha\beta = 1$ and $\alpha < \frac{1}{2}$;*
- *In the AoF sense if $\alpha\beta = 1$ and $\alpha < 0.316$; otherwise, see Section 4.4;*
- *In the EC sense if $\alpha\beta = 1$ and $\alpha < 0.423$; otherwise, see Section 4.4.*

Thus, in the case $\alpha\beta = 1$, the IPL model is in the full HRR if $\alpha < 0.316$, in the strong HRR if $0.316 < \alpha < 0.423$, in the weak HRR if $0.423 < \alpha < \frac{1}{2}$, and does not experience the HRR if $\alpha > \frac{1}{2}$.

Proof. From the definition given in [54], it is clear that the IPL channel is said to be in the hyper-Rayleigh regime (in the AoF sense, OP sense, or EC sense) if the following corresponding conditions are satisfied:

$$\mathrm{AoF}^{Ray} < \mathrm{AoF}^{IPL}, \tag{46a}$$

$$\mathrm{P}_{\mathrm{out}}^{Ray}(\gamma_{th})\big|_{\bar{\gamma}\to\infty} < \mathrm{P}_{\mathrm{out}}^{IPL}(\gamma_{th})\big|_{\bar{\gamma}\to\infty}, \tag{46b}$$

$$\overline{C}^{Ray}\,\big|_{\bar{\gamma}\to\infty} > \overline{C}^{IPL}\,\big|_{\bar{\gamma}\to\infty}. \tag{46c}$$

This is equivalent to (recollecting the results from Theorems 2, 3, and 5):

$$\Gamma(\alpha)\frac{\Gamma\Big(1-\frac{2}{\beta}\Big)\Gamma\Big(\alpha+\frac{2}{\beta}\Big)}{\Big(\Gamma\Big(1-\frac{1}{\beta}\Big)\Gamma\Big(\alpha+\frac{1}{\beta}\Big)\Big)^2} > 2, \tag{47a}$$

$$(\mho_{\alpha,\beta})^{\alpha}\left(\frac{\gamma_{th}}{\bar{\gamma}}\right)^{\alpha\beta-1} < 1, \tag{47b}$$

$$\frac{1}{\beta \ln 2}\Big(G_e(1-\beta) + \psi^{(0)}(\alpha) - \ln \mho_{\alpha,\beta}\Big) < 0. \tag{47c}$$

Starting from (47b) and noting that if $\alpha\beta \geq 1$ for a sufficiently high SNR (i.e., the HRRs are defined asymptotically for $\lim_{\bar{\gamma}\to\infty}(\cdot)$), the inequality holds true, i.e., the channel is in the HRR (in the OP sense). Thus, if $\alpha\beta < 1$, the IPL model experiences at least the weak HRR.

If $\alpha\beta = 1$, the HRR requires $(\mho_{\alpha,\beta})^{\alpha}$ (which is called "power offset" in [54]) to be less than one. Assuming the definition of $\mho_{\alpha,\beta}$ (see (6)) and substituting $\beta = \frac{1}{\alpha}$ after some simplifications yields

$$\left(\mho_{\alpha,\frac{1}{\alpha}}\right)^{\alpha} = \frac{\Gamma(1-\alpha)\Gamma(2\alpha)}{\Gamma(\alpha)} < 1. \tag{48}$$

Noting that $\beta > 1 \Rightarrow \alpha < 1$ (since the definition does not require the second-order moment), the range of α satisfying (48) is $\alpha \in \left(0, \frac{1}{2}\right)$.

Concerning (47a) and (47c), in the case of $\beta = \frac{1}{\alpha}$, one obtains

$$\Gamma(\alpha)\frac{\Gamma(1-2\alpha)\Gamma(3\alpha)}{(\Gamma(1-\alpha)\Gamma(2\alpha))^2}>2, \tag{49a}$$

$$G_e(\alpha-1) + \ln\left(\frac{\Gamma(1-\alpha)\Gamma(2\alpha)}{\Gamma(\alpha)}\right) - \alpha\psi^{(0)}(\alpha)>0. \tag{49b}$$

Although the expressions cannot be solved in closed form, a numerical solution is possible and yields the ranges $\alpha \in (0, 0.316)$ for (49a) and $\alpha \in (0, 0.423)$ for (49b), and thus Lemma 2 holds.

Otherwise, if $\alpha\beta \neq 1$, no closed-form or numerical solution connecting α and β exists. In this case, the HRRs must be analyzed numerically for the whole range of parameter values. The detailed numerical analysis is given in Section 4.4. □

4. Simulation and Results' Analysis

To support the analytical derivations and verify the results, simulations were performed. The results evaluated using the obtained exact expressions (see Lemmas 1 and 2 and Theorems 1–5) and plotted with solid lines in Figures 1–8 were tested against those obtained via a Monte Carlo simulation with 10^8 samples (plotted with markers) and supplemented, where applicable, with high-SNR asymptotics (plotted with dashed blue curves).

The channel parameters ($\bar{\gamma}$, α, and β) were chosen to cover all possible scenarios, including heavy and light fading. It must be pointed out that for all results, $\alpha > 0$, since no restrictions are imposed on it. Therefore, in most plots, its lowest value was chosen to be 0.1, except for the hyper-Rayleigh analysis, where the smallest value of α was 0.01. Conversely, the lowest value of β was chosen differently depending on the analyzed characteristics: if its evaluation required the second-order moment, β was greater than two (e.g., the AoF, ABER, EC, and HRR); otherwise, $\beta > 1$ (e.g., the PDF, CDF, and MGF). As mentioned earlier, since the IPL model (as well as the baseline Lomax model) belongs to the heavy-tailed distributions and exhibits various HHRs, the simulation specifically focused on the lower values of the parameters.

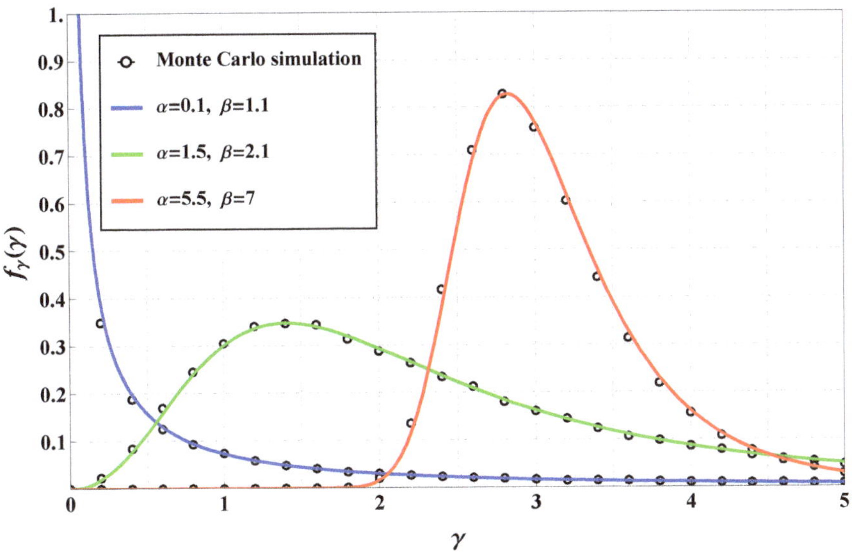

Figure 1. PDF comparison of numerical evaluation and numerical simulation for various channel parameters.

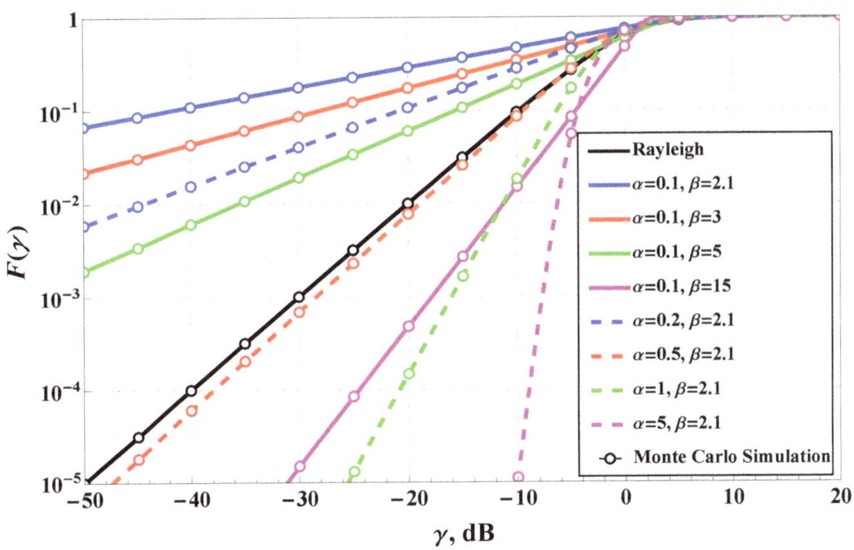

Figure 2. CDF comparison for various channel parameters.

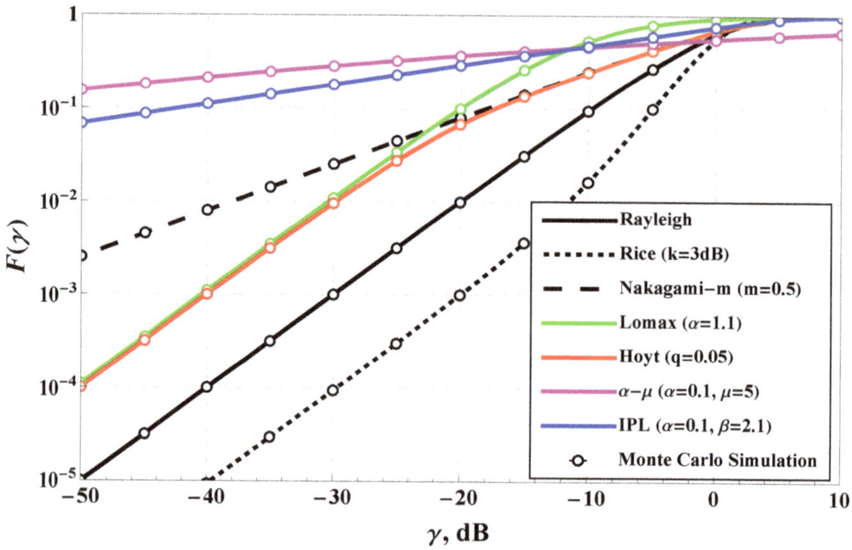

Figure 3. IPL positioning among various fading channel models.

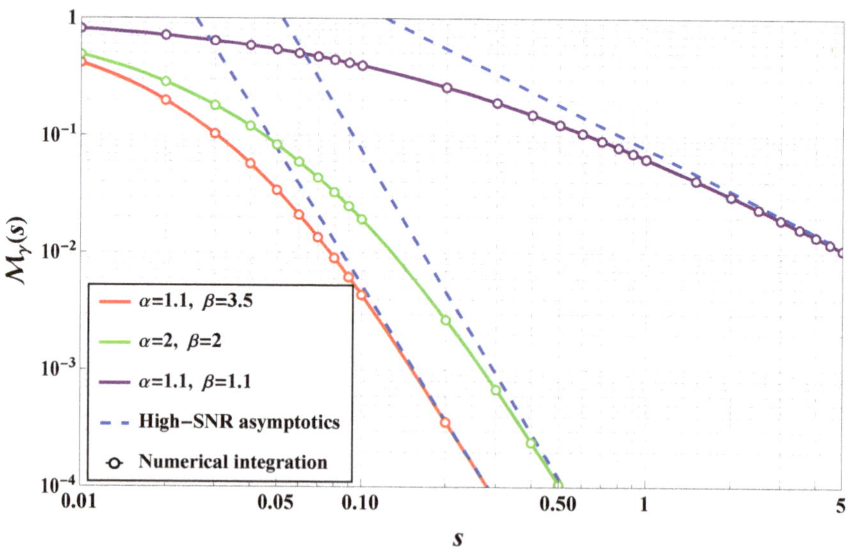

Figure 4. MGF comparison of numerical evaluation and derived closed-form expression for various channel parameters.

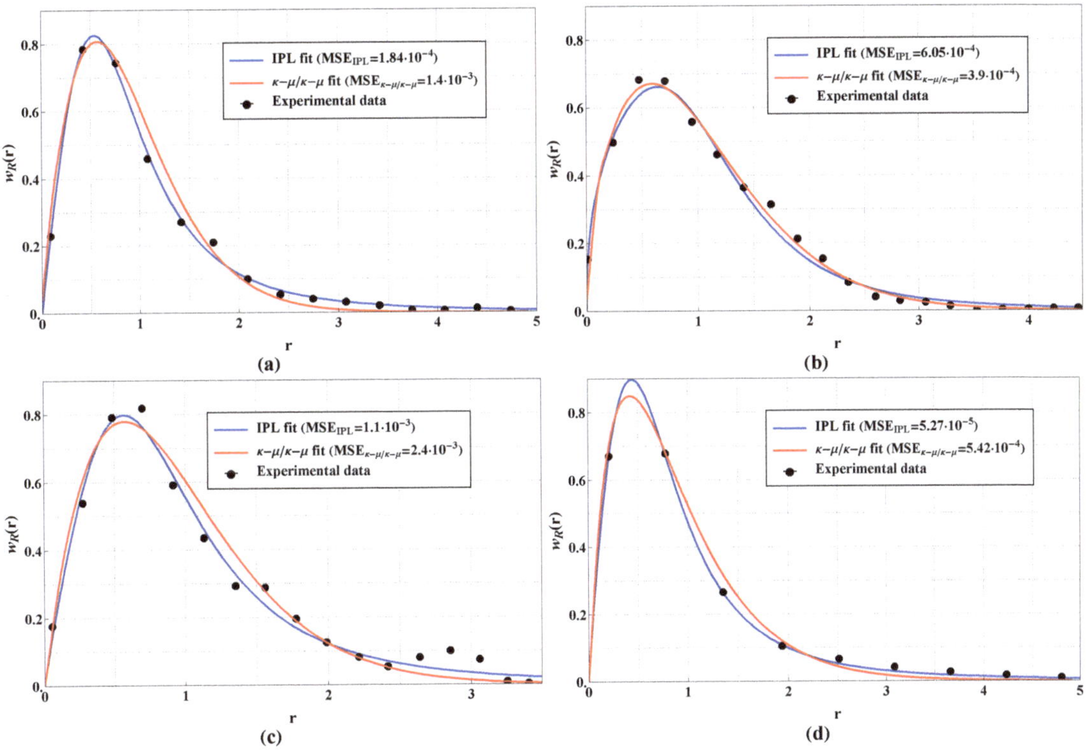

Figure 5. Empirical data from D2D communication experiment with fitted IPL model and $\kappa-\mu/\kappa-\mu$ model: (**a**) LOS indoor environment (see [56] Figure 12a), (**b**) NLOS indoor environment (see [56] Figure 12b), (**c**) LOS outdoor environment (see [56] Figure 12c), and (**d**) NLOS outdoor environment (see [56] Figure 12d).

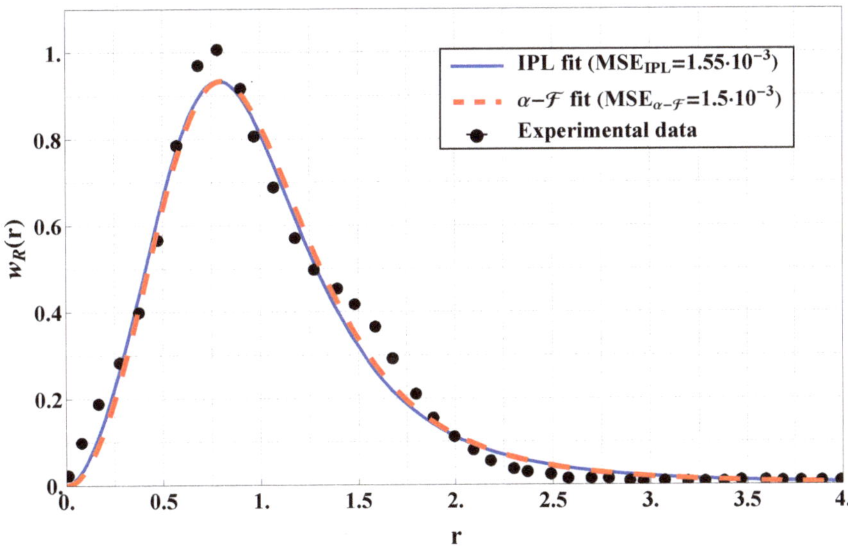

Figure 6. Empirical data from D2D communication experiment [57] with fitted IPL model and $\alpha-\mathcal{F}$ model (see [42]).

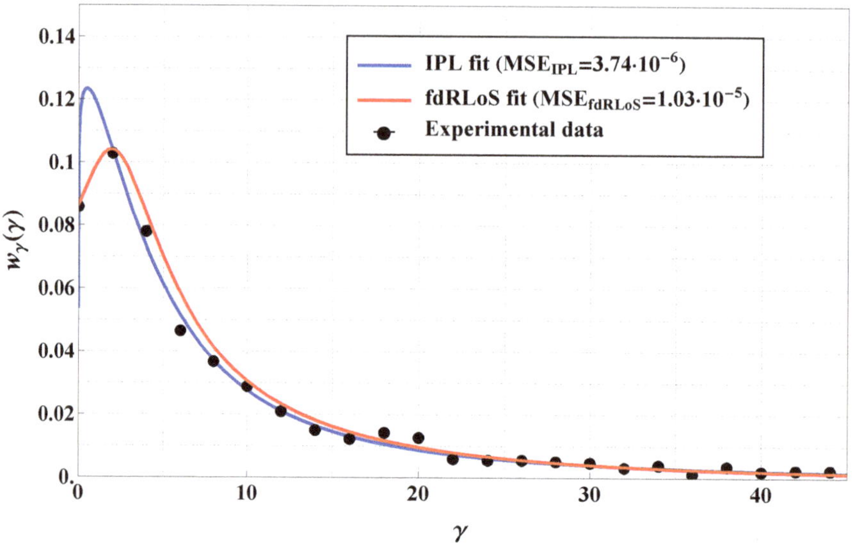

Figure 7. Experimental results from [58] fitted with the IPL and fdRLoS [59] models.

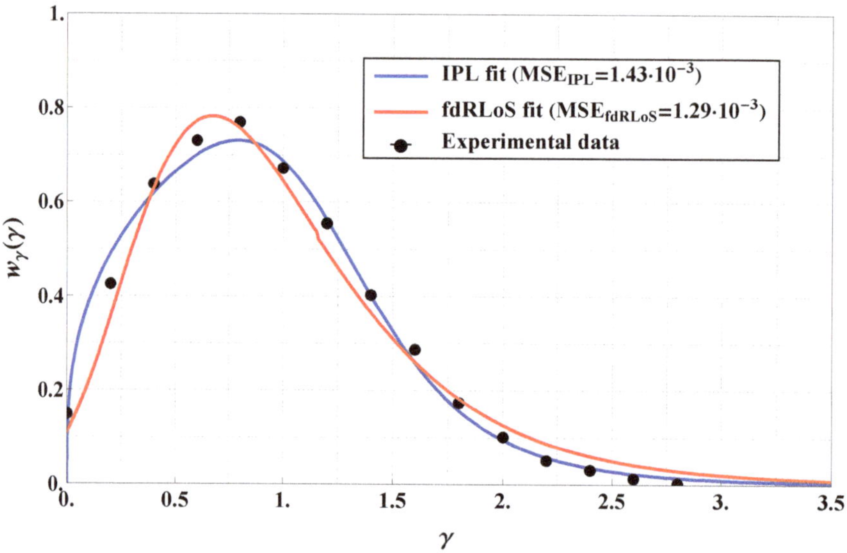

Figure 8. Experimental results from [60] fitted with the IPL and fdRLoS [59] models.

4.1. Model Verification and Positioning

First, to validate the correctness of the model, the simulated instantaneous SNR samples generated for specific channel parameters were compared with the theoretical PDF for the same parameter values. The results are presented in Figure 1 for three sets of α, β: $\alpha = 0.1, \beta = 1.1$; $\alpha = 1.5, \beta = 2.1$; and $\alpha = 5.5, \beta = 7$.

The plots show how the PDF varies with different channel conditions and demonstrate that a decrease in the channel parameters leads to a shift in the distribution mode toward the origin. This means that the smaller the values of (α, β), the heavier the fading (i.e., the more probable extremely small SNR values are). It must be highlighted that the Monte Carlo simulation data correspond closely with the theoretical PDF.

To better understand the impact of channel parameters, the CDF was plotted for various sets of (α, β) (see Figure 2), including the following:

- The increase in the shape parameter $\beta = (2.1; 3; 5; 15)$ for a fixed and small power-transformation parameter (i.e., $\alpha = 0.1$);
- The increase in the power-transformation parameter $\alpha = (0.1; 0.2; 0.5; 1; 5)$ for a fixed-shape parameter (i.e., $\beta = 2.1$).

To provide a baseline for comparison, Figure 2 includes the Rayleigh CDF.

It is clear that the proposed model can handle fading that is heavier (even much heavier) than Rayleigh (i.e., a small α and up to a moderate β), lighter than Rayleigh (see the curves with larger values of α), as well as cases very close to Rayleigh (see the curve for $\alpha = 0.5$ and $\beta = 2.1$).

As pointed out earlier, there are numerous classical models used in wireless communications to describe fading channels. It is interesting to see how the IPL model is positioned among them.

Figure 3 illustrates the results (in terms of the CDF) for common fading channel models, such as Rayleigh, Rice (with $k = 3\,\text{dB}$), Nakagami-m (with $m = 0.5$), Lomax (with $\alpha = 1.1$), Hoyt (with $q = 0.05$), and $\alpha - \mu$ (with $\alpha = 0.1, \mu = 5$). The parameters for Nakagami, Hoyt, Lomax, and $\alpha - \mu$ were chosen to correspond to heavy fading conditions. It can be seen that the presented IPL model with small parameter values demonstrates extremely deep fading (much heavier than the classical models). Combining the results depicted in Figures 2 and 3, one can state that by choosing a specific α and β, the IPL model can cover the fading regimes achieved with the classical models, demonstrating its high flexibility.

Figure 4 presents a comparison between the MGF obtained through numerical evaluation and the derived closed-form expression for different channel parameters. The parameters include $\alpha = 1.1, \beta = 3.5$; $\alpha = 2, \beta = 2$; and $\alpha = 1.1, \beta = 1.1$. The high-SNR asymptotic expressions are also depicted. The comparison validates the accuracy of the closed-form expression in capturing the behavior of the MGF across various channel conditions.

It must be pointed out that it is quite cumbersome to convert Monte Carlo simulation samples to the sampled MGF. Therefore, a different approach for validating the analytics was chosen: it was tested against the direct numerical integration in (10a).

4.2. Experimental Verification

Although the theoretical comparison of the IPL model with other well-known models demonstrated its high flexibility and usefulness, it does not justify its practical applicability. Therefore, further clarification is needed.

To resolve this issue, one should rely on real-life experimental data and synthetic data generated via certified models. The current research assumes both approaches.

To this extent, four datasets were considered:

1. Experimental data from a 5.8 GHz device-to-device (D2D) communication system with on-body devices forming a body area network (BAN) in indoor and outdoor line-of-sight and non-line-of-sight environments [56].
2. Experimental data from a 868.3 MHz two-user D2D system operating outdoors in a line-of-sight environment, with distances between the users ranging from 1 to 15 m [42,57].
3. Experimental measurements of a 4G LTE network functioning in a densely populated smart city [58].
4. Synthetic computer-generated datasets compliant with the COST 2100 model, employed for a channel multipath clustering assessment [60,61].

The first and second datasets presented experimental data from D2D communication between on-body devices. The assumed use cases involved a BAN with two users in indoor and outdoor environments, and the tests were conducted to assess the impact of the surrounding environment on link quality. In both datasets, the extracted data represented histogram values of the normalized envelope. Therefore, for fitting the IPL model, its SNR statistics (i.e., the PDF) were converted into envelope statistics using the standard random variable transformation technique.

The third dataset focuses on real-world network-performance indicators. Measurements were collected from an accessible 4G LTE network operating at a frequency of 1876.6 MHz with a bandwidth of 10 MHz. The tests involved a moving vehicle traveling up to 30 km/h and three static base stations over a distance of approximately 2 km. For this study, the SINR values were extracted from the site labeled "site 1" and converted from decibels to dimensionless quantities for analysis.

The fourth dataset is based on a simulated semi-urban environment at 5.3 GHz, also with a bandwidth of 10 MHz. This dataset was initially proposed for assessing the multipath clustering effects as outlined in the COST 2100 channel model. The extracted data, representing the relative power of different components from the power-delay profile, were normalized based on the sample variance, producing dimensionless values for further analysis.

For the first two datasets, the resulting data provided the sample PDF values. For the last two datasets, the extracted data were further processed using Wolfram Mathematica by generating the histogram list with fixed bins (obtained using the conventional Sturges binning). The results are presented in Figures 5–8 with black markers. All the data are accompanied by the fitting PDF curves of the corresponding models (i.e., $\kappa - \mu/\kappa - \mu$ for Figure 5, $\alpha - \mathcal{F}$ for Figure 6, and fdRLoS for Figures 7 and 8). The parameters for each channel were chosen in accordance with [42,56,59]. For each of the assumed datasets, the corresponding IPL PDF was obtained by selecting the channel model parameters (i.e., $\bar{\gamma}, \alpha, \beta$) that provided the best fit (in terms of the minimum mean squared error (MSE)). The resulting curves are depicted in Figures 5–8 as blue curves, with the corresponding MSE denoted in the legend.

It can be observed that in most cases, the IPL model produces a better fit (in terms of the MSE) than the best of the models applied to the current data. In cases where the IPL has a greater MSE than that of the initial model, the loss is generally insignificant for practical applications. Moreover, in these cases, it performs better than the other models referenced in [42,56,59] (i.e., only the best-fitting models were considered here). Furthermore, it should be highlighted that the IPL provides a much better fit in the distribution tails.

The analysis demonstrated that the IPL model can be successfully applied in all the considered scenarios. Thus, it supports the flexibility of the model and the versatility of its applicability.

4.3. IPL Model Performance: AoF, OP, ABER, and EC

Made certain of the correctness of the IPL model's statistical characteristics, further analysis focuses on its performance.

Figure 9 shows the relationship between the AoF and the channel parameters α and β. Different sets of parameters are considered, including $\alpha = 0.1, \beta = 2.1$; $\alpha = 0.1, \beta = 3.1$; $\alpha = 0.1, \beta = 4.1$; and $\alpha = 0.1, \beta = 5.1$. The graph demonstrates how the AoF changes with varying β values for a fixed α and vice versa. This analysis helps in understanding the dependence of the AoF on the channel parameters, providing insights into how different fading conditions impact the performance of wireless communication systems. Moreover, one can note that small values of α and β correspond to AoF > 1, which means that the channel exhibits the HHR in the AoF sense. The impact of α and β is unequal: the increase in α (for $\alpha > 1$) does not lead to a sufficient decrease in the AoF (i.e., there exists some threshold value of the AoF), while the increase in β (for $\beta > 3$) leads to an exponential decrease (linear on the logarithmic scale) in the AoF.

The dependence of the outage probability $P_{\text{out}}(\gamma)$ on the average signal-to-noise ratio $\bar{\gamma}$ is illustrated in Figure 10 for various channel parameters. The subplots demonstrate the impact of β for a fixed small $\alpha = 0.1$ (see Figure 10a), the effect of α for a fixed small $\beta = 2.1$ (see Figure 10b), and the impact of the threshold γ_{th} for fixed $\alpha = 0.5$ and $\beta = 2.1$. It can be seen that the derived high-SNR approximation for the OP is extremely accurate for a small power-transformation parameter, irrespective of the value of the shape parameter. Moreover, it is observed that a small α has a greater impact on the OP than β and γ_{th}.

Reasonable values of the OP (e.g., $P_{out} \sim 10^{-4}$ with $\beta = 2.1$) are achieved around 23.6 dB for $\alpha = 0.1$ and 11 dB for $\alpha = 2.1$. If $\alpha = 0.1$ is preserved, the same change in β (i.e., from $\beta = 2.1$ to $\beta = 4.1$) reduces γ (required to achieve $P_{out} \sim 10^{-4}$) by only about 5 dB (from 23.6 dB to 18.7 dB).

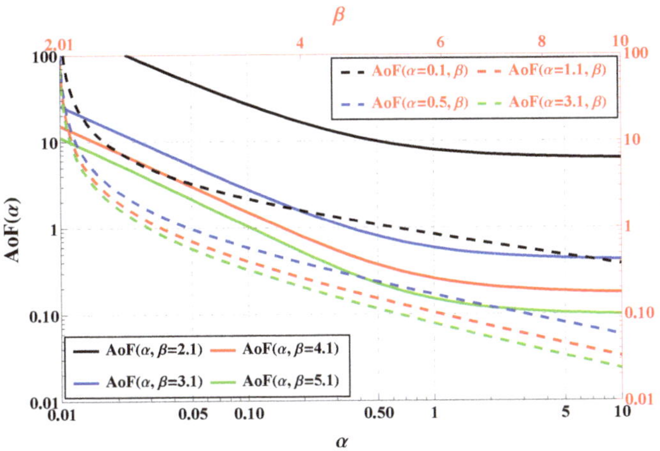

Figure 9. AoF vs. channel parameters (α, β).

Figure 11 displays the average bit error rate (ABER) as a function of $\bar{\gamma}$ for different modulations and channel parameters, including the impact of the constellation size for M-PSK (see Figure 11a) and M-QAM (see Figure 11b) modulations with $\alpha = 1, \beta = 3$, and the effect of fading parameters for QAM-4 (see Figure 11c).

The modulations were chosen to cover cases with both low-dimensional constellations (e.g., BPSK, QPSK, and QAM-4) and high-dimensional constellations (e.g., PSK-8 for PSK modulations, QAM-128, and QAM-1024). It is clear that reasonable ABER values ($\bar{P}_{er} \sim 10^{-6}$) are successfully achieved for a moderate $\bar{\gamma}$ (i.e., $\bar{\gamma} \sim$ 15–30 dB) only for a small-to-moderate M (not higher than 128).

The impact of fading parameters (see Figure 11c) was analyzed for five distinct sets of parameters: either a small α or small β (for a large second parameter), both parameters being either small or large, and an intermediate scenario (with both a moderate α and β). It can be seen that in the case of a small power-transformation parameter value, $\bar{P}_{er} \sim 10^{-6}$ cannot be achieved for any reasonable average SNR; thus, such a scenario (corresponding to deep fading) demonstrates an extremely poor link quality. In contrast, if both parameters are relatively high ($\alpha = 3, \beta = 10$), the required SNR can be as small as 15.5 dB. Again, as demonstrated in the previous plots, the high-SNR approximation becomes less accurate with a higher α.

Figure 12 presents the ergodic capacity C of the IPL channel as a function of the average SNR $\bar{\gamma}$ for different sets of channel parameters α and β. As a reference line, the ergodic capacity for a Rayleigh channel is also presented. It can be noted that small values of α correspond to cases where the capacity is smaller than that of a Rayleigh channel (i.e., the channel is in the HRR in the EC sense). For large values of the power-transformation parameter, the capacity is greater than Rayleigh's, even for a small β, indicating that the impact of α on the EC is more pronounced.

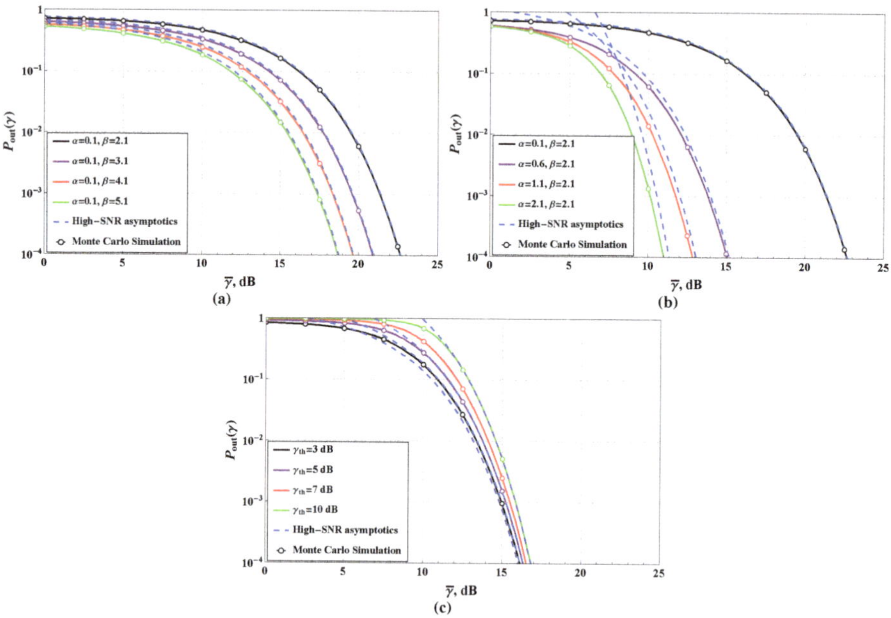

Figure 10. Outage probability vs. $\bar{\gamma}$ for various channel parameters: (**a**) the impact β for fixed small $\alpha = 0.1$, (**b**) the impact α for fixed small $\beta = 2.1$, and (**c**) the impact γ_{th} for fixed $\alpha = 0.5, \beta = 2.1$.

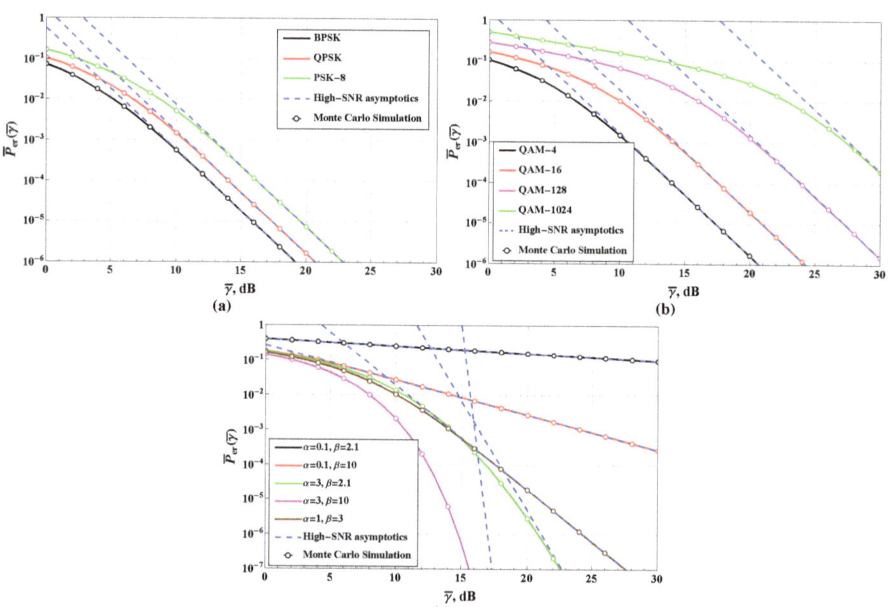

Figure 11. ABER vs. $\bar{\gamma}$ for modulations and channel parameters: (**a**) the impact of constellation size for M-PSK with $\alpha = 1, \beta = 3$, (**b**) the impact of constellation size for M-QAM with $\alpha = 1, \beta = 3$, and (**c**) the impact of fading parameters for QAM-4.

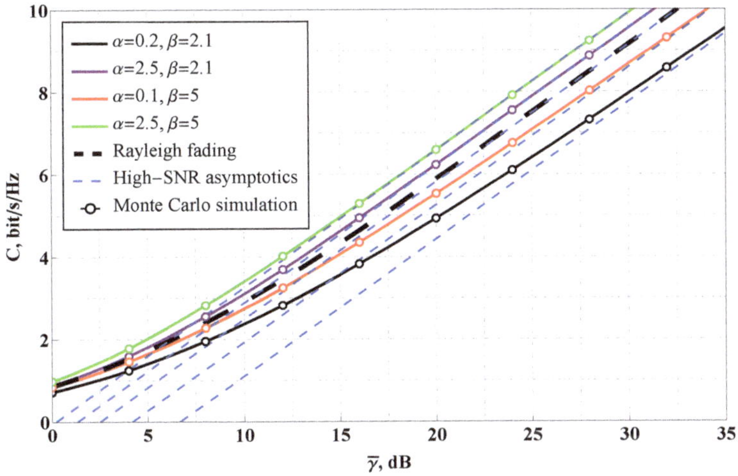

Figure 12. IPL channel ergodic capacity vs. $\bar{\gamma}$ for various channel parameters (α, β).

4.4. Hyper-Rayleigh Analysis

Combining the results of the EC, AoF, and OP analysis, it is clear that there are sets of parameters (α, β) that correspond to the HRR in a specific sense. Lemma 2 presented the regions of (α, β) for full, strong, weak, and no hyper-Rayleigh regime only for the case of $\alpha\beta = 1$. Otherwise, no unique closed-form relation exists; thus, the HRR must be analyzed numerically, and it is convenient to present the results in the form of a contour map as a function of (α, β). The results for the AoF (in logarithmic scale) are presented in Figure 13, and the residual EC (the difference between the EC of the IPL channel and the Rayleigh model) are presented in Figure 14.

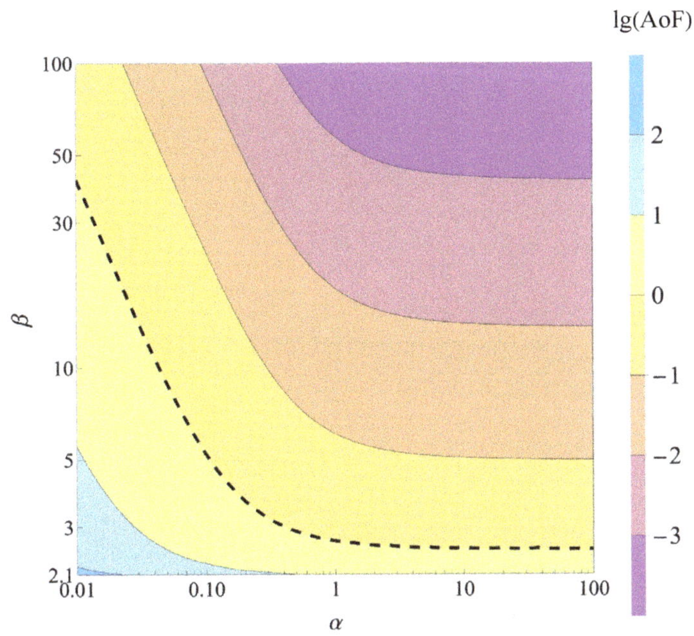

Figure 13. Logarithmically scaled AoF contour map for various channel parameters (α, β); black dashed line corresponds to Rayleigh fading.

It must be pointed out that the results for P_{out} are omitted here since they give trivial conditions (see Lemma 2) that can be expressed analytically in a closed form.

In both plots, the black dashed line corresponds to the Rayleigh fading condition (i.e., $\lg(\text{AoF}) = 0$ and $\Delta\bar{C} = 0$), separating the regions of hyper-Rayleigh fading and lighter-than-Rayleigh fading.

The combined analysis of Figures 13 and 14 demonstrates that the constraint implied by the ergodic capacity is stricter than that of the AoF. Recalling the result of Lemma 2, one can note that the constraint implied by the OP is stricter than by the EC. Thus, it can be finally stated that if the OP of the IPL model is worse than for Rayleigh, the channel experiences the full HRR; if the OP is better than Rayleigh's but the EC is worse, then the IPL model undergoes the strong HRR; if the EC is better than Rayleigh's but the AoF is greater than one, the channel experiences the weak HRR; otherwise, the IPL channel does not demonstrate hyper-Rayleigh behavior.

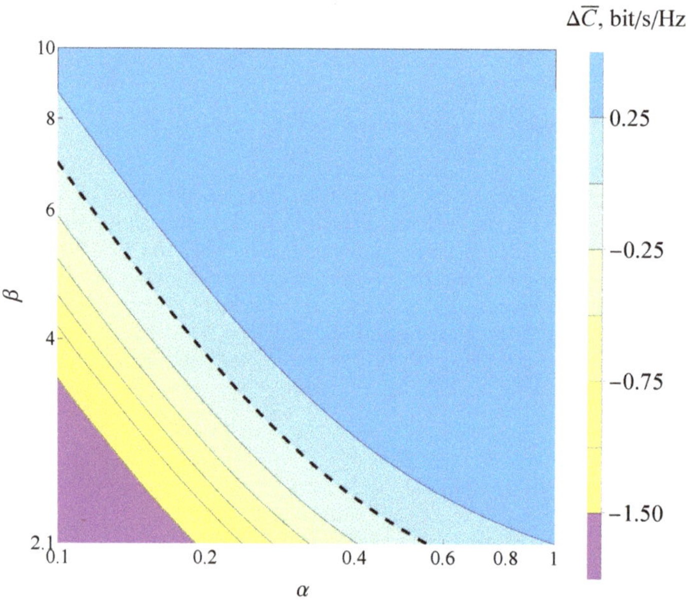

Figure 14. $\Delta\overline{C}$ contour map for various channel parameters (α, β); black dashed line corresponds to $\Delta\overline{C} = 0$, i.e., Rayleigh fading.

5. Conclusions

This research introduces and examines the inverse power Lomax fading channel model for wireless communications, demonstrating its versatility in describing a range of fading scenarios from severe to mild. The IPL model, defined by its probability density function, cumulative distribution function, and moment-generating function, captures the statistical characteristics of fading channels more flexibly than traditional models like Rayleigh, Rice, Nakagami-m, etc. This study provides exact analytical expressions for key performance metrics such as the amount of fading, outage probability, average bit error rate, and ergodic capacity. These metrics are crucial for evaluating the reliability and efficiency of wireless communication systems. The findings reveal that the IPL model can effectively operate under various fading conditions by adjusting its parameters, thereby encompassing the behavior of several well-known models. Additionally, this research confirms the accuracy of the derived expressions through extensive numerical analysis and statistical simulations, underscoring the practical applicability of the IPL model in real-world communication systems.

Funding: This work was supported by the Russian Science Foundation under grant 24-29-00516 (https://rscf.ru/en/project/24-29-00516/, accessed date: 7 August 2024).

Data Availability Statement: The original contributions presented in the study are included in the article.

Conflicts of Interest: The authors declare no conflicts of interest. The funders had no role in the design of the study; in the collection, analyses, or interpretation of data; in the writing of the manuscript; or in the decision to publish the results.

References

1. Paris, J.F. Advances in the Statistical Characterization of Fading: From 2005 to Present. *Int. J. Antennas Propag.* **2014**, *2014*, 258308. [CrossRef]
2. Shankar, P.M. *Fading and Shadowing in Wireless Systems;* Springer International Publishing: Berlin/Heidelberg, Germany, 2017. [CrossRef]
3. Khawaja, W.; Guvenc, I.; Matolak, D.W.; Fiebig, U.C.; Schneckenburger, N. A Survey of Air-to-Ground Propagation Channel Modeling for Unmanned Aerial Vehicles. *IEEE Commun. Surv. Tutorials* **2019**, *21*, 2361–2391. [CrossRef]
4. Yadav, P.; Kumar, S.; Kumar, R. A comprehensive survey of physical layer security over fading channels: Classifications, applications, and challenges. *Trans. Emerg. Telecommun. Technol.* **2021**, *32*, e4270. [CrossRef]
5. Liu, S.; Yu, X.; Guo, R.; Tang, Y.; Zhao, Z. THz channel modeling: Consolidating the road to THz communications. *China Commun.* **2021**, *18*, 33–49. [CrossRef]
6. Serghiou, D.; Khalily, M.; Brown, T.W.C.; Tafazolli, R. Terahertz Channel Propagation Phenomena, Measurement Techniques and Modeling for 6G Wireless Communication Applications: A Survey, Open Challenges and Future Research Directions. *IEEE Commun. Surv. Tutorials* **2022**, *24*, 1957–1996. [CrossRef]
7. Launay, F. *NG-RAN and 5G-NR: 5G Radio Access Network and Radio Interface*, 1st ed.; Wiley-ISTE: Hoboken, NJ, USA, 2021.
8. Jiang, W.; Luo, F.L. *6G Key Technologies: A Comprehensive Guide*; Wiley-IEEE Press: New York, NY, USA, 2022.
9. Alalewi, A.; Dayoub, I.; Cherkaoui, S. On 5G-V2X Use Cases and Enabling Technologies: A Comprehensive Survey. *IEEE Access* **2021**, *9*, 107710–107737. [CrossRef]
10. Chen, C.; Chen, X.; Das, D.; Akhmetov, D.; Cordeiro, C. Overview and Performance Evaluation of Wi-Fi 7. *IEEE Commun. Stand. Mag.* **2022**, *6*, 12–18. [CrossRef]
11. Reshef, E.; Cordeiro, C. Future Directions for Wi-Fi 8 and Beyond. *IEEE Commun. Mag.* **2022**, *60*, 50–55. [CrossRef]
12. Mao, K.; Zhu, Q.; Qiu, Y.; Liu, X.; Song, M.; Fan, W.; Kokkeler, A.B.J.; Miao, Y. A UAV-Aided Real-Time Channel Sounder for Highly Dynamic Nonstationary A2G Scenarios. *IEEE Trans. Instrum. Meas.* **2023**, *72*, 1–15. [CrossRef]
13. Lyu, Y.; Wang, W.; Sun, Y.; Rashdan, I. Measurement-based fading characteristics analysis and modeling of UAV to vehicles channel. *Veh. Commun.* **2024**, *45*, 100707. [CrossRef]
14. Schmieder, M.; Eichler, T.; Wittig, S.; Peter, M.; Keusgen, W. Measurement and Characterization of an Indoor Industrial Environment at 3.7 and 28 GHz. In Proceedings of the 2020 14th European Conference on Antennas and Propagation (EuCAP), Copenhagen, Denmark, 15–20 March 2020; pp. 1–5. [CrossRef]
15. Musthafa, A.M.; Luini, L.; Riva, C.; Livieratos, S.N.; Roveda, G. A long-term experimental investigation on the impact of rainfall on short 6G D-band links. *Radio Sci.* **2023**, *58*, 1–10. [CrossRef]
16. Zhang, J.; Lin, J.; Tang, P.; Zhang, Y.; Xu, H.; Gao, T.; Miao, H.; Chai, Z.; Zhou, Z.; Li, Y.; et al. Channel Measurement, Modeling, and Simulation for 6G: A Survey and Tutorial. *arXiv* **2023**, arXiv:2305.16616. [CrossRef]
17. Qasem, N. Measurement and Simulation for Improving Indoor Wireless Communication System Performance at 2.4 GHz by Modifying the Environment. *IEEE Access* **2024**, *12*, 96660–96671. [CrossRef]
18. Huang, Y.; Xin, W.; Wang, X.; Wang, T.; Shi, Y.; Jiang, Z. A Comparative Study on Urban Micro-Cell Channel Measurements at 4.9 and 28 GHz. In Proceedings of the 2024 IEEE Wireless Communications and Networking Conference (WCNC), Dubai, United Arab Emirates, 21–24 April 2024; pp. 1–6. [CrossRef]
19. Ju, S.; Shakya, D.; Poddar, H.; Xing, Y.; Kanhere, O.; Rappaport, T.S. 142 GHz Sub-Terahertz Radio Propagation Measurements and Channel Characterization in Factory Buildings. *IEEE Trans. Wirel. Commun.* **2024**, *23*, 7127–7143. [CrossRef]
20. Fraidenraich, G.; Yacoub, M.D. The α-η-μ and α-κ-μ Fading Distributions. In Proceedings of the 2006 IEEE Ninth International Symposium on Spread Spectrum Techniques and Applications, Manaus, Brazil, 28–31 August 2006; pp. 16–20. [CrossRef]
21. Yacoub, M.D. The κ-μ distribution and the η-μ distribution. *IEEE Antennas Propag. Mag.* **2007**, *49*, 68–81. [CrossRef]
22. Yacoub, M.D. The α-μ Distribution: A Physical Fading Model for the Stacy Distribution. *IEEE Trans. Veh. Technol.* **2007**, *56*, 27–34. [CrossRef]
23. Rabelo, G.S.; Yacoub, M.D. The κ-μ Extreme Distribution. *IEEE Trans. Commun.* **2011**, *59*, 2776–2785. [CrossRef]
24. Paris, J.F. Statistical Characterization of κ-μ Shadowed Fading. *IEEE Trans. Veh. Technol.* **2014**, *63*, 518–526. [CrossRef]
25. Yacoub, M.D. The $\alpha - \eta - \kappa - \mu$ Fading Model. *IEEE Trans. Antennas Propag.* **2016**, *64*, 3597–3610. [CrossRef]
26. Sánchez, I.; López-Martínez, F.J. The Lomax Distribution for Wireless Channel Modeling: Theory and Applications. *IEEE Open J. Veh. Technol.* **2024**, *5*, 162–171. [CrossRef]
27. Badarneh, O.S.; Costa, D.B.d. Fluctuating Nakagami-m Fading Distribution. *IEEE Wirel. Commun. Lett.* **2024**, *13*, 959–963. [CrossRef]
28. Hassan, A.S.; Abd-Allah, M. On the Inverse Power Lomax Distribution. *Ann. Data Sci.* **2018**, *6*, 259–278. [CrossRef]
29. El-Sherpieny, E.S.A.; Almetwally, E.M.; Muhammed, H.Z. Progressive Type-II hybrid censored schemes based on maximum product spacing with application to Power Lomax distribution. *Phys. Stat. Mech. Its Appl.* **2020**, *553*, 124251. [CrossRef]

30. Shi, X.; Shi, Y. Inference for Inverse Power Lomax Distribution with Progressive First-Failure Censoring. *Entropy* **2021**, *23*, 1099. [CrossRef] [PubMed]
31. Hassan, A.S.; Al-Omar, A.I.; Ismail, D.M.; Al-Anzi, A. A new generalization of the inverse Lomax distribution with statistical properties and applications. *Int. J. Adv. Appl. Sci.* **2021**, *8*, 89–97. [CrossRef]
32. Bhattacharjee, S.; Giri, R.L.; Szymkowiak, M. Some results on characterization of distributions in reliability analysis. *Commun.-Stat.-Theory Methods* **2024**, 1–12. [CrossRef]
33. Akueson, G.A.H.; Bogninou, M.J.; Akossou, A.Y.J. Modified Alpha Power Transformed Inverse Power Lomax Distribution with appllications. **2024** . [CrossRef]
34. ZeinEldin, R.A.; ul Haq, M.A.; Hashmi, S.; Elsehety, M. Alpha Power Transformed Inverse Lomax Distribution with Different Methods of Estimation and Applications. *Complexity* **2020**, *2020*, 1–15. [CrossRef]
35. Ogunde, A.A.; Oseghale, I.O.; Nwanyibuife, O.B. The Type II Topp-Leone Inverse Power Lomax distribution with Simulation and Applications. *J. Mod. Appl. Stat. Methods* **2023**, *23, 1–28* . [CrossRef]
36. Lomax, K.S. Business Failures: Another Example of the Analysis of Failure Data. *J. Am. Stat. Assoc.* **1954**, *49*, 847–852. [CrossRef]
37. Lopez-Martinez, F.J.; Morales-Jimenez, D.; Martos-Naya, E.; Paris, J.F. On the Bivariate Nakagami-m Cumulative Distribution Function: Closed-Form Expression and Applications. *IEEE Trans. Commun.* **2013**, *61*, 1404–1414. [CrossRef]
38. Gvozdarev, A.S.; Artemova, T.K.; Manahov, R.; Mozzhukhin, S.; Veselkov, A. Energy-Based Spectrum Sensing Over the Severe Fading Channels. In Proceedings of the 2024 Systems of Signals Generating and Processing in the Field of on Board Communications, Moscow, Russian, 12–14 March 2024; pp. 1–5. [CrossRef]
39. Gvozdarev, A.S.; Artemova, T.K. Closed-form energy-based signal detection analysis in presence of a Lomax fading channel in full hyper-Rayleigh regime. *Phys. Commun.* **2024**, *66*, 102473. [CrossRef]
40. Ramírez-Espinosa, P.; López-Martínez, F.J. Composite Fading Models Based on Inverse Gamma Shadowing: Theory and Validation. *IEEE Trans. Wirel. Commun.* **2021**, *20*, 5034–5045. [CrossRef]
41. Agrawal, R.; Karmeshu; Sehgal, A. Composite Channel Model for Wireless Propagation with Wide-Range Signal Variation Using Rayleigh–Generalized Inverse Gaussian Distribution. *Iran. J. Sci. Technol. Trans. Electr. Eng.* **2021**, *46*, 213–223. [CrossRef]
42. Badarneh, O.S. The α-$\mathcal{F}$ Composite Fading Distribution: Statistical Characterization and Applications. *IEEE Trans. Veh. Technol.* **2020**, *69*, 8097–8106. [CrossRef]
43. Badarneh, O.S. The $\alpha - \eta - \mathcal{F}$ and $\alpha - \kappa - \mathcal{F}$ Composite Fading Distributions. *IEEE Commun. Lett.* **2020**, *24*, 1924–1928. [CrossRef]
44. Badarneh, O.S.; Muhaidat, S.; da Costa, D.B. The $\alpha - \eta - \kappa - \mathcal{F}$ Composite Fading Distribution. *IEEE Wirel. Commun. Lett.* **2020**, *9*, 2182–2186. [CrossRef]
45. Gvozdarev, A.S. Closed-Form Analysis of the α-Beaulieu-Xie Shadowed Fading Channel. *IEEE Wirel. Commun. Lett.* **2023**, *12*, 1623–1627. [CrossRef]
46. Olver, F.W.J. *NIST Handbook of Mathematical Functions*; Cambridge University Press: Cambridge, UK, 2010.
47. Simon, M.K.; Alouini, M.S. *Digital Communication over Fading Channels*; John Wiley & Sons, Inc.: Hoboken, NJ, USA, 2005. [CrossRef]
48. Atapattu, S.; Tellambura, C.; Jiang, H. MGF Based Analysis of Area under the ROC Curve in Energy Detection. *IEEE Commun. Lett.* **2011**, *15*, 1301–1303. [CrossRef]
49. Gvozdarev, A.S. The Generalized MGF Approach to Closed-Form Average Symbol Error Rate Calculation. *IEEE Commun. Lett.* **2021**, *25*, 1124–1128. [CrossRef]
50. Gvozdarev, A.S. A novel unified framework for energy-based spectrum sensing analysis in the presence of fading. *Sensors* **2022**, *22*, 1742. [CrossRef]
51. Mathai, A.; Saxena, R.; Haubold, H. *The H-Function: Theory and Applications*; Springer: New York, NY, USA, 2009.
52. Prudnikov, A.P.; Brychkov, Y.A.; Marichev, O.I. *Integrals and Series: More Special Functions*; Gordon and Breach Science Publishers: New York, NY, USA, 1990; Volume 3, p. 800.
53. Chauhan, P.S.; Kumar, S.; Jain, A.; Hanzo, L. An Asymptotic Framework for Fox's H-Fading Channel With Application to Diversity-Combining Receivers. *IEEE Open J. Veh. Technol.* **2023**, *4*, 404–416. [CrossRef]
54. Garcia-Corrales, C.; Fernandez-Plazaola, U.; Canete, F.J.; Paris, J.F.; Lopez-Martinez, F.J. Unveiling the Hyper-Rayleigh Regime of the Fluctuating Two-Ray Fading Model. *IEEE Access* **2019**, *7*, 75367–75377. [CrossRef]
55. Wang, Z.; Giannakis, G. A simple and general parameterization quantifying performance in fading channels. *IEEE Trans. Commun.* **2003**, *51*, 1389–1398. [CrossRef]
56. Bhargav, N.; da Silva, C.R.N.; Chun, Y.J.; Leonardo, E.J.; Cotton, S.L.; Yacoub, M.D. On the Product of Two $\kappa - \mu$ Random Variables and its Application to Double and Composite Fading Channels. *IEEE Trans. Wirel. Commun.* **2018**, *17*, 2457–2470. [CrossRef]
57. Cotton, S.L. Human Body Shadowing in Cellular Device-to-Device Communications: Channel Modeling Using the Shadowed $\kappa - \mu$ Fading Model. *IEEE J. Sel. Areas Commun.* **2015**, *33*, 111–119. [CrossRef]
58. Imoize, A.L.; Orolu, K.; Atayero, A.A.A. Analysis of key performance indicators of a 4G LTE network based on experimental data obtained from a densely populated smart city. *Data Brief* **2020**, *29*, 105304. [CrossRef]
59. Gvozdarev, A.; Artemova, T.; Morkovkin, A. The impact of modulation constellation type on the finite signal-to-noise ratio diversity gain in the presence of a multipath fading channel. *Probl. Inf. Transm.* **2024**, *60*, 71–93.

60. Blanza, J.; Teologo, A.J.; Materum, L. Datasets for Multipath Clustering at 285 MHz and 5.3 GHz Bands Based on COST 2100 MIMO Channel Model. In Proceedings of the 2019 International Symposium on Multimedia and Communication Technology (ISMAC), Quezon City, Philippines, 19–21 August 2019. [CrossRef]
61. Blanza, J.F.; Teologo, A.T.; Materum, L. Datasets for Multipath Clustering at 285 MHz and 5.3 GHz Bands Based on COST 2100 MIMO Channel Model. In Proceedings of the 2019 International Symposium on Multimedia and Communication Technology (ISMAC), Quezon City, Philippines, 19–21 August 2019; pp. 1–5. [CrossRef]

mathematics

Article

Intelligent Reflecting Surface-Assisted Wireless Communication Using RNNs: Comprehensive Insights

Rana Tabassum [1,2], Mohammad Abrar Shakil Sejan [3], Md Habibur Rahman [1,2], Md Abdul Aziz [1,2] and Hyoung-Kyu Song [1,2,*]

1 Department of Information and Communication Engineering, Sejong University, Seoul 05006, Republic of Korea; tanvi@sju.ac.kr (R.T.); habibur@sju.ac.kr (M.H.R.); aziz@sju.ac.kr (M.A.A.)
2 Department of Convergence Engineering for Intelligent Drone, Sejong University, Seoul 05006, Republic of Korea
3 Department of Electrical Engineering, Sejong University, Seoul 05006, Republic of Korea; sejan@sejong.ac.kr
* Correspondence: songhk@sejong.ac.kr

Abstract: By adjusting the propagation environment using reconfigurable reflecting elements, intelligent reflecting surfaces (IRSs) have become potential techniques used to improve the efficiency of wireless communication networks. In IRS-assisted communication systems, accurate channel estimation is crucial for optimizing signal transmission and achieving high spectral efficiency. As mobile data traffic continues to surge and the demand for high-capacity and low-latency wireless connectivity grows, IRSs are becoming pivotal technologies in the development of next-generation communication networks. IRSs offer the potential to revolutionize wireless propagation environments, improving network capacity and coverage, particularly in high-frequency wave scenarios where traditional signals encounter obstacles. Amidst this evolving landscape, machine learning (ML) emerges as a powerful tool to harness the full potential of IRS-assisted communication systems, particularly given the escalating computational complexity associated with deploying and operating IRSs in dynamic environments. This paper presents an overview of preliminary results for IRS-assisted communication using recurrent neural networks (RNNs). We first implement single- and double-layer LSTM, BiLSTM, and GRU techniques for an IRS-based communication system. In the next phase, we explore a hybrid approach, combining different RNN techniques, including LSTM-BiLSTM, LSTM-GRU, and BiLSTM-GRU, as well as their reverse configurations. These RNN algorithms were evaluated with respect to bit error rate (BER) and symbol error rate (SER) for IRS-enhanced communication. According to the experimental results, the BiLSTM double-layer model and the BiLSTM-GRU combination demonstrated the highest BER and SER accuracy compared to other approaches.

Keywords: intelligent reflecting surface; machine learning; recurrent neural network

MSC: 94A14

1. Introduction

The intelligent reflecting surface (IRS) is evaluated as a novel radio technology for enhancing wireless communication systems. To improve the spectral efficiency of upcoming wireless networks, significant attention has been given to IRS, a planar metasurface composed of multiple reflecting elements [1]. Communication systems can be greatly enhanced because the IRS proactively alters the wireless channel by enabling each reflecting element to adjust the incident signal with a desired phase shift [2]. This allows the IRS to create a virtual line-of-sight (LoS) link between user equipment (UE) and the base station (BS), particularly when obstacles block the direct path between them. This capability increases the degree of freedom (DoF) in terms of flexibility and reduces end-to-end latency for ultra-reliable low-latency communication (URLLC) applications [3]. Smartphones and other web-enabled devices have become essential tools for global communication, information sharing, and entertainment. The wireless communications sector is currently at an exciting juncture: while the fifth-generation (5G) technology has largely been standardized and commercialized, there is a growing focus on the sixth-generation (6G) wireless technology within both academia and industry. The Cisco annual internet report (2018–2023) has stated mobile connectivity is expected to reach over 70% of the world's population by 2023 [4]. Additionally, the mobile subscriber number is projected to increase from 5.1 billion in 2018 to 5.7 billion by 2023. This signifies a pivotal moment in the wireless sector, with strong momentum toward advancing to the next level of wireless technology to meet the increasing global demand for connectivity [5,6].

Due to increasingly aggressive frequency reuse, cellular networks have grown denser, leading to the development of intercellular synchronization technologies to manage this complexity [7]. However, the unpredictable nature of wireless transmission and the limited available spectrum continue to constrain network capacity. As a result, higher-frequency bands with abundant spectra are gradually gaining popularity. However, in urban areas, electromagnetic (EM) waves at these higher frequencies are more susceptible to blockage by structures such as buildings. To mitigate communication gaps and enhance network coverage, it requires more energy to deploy additional relays and BSs. Consequently, providing wireless service coverage using traditional cellular technologies becomes increasingly challenging. In response to the spectrum shortage faced by communication systems, IRSs have emerged as a key solution, offering the potential to achieve high spectrum and energy efficiency [8].

The utilization of machine learning (ML) techniques in wireless communications, particularly in the context of IRSs [9–12], has garnered significant attention due to their adaptability and the need to operate across broader search areas. In recent years, numerous researchers have explored diverse ML algorithms tailored to the communication sector in an effort to address challenges within this domain. These efforts aim to enable infrastructure to autonomously overcome various obstacles. Typically, ML methods involve learning parameters and optimization models based on input data to achieve specific objectives. Given the substantial volume of data involved, the efficiency and effectiveness of mathematical optimization procedures are crucial in determining the popularity and application of ML models in the current landscape [13].

By minimizing inter-user interference (IUI), improving non-line-of-sight connections, managing channel blockages, and expanding coverage, Intelligent reflecting surfaces (IRSs) have shown significant potential for enhancing wireless communications. Each IRS component can adjust the phase shifts and amplitude reflection coefficients, modifying the phase and amplitude of incoming signals to enable passive beamforming (BF). Unlike traditional MIMO systems, which focus on beamforming at the base station (BS) and user ends, IRS-assisted communication systems require a joint design of passive BF for IRSs and active BF for BS to effectively harness passive BF gains. Additionally, by employing large-scale passive reflecting components, IRS-assisted systems can enhance both energy efficiency (EE) and spectrum efficiency (SE) while remaining cost-effective and energy-efficient. Recent research has explored various aspects of IRS-supported wireless networks, including channel estimation, modulation and encoding, channel modeling, outage probability, SE analysis, symbol error rate (SER) probability, bit error rate (BER) probability, energy efficiency, weighted sum rate, and performance assessment. The extensive research on these topics reflects the enormous potential benefits offered by IRS-enabled environments.

In this paper, we present an overview of the in-depth results of IRS-assisted communication using RNNs. RNN algorithms can efficiently handle temporal dependencies, which can be used for time-varying systems like IRS-based wireless communication systems [14]. In addition, RNN algorithms can help obtain accurate channel state information for the IRS channel. One of the core tasks in IRS-based communication is beamforming, where signals are reflected in specific directions to maximize signal strength and minimize interference. RNNs can help optimize this process by continuously learning the optimal reflection angles and phase shifts over time. RNNs can reduce the computational time required to process data for IRS-based channels. Various RNN algorithms were implemented to evaluate performance in terms of BER and SER probability for IRS-enhanced communication. The types of used RNN algorithms are shown in Figure 1. We examine single-layer and double-layer LSTM, BiLSTM, and gated recurrent units (GRU) models for IRS-based communication. This experiment aims to assess the effectiveness of each RNN model and provide a comparative analysis of the results. In the next phase, we create hybrid models by combining two RNN techniques to observe their performance in IRS-based communication. Specifically, LSTM and BiLSTM are combined to form one hybrid model, and vice versa. Similarly, LSTM and GRU are combined to create another hybrid model, and vice versa. Lastly, we consider a combination of GRU and BiLSTM for IRS-based communication, and vice versa.

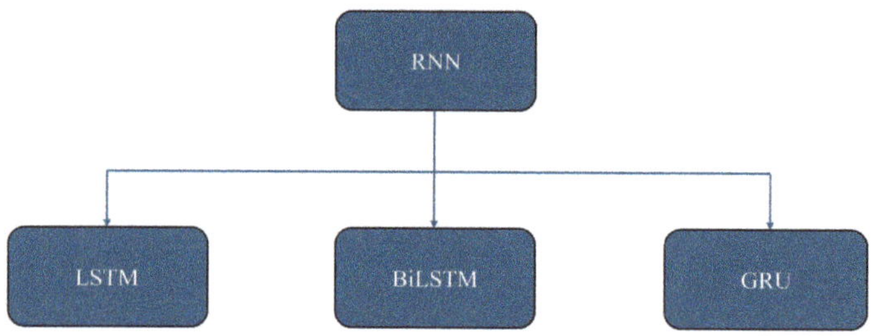

Figure 1. Types of RNN algorithms.

In this paper, the main contributions can be stated as follows:

- The goal of this experiment is to estimate the channel for intelligent reflecting surface-based wireless communication using RNN algorithms. Different combinations of RNN algorithms are considered for channel estimation.
- Rather than using traditional CE techniques like LS and MMSE, the suggested LSTM, BiLSTM, and GRU combined model can jointly estimate and identify the transmitted data from BS to UE.
- Comparative simulation results are presented to evaluate the effectiveness of the suggested system with respect to BER and SER measures.

The rest of the article is organized as follows: Section 3 presents the wireless communication system model with IRS, including the reflecting elements and the training data generation process. Section 4 describes the RNN models, while Section 5 details the proposed model using RNN techniques. Section 6 presents the simulation results, and Section 7 provides the conclusions.

2. Related Works

The utilization of machine learning (ML) techniques in wireless communications, particularly in the context of IRSs [9–12], has garnered significant attention due to their adaptability and the ability to operate across broader search areas. In recent years, numerous researchers have explored diverse ML algorithms tailored to the communication sector in an effort to address the challenges within this domain. These efforts aim to enable communication infrastructure to autonomously overcome various obstacles. Typically, ML methods involve learning parameters and optimization models from input data to achieve specific objectives. Given the substantial volume of data involved, the efficiency and effectiveness of mathematical optimization procedures play a crucial role in determining the popularity and application of ML models in the current landscape [13]. Today, ML is applied to solve many aspects of IRS-based communication, with key areas of application including channel estimation, beamforming, energy efficiency, resource management, signal detection, and security [15]. Various approaches have been adopted to address different problems, such as supervised learning, unsupervised learning, reinforcement learning, federated learning, graph learning, transfer learning, hierarchical learning, and meta-learning [16]. In supervised learning, models are trained with labeled data using algorithms such as deep neural networks, convolutional neural networks, decision trees, and support vector machines for IRS-based communication [17]. Unsupervised learning, on the other hand, is applied when input data lacks labels, and results are organized into clusters using methods like k-means or unsupervised neural networks. Reinforcement learning has gained particular attention due to its strong problem-solving capabilities. Common reinforcement learning algorithms include Q-learning, actor-critic learning, multi-agent learning, and double-deep Q-learning. Federated learning allows local models to share parameters in order to create a robust global model, which is particularly desirable for user privacy. Graph-based learning, using graph neural networks, has also been studied extensively. Transfer learning, which reuses existing knowledge to speed up the current learning process, is valued for achieving faster training results. Hierarchical learning divides complex tasks into sub-tasks and solves them from the root level. Finally, meta-learning enhances performance by utilizing the experience gained from prior learning tasks. In the literature, various studies have explored recurrent neural network (RNN) techniques for IRS-based communication. A stacked bidirectional LSTM (BiLSTM) architecture for IRS-assisted UAV communication systems was proposed in [12]. A long short-term memory (LSTM)-based neural network for IRS channel tracking and prediction was introduced in [18]. Additionally, Ref. [19] proposed a prediction method that considers the movement of wireless terminals in IRS-based communication as time series data, using LSTM for this purpose. The study in [20] proposed IRS-assisted NOMA network channel estimation using a convolutional neural network (CNN)-LSTM approach. The authors in [21] applied ML techniques to solve complex optimization problems in wireless communication systems. By combining a hybrid approach of ML-driven method and model-free optimization approach, the results show this approach can increase learning efficiency.

3. System Model

3.1. IRS Architecture and Working

When an electromagnetic (EM) wave encounters the boundary between two isotropic mediums, the relationship between the angles of incidence, reflection, and refraction is described by Snell's law. In standard cases, the angle of incidence equals the angle of reflection for smooth surfaces, as shown in Figure 2a. However, recent developments in metasurfaces allow modifying surface impedance to induce specific phase shifts between incident and scattered waves. By dividing the surface into many small elements, each programmed with a particular phase shift, the reflected EM wave can be steered to a different angle than predicted by Snell's law. As shown in Figure 2b, a fully configurable phase shift for each metasurface element can direct the reflected beam to any desired angle. In practice, achieving precise or continuous phase shifts is challenging due to control complexities, making signal processing and machine learning techniques useful for designing IRSs [22].

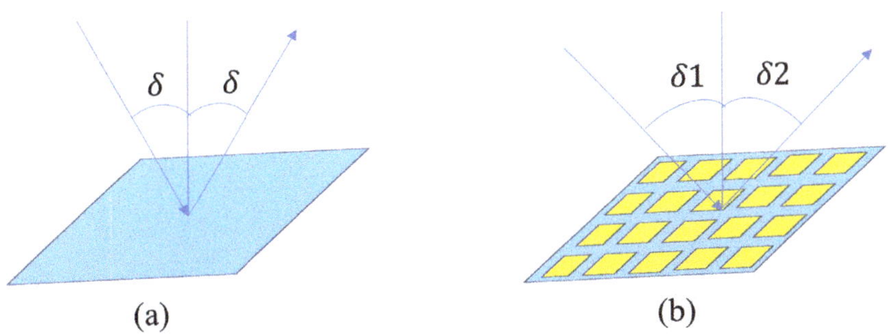

Figure 2. (**a**) The regular reflector; (**b**) IRS reflector.

IRSs are made up of a wide range of inexpensive passive reflecting components that are capable of adjusting the incident electromagnetic wave phase, amplitude, and even polarization. The propagation environment of wireless signals may be dynamically adjusted using these programmable surfaces. For a variety of purposes, including boosting signal strength, lowering interference, or covering regions with poor signals, IRSs in wireless communication systems work by reflecting and redirecting signals. An example of an IRS-aided wireless system in action is shown in Figure 3.

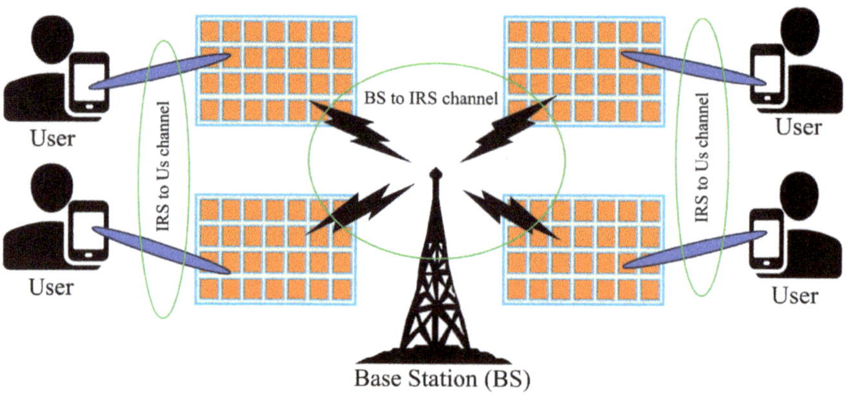

Figure 3. IRS working principle in wireless mobile communication system.

Here is how IRSs work in a typical wireless communication setup:

Signal Transmission: A signal is transmitted from a BS or user toward the IRS. *Surface Reflection:* The IRS components are adjusted to modify the incoming signal's phase and/or amplitude in order to direct it toward the intended target, which might be a BS or another user. *Signal Reception:* The modified signal reaches the receiver with enhanced quality, having avoided obstacles, reduced interference, and optimized the transmission path for improved signal strength and reliability.

In order to provide a smart propagation environment where wireless communication is boosted without the need for active, power-hungry components like amplifiers, IRSs may be managed in real-time to react to changing environmental circumstances, such as user movement or barriers. Key motivations behind using IRSs in wireless communication are listed as follows:

- *Improved signal coverage:* IRSs may greatly increase the coverage area, especially in areas with many impediments, by reflecting signals around objects like walls or buildings that often impede wireless signals.
- *Energy efficiency:* IRSs may reflect signals with low energy usage as it is made up of passive components that do not require active, power-hungry transmitters. This lowers the total energy consumption of wireless networks.
- *Reduced interference:* IRSs may reduce interference by carefully rerouting communications away from places or devices that cause interference. This is especially helpful when there are lots of devices interacting at once in congested areas.
- *Enhanced spectral efficiency:* IRSs aid in the effective use of the available spectrum by modifying the wireless environment and boosting the data rates and overall capacity of wireless networks.
- *Cost-effective deployment:* As an IRS is passive, it is less expensive to install and operate than power-hungry relays or new base stations. This makes it a more cost-effective option overall.
- *Support for 5G and beyond:* By improving communication at high-frequency bands like millimeter waves (mmWave), which are vulnerable to high route loss and limited penetration, IRSs can supplement 5G technology.

The common perception of relays is that they are active components that require a separate power supply to function. The devices are furnished with active electronic components, including mixers, low-noise amplifiers for receiving, power amplifiers for transmission, and digital-to-analog converters (DACs and ADCs). Decoder-and-forward (DF) and amplify-and-forward (AF) relays usually need a number of electrical components. This means that relays to implement multiple-antenna systems in millimeter and submillimeter wave frequency bands may be expensive and energy-intensive [23]. On the other hand, IRSs are composite material layers composed of printed metallic or dielectric patches on a grounded dielectric substrate. Low-power, low-complexity electronic circuits (switches or varactors) guarantee their configurability [24]. Since discrete power amplifiers, mixers, and DACs/ADCs are typically not needed, IRSs are intended to be less complicated than relays, especially when IRSs are implemented in large quantities and with low-cost big-area electronics [25]. The additive noise that exists in relays and degrades the functionality of traditional relaying methods is caused by the active electrical components employed in relays [25]. However, additive noise does not affect IRSs that exhibit aberrant reflector behavior. Nevertheless, phase noises might interfere with them [26]. Furthermore, they are unable to amplify or renew the impulses if they are almost inactive [27]. The spectral efficiency of relay-assisted systems depends on the duplexing method. A half-duplex (HD) reduces the rate by half but can improve the signal-to-noise ratio. A full-duplex (FD) avoids this but introduces interference from simultaneous transmissions. IRSs bypass these issues, as they do not face HD constraints or self-interference and can enhance signal combinations by optimizing their reflection properties [25].

3.2. IRS System Structure

The IRS-assisted communication model consists of a transmitter, a receiver, and IRS-reflecting elements. In this study, we consider IRS-assisted communication as shown in Figure 4. We consider each user, I, to have a single antenna, and the serving base station (BS) has a uniform linear array of L antennas. The number of IRS elements is considered as N. The received signal for the i-th ($i = 1, 2, 3, ..., I$) user from the IRS can be expressed as follows:

$$y_i = \mathbf{H}_{z,i}^H \Phi \mathbf{H}\mathbf{x} + n_i, \tag{1}$$

where y_i is the received signal at the i-th user, $\mathbf{x} \in \mathbb{C}^{L\times 1}$ is the transmitted signal, $\mathbf{H} \in \mathbb{C}^{N\times L}$ is the channel matrix from the BS to IRS, $\mathbf{H}_{z,i}^H$ is the channel from the IRS to the i-th user, and $n_i \sim \mathcal{CN}(0, \sigma^2)$ is the additive white Gaussian noise (AWGN) at the i-th user. Again, $\Phi = diag(\mathbf{d}) \in {}^{N\times N}$ represents a diagonal matrix that contains phase shift values of reflective elements of the IRS, where $\mathbf{d} = [\beta_1 e^{j\phi_1}, \beta_2 e^{j\phi_2}, ..., \beta_N e^{j\phi_N}]^T \in \mathbb{C}^{N\times 1}$, and $\phi_N \in [0, 2\pi]$ and $\beta_N \in [0, 1]$ denote the phase coefficient and amplitude of the i-th reflective element, respectively. For the experiment in this study, we consider $\beta_N = 1$. We also consider that the direct link between BS and the user exists and is expressed as $\mathbf{H}_d \in \mathbb{C}^{L\times 1}$. The Φ is considered as a diagonal matrix, i.e., $\Phi = diag(\mathbf{d})$. The channel from the BS to the user via the IRS can be transformed as follows:

$$\mathbf{H}_{z,i}^H \Phi \mathbf{H} = \mathbf{H}_{z,i}^H diag(\mathbf{d})\mathbf{H} = \mathbf{d}^T diag(\mathbf{H}_{z,i}^H)\mathbf{H}, \tag{2}$$

$$\mathbf{H}_c = \mathbf{H}_{z,i}^H \Phi \mathbf{H}, \tag{3}$$

where $\mathbf{H}_c$ is the cascaded channel for the i-th user and $\mathbf{H}_c \in^{N\times L}$, which relies on the downlink channel state information. We can write the total channel using both the cascaded channel and direct channel as follows:

$$\mathbf{y}_{ti} = (\mathbf{H}_c + \mathbf{H}_d)\mathbf{x} + \mathbf{n}_i, \tag{4}$$

where the direct communication link is $\mathbf{H}_d$ from the BS to the i-th user.

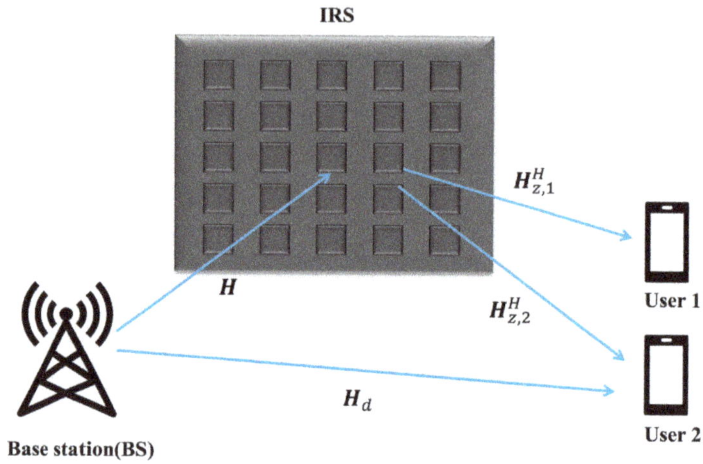

Figure 4. IRS-based communication system.

The three-dimensional (3D) Saleh–Valenzuela channel model is adapted for millimeter-wave (mmWave) propagation [28]. This method is a statistical channel model in a multipath propagation scenario. The channel model for mmWave propagation can be expressed as follows [10]:

$$\mathbf{H}_{mm} = \sqrt{\frac{N}{T}} \sum_{t=1}^{T} \eta_t \mathbf{a}(\rho_t^H, \phi_t^H), \tag{5}$$

where $\mathbf{H}$ is the channel vector, the t-th path complex gain is represented as η_t, and the total number of paths is T. The elevation angle of departure and azimuth angle of departure for transmission signals are ϕ_t^H and ρ_t^H, respectively. The array response vector is expressed as $\mathbf{a}(\rho_t^H, \phi_t^H)$. We can express the array response vector as follows:

$$\mathbf{a}(\rho, \phi) = \frac{1}{\sqrt{N}} [e^{-j2\pi d \sin(\rho)\cos(\phi)\mathbf{n}_1/\lambda}] \otimes [e^{-j2\pi d \sin(\phi)\mathbf{n}_2/\lambda}], \tag{6}$$

where λ is the carrier wavelength, d is the antenna spacing, $\mathbf{n}_1 = [0, 1, ..., N_1 - 1]$, $\mathbf{n}_2 = [0, 1, ..., N_2 - 1]$, and $(N = N_1 \times N_2)$. For proper communication, the antenna spacing satisfies the condition of $d = \lambda/2$. Again, the BS channel $\mathbf{H}$ to the IRS channel can be expressed as follows:

$$\mathbf{H} = \sqrt{\frac{LN}{T_1}} \sum_{t_1=1}^{T_1} \eta_{t_1} \mathbf{b}(\rho_{t_1}^{H_r}, \phi_{t_1}^{H_r}) \mathbf{a}^H(\rho_{t_1}^{H_t}, \phi_{t_1}^{H_t})^T, \tag{7}$$

where T_1 represents the number of paths between the BS and IRS, η_1 stands for the complex gain of the paths available from the BS to IRS, $\mathbf{b}(\rho_{t_1}^{H_r}, \phi_{t_1}^{H_r})$ is the steering vector related to IRS elements, and $\mathbf{a}^H(\rho_{t_1}^{H_t}, \phi_{t_1}^{H_t})$ is the steering vector connected to the BS for the t_1-th path. To define the channel between the IRS and the user, the channel matrix can be described as follows:

$$\mathbf{H}_{z,i}^H = \sqrt{\frac{N}{T_2}} \sum_{t_2=1}^{T_2} \eta_{t_2} \mathbf{a}^H(\rho_{t_2}^{H_t}, \phi_{t_2}^{H_t}), \tag{8}$$

where the complex gain of paths is represented as η_{t_1}; T_2 is the number of paths between the IRS and the user, $\rho_{t_1}^{H_t}$ is the departure angle in the azimuth direction, and $\rho_{t_1}^{H_t}$ is the elevation angle. The array response vector is represented as $\mathbf{a}^H(\rho_{t_1}^{H_t}, \phi_{t_1}^{H_t})$. Thus, the total channel can be written as follows:

$$\mathbf{H}_c = \sqrt{\frac{LN}{T_1 T_2}} \sum_{t_1=1}^{T_1} \sum_{t_2=1}^{T_2} \eta_{t_1}\eta_{t_2} diag(\mathbf{a}^H(\rho_{t_2}^{H_t}, \phi_{t_2}^{H_t}))\mathbf{b}(\rho_{t_1}^{H_r}, \phi_{t_1}^{H_r})\mathbf{a}^H(\rho_{t_1}^{H_t}, \phi_{t_1}^{H_t})^T. \quad (9)$$

3.3. Data Generation for Training from the IRS Network

The data generation and model training processes are illustrated in Figure 5. In the first phase, a random binary data sequence is generated for transmission. The generated bits are then transformed into OFDM symbols, using 128 subcarriers with quadrature phase shift keying (QPSK) modulation for symbol generation. The data symbols are transmitted through a simulated transmission channel that considers IRS reflection. The channel state information (CSI) is acquired using the Rayleigh fading model. Pilot signals are transmitted through the IRS channel, and the CSI is estimated based on the received data. Once the CSI is obtained, the training signals are transmitted through the channel and used for model training. The channel simulation parameters are listed in Table 1. The direct link is also considered when the received signal is calculated. Upon reception, the signal is demodulated, and the data symbols are retrieved. Sixteen unique symbols are used, and each symbol is assigned as a label value for classification. The real and imaginary parts of the symbols are separated to generate training data, as complex numbers cannot be directly used in the machine learning training process. A total of 64,000 samples were generated for the training and testing process. Of these, 512,000 data symbols were used for training, and 128,000 symbols were used for testing. Two separate datasets were saved as training and test datasets for model training and evaluation. Each model was then trained using RNN techniques, and the training progress was monitored. Hyperparameters were tuned to optimize performance and improve model accuracy. After training, the test data were used to evaluate each model's performance on unseen data and were measured as test accuracy. Following this step, the model could be used for IRS-based channel prediction.

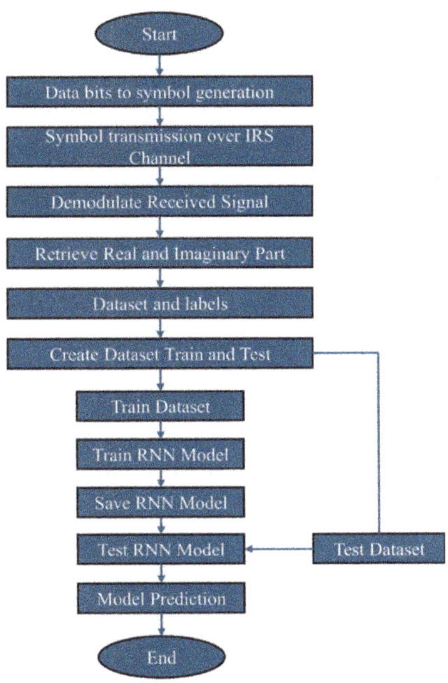

Figure 5. Training data generation and training process flowchart.

Table 1. Simulation parameters.

Parameter	Value
Number of IRS elements	512
Number of paths of BS to IRS	2
Number of paths of IRS to user	4
Antenna spacing for transmitter	0.5λ
IRS element spacing	0.5λ
Transmitter antenna	2
Receiver antenna	2
OFDM subcarrier	128
Modulation	QPSK
Noise for channel	AWGN

4. Recurrent Neural Network Models

This section briefly describes the fundamentals of LSTM, Bi-LSTM, and GRU approaches. After that, the proposed model descriptions are given.

4.1. LSTM Working Structure

Specialized memory cells with gating mechanisms are incorporated into LSTM networks, enabling them to store, update, and retrieve data selectively over long durations. Because of their distinct architecture, LSTM networks are better able to learn complex temporal patterns and relationships than ordinary RNNs, which helps them overcome some of their limitations. The input gate, forget gate, output gate, and cell state are the main parts of an LSTM unit. They manage information flow and maintain pertinent data throughout several time steps. LSTM networks are powerful tools for jobs requiring long-range temporal modeling and prediction because they can capture complex dependencies and context within sequential data by efficiently controlling the flow of information through the network. Figure 6a shows the internal structure of an LSTM cell. The gates of the LSTM can be mathematically expressed as follows:

$$i_t = \sigma(W_{xi}x_t + W_{hi}h_{t-1} + W_{ci}c_{t-1} + b_i) \tag{10}$$

$$f_t = \sigma(W_{xf}x_t + W_{hf}h_{t-1} + W_{cf}c_{t-1} + b_f) \tag{11}$$

$$g_t = \tanh(W_{xg}x_t + W_{hg}h_{t-1} + b_g) \tag{12}$$

$$c_t = f_t \odot c_{t-1} + i_t \odot g_t \tag{13}$$

$$o_t = \sigma(W_{xo}x_t + W_{ho}h_{t-1} + W_{co}c_t + b_o) \tag{14}$$

$$h_t = o_t \odot \tanh(c_t), \tag{15}$$

where at time step t, the input is x_t, at the $t-1$ time step, the hidden state is expressed as h_{t-1}, from the previous time step, c_{t-1} is the cell state, the input gate vector is i_t, the forget gate vector is f_t, the update gate vector is g_t, the output gate vector is o_t, σ is the sigmoid activation function, element-wise multiplication is represented as $\odot$, the weight matrix is W, and the bias vector is b. The input layer is connected to the LSTM layer, and the data then move through a fully connected layer and a dropout layer. Lastly, a classification layer is applied to determine the class that the input data belong to. The model can efficiently process sequential data, identify pertinent features, and produce precise predictions due to its architecture.

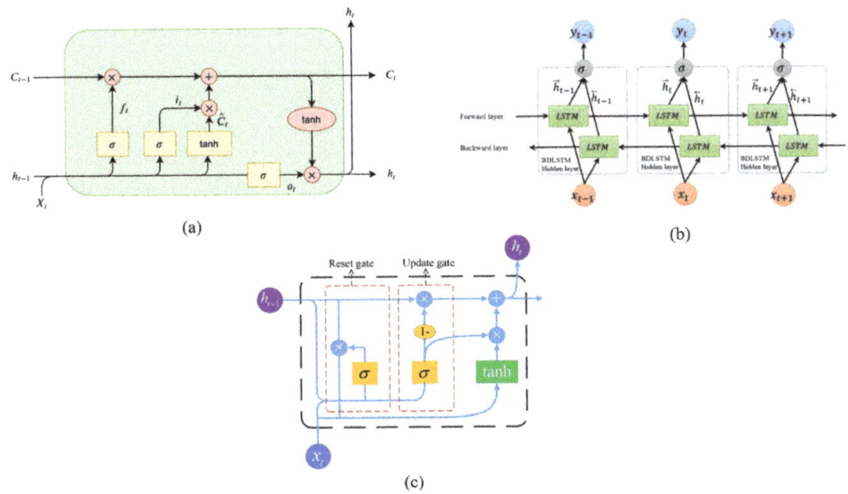

Figure 6. (**a**) Internal structure of LSTM [29], (**b**) internal structure of BiLSTM [30], (**c**) internal structure of GRU [31].

4.2. BiLSTM Working Structure

BiLSTM networks are extensions of the traditional LSTM architecture designed to store dependencies in sequential data in both directions. BiLSTMs employ two distinct LSTM layers (one for forward-processing input sequences and another for backward-processing) instead of traditional LSTMs, which process input sequences sequentially from the past to the future. The network's bidirectional processing allows future and past contexts, which facilitates a more full understanding of the input sequence. The internal organization of the BiLSTM is depicted in Figure 6b. For every time step in a BiLSTM, the output of each LSTM layer is merged, combining data from both directions. This concatenated form, which incorporates context from the past and future, represents the input sequence as a whole. Tasks requiring context from both sides, like named entity recognition, sentiment analysis, and part-of-speech tagging, are especially well-suited for BiLSTMs. Because BiLSTMs are bidirectional, they are effective at capturing contextual information and long-range relationships, which improves their performance in a variety of sequence modeling applications. But keep in mind that because BiLSTMs must process input sequences in both ways, they may add to the computational complexity. However, they are a popular option in many sequence prediction and natural language processing applications due to their capacity to take advantage of bidirectional context. BiLSTM can be expressed as follows [32]:

$$\overrightarrow{h_f} = \sigma(W_f S_t + W_f h_{t-1} + b_f), \tag{16}$$

$$\overleftarrow{h_r} = \sigma(W_r S_t + W_r h_{t+1} + b_r), \tag{17}$$

where the activation function is expressed as σ, the forward and backward time steps are represented as $t-1$ and $t+1$, the transmitted signal is S_t, the hidden state of the previous step is h_{t-1}, the hidden state of next state is h_{t+1}, the weights for the forward and reverse direction are W_f and W_r, learnable biases of forward and reverse directions are b_f and b_r, $h_f \rightarrow$ is the forward direction of network output, and $h_r \leftarrow$ is the backward direction of the LSTM network output, respectively. A dropout layer comes after the BiLSTM layer receives the input data. The input data are then classified using a fully connected layer and a classification layer. This architecture improves the model's capacity to effectively minimize interference by allowing it to capture context information from the past and the future.

4.3. GRU Working Structure

RNN architectures, such as the vanishing gradient problem and the incapacity to accurately capture long-term dependencies, are addressed by the GRU. Because GRUs are made to update and reset their internal state selectively, they can avoid consuming superfluous memory and retain pertinent information across longer sequences. The update gate and the reset gate are two of the main parts of the GRU design. Figure 6c illustrates GRU's internal organization. At every time step, these gates decide which data to keep and which to delete, controlling the flow of information through the network. At every time step, these gates decide which data to keep and which to delete, controlling the flow of information through the network. In addition, GRUs have a candidate state calculation phase that uses the reset gate and the current input to create a new candidate state. The final concealed state for the current time step is then created by combining the output of the update gate with this candidate state. GRUs have several advantages over conventional RNNs, one of which is their reduced parameter requirements and improved ability to capture long-term dependencies. Because of this, GRUs are especially well suited for sequential data tasks including time series prediction, machine translation, and language modeling. Furthermore, because of their more straightforward architecture, GRUs can be trained more quickly than LSTMs, which makes them a preferred option for many recurrent neural network applications. The GRU calculation procedure is explained using the formulas as follows [33]:

$$U_t = \sigma(G_U X_t + O_U H_{t-1} + b_U) \tag{18}$$
$$R_t = \sigma(G_R X_t + O_R H_{t-1} + b_R) \tag{19}$$
$$\widehat{H}_t = \tanh(G_H X_t + O_H (R_t \otimes H_{t-1}) + b_H) \tag{20}$$
$$H_t = (1 - U_t) \otimes H_{t-1} + U_t \otimes \widehat{H}_t, \tag{21}$$

where the biases are expressed as, respectively, b_U, b_R, and b_H. Again, G_U, G_R, G_H, O_U, O_R, and O_H are weight matrices. $\sigma(c) = (1 + e^{-c})^{-1}$ is used to calculate the gate activation function and expressed as a sigmoid function. The output of the hidden layer at the current instant is denoted as H_t, while the input state X_t and the output of the hidden layer at the preceding instant are combined to yield $\widehat{H}_t$. The state activation function is computed via the hyperbolic tangent function, (tanh). This uniformity makes it possible to assess the effectiveness of various RNN techniques for interference management fairly.

5. Proposed Model Architecture

We consider different RNN approaches for this study to estimate the channel performance for IRS-based communication. Two different approaches are considered for the experiment. First, the RNN techniques, namely LSTM, BiLSTM, and GRU, are tested with a single layer and double layer. Figure 7 shows the model architecture for a single RNN approach. To make a fair comparison, we keep the overall architecture similar for each technique. Each layer starts with a sequence input layer followed by an RNN layer and a dropout layer. Another RNN layer and dropout layer are added for the double layer. After that, layer normalization is applied and a fully connected layer is added. Finally, softmax and classification layers are used to classify the input data. LSTM, BiLSTM, and GRU single and double-layer architectures are shown in Figure 7a–c. In the next step, a hybrid approach is applied to see the performance of each combination. We consider combining two approaches, i.e., LSTM-BiLSTM, LSTM-GRU, and BiLSTM-GRU, and vice-versa. The model architectures for the LSTM and BiLSTM layers are shown in Figure 8a. The positions of the LSTM and BiLSTM layers are interchanged to observe the model accuracy effect. Similarly, the LSTM and GRU layer-based model architectures are shown in Figure 8b. The LSTM and GRU layers are switched for model accuracy observations. Lastly, BiLSTM and GRU model architectures are presented in Figure 8c. Here, the positions of the BiLSTM layer and the GRU layer are changed for model accuracy improvements. The other parameter for model training is presented in Table 2.

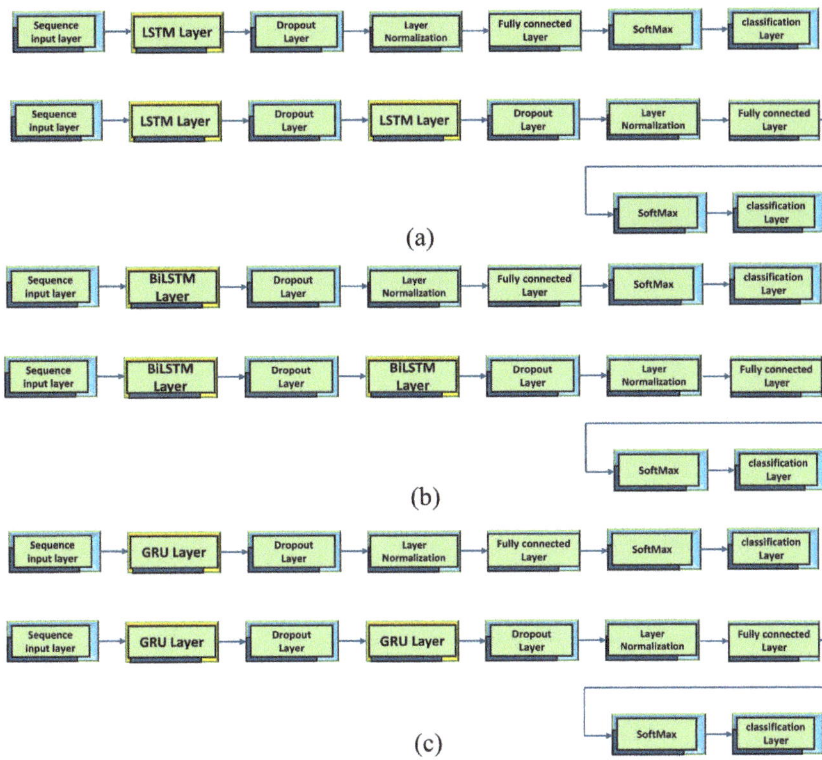

Figure 7. (**a**) LSTM model for the single layer and double layer, (**b**) BiLSTM model for the single layer and double layer, (**c**) GRU model for the single layer and double layer.

Table 2. Machine learning parameters.

Parameter	Value
Minibatch size	200
Maximum number of episodes	25
Learning rate	0.001
Hidden units	200
Number of classes	16
Dropout rate	0.02
Optimizer	ADAM
Gradient threshold	1
Validation frequency	50
Shuffle	every episode
Iteration per episode	2560

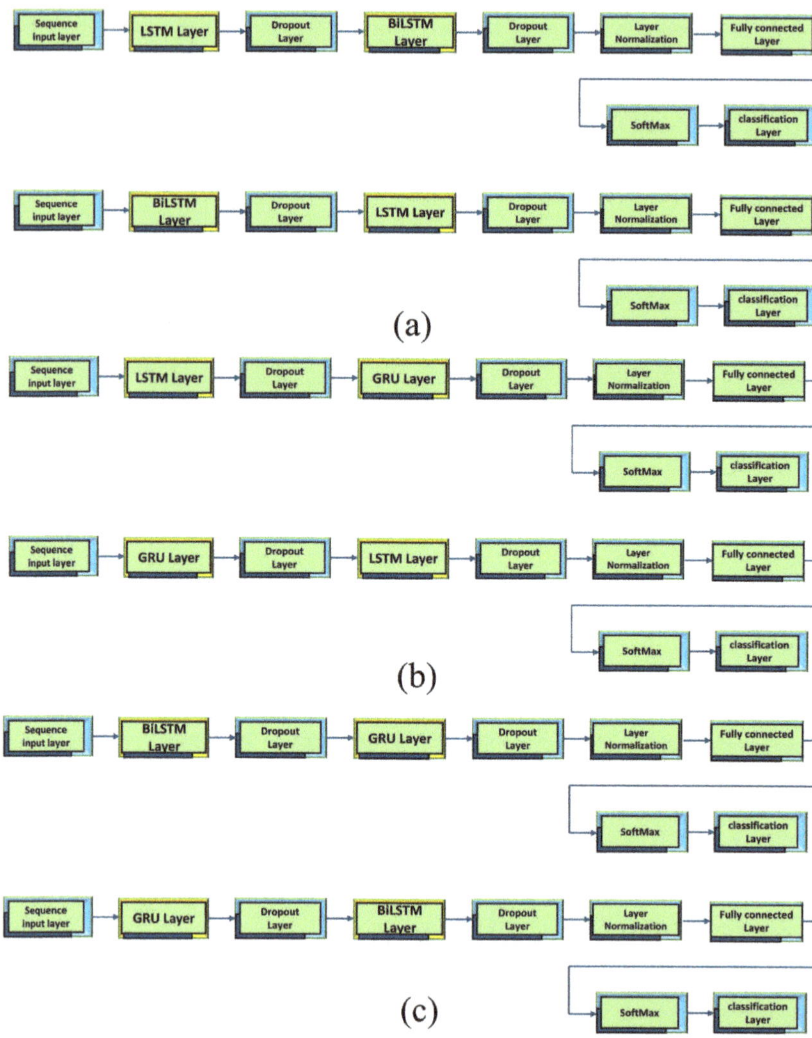

Figure 8. (**a**) LSTM and BiLSTM combination model, (**b**) LSTM and GRU combination model, (**c**) BiLSTM and GRU combination model.

6. Results

6.1. Model Training

Each model is trained using the generated dataset. First, we examined the LSTM model with the generated dataset under different hyperparameter setups. Table 2 lists the optimal hyperparameters for the best training results. Figure 9a shows the training progress of the LSTM with single and double layers. Both models exhibit a similar trend in reducing loss; however, the single-layer LSTM model reaches stable accuracy sooner than the double-layer model. For the BiLSTM model, the trend of the loss and accuracy curves during the training phase is shown in Figure 9b. As the number of layers increases, the training accuracy remains almost unchanged. The accuracy and loss of GRU-based models are depicted in Figure 9c, where the double-layer GRU achieves slightly higher accuracy than the single-layer GRU model.

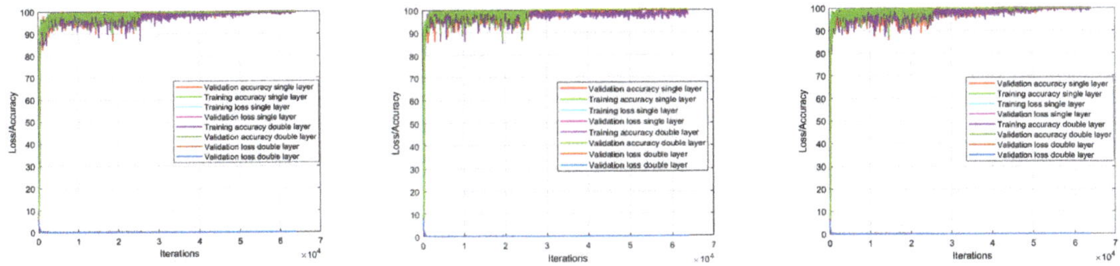

(a) (b) (c)

Figure 9. Model training summary; (**a**) LSTM single and double layer, (**b**) BiLSTM single and double layer, and (**c**) single and double layers.

Figure 10 shows the training and loss curves for the hybrid models. The training curves for the LSTM-BiLSTM model are presented in Figure 10a. The validation accuracy for the LSTM-BiLSTM model is lower than that of the BiLSTM-LSTM model, indicating that the BiLSTM-LSTM model can extract more features from the training data and is expected to yield better results compared to the LSTM-BiLSTM model. Figure 10b displays the training curves for the LSTM-GRU and GRU-LSTM models. At the beginning of the training process, the LSTM-GRU model shows instability, but by the end of the training, it becomes more stable and comparable to the GRU-LSTM model. The BiLSTM-GRU and GRU-BiLSTM models are shown in Figure 10c. Both models exhibit similar trends and are expected to have comparable prediction results.

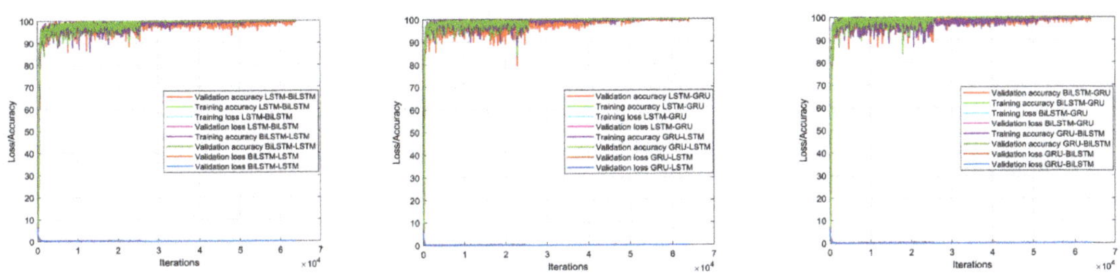

(a) (b) (c)

Figure 10. Model training summary (**a**) LSTM-BiLSTM model and BiLSTM-LSTM model, (**b**) LSTM-GRU model and GRU-LSTM model, and (**c**) BiLSTM-GRU model and GRU-BiLSTM model.

6.2. Results of Individual RNN Techniques

We first describe the results of the RNN techniques with single and double layers. The BER performance of the single-layer LSTM is shown in Figure 11. At an SNR value of 30 dB, the BER reaches 8.18×10^{-4}. However, when the number of layers is increased, the performance decreases to 9.47×10^{-4}. This decline is likely due to the feature extraction process being less effective in the double-layer configuration compared to the single-layer setup. A similar trend is observed in Figure 12, which depicts the SER for both the single-layer and double-layer LSTM models. The single-layer BiLSTM, on average, performs better than the LSTM, with the best performance achieved by the double-layer BiLSTM, as shown in Figures 11 and 12. At 30 dB SNR, the BER is 2.89×10^{-4}, and the SER is 9.95×10^{-4}. For the single-layer GRU, the BER performance is better than both the LSTM and the single-layer BiLSTM. In contrast, the double-layer GRU shows better performance across all SNR ranges, with the second-highest BER performance, as GRU is effective in feature extraction.

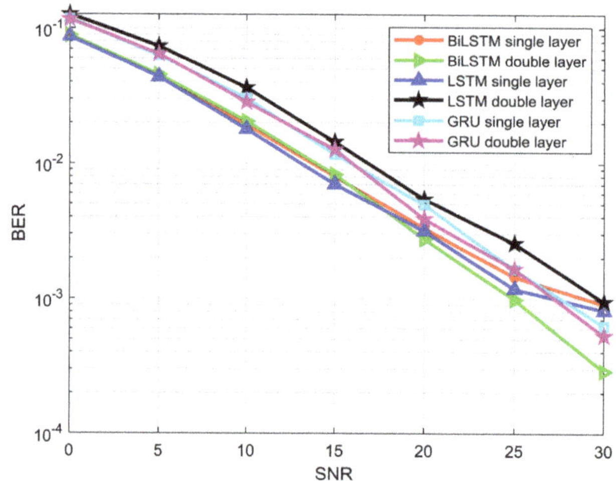

Figure 11. BER of LSTM, BiLSTM, and the GRU model for single and double layers.

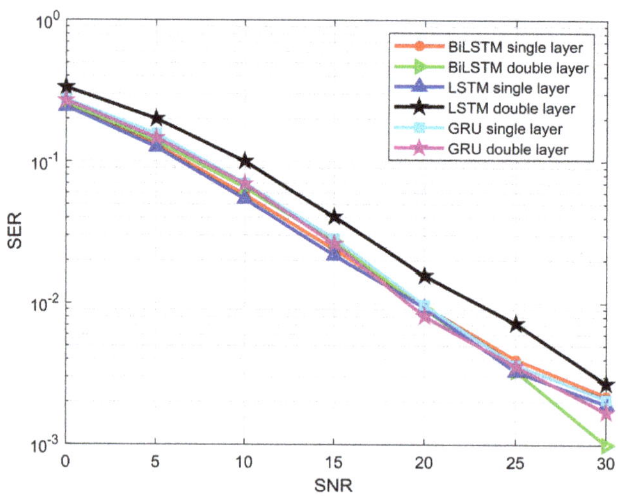

Figure 12. SER of LSTM, BiLSTM, and the GRU model for single and double layers.

6.3. Hybrid Model Performances

The combination of BiLSTM and LSTM was experimented with in the first step. The input data are first passed through a BiLSTM layer and then through an LSTM layer. The BER results for the BiLSTM-LSTM combination are shown in Figure 13, with a value of 2.08×10^{-4} at an SNR of 30 dB. However, when the layer order is reversed, the BER increases to 2.64×10^{-4}. This indicates that IRS channel data demodulation is more accurate with the BiLSTM-LSTM combination. A similar trend is observed in Figure 14, which shows the SER performance for both BiLSTM-LSTM and LSTM-BiLSTM layers. Next, we consider the LSTM-GRU and GRU-LSTM layer combinations for IRS communication. As seen in Figure 13, the LSTM-GRU layer combination achieves BER of 2.38×10^{-4}, while the GRU-LSTM combination has a BER of 2.86×10^{-4} at 30 dB SNR. This indicates better performance for the LSTM-GRU combination compared to GRU-LSTM, as LSTM can extract primary features more accurately, and GRU can be applied for classification. This combination demonstrates better accuracy compared to LSTM-BiLSTM. The SER for the LSTM and

GRU combination is shown in Figure 14. The final considered combination is between BiLSTM and GRU. Both combinations yield excellent results for BER and SER. At 30 dB SNR, the GRU-BiLSTM combination has a BER of 9.72×10^{-5}, while the BiLSTM-GRU combination achieves a BER of 9.45×10^{-5}. Among all the combinations, BiLSTM-GRU demonstrates the best BER performance. Since BiLSTM processes data in both forward and backward directions, its feature extraction capabilities are enhanced. Additionally, the GRU layer helps extract important features, leading to high classification accuracy. Thus, the BiLSTM-GRU-based model achieves more accurate results compared to other combinations.

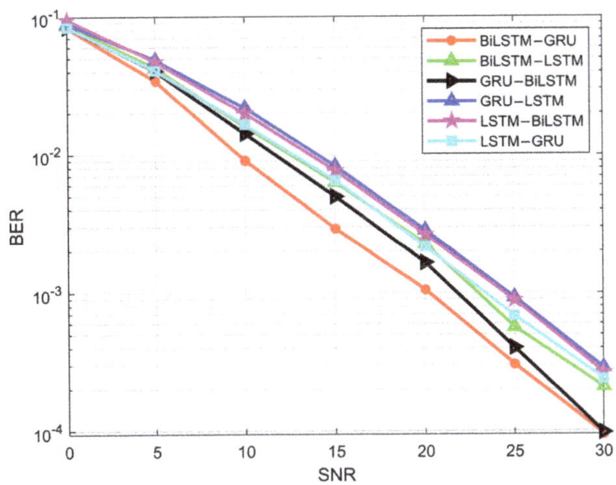

Figure 13. BER of LSTM, BiLSTM, and GRU hybrid models.

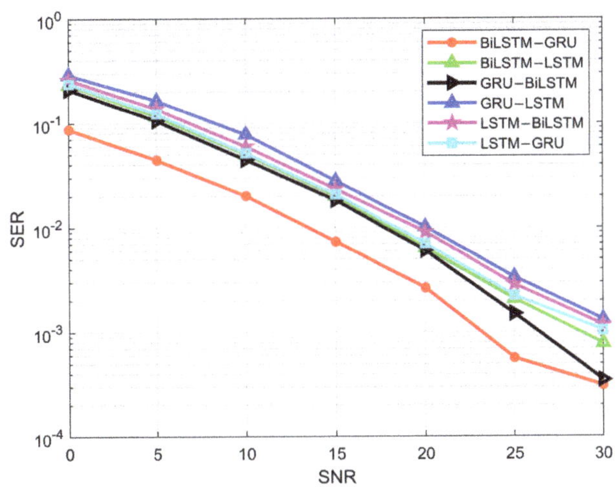

Figure 14. SER of LSTM, BiLSTM, and GRU hybrid models.

We also compared the performance of the proposed RNN architecture with a traditional wireless communication system without IRS reflection. For this experiment, we adopted three hybrid models, BiLSTM-GRU, GRU-BiLSTM, and LSTM-BiLSTM, as these models perform well in IRS reflection channels. As shown in Figure 15, all models demonstrate good performance at certain SNRs, but the overall performance is not satisfactory.

Specifically, in the high SNR range, the system without IRS fails to achieve low BER compared to the IRS-based system. This result highlights the superior BER performance of IRS-based systems.

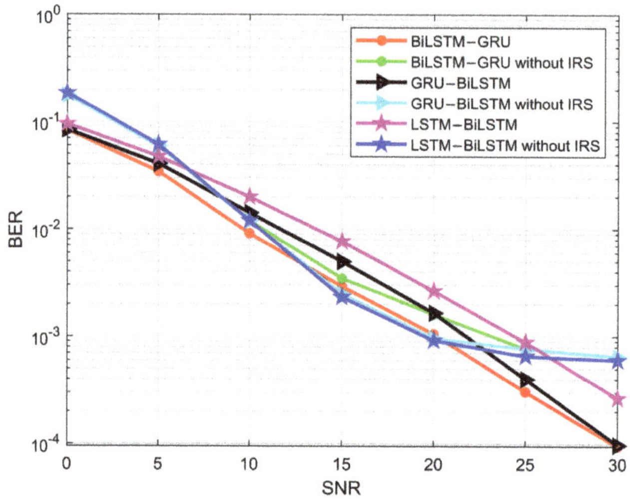

Figure 15. Comparison of the wireless communication channel performance with and without IRS using RNN hybrid models.

BiLSTM-GRU model's performance was also evaluated with other studies in the literature. Figure 16 compares the performance with other similar studies. The proposed BiLSTM-GRU model can perform well as compared to TCN [34] and traditional demodulation techniques. However, it has a similar performance to Demod-CNN [35] as CNNs are very powerful in classifying data.

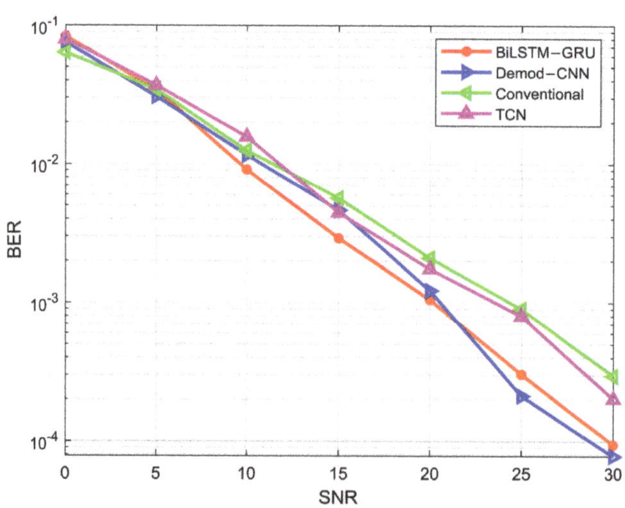

Figure 16. Comparison with different deep learning models with the proposed BiLSTM-GRU model.

7. Conclusions

In this study, we considered an IRS-based wireless communication system and conducted a performance analysis using RNN approaches. We examined both the IRS-reflected signal and the direct signal for channel estimation. To apply machine learning, we specifically focused on RNN techniques, namely LSTM, BiLSTM, and GRU, for estimating the IRS channel. First, each individual RNN technique was studied with single and double layers. In this setup, the BiLSTM with double layers showed the best BER performance compared to LSTM and GRU. In the next step, a hybrid approach was considered by combining each of the RNN techniques. The experimental results demonstrated that individual RNN techniques are capable of learning data patterns from the training process. Among the RNN techniques, the BiLSTM double-layer model achieved the highest accuracy compared to the LSTM and GRU models. In the hybrid approach, the BiLSTM-GRU combination provided the best performance in terms of BER and SER. We also compared our results with previous algorithms, and the outcomes are comparable to other studies. When the CSI and channel parameters are varied, the model must be adjusted to achieve optimal results. Retraining the model with newly generated data is necessary. Additionally, the model architecture may need to be restructured to reduce the BER. This study can be extended to develop larger models that account for multiple IRS reflection scenarios.

Author Contributions: Conceptualization, R.T. and M.A.S.S.; methodology, R.T., M.A.S.S.,and M.A.A.; software, R.T., M.H.R. and M.A.A.; validation, R.T. and M.H.R.; formal analysis, R.T., M.A.S.S., M.H.R. and M.A.A.; investigation, R.T., M.A.S.S. and M.A.A.; resources, H.-K.S.; data curation, R.T. and M.A.S.S.; writing—original draft preparation, R.T.; writing—review and editing, R.T., M.A.S.S., M.H.R., M.A.A. and H.-K.S.; visualization, R.T., M.A.S.S., M.H.R. and M.A.A.; supervision, H.-K.S.; project administration, H.-K.S.; funding acquisition, H.-K.S. All authors have read and agreed to the published version of the manuscript.

Funding: This work was supported in part by the Institute of Information and Communications Technology Planning and Evaluation (IITP) grant funded by the Korean Government [Ministry of Science and ICT (MSIT)], South Korea, under the Metaverse Support Program to Nurture the Best Talents (grant no. IITP-2024-RS-2023-00254529). It was also supported in part by the Basic Science Research Program through the National Research Foundation of Korea (NRF), funded by the Ministry of Education (grant no. 2020R1A6A1A03038540); and in part by the IITP (Institute of Information & Communications Technology Planning & Evaluation)–ITRC (Information Technology Research Center) grant, funded by the Korean government (Ministry of Science and ICT), under IITP-2024-RS-2024-00437191.

Data Availability Statement: The data will be made available by the authors upon request.

Conflicts of Interest: The authors declare no conflicts of interest.

References

1. Wu, Q.; Zhang, R. Towards smart and reconfigurable environment: Intelligent reflecting surface aided wireless network. *IEEE Commun. Mag.* **2019**, *58*, 106–112. [CrossRef]
2. Alexandropoulos, G.C.; Lerosey, G.; Debbah, M.; Fink, M. Reconfigurable intelligent surfaces and metamaterials: The potential of wave propagation control for 6G wireless communications. *arXiv* **2020**, arXiv:2006.11136.
3. Mishra, D.; Johansson, H. Channel estimation and low-complexity beamforming design for passive intelligent surface assisted MISO wireless energy transfer. In Proceedings of the ICASSP 2019—2019 IEEE International Conference on Acoustics, Speech and Signal Processing (ICASSP), Brighton, UK, 12–17 May 2019; pp. 4659–4663.
4. Alwazani, H.; Kammoun, A.; Chaaban, A.; Debbah, M.; Alouini, M.S. Intelligent reflecting surface-assisted multi-user MISO communication: Channel estimation and beamforming design. *IEEE Open J. Commun. Soc.* **2020**, *1*, 661–680.
5. Guo, H.; Lau, V.K. Uplink cascaded channel estimation for intelligent reflecting surface assisted multiuser MISO systems. *IEEE Trans. Signal Process.* **2022**, *70*, 3964–3977. [CrossRef]
6. He, Z.Q.; Yuan, X. Cascaded channel estimation for large intelligent metasurface assisted massive MIMO. *IEEE Wirel. Commun. Lett.* **2019**, *9*, 210–214. [CrossRef]
7. Wang, P.; Fang, J.; Duan, H.; Li, H. Compressed channel estimation for intelligent reflecting surface-assisted millimeter wave systems. *IEEE Signal Process. Lett.* **2020**, *27*, 905–909. [CrossRef]

8. Lin, T.; Yu, X.; Zhu, Y.; Schober, R. Channel estimation for intelligent reflecting surface-assisted millimeter wave MIMO systems. In Proceedings of the GLOBECOM 2020—2020 IEEE Global Communications Conference, Taipei, Taiwan, 7–11 December 2020; pp. 1–6.
9. Sejan, M.A.S.; Rahman, M.H.; Shin, B.S.; Oh, J.H.; You, Y.H.; Song, H.K. Machine learning for intelligent-reflecting-surface-based wireless communication towards 6G: A review. *Sensors* **2022**, *22*, 5405. [CrossRef]
10. Wei, X.; Shen, D.; Dai, L. Channel estimation for RIS assisted wireless communications—Part II: An improved solution based on double-structured sparsity. *IEEE Commun. Lett.* **2021**, *25*, 1403–1407. [CrossRef]
11. Shtaiwi, E.; Zhang, H.; Vishwanath, S.; Youssef, M.; Abdelhadi, A.; Han, Z. Channel estimation approach for RIS assisted MIMO systems. *IEEE Trans. Cogn. Commun. Netw.* **2021**, *7*, 452–465. [CrossRef]
12. Yu, J.; Liu, X.; Gao, Y.; Zhang, C.; Zhang, W. Deep learning for channel tracking in IRS-assisted UAV communication systems. *IEEE Trans. Wirel. Commun.* **2022**, *21*, 7711–7722. [CrossRef]
13. Song, H.; Zhang, M.; Gao, J.; Zhong, C. Unsupervised learning-based joint active and passive beamforming design for reconfigurable intelligent surfaces aided wireless networks. *IEEE Commun. Lett.* **2020**, *25*, 892–896. [CrossRef]
14. Lin, T.; Horne, B.G.; Giles, C.L. How embedded memory in recurrent neural network architectures helps learning long-term temporal dependencies. *Neural Netw.* **1998**, *11*, 861–868. [CrossRef] [PubMed]
15. Faisal, K.; Choi, W. Machine learning approaches for reconfigurable intelligent surfaces: A survey. *IEEE Access* **2022**, *10*, 27343–27367. [CrossRef]
16. Zhou, H.; Erol-Kantarci, M.; Liu, Y.; Poor, H.V. A Survey on Model-Based, Heuristic, and Machine Learning Optimization Approaches in RIS-Aided Wireless Networks. *IEEE Commun. Surv. Tutor.* **2024**, *26*, 781–823. [CrossRef]
17. Zhai, W.; Wang, X.; Greco, M.S.; Gini, F. CNN based Sparse IRS Design for Channel Estimation in Assisted Uplink Communications. In Proceedings of the 2023 IEEE International Radar Conference (RADAR), Sydney, Australia, 6–10 November 2023; pp. 1–6. [CrossRef]
18. Wei, Y.; Zhao, M.M.; Liu, A.; Zhao, M.J. Channel Tracking and Prediction for IRS-Aided Wireless Communications. *IEEE Trans. Wirel. Commun.* **2023**, *22*, 563–579. [CrossRef]
19. Tsuchiya, Y.; Suga, N.; Uruma, K.; Fujisawa, M. LSTM-based Spectral Efficiency Prediction by Capturing Wireless Terminal Movement in IRS-Assisted Systems. In Proceedings of the 2022 IEEE 95th Vehicular Technology Conference: (VTC2022-Spring), Helsinki, Finland, 19–22 June 2022; pp. 1–5. [CrossRef]
20. Nguyen, C.; Hoang, T.M.; Cheema, A.A. Channel estimation using CNN-LSTM in RIS-NOMA assisted 6G network. *IEEE Trans. Mach. Learn. Commun. Netw.* **2023**, *1*, 43–60. [CrossRef]
21. Gong, S.; Lin, J.; Ding, B.; Niyato, D.; Kim, D.I.; Guizani, M. When optimization meets machine learning: The case of IRS-assisted wireless networks. *IEEE Netw.* **2022**, *36*, 190–198. [CrossRef]
22. Dajer, M.; Ma, Z.; Piazzi, L.; Prasad, N.; Qi, X.F.; Sheen, B.; Yang, J.; Yue, G. Reconfigurable intelligent surface: Design the channel—A new opportunity for future wireless networks. *Digit. Commun. Netw.* **2022**, *8*, 87–104. [CrossRef]
23. Rappaport, T.S.; Xing, Y.; Kanhere, O.; Ju, S.; Madanayake, A.; Mandal, S.; Alkhateeb, A.; Trichopoulos, G.C. Wireless communications and applications above 100 GHz: Opportunities and challenges for 6G and beyond. *IEEE Access* **2019**, *7*, 78729–78757. [CrossRef]
24. Yu, N.; Genevet, P.; Kats, M.A.; Aieta, F.; Tetienne, J.P.; Capasso, F.; Gaburro, Z. Light propagation with phase discontinuities: Generalized laws of reflection and refraction. *Science* **2011**, *334*, 333–337. [CrossRef]
25. Di Renzo, M.; Ntontin, K.; Song, J.; Danufane, F.H.; Qian, X.; Lazarakis, F.; De Rosny, J.; Phan-Huy, D.T.; Simeone, O.; Zhang, R.; et al. Reconfigurable intelligent surfaces vs. relaying: Differences, similarities, and performance comparison. *IEEE Open J. Commun. Soc.* **2020**, *1*, 798–807. [CrossRef]
26. Qian, X.; Di Renzo, M.; Liu, J.; Kammoun, A.; Alouini, M.S. Beamforming through reconfigurable intelligent surfaces in single-user MIMO systems: SNR distribution and scaling laws in the presence of channel fading and phase noise. *IEEE Wirel. Commun. Lett.* **2020**, *10*, 77–81. [CrossRef]
27. Basar, E.; Di Renzo, M.; De Rosny, J.; Debbah, M.; Alouini, M.S.; Zhang, R. Wireless communications through reconfigurable intelligent surfaces. *IEEE Access* **2019**, *7*, 116753–116773. [CrossRef]
28. Busari, S.A.; Huq, K.M.S.; Mumtaz, S.; Dai, L.; Rodriguez, J. Millimeter-wave massive MIMO communication for future wireless systems: A survey. *IEEE Commun. Surv. Tutor.* **2017**, *20*, 836–869. [CrossRef]
29. Medium. Long Short-Term Memory (LSTM). Available online: https://medium.com/@saba99/long-short-term-memory-lstm-fffc5eaebfdc (accessed on 29 August 2024).
30. Terminology, Z. BiLSTM (Bidirectional Long Short-Term Memory). Available online: https://computer-nerd-zettelkasten.tistory.com/entry/BiLSTM-Bidirectional-Long-Short-Term-Memory (accessed on 29 August 2024).
31. Chen, J.; Huang, X.; Jiang, H.; Miao, X. Low-cost and device-free human activity recognition based on hierarchical learning model. *Sensors* **2021**, *21*, 2359. [CrossRef] [PubMed]
32. Rahman, M.H.; Sejan, M.A.S.; Aziz, M.A.; You, Y.H.; Song, H.K. HyDNN: A Hybrid Deep Learning Framework Based Multiuser Uplink Channel Estimation and Signal Detection for NOMA-OFDM System. *IEEE Access* **2023**, *11*, 66742–66755. [CrossRef]
33. Rahman, M.H.; Sejan, M.A.S.; Aziz, M.A.; Kim, D.S.; You, Y.H.; Song, H.K. Deep Convolutional and Recurrent Neural-Network-Based Optimal Decoding for RIS-Assisted MIMO Communication. *Mathematics* **2023**, *11*, 3397. [CrossRef]

34. Sejan, M.A.S.; Rahman, M.H.; Aziz, M.A.; You, Y.H.; Song, H.K. Temporal Neural Network Framework Adaptation in Reconfigurable Intelligent Surface-Assisted Wireless Communication. *Sensors* **2023**, *23*, 2777. [CrossRef]
35. Sejan, M.A.S.; Rahman, M.H.; Song, H.K. Demod-CNN: A Robust Deep Learning Approach for Intelligent Reflecting Surface-Assisted Multiuser MIMO Communication. *Sensors* **2022**, *22*, 5971. [CrossRef]

mathematics

MDPI

Article

Interference Management for a Wireless Communication Network Using a Recurrent Neural Network Approach

Mohammad Abrar Shakil Sejan [1,2], Md Habibur Rahman [1,2], Md Abdul Aziz [1,2], Rana Tabassum [1,2], Young-Hwan You [2,3], Duck-Dong Hwang [4] and Hyoung-Kyu Song [1,2,*]

1 Department of Information and Communication Engineering, Sejong University, Seoul 05006, Republic of Korea; sejan@sejong.ac.kr (M.A.S.S.); habibur@sju.ac.kr (M.H.R.); aziz@sju.ac.kr (M.A.A.); tanvi@sju.ac.kr (R.T.)

2 Department of Convergence Engineering for Intelligent Drone, Sejong University, Seoul 05006, Republic of Korea; yhyou@sejong.ac.kr

3 Department Computer Engineering, Sejong University, Seoul 05006, Republic of Korea

4 Department of Electronics and Communication Engineering, Sejong University, Seoul 05006, Republic of Korea; duckdonh@sejong.ac.kr

* Correspondence: songhk@sejong.ac.kr

Abstract: Wireless communication technologies have profoundly impacted the interconnectivity of mobile users and terminals. Nevertheless, the exponential increase in the number of users poses significant challenges, particularly in interference management, which is a major concern in wireless communication. Machine learning (ML) approaches have emerged as powerful tools for solving various problems in this domain. However, existing studies have not fully addressed the problem of interference management for wireless communication using ML techniques. In this paper, we explore the application of recurrent neural network (RNN) approaches to address co-channel interference in wireless communication. Specifically, we investigate the effectiveness of long short-term memory (LSTM), bidirectional LSTM (Bi-LSTM), and gated recurrent unit (GRU) network architectures in two different network settings. The first network comprises 10 connected devices, while the second network involves 20 devices. Our experimental results demonstrate that Bi-LSTM outperforms LSTM and GRU in terms of mean squared error, normalized mean squared error, and sum rate. While LSTM and GRU produce similar results, LSTM exhibits a marginal advantage over GRU. In addition, a combined RNN approach is also studied, and it can provide better results in dense networks.

Keywords: interference management; wireless network; deep learning; recurrent neural network

MSC: 94A13

1. Introduction

Present-day wireless communication is the cornerstone of our interconnected world, facilitating seamless connectivity and information exchange across a multitude of devices and applications. From smartphones and tablets to Internet of Things (IoT) sensors and autonomous vehicles, the proliferation of wireless technologies has revolutionized how we communicate, work, and interact with our surroundings [1]. The fifth generation (5G) communication technology allows wireless communication by employing massive multiple-input–multiple-output (MIMO), ultra-dense network, and device-to-device communication [2]. However, amidst this exponential growth in wireless connectivity and high-density data connectivity, the efficient management of interference within wireless networks has emerged as a paramount challenge, dictating the quality and reliability of wireless experiences. Interference, in the context of wireless communication, refers to the phenomenon inwhich signals from multiple devices or networks overlap and disrupt each other, leading to signal degradation, reduced throughput, and impaired performance [3,4]. This interference can arise from various sources, including co-channel interference, adjacent channel interference, and inter-symbol interference, each presenting unique hurdles to overcome in the quest for seamless wireless connectivity [4]. The escalating demand for wireless services, coupled with the finite nature of the radio spectrum, underscores the urgency of devising effective strategies for interference management in wireless networks. With the advent of technologies like 5G and beyond, which promise unprecedented data rates, ultra-low latency, and massive connectivity, the importance of mitigating interference becomes even more pronounced [5]. Thus, next-generation networks introduce new complexities, such as denser deployments of devices, heterogeneous architectures for networks, and diverse quality-of-service requirements, further accentuating the need for robust interference management solutions.

The solution to interference in wireless networks necessitates a multidisciplinary approach, drawing upon insights from signal processing, information theory, network optimization, and regulatory frameworks [6]. Advanced signal processing techniques, including adaptive filtering, beamforming, and interference cancellation, play a crucial role in mitigating interference and improving spectral efficiency [7–9]. Resource allocation algorithms, such as dynamic spectrum access and cognitive radio, enable the efficient utilization of available spectrum resources, minimizing interference while maximizing network capacity [10]. Recent studies in the literature have tried to solve the interference problem. The authors of [11] used graph representation learning for interference management in wireless networks. An actor–critic graph representation learning algorithm was developed to train neural networks to construct the optimal graph representing the impact of interference in a wireless network. The study reported in [12] employed a rate-splitting multiple-access precoding method in a multiple-antenna interference channel using deep reinforcement learning. The authors of [13] presented a study on satellite–ground integrated network resource interference management by analyzing coverage overlap and beneficially scheduling both beam-domain and power-domain resources. A deep Q network was designed using LSTM to predict the channel state information of satellite links. Reconfigurable intelligent surface-aided proactive mobile network interference mitigation was proposed in [14]. This study focused on developing an advantage actor–critic-based method for the interference problem by designing action space, state space, and a reward function. To mitigate RF channel interference, a deep learning approach was proposed in [15]. The authors modified the VGGNet-16 by adding a new convolution layer with 1280 filters, which enhanced the performance of the wireless communication system. Moreover, spectrum management policies and regulatory frameworks play a pivotal role in shaping the landscape of interference management, balancing the interests of various stakeholders, and ensuring fair and equitable access to the radio spectrum. As wireless networks continue to evolve and diversify, driven by emerging technologies like IoT, edge computing, and machine learning (ML), the quest for effective interference management remains a dynamic and ongoing endeavor.

In recent years, ML has been very successful in solving many complex problems across various domains. ML can be applied efficiently in various tasks, including computer vision, image interpretation, text classification, human behavior and identity recognition, fraud detection, recommendation systems, drug discovery, visual art generation, and natural language processing [16–23]. In addition, ML has significantly impacted wireless communication, revolutionizing various aspects of network operation, optimization, and management [24–26]. In particular, ML algorithms based on deep learning offer powerful tools for extracting valuable insights from the vast amount of data generated by wireless networks. Deep learning algorithms demonstrably outperform shallow machine learning algorithms in various ways. Key advantages include feature learning, superior performance on complex tasks like image and speech recognition, natural language processing, and time series forecasting, the ability to handle increasingly large and intricate datasets with complex relationships, and the facilitation of end-to-end learning, transfer learning, and the efficient handling of sequential data [27]. By analyzing historical usage patterns and real-time network conditions, ML algorithms can dynamically allocate spectrum resources, optimize transmission parameters, and mitigate interference, thereby enhancing spectral efficiency and network capacity [28,29]. Moreover, ML techniques enable intelligent predictive maintenance and fault detection, allowing network operators to proactively identify and address potential issues before they affect service quality [30]. Additionally, ML-based approaches facilitate the design of adaptive and self-optimizing networks that can autonomously reconfigure themselves to adapt to changing environmental conditions and user requirements. ML offers an efficient and autonomous approach to determine the nonlinear mapping relationship between fatigue life and numerous variables, leveraging available experimental data [31]. Overall, the integration of machine learning techniques into wireless communication holds immense promise for improving network performance, reliability, and user experience [32]. A class of neural networks known as recurrent neural networks (RNNs) was created to process sequential input efficiently by preserving memory over time steps [33]. Unlike traditional feed-forward neural networks, RNNs possess feedback connections that allow information to persist and flow through the network's hidden states. This inherent memory capability makes RNNs particularly well-suited for tasks involving sequential data, such as natural language processing, time series analysis, and speech recognition [34]. The architecture of an RNN consists of recurrent connections between neurons, enabling the network to capture dependencies and temporal patterns within sequences. However, traditional RNNs suffer from the vanishing gradient problem, limiting their ability to capture long-term dependencies. To solve this problem, various extensions of RNNs have been proposed, including long short-term memory (LSTM), bidirectional LSTM, and gated recurrent unit (GRU) networks. These variants incorporate sophisticated gating mechanisms that regulate the flow of information through the network, allowing them to effectively capture and propagate information over long sequences. RNNs have demonstrated remarkable success in a wide range of applications, including language modeling, time series prediction, machine translation, and sentiment analysis [35,36]. Their ability to model sequential data dynamics and capture contextual information makes them indispensable tools in the field of deep learning.

ML techniques have already been widely adopted in previous studies to address interference-related problems in wireless communication. For instance, the authors of [37] applied a convolutional neural network (CNN) and LSTM to form a convolutional LSTM autoencoder to retrieve corrupted data. The proposed model demonstrated improved symbol error rate performance in interference-affected signals after neural network training. Similarly, [38] utilized field-programmable gate arrays (FPGAs) as wireless receivers and employed a neural-network-based model for data demodulation. In [39], a CNN-based interference reduction method was introduced to classify received signals, achieving a high true positive rate with the deployment of two convolutional layers. Two convolutional layers were deployed to classify the signal label with a good true positive rate. Moreover, promising results were reported in [40] regarding CNN-based techniques for denoising and interference mitigation in radar processing. The study experimented with three models, one with three layers and the other two with a bottleneck-based structure, which reduces the number of channels. To mitigate interference in radar signals, another CNN approach was proposed in [41], in which a single convolutional layer and two batch normalization layers were utilized to create the model. A single convolutional layer and two batch normalization layers were used to create the model. Additionally, RNN-based approaches have been explored in previous research. In Ref. [42], the authors applied a multi-layer GRU-based model, demonstrating high performance under various interference conditions with low processing time. Moreover, Ref. [43] utilized a derivative-free Kalman filter-based RNN to examine interference reduction methods for the global positioning system (GPS). An adaptive unscented Kalman filter-based RNN (UKF-RNN) was used to lessen pulsed, swept, continuous wave interference (CWI), and multi-tone CWI. Furthermore, Ref. [42] suggested an RNN with self-attention to reduce radar interference caused by frequency-modulated continuous waveforms (FMCWs) and orthogonal frequency-division multiplexing (OFDM). Previous studies in the literature focused on diverse problems in wireless communication, and only a few studies considered the mitigation of interference for wireless communication. Additionally, the results produced by previous studies can be improved by applying new machine learning approaches. RNNs can be a promising candidate for examining and reducing interference in the channel. There is a research gap in considering the performance of the RNN techniques in interference management, and this gap needs further research.

In this paper, we model a co-channel interference network and simulate a real-time environment for data extraction. We utilize RNN machine learning approaches to measure the performance of the channel under interference. Each of the techniques, LSTM, BiLSTM, GRU, and a combined approach are employed to enhance the performance of interference-based communication.

The remainder of the paper is organized as follows: Section 2 details the wireless network interference structure, Section 3 describes RNN approaches, Section 4 outlines the research results and discussion, and Section 5 provides concluding remarks.

2. Wireless Network Interference

In any network system, the communication system is susceptible to signal or noise interference. Figure 1 depicts a typical wireless communication scenario. As shown in the figure, region A contains numerous users, while region B has fewer users. This discrepancy suggests the potential for interference in region A, which could diminish the user experience.

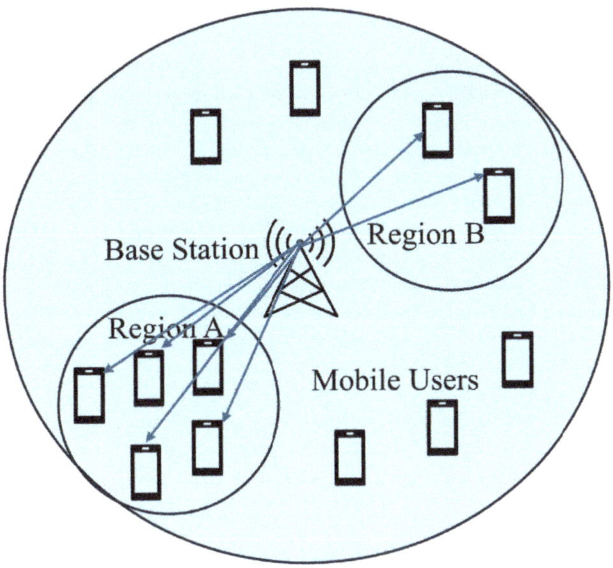

Figure 1. Interference example in a typical network between the base station and mobile users, where interference is prominent in region A and less interference in region B.

To solve the interference problem, we can consider the following optimization problem:

$$\begin{array}{ll} \text{minimize} & p(e;z), \\ \text{subject to} & e \in E \end{array} \tag{1}$$

where $e : \mathbb{R}^n \to \mathbb{R}$ is a continuous nonconvex objective function, e is the problem parameter, $E \in \mathbb{R}^n$ is the feasible region, and $z \in \mathbb{R}^n$ is a vector of the problem parameter. The original signal, which is received by the receiver, is added with other signals that cause signal interference. The communication links between different users and the base station cause interference. The throughput expression using Shannon's formula can be expressed as follows:

$$T_i = B\log_2(1 + SINR_i) \tag{2}$$

where T_i is the throughput for the i-th pair of communication, B is the bandwidth of the channel, and $SINR_i$ is the signal-to-interference-plus-noise ratio. Again, the $SNIR_i$ can be expressed as follows [44]:

$$SINR_i = \frac{|h_{ii}|^2 p_i}{\sum_{i\neq j} |h_{ij}|^2 p_j + \delta_i^2}, \tag{3}$$

where $h_{ii} \in \mathbb{C}$ is the direct channel between transmitter i and receiver i, $h_{ij} \in \mathbb{C}$ is the interference channel from transmitter i to receiver j, p_i is the power of the transmitter, p_j is the other transmitter power in the network, and δ_i^2 is the noise power at the receiver. Our goal is to retrieve the original signal from the corrupted signal by interference and solve Equation (1).

3. Method with RNN for Interference Management

3.1. Dataset Generation

We consider a network with N devices, each connected to a base station. The total signal received by a device is the sum of the direct signal and the interference signal. Initially, we calculate the direct signal using the diagonal parts of the channel matrix. Subsequently, the interference signal is determined by computing the cross channel, excluding the diagonal parts of the channel matrix. In the training data, we apply the interference signal to reveal the original signal; in turn, this can minimize interference. To simulate interference, we conducted experiments using two different network sizes. The first network comprised 10 devices operating simultaneously, while the second network involved 20 devices operating concurrently. We opted for these network sizes to mitigate network complexity. Initially, a simple RNN structure was employed, which can later be scaled up for a larger number of devices. Each device generated 20,000 samples, with a noise variance assumed to be 1 for this experiment. Additionally, each device generated 20 different feature values, which served as input for the RNN algorithm. Furthermore, each device had a unique label for classification purposes. The input data dimensions for the training sets were 20,000 samples $\times$ 10 features $\times$ 20 dimensions per feature (for the 10-device network) and 20,000 samples $\times$ 20 features $\times$ 20 dimensions per feature (for the 20-device network). To evaluate the models, we generated separate test sets of 2000 samples each, maintaining the same network configurations (10 or 20 features, 20 dimensions per feature) as the training data. The parameters for the considered network are listed in Table 1. After the data are received through the channel, they are utilized for training purposes. Algorithm 1 outlines the process of data generation for training and testing.

Table 1. Parameter list for data generation.

Parameter	Value
Number of devices	20, 10
Training samples	20,000
Noise variance	1
Feature of each device	20
Label for each device	1

Algorithm 1 Training Data Generation Process

1: Initialize data: number of users N, number of features q, number of labels for each user s, total number of data generation t.
2: Calculate h_{ii} and h_{ij} for each user.
3: Employing simulated channel the training data is passed. $y = xh_{ii} + xh_{ij} + n$.
4: The received data y is saved for training.
5: Separate data is generated for testing process.

3.2. Details of RNN Models

In this section, we describe the RNN models considered for inference reduction. After generating the training data, we employ the models for training.

3.2.1. LSTM

LSTM networks incorporate specialized memory cells with gating mechanisms, allowing them to selectively store, update, and retrieve information over extended sequences. This unique architecture enables LSTM networks to overcome the limitations of standard RNNs by facilitating the learning of complex temporal patterns and dependencies. The key components of an LSTM unit include the input gate, forget gate, output gate, and cell state, which are each responsible for controlling the flow of information and preserving relevant information over multiple time steps. By effectively managing the flow of information through the network, LSTM networks can capture intricate dependencies and context within sequential data, making them powerful tools for tasks requiring long-range temporal modeling and prediction. The gates of LSTM can be expressed as follows:

$$i_t = \sigma(W_{xi}x_t + W_{hi}h_{t-1} + W_{ci}c_{t-1} + b_i) \tag{4}$$

$$f_t = \sigma(W_{xf}x_t + W_{hf}h_{t-1} + W_{cf}c_{t-1} + b_f) \tag{5}$$

$$g_t = \tanh(W_{xg}x_t + W_{hg}h_{t-1} + b_g) \tag{6}$$

$$c_t = f_t \odot c_{t-1} + i_t \odot g_t \tag{7}$$

$$o_t = \sigma(W_{xo}x_t + W_{ho}h_{t-1} + W_{co}c_t + b_o) \tag{8}$$

$$h_t = o_t \odot \tanh(c_t), \tag{9}$$

where x_t is the input at time stop t, h_{t-1} is the hidden state from the previous time step, c_{t-1} is the cell state from the previous time step, i_t is the input gate vector, f_t is the forget gate vector, g_t is the update gate vector, o_t is the output gate vector, σ is the sigmoid activation function, $\odot$ represents element-wise multiplication, W is the weight matrix, and b is bias vector. The architecture of the LSTM-based machine learning model is depicted in Figure 2. The LSTM layer is connected to the input layer, after which the data pass through a dropout layer and a fully connected layer. Finally, a classification layer is utilized to predict the correct class of the input data. This architecture allows the model to effectively process sequential data, extract relevant features, and make accurate predictions.

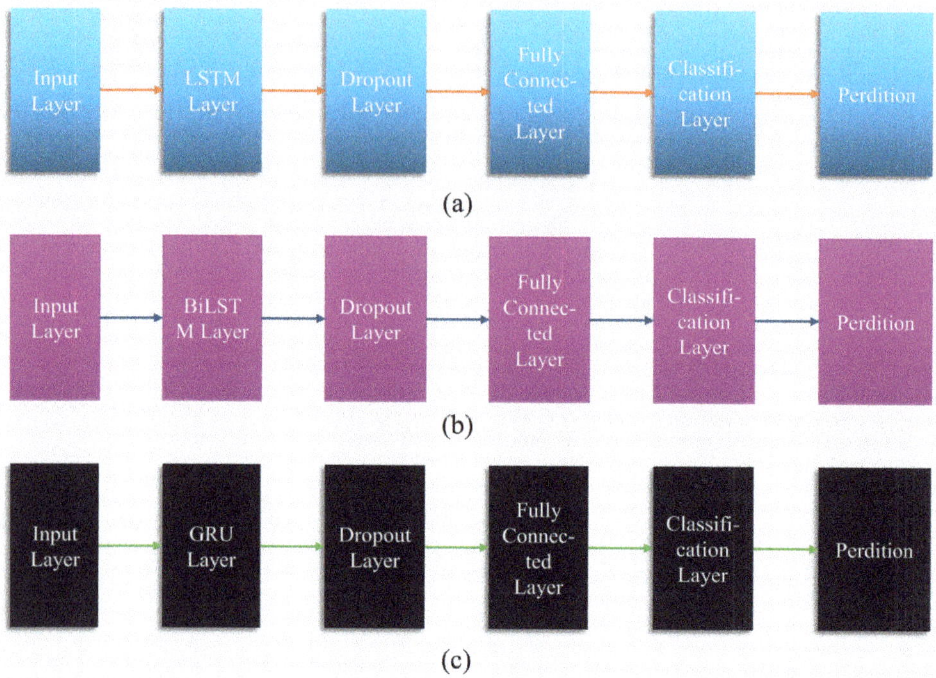

Figure 2. Model architecture for training data; (**a**) LSTM ML model architecture for interference reduction; (**b**) the BiLSTM ML model architecture for interference reduction; (**c**) the GRU ML model architecture for interference reduction.

3.2.2. BiLSTM

BiLSTM networks are an extension of the traditional LSTM architecture, designed to capture dependencies in both the forward and backward directions within sequential data. Unlike standard LSTMs, which process input sequences sequentially from past to future, BiLSTMs incorporate two separate LSTM layers: one processing the input sequence in the forward direction and the other processing it in the backward direction. This bidirectional processing allows the network to capture contextual information from both past and future contexts, enabling a more comprehensive understanding of the input sequence. In a BiLSTM, the output of each LSTM layer is concatenated at each time step, effectively merging information from both directions. This concatenated representation contains information about the input sequence as a whole, incorporating both past and future contexts. BiLSTMs are particularly effective in tasks where context from both directions is crucial, such as part-of-speech tagging, named entity recognition, and sentiment analysis. The bidirectional nature of BiLSTMs makes them powerful tools for capturing long-range dependencies and contextual information, leading to improved performance in various sequence modeling tasks. However, it is worth noting that BiLSTMs may introduce additional computational complexity due to the need to process input sequences in both directions. Nonetheless, their ability to leverage bidirectional context makes them a popular choice in many natural language processing and sequence prediction applications. The mathematical expression of BiLSTM can be expressed as follows [45]:

$$\overrightarrow{h_f} = \sigma(W_f S_t + W_f h_{t-1} + b_f), \tag{10}$$

$$\overleftarrow{h_r} = \sigma(W_r S_t + W_r h_{t+1} + b_r), \tag{11}$$

where σ is the activation function, the time steps of forward and backward represent $t-1$ and $t+1$, S_t is the transmitted signal, the hidden state of previous and the next are h_{t-1} and h_{t+1}, the weights and learnable bias of both directions are Wf and Wr and bf and br, and $hf \rightarrow$ and $h3r \leftarrow$ are the forward- and backward-direction LSTM network outputs, respectively. The model architecture for BiLSTM to tackle interference is illustrated in Figure 2b. The input data are fed into the BiLSTM layer, followed by a dropout layer. Subsequently, a fully connected layer and a classification layer are employed to classify the input data. This architecture enables the model to capture both past and future context information, thus enhancing its ability to mitigate interference effectively.

3.2.3. GRU

GRU is a type of RNN architecture that addresses some limitations of traditional RNNs, such as the vanishing gradient problem and the inability to capture long-term dependencies effectively. GRUs are designed to selectively update and reset their internal state, allowing them to retain relevant information over longer sequences while avoiding unnecessary memory consumption. The GRU architecture consists of several key components, including the update gate and the reset gate. These gates regulate the flow of information through the network, determining which information to retain and which to discard at each time step. GRUs also feature a candidate state computation step, which generates a new candidate state based on the current input and the reset gate. This candidate state is then combined with the update gate's output to produce the final hidden state for the current time step. One of the key advantages of GRUs over traditional RNNs is their ability to capture long-term dependencies more effectively while requiring fewer parameters. This makes GRUs particularly well suited for tasks involving sequential data, such as language modeling, machine translation, and time series prediction. Additionally, GRUs are generally faster to train than LSTMs due to their simpler architecture, making them a popular choice in many applications requiring recurrent neural networks. The GRU computation procedure is explained using the formulas as follows [46]:

$$U_t = \sigma(W_U X_t + V_U H_{t-1} + b_U) \tag{12}$$

$$R_t = \sigma(W_R X_t + V_R H_{t-1} + b_R) \tag{13}$$

$$\widehat{H}_t = \tanh(W_H X_t + V_H (R_t \otimes H_{t-1}) + b_H) \tag{14}$$

$$H_t = (1 - U_t) \otimes H_{t-1} + U_t \otimes \widehat{H}_t, \tag{15}$$

where the biases are b_U, b_R, and b_H. The gate activation function is calculated using the sigmoid function as $\sigma(c) = (1 + e^{-c})^{-1}$. Again, W_U, W_R, W_H, V_U, V_R, and V_H are weight matrices. While the input state X_t and the output of the hidden layer at the previous instant are merged to generate $\widehat{H_t}$, the output of the hidden layer at the current instant is represented by H_t. The hyperbolic tangent function known as (tanh) is in charge of calculating the state activation function. As depicted in Figure 2c, the GRU-based machine learning model architecture is connected to the input data. The same architecture is followed as previously used in LSTM and BiLSTM for the sake of the best comparison. This consistency allows for a fair evaluation of the performance of different RNN approaches in interference management.

4. Experiment Result and Discussion

4.1. Training and Testing of the Models

We have selected two different network sizes to test various RNN approaches. The first network comprises 20 active devices, while the second network has 10 active devices. The parameter list for machine learning is provided in Table 2. To minimize model complexity, we opted for a single hidden layer in each RNN architecture. To determine the optimal number of hidden layers for each approach, we started with 10 hidden units and gradually increased this number to 100. However, for hidden units below 50, we observed significant fluctuations in validation accuracy and loss, leading to unstable results. Conversely, when the number of hidden units exceeded 50, validation accuracy began to decrease. Consequently, 50 hidden units were chosen as they provided stable results along with high validation accuracy. Additionally, we only considered one optimizer and one loss function for the experiment. To describe the internal structure and complexity of LSTM, BiLSTM, and GRU, we can refer to their formulations. The complexity of LSTM is expressed as $(t_s \times t_h \times (4t_i + 4t_h + 3))$ [46,47], where t_i represents the number of features in the input vector, t_s denotes the size of the input time sequence, and t_h indicates the number of hidden units. For BiLSTM, the complexity is represented as $(t_s \times t_h \times 2(4t_i + 4t_h + 3))$. Similarly, the complexity of GRU is expressed as $(t_s \times t_h \times (3t_i + 3t_h + 3))$.

Table 2. Parameter list for machine learning.

Parameter	Value
RNN approach	LSTM, BiLSTM, GNN, Combied
Layers	Single layer
Hidden units LSTM	50
Hidden units BiLSTM	50
Hidden units GRU	50
Training epochs	150
Learning rate	0.001
Number of iterations	200
Learning rate decay	None
Optimizer	Adam
Loss function	Mean squared error

Figure 3 illustrates the training and validation accuracy, as well as loss, for different RNN approaches. Specifically, Figure 3a displays the performance of LSTM for the 10-device network. The training accuracy for the LSTM algorithm reaches 81.74%, with a validation accuracy of 80.28%. Additionally, the training loss decreases to 0.089, and the validation loss reaches 0.1041 after 150 episodes. The proposed LSTM network's training is completed in 150 episodes, after which the model's learning accuracy converges and remains stable. Therefore, the training episodes were kept at 150 for all algorithms to ensure a fair comparison. Figure 3b depicts the training accuracy and loss for BiLSTM in the 10-device network. The training accuracy for BiLSTM reaches 86.12%, and the validation accuracy reaches 82.62%. In contrast, the training loss reduces to 0.053, and the validation loss decreases to 0.1040. This indicates that BiLSTM has a higher performance in terms of both accuracy and loss reduction compared to LSTM. BiLSTM demonstrates slightly better performance in training accuracy compared to LSTM in terms of validation accuracy. In the case of GRU, the training accuracy and loss are presented in Figure 3c. GRU has shown a training accuracy of 81.31% and a validation accuracy of 79.79% after 150 episodes. The training loss for GRU is 0.093, and the validation loss is 0.1051, indicating a lower performance compared to BiLSTM and LSTM. In the subsequent phase, a 20-device network was considered for training. Figure 3d illustrates the training progress of LSTM, where both the accuracy and loss converge after 20 episodes. The training accuracy for the 20-device network is 79.35%, and the validation accuracy is 78.14% after 150 episodes. The training loss for the LSTM in the 20-device network is 0.087, and the validation loss is 0.096. Similarly, in the case of BiLSTM, as shown in Figure 3e, the training accuracy reaches 81.18% and the validation accuracy is 78.48% after 150 episodes. The training loss for BiLSTM decreases to 0.0723, and the validation loss is 0.0958 at the end of 150 episodes. For GRU, the training accuracy is 78.79% and the validation accuracy is 77.99%, as depicted in Figure 3f. Additionally, the training loss for GRU is 0.091 and the validation loss is 0.0966 at the end of training.

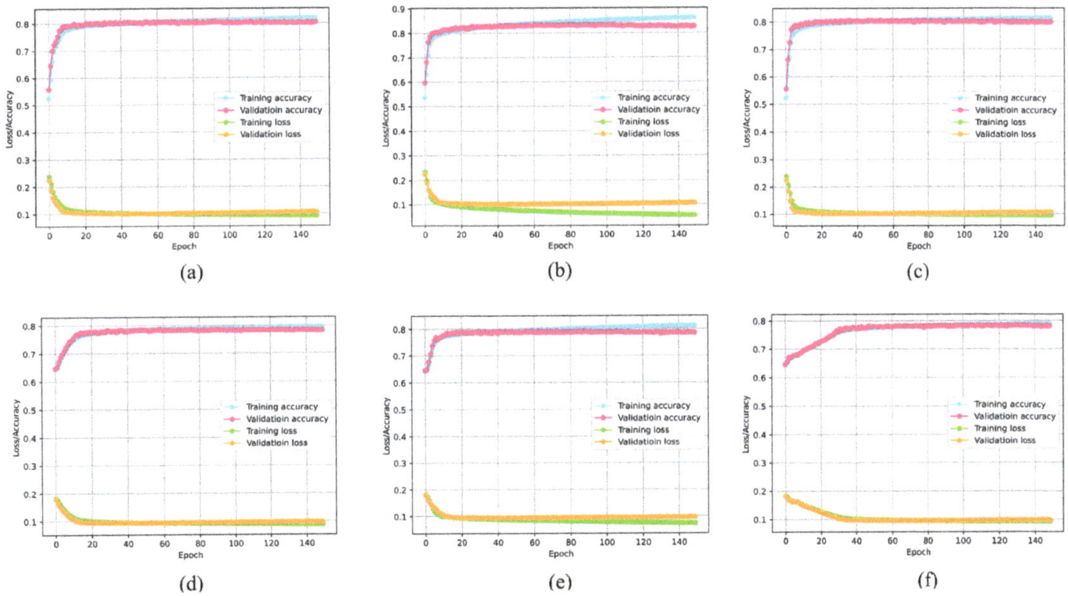

Figure 3. Accuracy and loss curve for different RNN techniques during model training; (**a**) LSTM training accuracy and loss with 10-device network; (**b**) BiLSTM training accuracy and loss with 10-device network; (**c**) GRU training accuracy and loss with 10-device network; (**d**) LSTM training accuracy and loss with 20-device network; (**e**) BiLSTM training accuracy and loss with a 20-device network; and (**f**) GRU training accuracy and loss with 20-device network.

4.2. Model Performance in Wireless Network

After passing the data through a noisy channel, we recovered the data using the RNN approach. To evaluate the quality of the received signal, we calculated the mean squared error (MSE) and normalized mean squared error (NMSE) in the initial phase. Subsequently, in the next phase, we calculated the sum rate for the received signal to assess the performance of each of the RNN algorithms. Initially, we analyzed the results of the mean squared error for the three RNN approaches across two different network sizes.

Figure 4 displays the MSE for BiLSTM, LSTM, and GRU against the episode number during training for the 10-device network. The MSE for LSTM decreases as the number of episodes increases, with a mean and standard deviation of 0.00522 ± 0.0007. GRU shows a similar trend to LSTM with a slightly higher MSE, with a mean and standard deviation of 0.00534 ± 0.00063. BiLSTM exhibits better performance compared to both LSTM and GRU, with a mean and standard deviation for MSE of 0.0043 ± 0.00067. BiLSTM outperforms LSTM and GRU, having the lowest mean value among the three and showing a consistent downward curve in MSE, as illustrated in Figure 4.

Figure 5 illustrates the MSE performance for the 20-device network. Here, all approaches exhibit a decreasing trend in MSE as the number of episodes increases. LSTM achieves a mean and standard deviation of 0.00488 ± 0.0062, with a slight upward trend towards the end of the episodes. GRU demonstrates a similar trend to LSTM, with a slightly larger mean MSE of 0.00523 ± 0.000989. According to the results presented in Figure 5, the MSE mean and standard deviation for BiLSTM are 0.00469 ± 0.000563. For the 20-device network, the MSE trend is similar for all three approaches, but BiLSTM exhibits comparatively better results than LSTM and GRU, as shown in Figure 5. Although the difference in MSE values is small, it can have a performance impact in delay-sensitive networks.

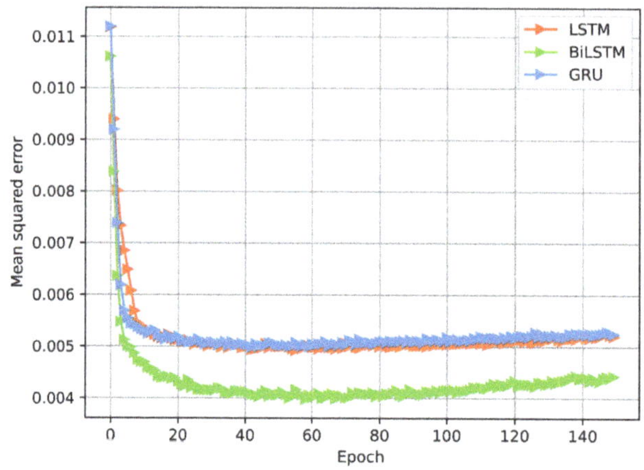

Figure 4. Mean squared error for a 10-device network.

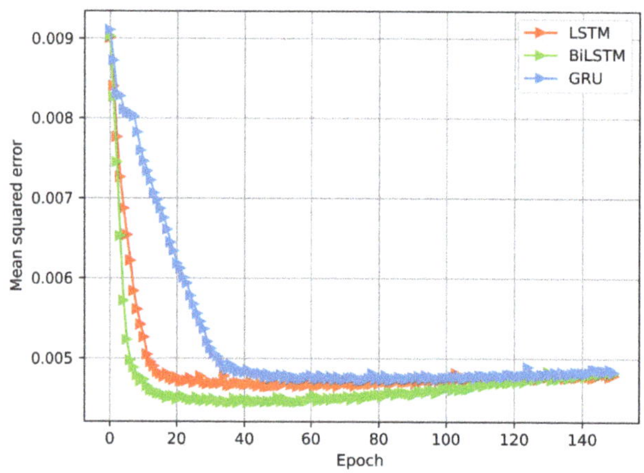

Figure 5. Mean squared error for 20-device network.

Additionally, we calculated the NMSE for both networks. The NMSE is calculated as follows:

$$NMSE = \frac{(\sum_{i=1}^{M} y_{estimated}(i) - y_{desired}(i))^2}{\sum_{i=1}^{M} y_{desired}(i)}, \tag{16}$$

where $y_{estimated}(i)$ is the i-th estimation from the ML model and $y_{desired}(i)$ is the ground truth.

The corresponding results are depicted in Figure 6 for the 10-device network. In the case of NMSE, LSTM and GRU show similar results, as depicted in Figure 6. The mean and standard deviation for the NMSE of LSTM are 0.278 ± 0.037. For GRU, the mean and standard deviation are 0.279 ± 0.033, which are similar to those of LSTM. However, BiLSTM exhibits better performance than LSTM and GRU, as indicated by its curve being closer to the lowest NMSE value. For BiLSTM, the mean and standard deviation are 0.231 ± 0.036. Thus, for the 10-device network, BiLSTM demonstrates the best performance among the three approaches, as shown in Figure 6.

Figure 7 shows the NMSE results for the 20-device network. In this case, the mean and standard deviation for LSTM are 0.4156 ± 0.52 and for GRU are 0.4452 ± 0.0828. BiLSTM again shows better performance, with a mean and standard deviation of 0.4010 ± 0.04673.

For the 20-device network, BiLSTM shows overall better results compared to GRU and LSTM, as shown in Figure 7. While LSTM and GRU exhibit similar results, LSTM has slightly higher performance. In both the 10-device and 20-device network sizes, the NMSE exhibits a trend similar to that of MSE, reinforcing the superior performance of BiLSTM in reducing the error in interference-prone wireless communication channels.

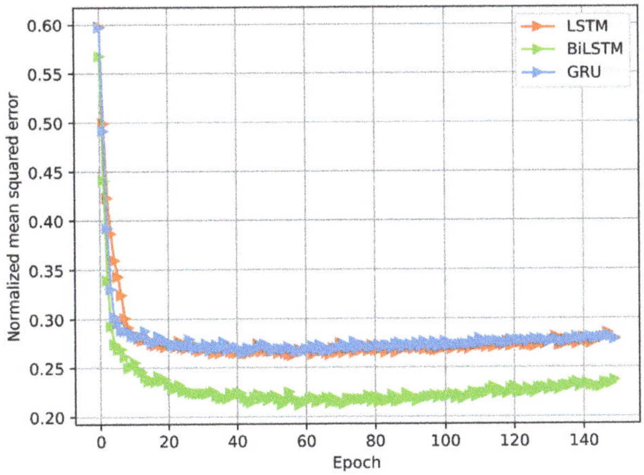

Figure 6. Normalized mean squared error for the 10-device network.

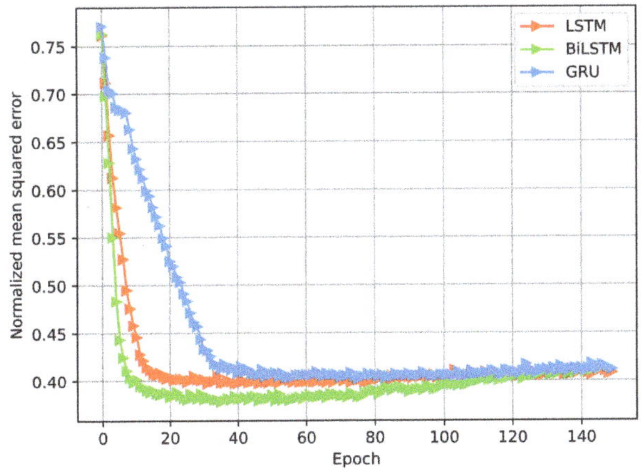

Figure 7. Normalized mean squared error for the 20-device network.

To assess the performance of complex RNN structures, we designed a model incorporating an LSTM layer, a GRU layer, and a BiLSTM layer. The input data first undergo processing by the LSTM layer, followed by a dropout layer with a retention rate of 85% (1–0.15). Similarly, the output from the LSTM layer feeds into the GRU layer, again followed by a dropout layer. Finally, the BiLSTM layer receives the processed data from the GRU layer, with another dropout layer applied. The output of the final dropout layer connects to a fully connected layer. As illustrated in Figure 8, for the 10-device network, the combined approach outperforms both the single-LSTM and -GRU models. However, the BiLSTM model still achieves superior performance compared to the combined model. Interestingly, when the number of devices in the network increases, the combined approach demonstrates significant improvement, surpassing the performance of single-layer models in terms of MSE, as shown in Figure 9.

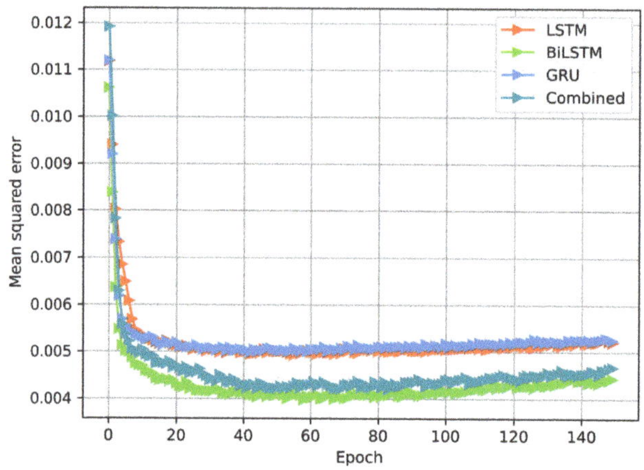

Figure 8. MSE performance comparison of a combined model (LSTM + GRU + BILSTM) with a single-layer model for the 10-device network.

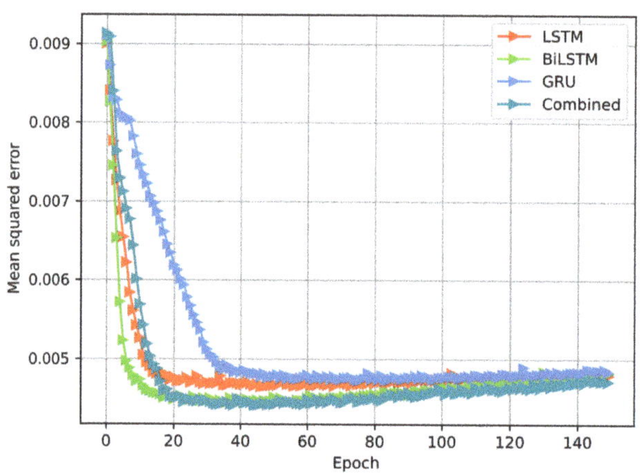

Figure 9. MSE performance comparison of a combined model (LSTM + GRU + BILSTM) with a single-layer model for the 20-device network.

Figure 10 presents the maximum achievable sum rate for the 10-device network. It is evident from the figure that BiLSTM achieves a higher sum rate compared to the other two approaches, while LSTM and GRU demonstrate similar sum rates. Specifically, LSTM slightly outperforms GRU, with mean sum rates of 1.7289 and 1.7264, respectively. In the case of the 20-device network shown in Figure 11, the overall performance of all approaches improves, and each approach approaches the maximum sum rate. In this experiment, we observed the better performance of BiLSTM, as it is closer to the maximum sum rate. LSTM secures the second position, with its curve being above that of GRU.

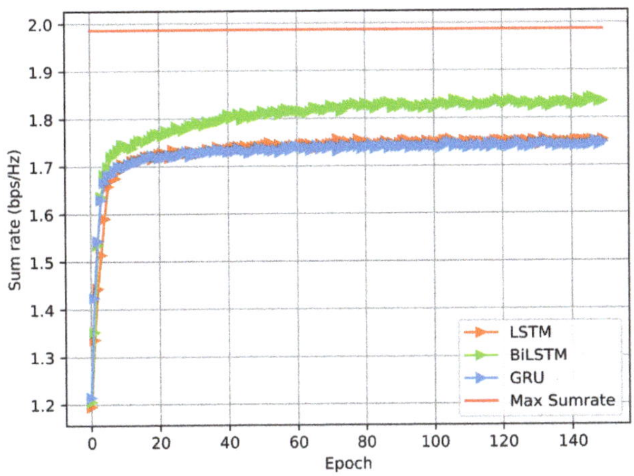

Figure 10. Sum rate achieved by different RNN approaches for the 10-device network.

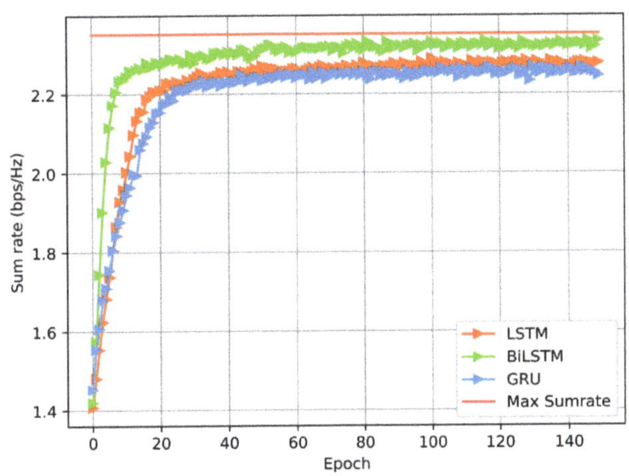

Figure 11. Sum rate achieved by different RNN approaches for the 20-device network.

Figure 12 compares the MSE of the proposed RNN approach with the study conducted in [44,48]. In the study by Sun et al. [44], deep neural networks were adopted to reduce interference in the wireless channel. Additionally, Chun et al. [48] proposed a deep learning model based on successive interference cancellation. The comparison reveals that the proposed RNN approach achieves a lower MSE in the validation set compared to the approaches described in [44,48].

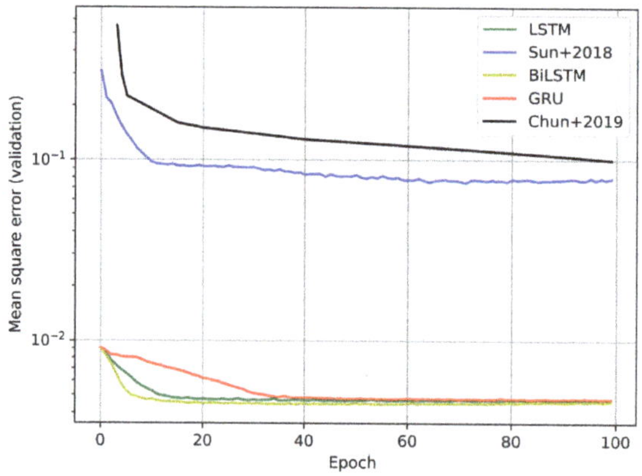

Figure 12. MSE comparison of different RNN approaches and other methods [44,48].

5. Conclusions

In this paper, we investigated the application of recurrent neural networks, specifically LSTM, BiLSTM, and GRU, to mitigate the effects of co-channel interference in wireless communication channels. We modeled two network scenarios: one with 10 devices and another with 20 devices. Data from each network were transmitted with controlled interference, and the received, corrupted data were used to train the RNN models. Our experiments revealed that BiLSTM achieved superior performance compared to LSTM and GRU, likely due to its enhanced ability to extract relevant features from the training data. Notably, LSTM and GRU exhibited similar results across both network sizes. However, a key limitation of this study is the potential performance degradation when the number of devices increased significantly. This may be attributed to the growing complexity of the interference environment and the limitations of the training data size. Future research directions can explore various avenues. As the number of devices in a network is expected to grow exponentially, the investigation of the performance of these RNN models with a massive swarm of devices would be valuable. Additionally, real-world communication channels experience dynamic interference due to environmental factors. Therefore, the incorporation of the dynamic changes in the interference into the training data could enhance the model's robustness. Furthermore, exploring more complex interference structures and new wireless communication channel models, such as intelligent reflecting surface-based communication and massive MIMO, could provide promising avenues for future research on interference mitigation using RNNs.

Author Contributions: Conceptualization, M.A.S.S. and M.H.R.; methodology, M.A.S.S., R.T. and Y.-H.Y.; software, M.A.S.S., M.H.R., M.A.A. and R.T.; validation, M.A.S.S., D.-D.H. and M.H.R.; formal analysis, M.A.S.S., M.H.R. and M.A.A.; investigation, M.A.S.S., R.T. and M.A.A.; resources, H.-K.S.; data curation, M.A.S.S. and R.T.; writing—original draft preparation, M.A.S.S.; writing—review and editing, M.A.S.S., M.H.R., M.A.A. and H.-K.S.; visualization, M.A.S.S., M.H.R. and M.A.A.; supervision, H.-K.S., Y.-H.Y. and D.-D.H.; project administration, H.-K.S.; funding acquisition, H.-K.S. All authors have read and agreed to the published version of the manuscript.

Funding: This work was supported by the Institute of Information & Communications Technology Planning & Evaluation (IITP) under the Metaverse support program to nurture the best talents (IITP-2024-RS-2023-00254529). The grant was funded by the Korean government (MSIT) and in part by the Basic Science Research Program through the National Research Foundation of Korea (NRF) funded by the Ministry of Education (2020R1A6A1A03038540), and it was also funded in part by the National Research Foundation of Korea (NRF) grant funded by the Korea Government (MSIT) (RS-2023-00219051).

Data Availability Statement: The data will be made available by the authors on request.

Conflicts of Interest: The authors declare no conflicts of interest.

References

1. Wu, Q.; Li, G.Y.; Chen, W.; Ng, D.W.K.; Schober, R. An overview of sustainable green 5G networks. *IEEE Wirel. Commun.* **2017**, *24*, 72–80. [CrossRef]
2. Duan, W.; Gu, J.; Wen, M.; Zhang, G.; Ji, Y.; Mumtaz, S. Emerging Technologies for 5G-IoV Networks: Applications, Trends and Opportunities. *IEEE Netw.* **2020**, *34*, 283–289. [CrossRef]
3. Alzubaidi, O.T.H.; Hindia, M.N.; Dimyati, K.; Noordin, K.A.; Wahab, A.N.A.; Qamar, F.; Hassan, R. Interference challenges and management in B5G network design: A comprehensive review. *Electronics* **2022**, *11*, 2842. [CrossRef]
4. Siddiqui, M.U.A.; Qamar, F.; Ahmed, F.; Nguyen, Q.N.; Hassan, R. Interference management in 5G and beyond network: Requirements, challenges and future directions. *IEEE Access* **2021**, *9*, 68932–68965. [CrossRef]
5. Dangi, R.; Lalwani, P.; Choudhary, G.; You, I.; Pau, G. Study and investigation on 5G technology: A systematic review. *Sensors* **2021**, *22*, 26. [CrossRef] [PubMed]
6. Goldsmith, A. *Wireless Communications*; Cambridge University Press: Cambridge, UK, 2005.
7. Wang, J. CFAR-based interference mitigation for FMCW automotive radar systems. *IEEE Trans. Intell. Transp. Syst.* **2021**, *23*, 12229–12238. [CrossRef]
8. Qaisar, Z.H.; Irfan, M.; Ali, T.; Ahmad, A.; Ali, G.; Glowacz, A.; Glowacz, W.; Caesarendra, W.; Mashraqi, A.M.; Draz, U.; et al. Effective beamforming technique amid optimal value for wireless communication. *Electronics* **2020**, *9*, 1869. [CrossRef]
9. Zhang, J.; Zhang, J.; Björnson, E.; Ai, B. Local partial zero-forcing combining for cell-free massive MIMO systems. *IEEE Trans. Commun.* **2021**, *69*, 8459–8473. [CrossRef]
10. Zambianco, M.; Verticale, G. Interference minimization in 5G physical-layer network slicing. *IEEE Trans. Commun.* **2020**, *68*, 4554–4564. [CrossRef]
11. Gu, Z.; Vucetic, B.; Chikkam, K.; Aliberti, P.; Hardjawana, W. Graph Representation Learning for Contention and Interference Management in Wireless Networks. *IEEE/ACM Trans. Netw.* **2024**, 1–16. [CrossRef]
12. Irkicatal, O.N.; Ceran, E.T.; Yuksel, M. Deep Reinforcement Learning Enhanced Rate-Splitting Multiple Access for Interference Mitigation. *arXiv* **2024**, arXiv:2403.05974.
13. Ding, X.; Lei, Y.; Zou, Y.; Zhang, G.; Hanzo, L. Interference Management by Harnessing Multi-Domain Resources in Spectrum-Sharing Aided Satellite-Ground Integrated Networks. *IEEE Trans. Veh. Technol.* **2024**, 1–16. [CrossRef]
14. Wang, Y.; Sun, M.; Cui, Q.; Chen, K.C.; Liao, Y. RIS-aided proactive mobile network downlink interference suppression: A deep reinforcement learning approach. *Sensors* **2023**, *23*, 6550. [CrossRef] [PubMed]
15. Gul, O.M.; Kulhandjian, M.; Kantarci, B.; Touazi, A.; Ellement, C.; D'amours, C. Secure industrial iot systems via rf fingerprinting under impaired channels with interference and noise. *IEEE Access* **2023**, *11*, 26289–26307. [CrossRef]
16. Voulodimos, A.; Doulamis, N.; Doulamis, A.; Protopapadakis, E. Deep learning for computer vision: A brief review. *Comput. Intell. Neurosci.* **2018**, *2018*, 7068349. [CrossRef]
17. Lee, H.; Huang, C.; Yune, S.; Tajmir, S.H.; Kim, M.; Do, S. Machine friendly machine learning: Interpretation of computed tomography without image reconstruction. *Sci. Rep.* **2019**, *9*, 1–9. [CrossRef] [PubMed]
18. Kadhim, A.I. Survey on supervised machine learning techniques for automatic text classification. *Artif. Intell. Rev.* **2019**, *52*, 273–292. [CrossRef]
19. Seota, S.B.W.; Klein, R.; Van Zyl, T. Modeling e-behaviour, personality and academic performance with machine learning. *Appl. Sci.* **2021**, *11*, 10546. [CrossRef]
20. Dornadula, V.N.; Geetha, S. Credit card fraud detection using machine learning algorithms. *Procedia Comput. Sci.* **2019**, *165*, 631–641. [CrossRef]
21. Khanal, S.S.; Prasad, P.; Alsadoon, A.; Maag, A. A systematic review: Machine learning based recommendation systems for e-learning. *Educ. Inf. Technol.* **2020**, *25*, 2635–2664. [CrossRef]
22. Santos, I.; Castro, L.; Rodriguez-Fernandez, N.; Torrente-Patino, A.; Carballal, A. Artificial neural networks and deep learning in the visual arts: A review. *Neural Comput. Appl.* **2021**, *33*, 121–157. [CrossRef]
23. Haque, R.; Islam, N.; Islam, M.; Ahsan, M.M. A comparative analysis on suicidal ideation detection using NLP, machine, and deep learning. *Technologies* **2022**, *10*, 57. [CrossRef]
24. Hu, S.; Chen, X.; Ni, W.; Hossain, E.; Wang, X. Distributed Machine Learning for Wireless Communication Networks: Techniques, Architectures, and Applications. *IEEE Commun. Surv. Tutor.* **2021**, *23*, 1458–1493. [CrossRef]
25. Sejan, M.A.S.; Rahman, M.H.; Shin, B.S.; Oh, J.H.; You, Y.H.; Song, H.K. Machine learning for intelligent-reflecting-surface-based wireless communication towards 6G: A review. *Sensors* **2022**, *22*, 5405. [CrossRef] [PubMed]
26. Rahman, M.H.; Sejan, M.A.S.; Aziz, M.A.; Tabassum, R.; Baik, J.I.; Song, H.K. A Comprehensive Survey of Unmanned Aerial Vehicles Detection and Classification Using Machine Learning Approach: Challenges, Solutions, and Future Directions. *Remote Sens.* **2024**, *16*, 879. [CrossRef]

27. Ahmed, S.F.; Alam, M.S.B.; Hassan, M.; Rozbu, M.R.; Ishtiak, T.; Rafa, N.; Mofijur, M.; Shawkat Ali, A.; Gandomi, A.H. Deep learning modelling techniques: Current progress, applications, advantages, and challenges. *Artif. Intell. Rev.* **2023**, *56*, 13521–13617. [CrossRef]
28. Du, J.; Jiang, C.; Wang, J.; Ren, Y.; Debbah, M. Machine learning for 6G wireless networks: Carrying forward enhanced bandwidth, massive access, and ultrareliable/low-latency service. *IEEE Veh. Technol. Mag.* **2020**, *15*, 122–134. [CrossRef]
29. Aziz, M.A.; Rahman, M.H.; Sejan, M.A.S.; Baik, J.I.; Kim, D.S.; Song, H.K. Spectral Efficiency Improvement Using Bi-Deep Learning Model for IRS-Assisted MU-MISO Communication System. *Sensors* **2023**, *23*, 7793. [CrossRef] [PubMed]
30. Hsu, J.Y.; Wang, Y.F.; Lin, K.C.; Chen, M.Y.; Hsu, J.H.Y. Wind turbine fault diagnosis and predictive maintenance through statistical process control and machine learning. *IEEE Access* **2020**, *8*, 23427–23439. [CrossRef]
31. Gao, J.; Heng, F.; Yuan, Y.; Liu, Y. A novel machine learning method for multiaxial fatigue life prediction: Improved adaptive neuro-fuzzy inference system. *Int. J. Fatigue* **2024**, *178*, 108007. [CrossRef]
32. Zhu, G.; Liu, D.; Du, Y.; You, C.; Zhang, J.; Huang, K. Toward an intelligent edge: Wireless communication meets machine learning. *IEEE Commun. Mag.* **2020**, *58*, 19–25. [CrossRef]
33. Salehinejad, H.; Sankar, S.; Barfett, J.; Colak, E.; Valaee, S. Recent advances in recurrent neural networks. *arXiv* **2017**, arXiv:1801.01078.
34. Cossu, A.; Carta, A.; Lomonaco, V.; Bacciu, D. Continual learning for recurrent neural networks: An empirical evaluation. *Neural Netw.* **2021**, *143*, 607–627. [CrossRef] [PubMed]
35. Baliyan, A.; Batra, A.; Singh, S.P. Multilingual sentiment analysis using RNN-LSTM and neural machine translation. In Proceedings of the 2021 8th International Conference on Computing for Sustainable Global Development (INDIACom), New Delhi, India, 17–19 March 2021; pp. 710–713.
36. Onan, A. Bidirectional convolutional recurrent neural network architecture with group-wise enhancement mechanism for text sentiment classification. *J. King Saud Univ.-Comput. Inf. Sci.* **2022**, *34*, 2098–2117. [CrossRef]
37. Zhou, Y.; Samiee, A.; Zhou, T.; Jalali, B. Deep learning interference cancellation in wireless networks. *arXiv* **2020**, arXiv:2009.05533.
38. Bhatia, A.; Robinson, J.; Carmack, J.; Kuzdeba, S. FPGA implementation of radio frequency neural networks. In Proceedings of the 2022 IEEE 12th Annual Computing and Communication Workshop and Conference (CCWC), Las Vegas, NV, USA, 26–29 January 2022; pp. 0613–0618.
39. Grunau, S.; Block, D.; Meier, U. Multi-label wireless interference identification with convolutional neural networks. *arXiv* **2018**, arXiv:1804.04395.
40. Rock, J.; Roth, W.; Toth, M.; Meissner, P.; Pernkopf, F. Resource-efficient deep neural networks for automotive radar interference mitigation. *IEEE J. Sel. Top. Signal Process.* **2021**, *15*, 927–940. [CrossRef]
41. Rock, J.; Toth, M.; Messner, E.; Meissner, P.; Pernkopf, F. Complex signal denoising and interference mitigation for automotive radar using convolutional neural networks. In Proceedings of the 2019 22th International Conference on Information Fusion (FUSION), Ottawa, ON, Canada, 2–5 July 2019; pp. 1–8.
42. Mun, J.; Kim, H.; Lee, J. A deep learning approach for automotive radar interference mitigation. In Proceedings of the 2018 IEEE 88th Vehicular Technology Conference (VTC-Fall), Chicago, IL, USA, 27–30 August 2018; pp. 1–5.
43. Wei-Lung, M. Gps interference mitigation using derivative-free kalman filter-based rnn. *Radioengineering* **2016**, *25*, 519.
44. Sun, H.; Chen, X.; Shi, Q.; Hong, M.; Fu, X.; Sidiropoulos, N.D. Learning to optimize: Training deep neural networks for interference management. *IEEE Trans. Signal Process.* **2018**, *66*, 5438–5453. [CrossRef]
45. Rahman, M.H.; Sejan, M.A.S.; Aziz, M.A.; You, Y.H.; Song, H.K. HyDNN: A Hybrid Deep Learning Framework Based Multiuser Uplink Channel Estimation and Signal Detection for NOMA-OFDM System. *IEEE Access* **2023**, *11*, 66742–66755. [CrossRef]
46. Rahman, M.H.; Sejan, M.A.S.; Aziz, M.A.; Kim, D.S.; You, Y.H.; Song, H.K. Deep Convolutional and Recurrent Neural-Network-Based Optimal Decoding for RIS-Assisted MIMO Communication. *Mathematics* **2023**, *11*, 3397. [CrossRef]
47. Freire, P.; Srivallapanondh, S.; Spinnler, B.; Napoli, A.; Costa, N.; Prilepsky, J.E.; Turitsyn, S.K. Computational Complexity Optimization of Neural Network-Based Equalizers in Digital Signal Processing: A Comprehensive Approach. *J. Light. Technol.* **2024**, 1–25. [CrossRef]
48. Chun, C.J.; Kang, J.M.; Kim, I.M. Deep learning-based joint pilot design and channel estimation for multiuser MIMO channels. *IEEE Commun. Lett.* **2019**, *23*, 1999–2003. [CrossRef]

mathematics

MDPI

Article

Deep Bidirectional Learning Based Enhanced Outage Probability for Aerial Reconfigurable Intelligent Surface Assisted Communication Systems

Md Habibur Rahman [1,2], Mohammad Abrar Shakil Sejan [1,2], Md Abdul Aziz [1,2], Rana Tabassum [1,2] and Hyoung-Kyu Song [1,2,*]

[1] Department of Information and Communication Engineering, Sejong University, Seoul 05006, Republic of Korea; habibur@sju.ac.kr (M.H.R.); sejan@sejong.ac.kr (M.A.S.S.); aziz@sju.ac.kr (M.A.A.); tanvi@sju.ac.kr (R.T.)

[2] Department of Convergence Engineering for Intelligent Drone, Sejong University, Seoul 05006, Republic of Korea

* Correspondence: songhk@sejong.ac.kr

Abstract: The reconfiguration of wireless channels with reconfigurable reflecting surface (RIS) technology offers new design options for future wireless networks. Due to its high altitude and increased probability of establishing line-of-sight linkages with ground source/destination nodes, aerial RIS (ARIS) has greater deployment flexibility than traditional terrestrial RIS. It also provides a wider-view signal reflection. To leverage the advantages of ARIS-enabled systems, this paper defines air-to-ground linkages via Nakagami-m small-scale fading and inverse-Gamma large-scale shadowing, considering realistic composite fading channels. To construct a tight approximate closed-form formula for the outage probability (OP), a new mathematical framework is proposed. Additionally, a deep-learning-based system called the BiLSTM model is deployed to evaluate OP performance in the 3D spatial movement of the ARIS system. In the offline phase, the proposed model is trained with real-value channel state estimation sets and enhances OP performance in the online phase by learning channel information in a bidirectional manner. Simulation results demonstrate that the proposed BiLSTM model outperforms all other models in analyzing OP for the ARIS system.

Keywords: aerial RIS system; deep learning; BiLSTM; outage probability analysis; Namkagami-m fading

MSC: 94A14

1. Introduction

As the fifth-generation (5G) wireless communication network unfolds globally, the exploration of the sixth-generation (6G) wireless network is now underway. 6G aims to establish pervasive connectivity with data rates surpassing 100 to 1000 times those of 5G, reaching an impressive 1 Terabyte per second [1,2]. For this purpose, prominent wireless transmission technologies like Terahertz communication and ultra-massive multiple-input multiple-output have become focal points of extensive research attention [3]. Flexible and mobile wireless smart communication environments are encountering challenges in sustaining direct line-of-sight connections between mobile devices or IoT devices and base stations, particularly in obstructed or densely populated urban settings. A potential solution to overcome these obstacles is the utilization of reconfigurable intelligent surfaces (RIS) [4]. RIS operates by manipulating the propagation of wireless signals and enhancing or suppressing them through reflective or refractive mechanisms. The recent advancements in smart radio systems, which enable the control of wireless signal operations, have paved the way for exploring novel applications of RIS in advanced wireless communication networks beyond the realms of 5G and into the domain of 6G. Various strategies have initially employed stationary RIS installed on buildings and utilized unmanned aerial vehicles (UAVs) as relays to enhance different communication metrics, including transmission rate, throughput, age of information, energy efficiency, sum secrecy rate, and signal-to-interference-plus-noise ratio, primarily in non-adversarial scenarios [5–8].

The research focus has shifted towards terrestrial RIS-assisted communication, primarily used within interior walls, ceilings, or building facades. However, this architecture presents several fundamental challenges. Firstly, determining optimal RIS installation locations can be complex due to factors like site rent, urban environment effects, and property-owner permissions. **Comprehensive RIS placement:** Using the strategy depicted in Figure 1, it can be identified by a set of all potential spots in the layout that are organized in a grid, denoted as G_R, with a predefined distance, d_R, separating each position. The orientation of the RIS panels is equally distributed within the range $[0, 2\pi]$ at every position in G_R. It is crucial to note that RIS grid coordinates lying inside any prohibited regions are not considered when placing the RIS, as the line of sight (LOS) for the base station (BS)-RIS link would be obstructed. After determining the placement of RIS within each blocked region, the shortest distance is calculated between each identified RIS position and the boundaries of the corresponding blocked region [9].

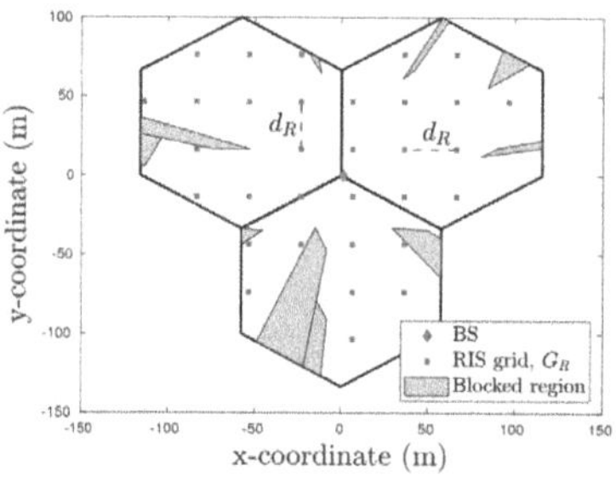

Figure 1. RIS placement grid inside an urban cell's three hexagonal cell sectors and around a single macrosite, divided by a predetermined distance [9].

Site rent: It is assumed that a telecom operator wishes to improve signal coverage in a highly populated metropolitan region by implementing RIS technology. When installing RIS panels on utility poles or buildings, they must contend with the expense of site leasing. Deploying RIS in other regions may not be as feasible due to high site leasing expenses. To reduce these expenses, companies could have to bargain with property owners or look into different deployment techniques. **Effects of the urban environment:** RIS technology is being implemented in a busy city center characterized by tall buildings and winding streets. The urban setting adds complications such as restricted line-of-sight chances, multipath propagation, and signal interference from nearby structures. These surrounding conditions may have an impact on how well RIS deployment works. To determine the best places for RIS installations to reduce the impacts of urban morphology on signal transmission, sophisticated modeling approaches like ray-tracing simulations could be required. **Permissions from property owners:** It is assumed that a telecom company wants to enhance network coverage by installing RIS panels on private properties. However, it becomes logistically difficult to obtain property owners' consent to put the panels on their buildings or other structures. Some property owners can be worried about liability, possible structural damage, or aesthetics. Deployment plans can be postponed or stopped completely in the absence of sufficient authorization. RIS installation approvals may need to be obtained through revenue-sharing or incentive agreements, property-owner negotiations, and resolving their concerns. Secondly, the RIS positioned on building walls or facades only serves terminals on one side, limiting coverage as shown in Figure 2a. Thirdly, radio signals undergo multiple reflections in urban environments, leading to signal scattering and attenuation even with RIS passive beamforming which is shown in Figure 3a. To address these issues, this paper proposes a novel three-dimensional (3D) wireless network enabled by aerial RIS (ARIS), where RIS is deployed on aerial platforms such as balloons or UAVs for intelligent reflection from the sky. ARIS offers several advantages over traditional terrestrial RIS. Firstly, it establishes line-of-sight (LoS) communication with ground nodes more effectively due to its elevated position [10], enhancing channel strength and enabling flexible optimization of aerial platform location or trajectory. Secondly, ARIS enables panoramic/full-angle reflection, allowing it to theoretically assist in reflecting signals between any pair of ground-based nodes shown in Figure 2b, unlike terrestrial RIS limited to half of the available space. Lastly, ARIS with its high probability of LoS linkages typically achieves required signal modification by a single reflection even in complex urban environments (see Figure 3b), significantly reducing signal power loss compared to terrestrial RIS. For example, the advantages of ARIS over traditional terrestrial RIS are an important factor in enabling future wireless communication non-orthogonal multiple access (NOMA) systems. The authors in [11] showed, through a comparison of ARIS-NOMA with the outcomes of terrestrial RIS-NOMA and ADF-NOMA deployment, that the energy efficiency curves versus transmit power for various reflecting element counts and their corresponding ARIS-NOMA, terrestrial RIS-NOMA, and ADF-NOMA schemes are depicted in Figure 4. It is evident that ARIS-NOMA systems may achieve higher levels of energy efficiency compared to conventional terrestrial RIS deployment. The analysis indicates that greater performance benefits can be obtained with less transmit power by investigating aerial mobility than with traditional terrestrial RIS-NOMA.

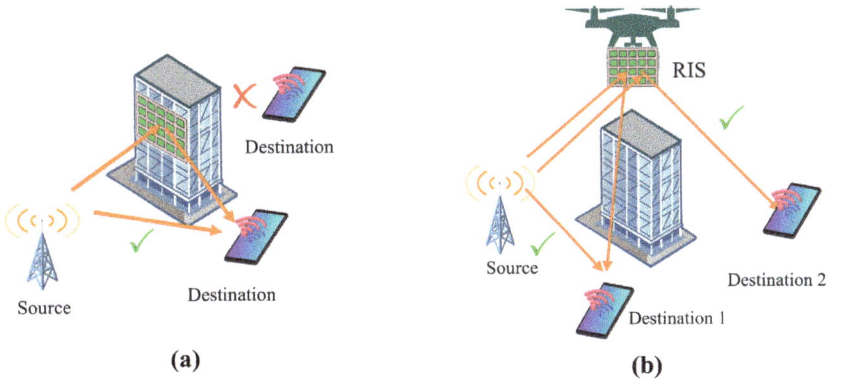

Figure 2. The comparison of reflection; (**a**) half-space reflection through terrestrial RIS; (**b**) panoramic/full-angle reflection of ARIS.

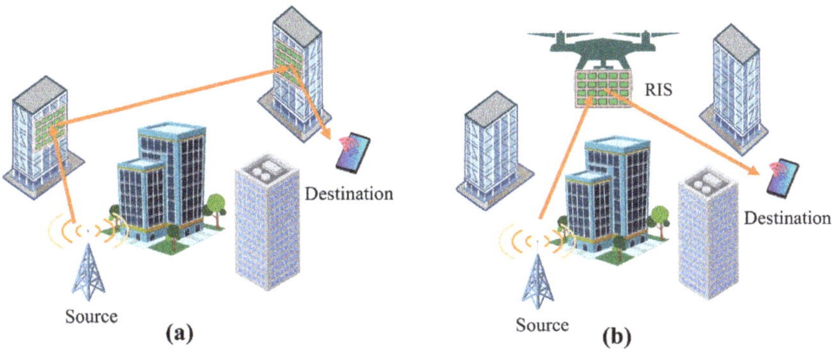

Figure 3. The comparison of lower the number of reflections; (**a**) terrestrial RIS; (**b**) ARIS.

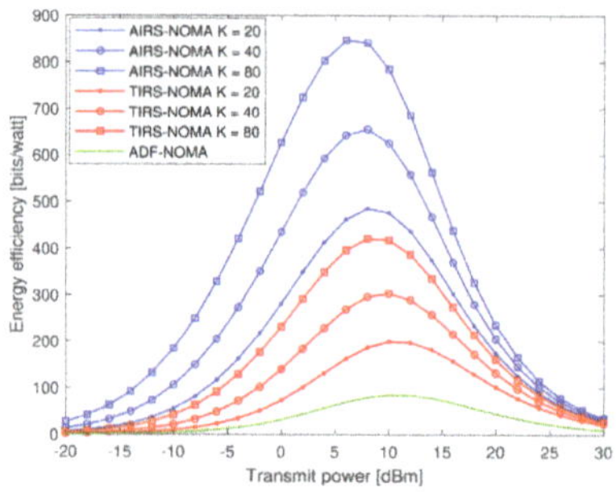

Figure 4. The transmit power versus energy efficiency curves for varying numbers of reflecting elements in the ARIS-NOMA, terrestrial RIS-NOMA, and ADF-NOMA systems [11].

In recent years, deep learning (DL)-based techniques have gained attention in the ARIS-assisted system due to the prominent benefits of performance increment [12–17]. In ref. [12], the authors suggested the utilization of the deep deterministic policy gradient as a model-free off-policy reinforcement learning (RL) method to maximize the MU-MISO system's sum rate. This is achieved by the cooperative optimization of the phase shifters of the passive reflecting components at the ARIS and the active beamforming matrix at the ground base station. In the context of a UAV eavesdropper, an efficient ARIS-assisted PLS technique in [13] maximizes the average downlink secrecy rate of several terrestrial ground terminals using a DDPG. The authors in [14] maximized the average secrecy rate by jointly investigating the reflection of RISs and the flying trajectory of the ARIS by deploying an RL model.

In the progression towards 6G technology, non-terrestrial communication methods are acknowledged as dependable tools for establishing connections in remote and challenging terrains. However, the dynamic nature of spatial movements necessitates the development of novel fading models that can accurately depict both the downlink and uplink channels in air-to-ground (A2G) communication scenarios. Research on A2G communication, particularly focusing on UAV-aided dual-hop relaying and outage probability (OP) analysis, has gained significant attention due to its potential applications in various fields like surveillance, disaster management, and communication in remote areas. A2G channel modeling for UAV-assisted communication was discussed in [18–22]. The authors in [18] presented the ideal height of the UAV in terms of communication coverage in the A2G channel, as well as the ideal height and location of UAVs while acting as relays. The authors of [19] examined how the interfering node affects dependable UAV communications. They calculated the OP in closed form for all feasible situations (A2G, G2A, G2G, A2A) of main and interference connections after describing the channel model impacted by the communication link's elevation angle. The study in [20] developed a LOS probability expression for A2G channels using urban scenario statistics and geometry. Their simplified parametric model expedites predictions and aids in channel modeling. Leveraging this analytical model, they trained a back-propagation neural network for parameter generation. Simulation results confirmed its versatility across altitudes and agreement with Monte Carlo data. In a densely populated urban setting, the authors in [21] calculated the likelihood of a loss of service for A2G connection based on the drone's height and the typical building height. Additionally, the study [22] discussed UAV-based relaying in urban contexts in conjunction with energy harvesting. They considered A2G channels which are classified as either shadowing Rayleigh fading or shadowed Rician fading. The authors created a closed-form expression for OP and a cooperative channel model for a space-to-ground network. A 3D UAV stochastic mixed mobility (MM) model mechanism allows a UAV to move either vertically by adhering to the random waypoint mobility (RWPM) model or experiencing a spatial digression determined by the random movement model for raw walk [23,24]. The following mechanisms, such as trajectory control and altitude control, define the 3D

UAV movement process. The analytically tractable MM model may be used to evaluate ARIS network performance. However, UAVs are more likely to travel in a uniform spatial manner. As a result, developing an efficient 3D mobility model for UAV movement that accounts for both of the aforementioned mechanisms remains very challenging. Furthermore, the 3D mobility model's steady-state distribution of UAV positions must be tractable enough to allow for UE performance monitoring inside the network [24]. Therefore, this study presents a mixed mobility model for the 3D UAV movement process, with the RWPM models characterizing the vertical and spatial motions. Next, the coverage probability of the interference received at a reference ground destination under inverse-Gamma shadowing and Nakagami-m fading is assessed based on the suggested 3D mixed mobility model. In the study [25], the OP of UAV-aided dual-hop relaying was investigated, taking into account 3D spatial movements. Another recent work [26] utilized empirical data from A2G trials to model A2G channels, employing Nakagami-m multipath fading and inverse-Gamma shadowing. Deep neural networks (DNNs) have shown promising effectiveness in predicting the secrecy performance of A2G communications and OP in ARIS-assisted wireless communication networks. By leveraging their ability to learn complex patterns from data, DNNs can capture the intricate relationships between various parameters affecting secrecy and OP. DNN represents the connection between system characteristics and the associated secure OP, which can accurately forecast the secure OP's performance for a range of interesting real-world scenarios [15]. This enables more accurate predictions, aiding network optimization and resource allocation. In a different approach [15], DNN were employed to predict the secrecy performance of A2G communications. Furthermore, considering the ARIS's 3D spatial mobility, the authors in [27] constructed a DNN that can precisely anticipate the OP. In order to obtain consistent performance and a notable improvement in secure performance, in [16], authors suggested a DNN-based method that additionally takes use of ARIS in the non-orthogonal multiple access assisted wireless system. In ref. [17], the authors suggested the use of ARIS in a relaying system to preserve the accuracy of data on distant Internet of Things wireless networks using the RL model. The goal is to reduce the predicted cumulative age of information by optimizing the RIS components' phase-shift matrix, communication schedule, and UAV elevation.

In a feed-forward neural network execution, RNN is capable of processing input sequences of different durations across time. RNNs remember past occurrences and use knowledge to predict future results. To regulate access to memory cells, an RNN variation called LSTM is created using certain gating approaches [28]. Sequencing problems are often solved using RNNs such as LSTM. Unlike the hidden unit in a typical RNN design, the LSTM has a gate architecture that allows it to identify and store important information while rejecting irrelevant information. The BiLSTM network is better at sequence classification than LSTM as it allows for bidirectional data flow. By using information from both previous (backward) and subsequent (forward) events concurrently, BiLSTM outperforms LSTM in terms of accuracy. This architecture improves the flow of data in the BiLSTM network, which improves learning performance by capturing more characteristics in the input data [29]. For RIS-enabled multi-user multiple-input single-output downlink orthogonal frequency division multiplexing systems, in ref. [30], authors suggested a BiLSTM-model-based channel estimation and signal detection. By capturing more characteristics from the input data, this architecture improves learning performance as the data flow inside the BiLSTM network is optimized [31]. According to the study [32], the experiment's results show that BiLSTM extracts features more effectively than LSTM. A stacked BiLSTM architecture for RIS-assisted unmanned aerial vehicle communication systems was previously reported by [33]. The authors in [34] proposed a BiLSTM-based massive MIMO system for channel estimation, where the proposed model effectively reduces the pilot overhead. BiLSTM has difficulties with hyperparameter tweaking, data reliance, and complexity. However, BiLSTM networks provide benefits for OP analysis by capturing bidirectional information flow and sequential context. Furthermore, the BiLSTM is promising for OP estimation problems due to its automated feature representation and context awareness.

Motivated by the above research work and the effective advantages of the BiLSTM model in wireless systems, in this paper, we suggest an effective ARIS-assisted wireless communication network that is susceptible to a combined fading channel model, in which the large-scale shadowing is distributed according to an inverse-Gamma distribution, and the small-scale fading is governed by a Nakagami-m distribution. We examine the ARIS-assisted method's outage efficiency in such a realistic fading channel scenario. In order to provide a tight approximated closed-form formulation for the OP, we put forth a novel mathematical framework. Next, we examine a mobile setting in which the RWMM is used to simulate the 3D spatial movement of ARIS. Because the locations of the ARIS are represented as random variables in such a mobile framework, the outage investigation becomes intractable. Therefore, in the offline phase, a BiLSTM is employed, and trained by pre-collected OP data as well as related channel state information (CSI) to forecast the OP. The BiLSTM model effectively trains the channel data in the forward and backward directions and enhances the learning efficiency of the model. The Adam optimization algorithm is utilized to effectively lessen the loss. In the online phase, the trained model is deployed for testing using the test data and it is found that during testing the proposed model effectively performs the OP analysis. The OP performance of the proposed model outperforms the other models which are shown in the simulation results.

2. Methods

2.1. System Model

In this study, we employ a low-complexity system assuming a source/transmitter and a destination/receiver equipped with a single antenna. In addition, it is considered that the direction channel of source to destination is not applicable in this system due to the ground obstacles. Alternatively, the source-to-destination channel is connected by ARIS, T, and can work passively to reflect signals from source to destination. The designed system is illustrated in Figure 5. Consider that the ARIS consists of N reflective elements and the complex channel coefficient vectors between the source and ARIS and ARIS and destination is $\mathbf{h}_{\mathbf{ST}} \in \mathbb{C}^{N\times 1}$ and $\mathbf{h}_{\mathbf{TD}} \in \mathbb{C}^{N\times 1}$, respectively. Through phase-shift matrix $\mathbf{\Psi} = \kappa \mathrm{diag}(e^{j\phi_1}, \ldots, e^{j\phi_N})$, the properties of ARIS are defined where $\phi_t \in [0, 2\pi]$, $t = 1, \ldots, N$, is the phase-shift angle in the ARIS elements of t, and the reflection coefficient is $\kappa \in [0, 1]$ [35]. The transmit signal from source to destination is defined as $\mathbb{E}[|s|^2] = 1$. Therefore, the received signal at the destination can be formulated as follows:

$$y = \sqrt{P} \sum_{t=1}^{N} [\mathbf{h}_{\mathrm{ST}}]_t \kappa e^{j\phi_t} [\mathbf{h}_{\mathrm{TD}}]_t s + w, \tag{1}$$

where the transmission signal power of the source is P and the additive white Gaussian noise (AWGN) at the destination is $\mathbf{w} \sim \mathcal{CN}(0, \sigma^2)$ with zero mean and σ^2 variance.

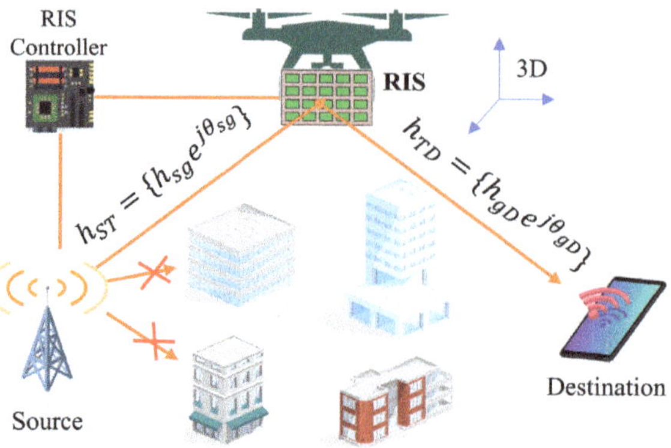

Figure 5. The architecture of ARIS-assisted communication scenario.

2.2. Channel Model

For the channel model, it is assumed that $[\mathbf{h_{ST}}]_t \triangleq \tilde{\mathbf{h}}_{St}$ and $[\mathbf{h_{TD}}]_t \triangleq \tilde{\mathbf{h}}_{tD}$. The expression of polar complex channel coefficient $\tilde{h}_{ccc}$ is as $\tilde{h}_{ccc} = h_{ccc}e^{j\theta_{ccc}}$, with $ccc \in \{St, tD\}$ where $h_{ccc} = |\tilde{h}_{ccc}|$ and $\theta_{ccc} \in [0, 2\pi]$. In a real-world composite fading channel, the small-scale fading, denoted by G_{ccc}, is represented by a Nakagami-m random variable, while the large-scale shadowing effects, denoted by L_{ccc}, follow an inverse-Gamma random variable. The composite fading channel, h_{ccc}, is then expressed as the product of these two components: $h_{ccc} = L_{ccc}G_{ccc}$. In more detail, G_{ccc}'s Nakagami-m probability density function (PDF) is provided as follows [27,36]:

$$f_{G_{ccc}}(x; m, \sigma) = 2\left(\frac{m}{\sigma}\right)^m \frac{1}{\xi(m)} x^{(2m-1)} e^{\frac{-m}{\sigma}x^2}, x > 0, \tag{2}$$

where the parameter of $m > 0$ is the shape parameter, indicating the severity of fading, and $\sigma > 0$ is the spread parameter of the distribution. However, if the L_{ccc} is given, the inverse-Gamma PDF is formulated as follows [27]:

$$f_{L_{ccc}}(x; \zeta, \delta) = \frac{\delta^{\zeta}}{\xi(\zeta)} x^{(-\zeta-1)} e^{\frac{-\delta}{x}}, x > 0, \tag{3}$$

where $\zeta > 1$ is the shaping parameter of the distribution, related to the severity of the shadowing, i.e., lower values of ζ result in lighter shadowing conditions, δ denotes the scaling parameter.

However, from (1), the proposed system's achievable capacity can be expressed as follows:

$$A_c = \max_{\phi_1,\ldots,\phi_N} \log_2\left(1 + \frac{P_S}{\tau^2}\left|\kappa \sum_{t=1}^{N} [\mathbf{h}_{\mathrm{ST}}]_t e^{j\phi_t} [\mathbf{h}_{\mathrm{TD}}]_t\right|^2\right). \tag{4}$$

2.3. Analysis of Outage Performance

This section examines the OP of the suggested communication system because it is typical practice to assess a communication system's performance and dependability using this measurement. A possibility that the system's instant mutual information would fall short of a predetermined goal spectral efficiency A_{cth}, [b/s/Hz], is known as the Op of the system. This can be stated mathematically as follows:

$$\mathrm{P_{out}} = \Pr(A_c < A_{\mathrm{cth}}) = \Pr(\eta < \eta_{\mathrm{th}}), \tag{5}$$

where $\eta_{\mathrm{th}} \triangleq 2^{A_{\mathrm{cth}}} - 1$ and at the destination, and η represents the end-to-end received signal-to-noise ratio (SNR), which can be formulated as follows:

$$\eta = \max_{\phi_1,\ldots,\phi_N} \bar{\eta}\left|\kappa \sum_{t=1}^{N} L_{St}G_{St}e^{j(\phi_t+\theta_{st}+\theta_{tD})}G_{tD}L_{tD}\right|^2, \tag{6}$$

where κ is the fixed amplitude reflection coefficient of ARIS, $L_t = L_{St}, L_{tD}$ is expressed as the end-to-end fading with respect to the one RIS element and $G_t = G_{St}, G_{tD}$ is expressed as the end-to-end shadowing with respect to the one RIS element, and the average transmit SNR [dB] is shown by $\bar{\eta} = P/\tau^2$. To solve (5), we first build up the phase-shift matrix, or $\mathbf{\Psi}^*$, in the best possible way so that (6) maximizes the end-to-end SNR. For each reflective element of ARIS, following the works [35,37,38], we configured the $\phi_t^* = -(\theta_{st} + \theta_{tD})\forall_t$. As a result, the OP in (5) can be rewritten as follows:

$$P_{\mathrm{out}} = \Pr\left(\bar{\eta}\kappa^2\left|\sum_{t=1}^{N} L_{St}G_{St}G_{tD}L_{tD}\right|^2 < \eta_{\mathrm{th}}\right). \tag{7}$$

In (7), derived the expression OP P_{out}, which is the probability that the received SNR falls below a certain threshold η_{th} in a wireless communication system. This expression is derived by analyzing the statistical properties of the received SNR, path loss, shadowing, and fading, resulting in an approximate closed-form expression for the OP P_{out} analysis. It is assumed that $W_t \triangleq L_{St}G_{St}G_{tD}L_{tD}, G_t \triangleq G_{St}G_{tD}, L_t \triangleq L_{St}L_{tD}$, and $\tilde{L}_t \triangleq \frac{1}{\sqrt{L_t}}$ and a [Gamma/(Gamma2)] distribution can be matched to the distribution of W_t, whose theoretical details are provided in Theorem 1. To achieve this, we first suggest a distribution matching for G_t and $\tilde{L}_t$, which is shown in Lemma 1. Assumed X is a Gamma random variable, and the PDF is provided as follows:

$$f_X(x;\mu,\varsigma) = \frac{1}{\varsigma^{\mu}\xi(\mu)} x^{(\mu-1)} e^{\frac{-x}{\varsigma}}, x > 0, \tag{8}$$

where it is indicated that $X \sim$ Gamma (μ,ς), where μ and ς stand for the shape and scale parameters, respectively.

However, Lemma 1 is given for the G_t and $\tilde{L}_t$ distributions, which are roughly matched to a Gamma distribution in the manner described as follows:

$$G_t \overset{\text{approx.}}{\sim} \text{Gamma}(m_G, \sigma_G/m_G), \tag{9}$$

$$\tilde{L}_t \overset{\text{approx.}}{\sim} \text{Gamma}(m_L, \sigma_L/m_L), \tag{10}$$

where $\sigma_G \triangleq \frac{\xi(m_s+1/2)\xi(m_d+1/2)}{\xi(m_s)\xi(m_d)} Y_G^{-1/2}$, $Y_G \triangleq \frac{m_s m_d}{\sigma_s \sigma_d}$, $m_G \triangleq \frac{\sigma_G^2}{\sigma_s\sigma_d - \sigma_G^2}$, $\sigma_L \triangleq \frac{\xi(\zeta_s+1/2)\xi(\zeta_d+1/2)}{\sqrt{\delta_s\delta_d}\xi(\zeta_s)\xi(\zeta_d)}$, $m_L \triangleq \frac{\sigma_L^2}{\left(\frac{\zeta_s\zeta_d}{\delta_s\delta_d}\right) - \sigma_L^2}$. The proof of Lemma 1 is described in [27].

$$F_{|Z|^2}(z) \approx \frac{1}{\xi(Nm_G+1)} z^{\frac{Nm_G}{2}} \sum_{\Xi_\tau,\forall\tau} \binom{N}{\tau_1,\ldots,\tau_K} \left[\prod_{k=1}^{K} \aleph_k^{v_k}\right] \Phi_2^{(K)}\left(m_G\tau_1,\ldots,m_G\tau_K; 1+Nm_G; -\frac{\sqrt{z}}{\varsigma_1},\ldots,-\frac{\sqrt{z}}{\varsigma_K}\right), \tag{11}$$

The closed-form formulas for the OPs, which are solely dependent on fading coefficients, are provided by Theorem 1. To determine the OP of the proposed ARIS-aided system, we offer a technological framework. Specifically, we employ the moment-matching approach to fit the distribution of the product of four separate random variables to that of a [Gamma/(Gamma2)] random variable, thus avoiding the emergence of parabolic cylinder functions. Gaussian–Laguerre quadrature is used to restructure the random variable and then convert it into a mixed Gamma random variable once more. The distribution of the sum of many converted mixed Gamma random variables is then obtained using the Laplace transform. We derive a tight approximate closed-form equation for the system OP with the help of our suggested technical framework [27].

Theorem 1. *Let consider* $Z \triangleq \sum_{t=1}^{N} W_t = \sum_{t=1}^{N} L_{St}G_{St}G_{tD}L_{tD}$, *for the cumulative distribution function (PDF) of* $|Z|^2$, *an approximate closed-form formula may be obtained, as in (11), top of the next page, where* $\binom{N}{\tau_1,\ldots,\tau_K} \triangleq \frac{N!}{\tau_1!\ldots\tau_K!}$, $\Xi_\tau \triangleq (\tau_1,\ldots,\tau_K)^{(\tau)}$ *signifies a potential pairing of* $\tau_1,\ldots,\tau_K$, *in which* $\tau_K \in \tau_1,\ldots,\tau_K$ *are positive integers that meet* $\sum_{k=1}^{K} \tau_k = N$.

The proof of the theorem is provided as follows:

Proof. Consider $\Lambda_G \triangleq \sigma_G/m_G$ and $\Lambda_L \triangleq \sigma_L/m_L$, and relying on Lemma 1 to determine the approximate PDF of the [Gamma/(Gamma2)] random variables of $W_t = G_t/\tilde{L}_t$ can be expressed as follows: □

$$f_{W_t}(y) \approx \frac{\Lambda_L^{-m_L}}{\xi(m_L)} \int_0^\infty x^{m_L+1} f_{G_t}\left(yx^2\right) e^{-\frac{x}{\Lambda_L}} dx. \quad (12)$$

We utilize the Gaussian–Laguerre (G-L) quadrature [39] to provide a tractable equation for the integral in (12) which is expressed as follows:

$$f_{W_t}(y) \approx \sum_{k=1}^{K} \psi_k \frac{y^{m_G-1}}{\xi(m_G)} e^{-\frac{y}{\varsigma_k}}, y > 0. \quad (13)$$

After normalization, the PDF of W_t may be obtained in approximate closed form as follows:

$$f_{W_r}(y) \approx \sum_{k=1}^{K} \aleph_k \frac{y^{m_G-1}}{\xi(m_G)} e^{-\frac{y}{\varsigma_k}}, \quad (14)$$

Therefore, the Laplace transform of W_t can be expressed as follows:

$$\mathcal{LP}_{W_r}(v) = \int_0^\infty e^{-vy} f_{W_t}(y) dy = \sum_{k=1}^{K} \aleph_k \left(\frac{1}{\varsigma_k} + v\right)^{-m_G}. \quad (15)$$

The CDF of Z may be stated as follows after a few derivation steps using the Laplace transform for Z.

$$F_Z(z) = \mathcal{LP}^{-1}\left\{\frac{1}{v}\left[\sum_{k=1}^{K} \aleph_k \left(\frac{1}{\varsigma_k} + v\right)^{-m_G}\right]^N ; v, z\right\}, \quad (16)$$

The inverse Laplace transform of $H(v)$ from the v-domain to the z-domain is shown by the notation $\mathcal{LP}^1 H(v), v, z$. (16) is reformulated using the multinominal theorem as follows:

$$\begin{aligned} F_Z(z) = &\sum_{\Xi_\tau, \forall\tau} \binom{N}{\tau_1, \ldots, \tau_K} \left[\prod_{k=1}^{K} \aleph_k^{\tau_k}\right] \\ &\times \mathcal{LP}^{-1}\left\{\frac{1}{v} \prod_{k=1}^{K} \left(\frac{1}{\varsigma_k} + v\right)^{-\tau_k m_G} ; v, z\right\}. \end{aligned} \quad (17)$$

Considering the Laplace transform's linearity feature, we obtain as follows:

$$\begin{aligned} \mathcal{LP}^{-1}&\left\{\frac{1}{v} \prod_{k=1}^{K} \left(\frac{1}{\varsigma_k} + v\right)^{-\tau_k m_G} ; v, z\right\} = \frac{1}{\xi(N m_G+1)} \\ &\times \mathcal{LP}^{-1}\left\{\frac{\xi(N m_G+1)}{v^{N m_G+1}} \prod_{k=1}^{K} \left(\frac{1}{\varsigma_k v} + 1\right)^{-\tau_k m_G} ; v, z\right\}, \end{aligned} \quad (18)$$

In addition, using [40], Equation (10) and following some minor mathematical adjustments or simplifications, the CDF of Z may be roughly expressed in closed form as follows:

$$\begin{aligned} F_Z(z) \approx &\frac{z^{N m_G}}{\xi(N m_G+1)} \sum_{\aleph_\tau, \forall\tau} \binom{N}{\tau_1, \ldots, \tau_K} \left[\prod_{k=1}^{K} \xi_k^{\tau_k}\right] \\ &\times \Phi_2^{(K)}\left(m_G\tau_1, \ldots, m_G\tau_K; N m_G + 1; -\frac{z}{\varsigma_1}, \ldots, -\frac{z}{\varsigma_K}\right). \end{aligned} \quad (19)$$

The CDF of $|Z|^2$ may be obtained as in (11) since $F(X^2) = F(X - \sqrt{X}), \quad X > 0$ and with this, Theorem 1 is completely proved.

Corollary 1. *By using Theorem 1, one may obtain a closed-form formula approximating the system OP as follows:*

$$P_{out} \approx F_{|Z|^2}\left(\eta_{th} / \left(\bar{\eta}\kappa^2\right)\right). \quad (20)$$

3. Proposed Deep Learning Model

In this study, it is assumed that the ARIS's 3D spatial movement is characterized by the RWMM [41]. Samples of its positions may be obtained from a homogeneous 3D Poisson process and the system model discussed in the preceding sections. For the purpose of explanation, we additionally assume that the ARIS moves spatially in a 3D cylinder, as shown in Figure 6. As a result, every element of the cylindrical coordinate of T can be produced from a uniform distribution which are as follows [27]: $\bar{\eta}[\text{dB}] = [5 - \epsilon_\eta, 5 + \epsilon_\eta]$, $\omega_T \sim \varkappa(0, 2\pi)$, $m_c = [2 - \epsilon_m, 2 + \epsilon_m]$, $\zeta_c = [2.5 - \epsilon_\zeta, 2.5 + \epsilon_\zeta]$, $N = [15/20/30 - \epsilon_N, 15/20/30 + \epsilon_N]$, $v = [2.7 - \epsilon_v, 2.7 + \epsilon_v]$, $r_T \sim 0.5\sqrt{\varkappa(0,1)}$, $h_T \sim \varkappa(0,1)$, $\delta_c = [1 - \epsilon_\delta, 1 + \epsilon_\delta]$, and $A_{\text{cth}} = [5 - \epsilon_{A_c}, 5 + \epsilon_{A_c}]$, respectively. It should be observed that distances among nodes in the suggested framework are now RVs due to the incorporation of 3D movement. This makes it impossible to derive the system OP specified in (7), mostly due to the multivariate confluent hypergeometric function. To get over this obstacle, we approach the task of locating the system OP in supervised learning. Specifically, we produce a data set that thoroughly describes the system under consideration. This dataset is used to train the generated BiLSTM model, enabling it to estimate the OP with high accuracy in different system configurations.

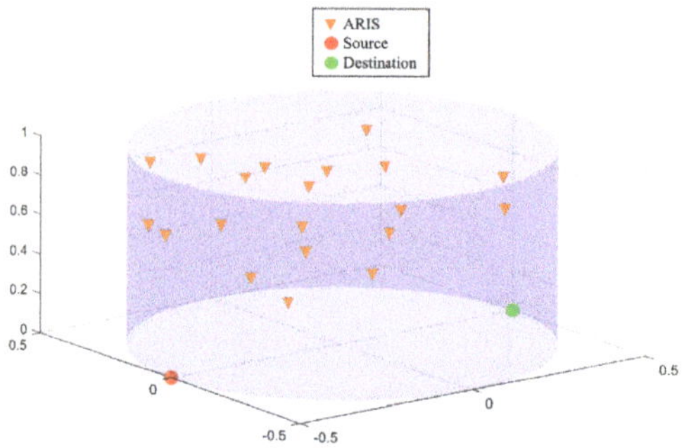

Figure 6. The architecture of the 3D spatial movement of the ARIS system.

3.1. 3D Modeling of Spatial Movement and Datasets Preparation

It is assumed that the azimuth, radial distance, and height of the cylindrical coordinate of T are represented by the symbols φ_T, r_T, and q_T, respectively. We take into consideration a unit normalization cylinder, that is, $\varphi_T \in [0, 2\pi]$, $r_T \in [0, 0.5]$, and $q_T \in [0, 1]$, without losing generality. This results in the conversion of R's 3D Cartesian coordinates, (x_T, y_T, z_T), to $x_T = r_T sin\varphi_T$, $y_T = r_T cos\varphi_T$, and $z_T = q_T$. Given the 3D Cartesian coordinates of the source and destination as (−0.5, 0, 0) and (0.5, 0, 0), respectively, one can compute the distance between two nodes using the following formula: $d_{ab} = \sqrt{(x_a - x_b)^2 + (y_a - y_b)^2 + (z_a - z_b)^2}$ where $\{a, b\} \in \{\text{source, destination, T}\}$.

In this study, the preparation of datasets, χ, is involved in the proposed ARIS system. Specifically, in the datasets, χ includes samples d_s of the information gathered from the AIRS system where the input–output connection could be $Data[d_s] = [\mathbf{F}[d_s], P_{d_j}^{d_s}]$, where $\mathbf{F}[d_s]$ represents the feature vector involving with all inputs characterized as shown in Table 1. However, this work has enough simulation samples to produce highly accurate outage performance estimates. After generating real-value CSI sets for each feature vector $\mathbf{F}[d_s]$, the simulation runs a Monte Carlo. To build the data set, we generate 20^3 samples in total, i.e., $Data[d_s], d_s = 1, ..., 20^3$, for utilizing the proposed model. After that, it splits the data set with ratios of 80%, 10%, and 10% for the training set Tr_χ, validation set Va_χ, and test set Te_χ. Through the mean-squared error (MSE), the accuracy of such a forecast may be assessed using (25).

Table 1. Inputs for the BiLSTM training and testing.

Parameters	Input Value
$\bar{\eta}[dB]$	−10∼20
φ_T	1
r_T	1
q_T	1
m_{St}	2.5
m_{tD}	2.5
ζ_c	2.5
N	15, 20, 30
v	1
h_T	1
δ_{St}	1
δ_{tD}	1
ζ_{St}	3
ζ_{tD}	3
A_{cth}	5

3.2. Proposed BiLSTM Model Structure and Mathematical Formulation

The BiLSTM network is proposed for the following objectives. Unidirectional LSTM networks execute sequences in the past without taking the future into account. This is because, since it only accepts input from the past, unidirectional LSTM only stores data from previous time steps [42]. Conversely, the BiLSTM consists of the unidirectional LSTM's forward (one from past to future) and backward (one from future to past) perspectives. As the input is bidirectional, it may leverage both sides of the data, providing more opportunities to extract training features and enhancing prediction accuracy. A schematic illustration of the internal cell architecture of the unidirectional LSTM and the BiLSTM is shown in Figure 7a,b, respectively. The proposed BiLSTM model consists of the input layer, BiLSTM layer, fully connected layer, dense layer, and output layer, respectively. The proposed model structure with different layers is shown in Figure 8. The mathematical formulation and operation of each of the layers are as follows:

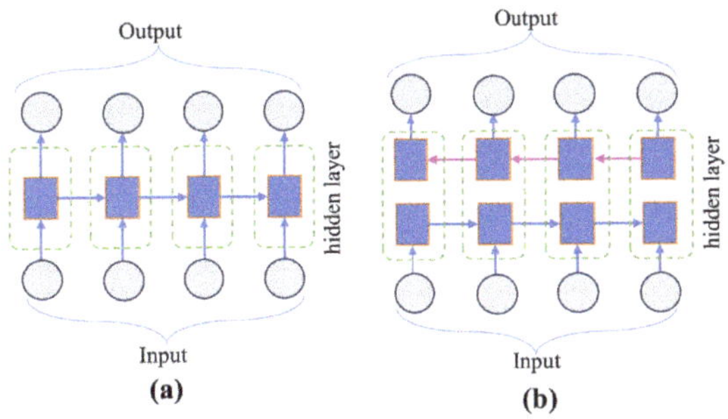

Figure 7. The structural difference of (**a**) LSTM and (**b**) BiLSTM.

Input Layer: The sequence of data with accompanying labels makes up the BiLSTM model input. The input sequence of features $\mathbf{F} = [s_1, s_2, s_3, \ldots, s_n]$ that contained the CSI is assumed. The dimension of 13×1 considers the size of the input features. The created dataset, which has the same number of features in the input data as the input size, is fed with the numerical values of sequence features together with their labels into the input layer of the BiLSTM.

BiLSTM Layer: The several gates such as input, output, and forget gates create the LSTM cell. In the proposed model, we have built up three BiLTM layers for obtaining the optimal training by the flow of the information in both directions. The BiLSTM layers in both directions flow can be calculated for an input signal of s at the current time step t as follows:

$$\overrightarrow{h_f} = \Theta(\mathcal{W}_f s_t + \mathcal{W}_f h_{t-1} + b_f), \tag{21}$$

$$\overleftarrow{h_r} = \Theta(\mathcal{W}_r s_t + \mathcal{W}_r h_{t+1} + b_r), \tag{22}$$

where the direction in both forward and reverse h_f and h_r is the corresponding output of the LSTM network, and Θ indicates the activation function. The future state is represented by h_{t+1}, and the past state is represented by h_{t-1}, respectively. The time steps of forward and backward are denoted by $t-1$ and $t+1$, and finally, for both directions, the weight and learnable bias are $\mathcal{W}_f$, $\mathcal{W}_r$ and b_f, b_r, respectively. As a result, the BiLSTM's output s_1 may be expressed as follows:

$$\begin{aligned} s_1 &= \Theta(\mathcal{W}_{s_1}\overrightarrow{h_f} \oplus \mathcal{W}_{s_1}\overleftarrow{h_r} + b_{s_1}) \\ &= \Theta(\mathcal{W}_{s_1}\tilde{h}_t + b_{s_1}), \end{aligned} \tag{23}$$

where the concentration of the hidden state in both directions of the BiLSTM is $\tilde{h}_t$, the learnable component of the BiLSTM output bias is b_{s_1}, and the output weights of the BiLSTM network are $\mathcal{W}_{s_1}$.

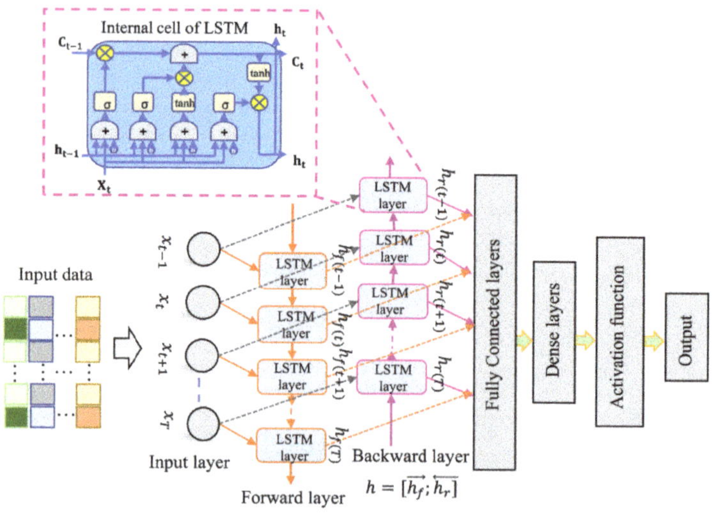

Figure 8. The proposed BiLSTM model structure for the ARIS system.

Fully Connected Layer: The fully connected layer is crucial because it performs OP tasks. In particular, the fully connected layer does the final distinction, and it processes the output of the BiLSTM layers. In the BiLSTM network, the fully linked layer operates independently on each time step. To get the optimal OP, a linear activation is explicitly specified for the dense output layer. Therefore, the predicted OP $(\hat{Y})$ can be expressed as follows:

$$(\hat{Y})(d_s) = f(L = l_{d_s}|S_1; (\mathcal{W}, b)), \tag{24}$$

where $\mathcal{W}$ and b stand for the weight and bias values, respectively; l_{d_s} is the computation result of the d_s-th samples of the input data; and finally, the linear activation function is $f(\cdot)$. However, the mean-squared error (MSE) of Π for the overall network is expressed as follows:

$$\Pi = \frac{1}{d_s}\sum_{q=1}^{d_s}(Y_q - \hat{Y}_q)^2, \tag{25}$$

where the predicted and target output OP of the q-th samples are $\hat{Y}_q$ and Y_q, respectively. The Adam optimization approach is used to lessen the loss [43].

3.3. Training and Testing of the Proposed Model

After the preparation of sufficient datasets and the design of the proposed BiLSTM model, we perform the training procedure in the offline phase. The algorithmic procedure of the training is described in the Algorithm 1. The sequence input train $Tr_{\mathcal{X}}$ and validation data $Va_{\mathcal{X}}$ are fed into the input layer and corresponding label as supervised training. The BiLSTM layers feed the input data with labels as a dimension of 13 × 1, and they train in a bidirectional manner. During training the model, a total of 50 epochs are performed, and the model achieves optimal knowledge, and we stop the training of the model. In addition, the learning rate of 0.01, minibatch of 32, and a hidden layer of 200 are set up for sufficient training of the model. In the testing phase, the trained model is loaded for inference by utilizing the test datasets $Te_{\mathcal{X}}$ and calculating the MSE according to the testing samples. The testing procedure of the proposed BiLSTM model algorithm is shown in Algorithm 2.

Algorithm 1 BiLSTM Training Algorithm

Initialize estimation function Θ, epochs E, training set $Tr_{\mathcal{X}}$.
Load training dataset, $Tr_{\mathcal{X}}$
Initialize the weights and biases of forward and backward as $(\mathcal{W}_f, \mathcal{W}_r)$ and (b_f, b_r)
Compute loss $\Pi(\mathcal{L})$
Optimization operation: $\mathcal{L}_{t+1} = \mathcal{L}_t - \frac{\mathcal{E}_t}{\sqrt{\hat{v}_t + \mathcal{Z}}} \cdot \hat{m}_t$
where m_t and v_t are first and second moment estimates of gradients $(\mathcal{L}_t)$ and $\mathcal{Z}$ is a small value to avoid division by zero.
for $e = 1 : E$ **do**
  for $q \in (1, \ldots, Tr_{\mathcal{X}})$ **do**
    Perform OP: $\hat{Y}_q = \Theta[\overrightarrow{h_f}(s); \overleftarrow{h_r}(s)]$
    Calculate loss by (25)
    Gradient descent: $\nabla_{\mathcal{L}_t}\Pi = \frac{\partial \Pi}{\partial \mathcal{L}_t}$
    Update parameters: $\mathcal{L}_t \leftarrow \mathcal{L}_t - \mathcal{E}_t \nabla_{\mathcal{L}_t}\Pi$
  end for
end for
Save the model as **BiLSTM.h5**

Algorithm 2 BiLSTM Testing Algorithm

Load the model as **BiLSTM.h5**
Input: Test dataset $Te_{\mathcal{X}}$
Output: Estimated outputs $\{\hat{\mathbf{Y}}_{\mathbf{q}_{\text{test}}}\}_{q=1}^{Ts_{\mathcal{X}}}$
Initialize RMSE = 0
for $q \in (1, \ldots, Ts_{\mathcal{X}})$ **do**
  for w : -10 to SNR **do**
    Perform estimation:
    $\hat{Y}_{q\text{test}} = \Theta[\overrightarrow{h_f}(s_{q\text{test}}); \overleftarrow{h_r}(s_{q\text{test}})]$
    Compute MSE for q th samples:
    $\Pi_q = \Pi = \frac{1}{d_s}\sum_{q=1}^{d_s}(Y_q - \hat{Y}_q)^2$,
    Compute RMSE $= \sqrt{\Pi}$
  end for
end for

3.4. Simulation Results

In this work, the simulation analysis is conducted in the Windows 11 operating system with the graphics processing unit. The program is written in the Keras-TensorFlow environment. An NVIDIA graphics card improves training performance. For the training of the model, we have considered (0–20) dB SNR during the generation of datasets. The Nakagami-m fading method considers the large-scale fading effect, represented as $\sigma_{ccc} = d_{ccc}^{\upsilon}$ for $ccc \in \{St, tD\}$, where υ represents the path-loss exponent, and the shape parameter m determines the fading severity. In IG shadowing, δ is normalized in relation to $\bar{\eta}$, and ζ denotes the intensity of shadowing. During the online testing phase, we consider (−10:2:20) dB SNR and obtain the optimal testing OP for the proposed system. The simulation is conducted in this study to evaluate the OP analysis against previous research and the suggested DL model. The DNN [27] and LSTM models are used to compare the efficacy of the suggested BiLSTM model. Acquiring knowledge about the channel parameter and other parameter adjustments enhances the efficacy of the proposed model.

To validate the proposed DL algorithm effectively, we analyze the MSE calculation for the suggested network, as seen in Figure 9, with different DL models such as LSTM and DNN. In this analysis, the training and validation of MSE BiLSTM, LSTM, and DNN models are calculated corresponding with different epoch numbers. It is seen from the figure that the proposed network has gained lower MSE with training and validation compared to the other models. For the case of the DNN model, the training and validation MSE with different epochs is gained with 10^{-4}, and the convergence is conducted very fast. In addition, the LSTM model shows almost the same MSE with DNN up to the epoch of 10, and after that, it outperforms the DNN model and gains 10^{-5} MSE. In contrast, the MSE of the proposed BiLSTM model outperforms the other models in terms of training and validation datasets. In the beginning, the curve for the BiLSTM model is not stable, and after the eigth epoch, it becomes stable. Furthermore, the convergence of the BILSTM model is slower than the other models and gains the MSE of 10^{-6}, which demonstrates the model's effectiveness in training.

During the training performance evaluation, we observed the iteration speeds for the proposed BiLSTM model compared to other models (DNN and LSTM). The iteration speed of the proposed BiLSTM model is 12 milliseconds (ms) per step, whereas the LSTM and DNN models showed 11 ms and 4 ms, respectively. Although the iteration speed of the proposed model is slower than the other models, it demonstrated a significant improvement in performance compared to the others. Additionally, it can be seen from Figure 7 that the convergence speed of the proposed model starts after the 25th epoch, which is almost the same as for the DNN and LSTM models. Furthermore, we have analyzed the computational complexity of the proposed model compared to other models, as shown in Table 2. From Table 2, it can be seen that the RMSE computation time is not significantly different from that of the other two models, as the proposed model has a large number of learnable parameters. Although the suggested HyDNN model has many parameters, the complexity can be reduced by leveraging GPU parallelization [44,45] during the real-time process. The suggested model has a higher complexity requirement than the others, but it performs better in estimation, and GPU parallel computing can shorten the calculation time.

Table 2. Computation complexity analysis.

Models	Learnable Parameters	RMSE Computation Time
BiLSTM	2,266,001	8 ms
LSTM	329,857	6 ms
DNN	67,969	2 ms

Figure 10, we have analyzed the OP versus transmit SNR in dB for the proposed system and compared it with the DNN model for different numbers of reflecting elements of $N = 15, 20$, and 30. It is evident from the figure that the increment of reflecting elements significantly reduces the power budget requirement for achieving the maximal OP performance. Compared to the DNN model, for $N = 15$, the proposed model gains almost 2.5 dB SNR; for $N = 20$, it gains almost 1.5 dB SNR; and finally, for $N = 20$, it gains almost 2.4 dB

SNR. In summary, the OP of the ARIS system is increased with different reflecting N for the proposed BiLSTM model compared with others.

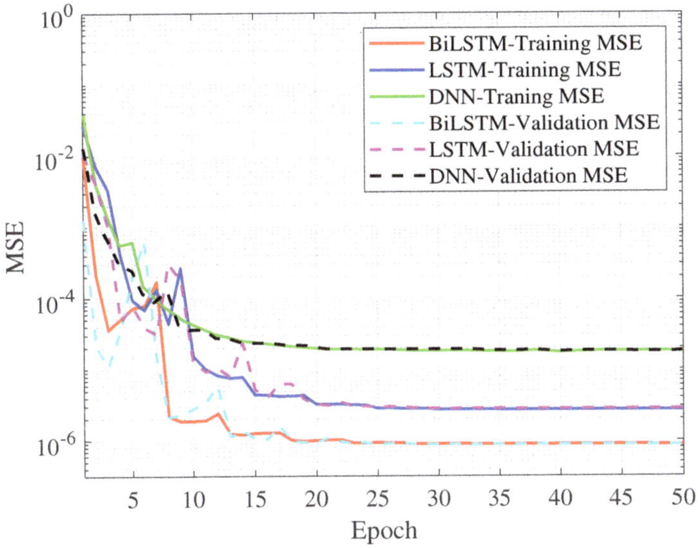

Figure 9. The MSE results of the proposed BiLSTM model compared with others models in terms of training and validation performances.

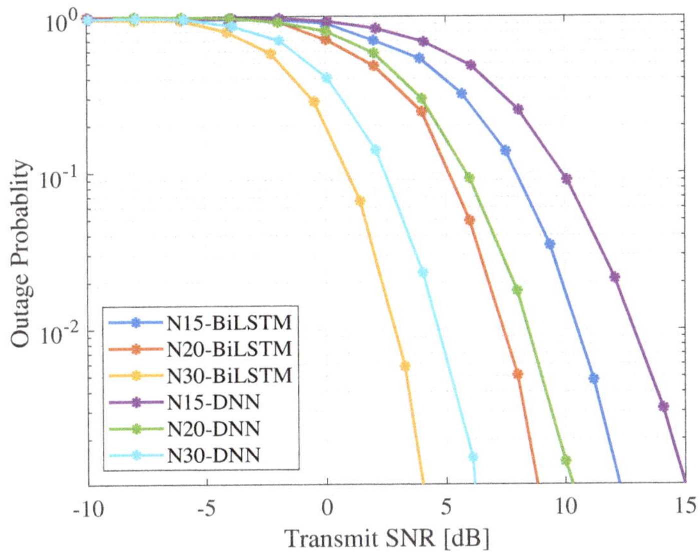

Figure 10. The analysis of OP for the proposed model compared with DNN for different ARIS element numbers of $N = 15, 20$, and 30.

The proposed BiLSTM OP performance is compared to the DNN model and the LSTM model based on the OP versus various SNR values in Figure 11. The proposed BILSTM-based system outperforms all other approaches, as the graph demonstrates. It appears that the suggested model works much better than other systems for SNRs between

−2 and 12 dB SNR and OP up to 10^{-5}. On the other hand, the LSTM model has OP of around 10^{-3} and SNR ranges of −2 to 12 dB. Furthermore, the DNN model has OP of about 10^{-4} and SNR ranges between −2 to 12. The suggested BiLSTM model has a significantly improved OP accuracy compared to the DNN and LSTM models. However, the LSTM model performs similarly to DNN models up to 6 dB SNR because of the lower SNR; at greater SNR, it performs a similar trend with the suggested model. The aforementioned simulation results demonstrate how the BiLSTM outperforms the other schemes in terms of performance. The aforementioned simulation results provide a summary: the suggested model may be a useful way to improve the OP for ARIS-based systems.

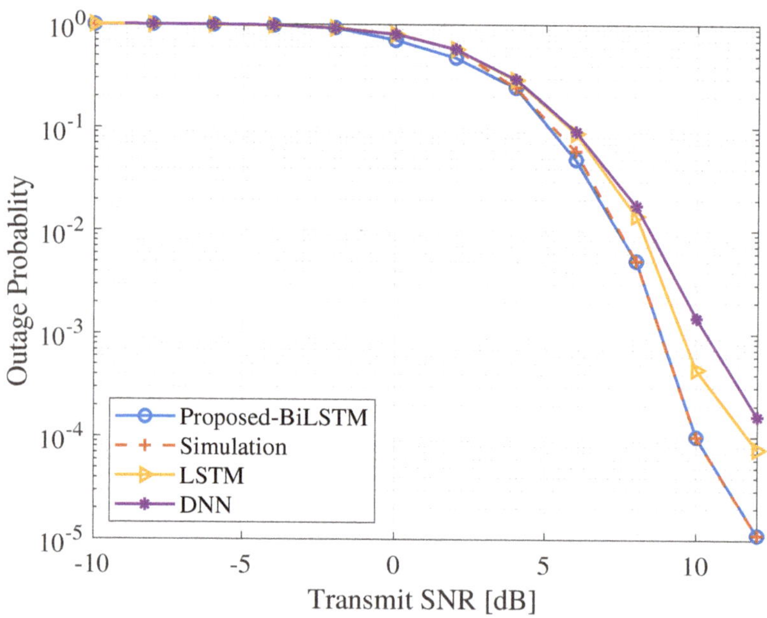

Figure 11. Simulation results of OP for the proposed BiLSTM with DNN and LSTM for the ARIS system.

4. Conclusions

In this paper, a DL network called the BiLSTM model is proposed to improve the OP in the 3D spatial excursion of the ARIS-assisted wireless communication system. The proposed model is configured with a BiLSTM layer that is capable of learning the ARIS system channel data in a bidirectional manner. Thus, the OP of the ARIS-based wireless system is enhanced with different SNR ranges and different reflecting elements. In the offline phase, the proposed model is trained using simulated data over the channel network assisted by ARIS. The effectiveness of the suggested model is next tested by retrieving the original data from the target terminal using the learned model, which is then implemented in the online phase. Thus, the sent data may be extracted to evaluate the OP performance. The well-known Adam optimization approach is used in order to evaluate the proposed model's error performance. The simulation results show that the suggested strategy outperforms other schemes in terms of MSE performance. Furthermore, in comparison to previous benchmarking models, the suggested BiLSTM model obtains an OP of 10^{-5} in 12 dB SNR. The simulation results clearly demonstrate that the suggested framework outperforms the DNN and LSTM models in terms of OP analysis. By training on long input sequence data utilizing a bidirectional framework, the proposed BiLSTM improves OP performance. The system's OP increases, and its training capacity is strengthened by adding bidirectional (forward and backward) tasks to the BiLSTM's hidden layers. A potential solution to improve the 6G communication system might be found in the suggested system. Further developments of the suggested paradigm might be used in more complex, sophisticated ARIS-assisted wireless communication systems. The limitation of the proposed system is that we did not consider the multiple antenna system for diverse wireless communication to analyze the OP. However, observing the effectiveness of the suggested approach could be the subject of further research. Furthermore, the proposed technology may be promising for AIRS-NOMA-OFDM systems and could be deployed in a hybrid DL model to improve system performance. Accurate channel estimation and feedback present issues in dynamic aerial situations. To improve RIS settings in real-time communication environments, several challenges must be addressed, and this can be a potential research direction [46].

Author Contributions: Conceptualization, M.H.R.; methodology, M.H.R.; software, M.H.R.; validation, M.H.R. and M.A.S.S.; formal analysis, M.H.R., M.A.S.S., M.A.A. and R.T.; investigation, M.H.R., M.A.S.S., M.A.A. and R.T.; resources, H.-K.S.; data curation, M.H.R. and M.A.S.S.; writing—original draft preparation, M.H.R.; writing—review and editing, M.H.R., M.A.S.S., M.A.A., R.T. and H.-K.S.; visualization, M.H.R. and M.A.S.S.; supervision, H.-K.S.; project administration, H.-K.S.; funding acquisition, H.-K.S. All authors have read and agreed to the published version of the manuscript.

Funding: This work was supported by the Institute of Information & Communications Technology Planning & Evaluation (IITP) under the metaverse support program to nurture the best talents (IITP-2024-RS-2023-00254529) grant funded by the Korean government (MSIT) and in part by the Basic Science Research Program through the National Research Foundation of Korea (NRF) funded by the Ministry of Education (2020R1A6A1A03038540) and in part by the MSIT, Korea, under the ITRC support program (IITP-2024-2021-0-01816) supervised by the IITP.

Institutional Review Board Statement: Not applicable.

Informed Consent Statement: Not applicable.

Data Availability Statement: The data will be made available by the authors on request.

Conflicts of Interest: The authors declare no conflicts of interest.

References

1. Latva-Aho, M.; Leppänen, K. *Key Drivers and Research Challenges for 6G Ubiquitous Wireless Intelligence*; 6G Flagship: Oulu, Finland, 2019.
2. Letaief, K.B.; Chen, W.; Shi, Y.; Zhang, J.; Zhang, Y.J.A. The roadmap to 6G: AI empowered wireless networks. *IEEE Commun. Mag.* **2019**, *57*, 84–90. [CrossRef]
3. Yang, P.; Xiao, Y.; Xiao, M.; Li, S. 6G wireless communications: Vision and potential techniques. *IEEE Netw.* **2019**, *33*, 70–75. [CrossRef]
4. Wu, Q.; Zhang, R. Towards smart and reconfigurable environment: Intelligent reflecting surface aided wireless network. *IEEE Commun. Mag.* **2019**, *58*, 106–112. [CrossRef]
5. Liu, X.; Liu, Y.; Chen, Y. Machine Learning Empowered Trajectory and Passive Beamforming Design in UAV-RIS Wireless Networks. *IEEE J. Sel. Areas Commun.* **2021**, *39*, 2042–2055. [CrossRef]
6. Guo, X.; Chen, Y.; Wang, Y. Learning-Based Robust and Secure Transmission for Reconfigurable Intelligent Surface Aided Millimeter Wave UAV Communications. *IEEE Wirel. Commun. Lett.* **2021**, *10*, 1795–1799. [CrossRef]
7. Zhao, J.; Yu, L.; Cai, K.; Zhu, Y.; Han, Z. RIS-aided ground-aerial NOMA communications: A distributionally robust DRL approach. *IEEE J. Sel. Areas Commun.* **2022**, *40*, 1287–1301. [CrossRef]
8. Wang, L.; Wang, K.; Pan, C.; Aslam, N. Joint trajectory and passive beamforming design for intelligent reflecting surface-aided UAV communications: A deep reinforcement learning approach. *IEEE Trans. Mob. Comput.* **2022**, *22*, 6543–6553. [CrossRef]
9. Goyal, A.; Gunturu, A.; Kumar Reddy Chavva, A.; Kim, H.; Kim, H.; Lee, S. Placement of Reconfigurable Intelligent Surfaces in Urban Cell For Improved Coverage—A Practical Approach. In Proceedings of the ICC 2023—IEEE International Conference on Communications, Rome, Italy, 28 May–1 June 2023; pp. 1074–1079. [CrossRef]
10. Zeng, Y.; Wu, Q.; Zhang, R. Accessing from the sky: A tutorial on UAV communications for 5G and beyond. *Proc. IEEE* **2019**, *107*, 2327–2375. [CrossRef]
11. Lima, B.K.S.; De Sena, A.S.; Dinis, R.; Benevides Da Costa, D.; Beko, M.; Oliveira, R.; Debbah, M. Aerial Intelligent Reflecting Surfaces in MIMO-NOMA Networks: Fundamentals, Potential Achievements, and Challenges. *IEEE Open J. Commun. Soc.* **2022**, *3*, 1007–1024. [CrossRef]

12. Abdalla, A.S.; Marojevic, V. DDPG learning for aerial RIS-assisted MU-MISO communications. In Proceedings of the 2022 IEEE 33rd Annual International Symposium on Personal, Indoor and Mobile Radio Communications (PIMRC), Kyoto, Japan, 12–15 September 2022; IEEE: Piscataway, NJ, USA, 2022; pp. 701–706.
13. Abdalla, A.S.; Marojevic, V. ARIS for safeguarding MISO wireless communications: A deep reinforcement learning approach. In Proceedings of the 2022 5th International Conference on Advanced Communication Technologies and Networking (CommNet), Marrakech, Morocco, 12–14 December 2022; IEEE: lPiscataway, NJ, USA, 2022; pp. 1–6.
14. Tang, X.; Jiang, T.; Liu, J.; Li, B.; Zhai, D.; Yu, F.R.; Han, Z. Secure Communication With UAV-Enabled Aerial RIS: Learning Trajectory With Reflection Optimization. *IEEE Trans. Intell. Veh.* **2023**, 1–10. [CrossRef]
15. Bao, T.; Zhu, J.; Yang, H.C.; Hasna, M.O. Secrecy outage performance of ground-to-air communications with multiple aerial eavesdroppers and its deep learning evaluation. *IEEE Wirel. Commun. Lett.* **2020**, *9*, 1351–1355. [CrossRef]
16. Dang, H.P.; Van Nguyen, M.S.; Do, D.T.; Nguyen, M.H.; Pham, M.T.; Kim, A.T. Secure Performance Analysis of Aerial RIS-NOMA-Aided Systems: Deep Neural Network Approach. *Electronics* **2022**, *11*, 2588. [CrossRef]
17. Samir, M.; Elhattab, M.; Assi, C.; Sharafeddine, S.; Ghrayeb, A. Optimizing age of information through aerial reconfigurable intelligent surfaces: A deep reinforcement learning approach. *IEEE Trans. Veh. Technol.* **2021**, *70*, 3978–3983. [CrossRef]
18. Al-Hourani, A.; Kandeepan, S.; Lardner, S. Optimal LAP altitude for maximum coverage. *IEEE Wirel. Commun. Lett.* **2014**, *3*, 569–572. [CrossRef]
19. Kim, M.; Lee, J. Outage probability of UAV communications in the presence of interference. In Proceedings of the 2018 IEEE Global Communications Conference (GLOBECOM), Abu Dhabi, United Arab Emirates, 9–13 December 2018; IEEE: Piscataway, NJ, USA, 2018; pp. 1–6.
20. Zhu, Q.; Bai, F.; Pang, M.; Li, J.; Zhong, W.; Chen, X.; Mao, K. Geometry-based stochastic line-of-sight probability model for A2G channels under urban scenarios. *IEEE Trans. Antennas Propag.* **2022**, *70*, 5784–5794. [CrossRef]
21. Al-Hourani, A.; Kandeepan, S.; Jamalipour, A. Modeling air-to-ground path loss for low altitude platforms in urban environments. In Proceedings of the 2014 IEEE Global Communications Conference, Austin, TX, USA, 8–12 December 2014; IEEE: Piscataway, NJ, USA, 2014; pp. 2898–2904.
22. Yang, L.; Chen, J.; Hasna, M.O.; Yang, H.C. Outage performance of UAV-assisted relaying systems with RF energy harvesting. *IEEE Commun. Lett.* **2018**, *22*, 2471–2474. [CrossRef]
23. Sharma, P.K.; Kim, D.I. Coverage probability of 3-D mobile UAV networks. *IEEE Wirel. Commun. Lett.* **2018**, *8*, 97–100. [CrossRef]
24. Sharma, P.K.; Kim, D.I. Random 3D mobile UAV networks: Mobility modeling and coverage probability. *IEEE Trans. Wirel. Commun.* **2019**, *18*, 2527–2538. [CrossRef]
25. Sharma, P.K.; Deepthi, D.; Kim, D.I. Outage probability of 3-D mobile UAV relaying for hybrid satellite-terrestrial networks. *IEEE Commun. Lett.* **2019**, *24*, 418–422. [CrossRef]
26. Bithas, P.S.; Nikolaidis, V.; Kanatas, A.G.; Karagiannidis, G.K. UAV-to-ground communications: Channel modeling and UAV selection. *IEEE Trans. Commun.* **2020**, *68*, 5135–5144. [CrossRef]
27. Do, T.N.; Kaddoum, G.; Nguyen, T.L.; da Costa, D.B.; Haas, Z.J. Aerial Reconfigurable Intelligent Surface-Aided Wireless Communication Systems. In Proceedings of the 2021 IEEE 32nd Annual International Symposium on Personal, Indoor and Mobile Radio Communications (PIMRC), Helsinki, Finland, 13–16 September 2021; pp. 525–530.
28. Hochreiter, S.; Schmidhuber, J. Long short-term memory. *Neural Comput.* **1997**, *9*, 1735–1780. [PubMed]
29. Rahman, M.H.; Sejan, M.A.S.; Yoo, S.G.; Kim, M.A.; You, Y.H.; Song, H.K. Multi-user joint detection using bi-directional deep neural network framework in NOMA-OFDM system. *Sensors* **2022**, *22*, 6994. [CrossRef] [PubMed]
30. Rahman, M.H.; Sejan, M.A.S.; Aziz, M.A.; Baik, J.I.; Kim, D.S.; Song, H.K. Deep Learning-Based Improved Cascaded Channel Estimation and Signal Detection for Reconfigurable Intelligent Surfaces-Assisted MU-MISO Systems. *IEEE Trans. Green Commun. Netw.* **2023**, *7*, 1515–1527.
31. Abduljabbar, R.L.; Dia, H.; Tsai, P.W. Unidirectional and bidirectional LSTM models for short-term traffic prediction. *J. Adv. Transp.* **2021**, *2021*, 5589075.
32. Siami-Namini, S.; Tavakoli, N.; Namin, A.S. The performance of LSTM and BiLSTM in forecasting time series. In Proceedings of the 2019 IEEE International Conference on Big Data (Big Data), Los Angeles, CA, USA, 9–12 December 2019; IEEE: Piscataway, NJ, USA, 2019; pp. 3285–3292.
33. Yu, J.; Liu, X.; Gao, Y.; Zhang, C.; Zhang, W. Deep learning for channel tracking in IRS-assisted UAV communication systems. *IEEE Trans. Wirel. Commun.* **2022**, *21*, 7711–7722.
34. Habibur Rahman, M.; Abrar Shakil Sejan, M.; Abdul Aziz, M.; Tabassum, R.; Baik, J.I.; Song, H.K. Deep Learning Based One Bit-ADCs Efficient Channel Estimation Using Fewer Pilots Overhead for Massive MIMO System. *IEEE Access* **2024**, *12*, 64823–64836. [CrossRef]
35. Björnson, E.; Özdogan, Ö.; Larsson, E.G. Intelligent reflecting surface versus decode-and-forward: How large surfaces are needed to beat relaying? *IEEE Wirel. Commun. Lett.* **2019**, *9*, 244–248.
36. Version, M. *MATLAB*, Version: R2021a, Update 2 64-Bit; MathWorks: Natick, MA, USA, 2021.
37. Basar, E.; Di Renzo, M.; De Rosny, J.; Debbah, M.; Alouini, M.S.; Zhang, R. Wireless communications through reconfigurable intelligent surfaces. *IEEE Access* **2019**, *7*, 116753–116773.
38. Van Chien, T.; Tu, L.T.; Chatzinotas, S.; Ottersten, B. Coverage probability and ergodic capacity of intelligent reflecting surface-enhanced communication systems. *IEEE Commun. Lett.* **2020**, *25*, 69–73.

39. Abramowitz, M.; Stegun, I.A. *Handbook of Mathematical Functions with Formulas, Graphs, and Mathematical Tables*; US Government Printing Office: Washington, DC, USA, 1968; Volume 55.
40. Lopez-Martinez, F.J.; Romero-Jerez, J.M.; Paris, J.F. On the calculation of the incomplete MGF with applications to wireless communications. *IEEE Trans. Commun.* **2016**, *65*, 458–469.
41. Ahuja, A.; Venkateswarlu, K.; Krishna, P.V. Stochastic characteristics and simulation of the random waypoint mobility model. *arXiv* **2012**, arXiv:1203.3920.
42. Khan, S.; Durrani, S.; Shahab, M.B.; Johnson, S.J.; Camtepe, S. Joint User and Data Detection in Grant-Free NOMA With Attention-Based BiLSTM Network. *IEEE Open J. Commun. Soc.* **2023**, *4*, 1499–1515. [CrossRef]
43. Kingma, D.P.; Ba, J. Adam: A method for stochastic optimization. *arXiv* **2014**, arXiv:1412.6980.
44. Rahman, M.H.; Sejan, M.A.S.; Aziz, M.A.; You, Y.H.; Song, H.K. HyDNN: A Hybrid Deep Learning Framework Based Multiuser Uplink Channel Estimation and Signal Detection for NOMA-OFDM System. *IEEE Access* **2023**, *11*, 66742–66755. [CrossRef]
45. Goodfellow, I.; Bengio, Y.; Courville, A. *Deep Learning*; MIT Press: Cambridge, MA, USA, 2016.
46. Arzykulov, S.; Celik, A.; Nauryzbayev, G.; Eltawil, A.M. Aerial RIS-Aided Physical Layer Security: Optimal Deployment and Partitioning. *IEEE Trans. Cogn. Commun. Netw.* **2024**, *early access*. [CrossRef]

mathematics

Article

Performance Analysis of RIS-Assisted SatComs Based on a ZFBF and Co-Phasing Scheme

Minchae Jung [1], Taehyoung Kim [2,*] and Hyukmin Son [3,*]

[1] Department of Electronics and Information Engineering, Sejong University, Seoul 05006, Republic of Korea; mcjung@sejong.ac.kr
[2] School of Electrical Engineering, Kookmin University, Seoul 02707, Republic of Korea
[3] Department of Electronic Engineering, Gachon University, Seongnam 13120, Republic of Korea
* Correspondence: th.kim@kookmin.ac.kr (T.K.); hson102@gachon.ac.kr (H.S.)

Abstract: In recent high-throughput satellite communication (SatCom) systems, the use of reconfigurable intelligent surfaces (RISs) has emerged as a promising solution to improve spectral efficiency and extend coverage in areas with limited terrestrial network access. However, the RIS may amplify the inter-beam interference (IBI) caused by multibeam transmission at the satellite, and multiple RISs can also cause inter-RIS interference (IRI) to terrestrial users. In this paper, the performance of the RIS-assisted SatCom system is asymptotically analyzed for both full and partial channel state information (CSI) scenarios. In particular, zero-forcing beamforming is considered as the active beamforming for data transmission, while the co-phasing scheme is considered as the passive beamforming for RIS reflection. Based on the asymptotic analyses, deterministic active and passive beamforming techniques using partial CSI are proposed that can gradually eliminate both IBI and IRI, ultimately achieving ideal performance. Simulation results validate the accuracy of asymptotic analyses and demonstrate the superiority of deterministic active and passive beamforming techniques using partial CSI. The simulation results also confirm that the proposed beamforming can achieve approximately 92.8% of the ideal performance, even though it only requires partial CSI.

Keywords: active and passive beamforming; co-phasing technique; full frequency reuse (FFR); reconfigurable intelligent surface (RIS); satellite communication; zero-forcing beamforming (ZFBF)

MSC: 94A05

1. Introduction

Recently, the demand for high-throughput satellite (HTS) systems has been increasing rapidly due to the growing number of Internet of Things (IoT) devices and emerging 6G applications [1–3]. However, existing satellite communication (SatCom) systems encounter limitations regarding signal attenuation and the ability to maintain a line-of-sight (LoS) communication link between the satellite and terrestrial user equipment (UE) caused by obstacles and shadowing. Recently, reconfigurable intelligent surfaces (RISs) have been proposed as a promising solution for supporting high-throughput SatCom systems [4–7]. RISs are passive reflecting surfaces that can control the phase, frequency, and amplitude of incoming radio frequency (RF) signals and reflect them to a desired destination [8–10]. Therefore, by using unmanned aerial vehicles (UAVs) equipped with RISs, it becomes possible to enhance the signal strength of satellites and expand the coverage of SatCom systems, especially in areas inaccessible due to obstacles and shadowing. Importantly, this can be achieved without consuming additional UAV battery power.

Previous studies [11–13] have focused on the challenges of employing UAVs as mobile relays in SatCom systems. The authors in [11] analyzed the performance of SatCom systems using UAVs equipped with multi-antenna arrays, operating as moving decode-and-forward relays. Another approach proposed in [12] explored blind beam-tracking techniques, where

UAVs equipped with large-scale antenna arrays acted as relays between satellites and terrestrial networks. Additionally, in [13], a relay architecture for UAVs operating between the ground and satellites was investigated. The studies in [5–7] examined the concept of RISs deployed on man-made structures, such as buildings and walls, to reflect or refract signals from satellites to the destinations. However, the assumption of UAVs as active relays consuming additional energy for signal processing, as presented in [11–13], is impractical due to the limited battery capacity of UAVs and the need for energy-efficient operation. Moreover, the deployment of RISs on the surface of ground structures, as assumed in [5–7], may not adequately complement existing SatCom systems or fully harness the potential for enhancing network capacity.

Meanwhile, to achieve high spectral efficiency and throughput in the upcoming multibeam HTS systems, full frequency reuse (FFR) is generally considered in SatCom systems [14]. This approach allows each beam to exploit all the available bandwidth, ensuring maximum resource utilization. However, the performance of FFR-based SatCom systems can be degraded by inter-beam interference (IBI) resulting from the overlapping side lobes between the multiple beam radiation patterns. In an RIS-assisted SatCom system, the FFR in the RIS–UE links may cause amplification of IBI, and multiple RISs may also result in inter-RIS interference (IRI) to terrestrial UEs. Moreover, acquiring full CSI for all RIS-related channels is challenging in practice [15]. Hence, integrating RIS into the SatCom system poses the inevitable challenge of mitigating IBI and IRI by relying on partial CSI of the desired channel. Recently, in [16], the performance of a two-layer RIS-assisted UAV communication system was analyzed. In [17], the authors jointly optimized user scheduling and phase shift in the simultaneously transmitting and reflecting (STAR)-RIS-assisted multicast SatCom systems. However, these studies did not consider the impact of IBI and IRI introduced by RIS on the performance of the considered system. In this paper, we conduct a performance analysis considering the impact of IBI and IRI, which are crucial factors in RIS-assisted SatCom systems.

To overcome the aforementioned limitations, we propose a deterministic zero-forcing beamforming (ZFBF) combined with the co-phasing technique that only requires partial CSI of the desired channel. In particular, the contributions of our work can be outlined as follows:

- First, we analyze the performance of the ZFBF combined with the co-phasing technique based on full channel state information (CSI) as the number of RIS elements increases towards infinity.
- To make the analysis more practical, we propose a novel deterministic ZFBF and co-phasing technique that only requires partial CSI while eliminating IBI and achieving near-optimal performance.
- Furthermore, we derive a closed-form expression for the asymptotic ergodic capacity of the proposed scheme.
- We then analyze the outage probability and demonstrate that the outage probability decreases as the number of RIS elements increases, thereby confirming the RIS's ability to provide reliable SatCom services in scenarios with a large number of RIS elements.

The simulation results validate the asymptotic optimality of the proposed scheme, showing that it closely approaches the ideal performance. Moreover, the outage probability exhibits a potential convergence to zero as the number of RIS elements increases, further supporting the reliability and efficiency of the proposed scheme.

Notations: Throughout the paper, boldface upper- and lower-case symbols represent matrices and vectors, respectively, and $\boldsymbol{I}_N$ and $\mathbf{1}_{M\times N}$ are, respectively, a size-N identity matrix and an $M \times N$ matrix of ones. The conjugate, transpose, inverse, and Hermitian transpose operators are $(\cdot)^*$, $(\cdot)^{\mathrm{T}}$, $(\cdot)^{-1}$, and $(\cdot)^{\mathrm{H}}$, respectively. The operator diag $(a_1, \cdots, a_n)$ is a diagonal operator that constructs a diagonal matrix from a vector $[a_1, \cdots, a_n]^{\mathrm{T}}$. The amplitude and phase of a complex number a are denoted by $|a|$ and $\angle a$, respectively. $\mathrm{E}[\cdot]$ and $\mathrm{Var}[\cdot]$ denote expectation and variance operators, respectively, and $\mathcal{O}(\cdot)$ denotes the big O notation. $\mathcal{L}^2\mathcal{N}(\eta, \sigma^2)$ denotes a double-log-normal distribution with mean η

and variance σ^2, and $\otimes$ and $\xrightarrow{\text{a.s.}}$, respectively, denote the Kronecker product and almost sure convergence.

2. System Model

We present an RIS-assisted SatCom system composed of a multibeam satellite, a single Earth station (ES), K UEs with single antennas, and K UAVs each equipped with an RIS consisting of N reflecting elements. As illustrated in Figure 1, the UAVs operate as mobile reflectors between the satellite and the UEs. The satellite generates K spot beams with K feeds; each beam is fed by a single feed, and the satellite's payload is considered to be transparent. The UAVs reflect the incident signals from the satellite using their RISs, and the phases of the incident wave are adjusted by N reflecting elements. It is assumed that the ES assigns one appropriate RIS to each UE, which means one RIS is allocated per UE, and a dedicated wireless control link is utilized to deliver a control signal from the ES to each RIS for the information of the reflection phase shift. In summary, the system operates as follows: (i) a satellite generating K beams receives signals from the ES and transmits signals to K UAVs. Subsequently, (ii) each UAV transmits a signal to each designated user through N reflecting elements. In this configuration, (iii) the UE receives IBI from the satellite through the RIS and simultaneously receives interference from adjacent RISs, referred to as IRI.

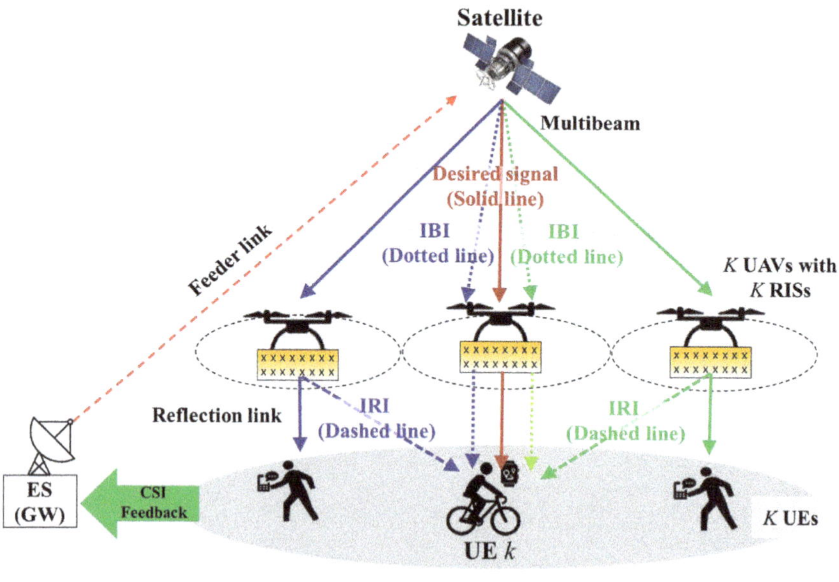

Figure 1. System model of the RIS-assisted SatCom system.

2.1. RIS-Assisted SatCom Channel Model

To achieve high spectral efficiency and throughput in multibeam HTS systems, FFR is generally considered in SatCom systems [18]. However, the performance of the FFR-based SatCom system can be degraded by co-channel interference among inter-beam UAVs, particularly for UAVs at the beam edge. Consequently, the RISs installed on those UAVs reflect the incident signals, including the IBI, to the UEs. This results in severe performance degradation due to the phase-aligned and amplified IBI. Thus, in order to mitigate IBI, we assume that beamforming (i.e., active beamforming) is performed at the ES. Additionally, we consider that the feeder link is ideal, similar to most related works [5–7]. Then, the received signal at UE k is obtained as:

$$y_k = \sqrt{\rho_k} \boldsymbol{f}_{k,k}^{\mathrm{H}} \boldsymbol{\Phi}_k \boldsymbol{D}_k \boldsymbol{B}_k \boldsymbol{w}_k x_k + n_k$$

$$+\sum_{i\neq k}^{K}(\underbrace{\sqrt{\rho_i}\boldsymbol{f}_{k,k}^{\mathrm{H}}\boldsymbol{\Phi}_k\boldsymbol{D}_k\boldsymbol{B}_k\boldsymbol{w}_i x_i}_{\text{IBI}}+\underbrace{\sqrt{\rho_i}\boldsymbol{f}_{i,k}^{\mathrm{H}}\boldsymbol{\Phi}_i\boldsymbol{D}_i\boldsymbol{B}_i\boldsymbol{w}_i x_i}_{\text{IRI}}), \tag{1}$$

where ρ_k and $\boldsymbol{w}_k \in \mathbb{C}^{K\times 1}$ are, respectively, the transmit signal-to-noise ratio (SNR) and the transmit beamforming vector for UE k, and $\boldsymbol{f}_{i,k} \in \mathbb{C}^{N\times 1}$ and $\boldsymbol{D}_k \in \mathbb{C}^{N\times N}$ are, respectively, the fading channels between RIS i and UE k and between the satellite and RIS k. Also, x_k and n_k are the transmit signal and additive complex Gaussian noise with zero mean and unit variance, respectively, and $\boldsymbol{\Phi}_k \in \mathbb{C}^{N\times N}$ is a reflection matrix that operates as a passive beamformer and can be controlled by the ES. The reflection matrix $\boldsymbol{\Phi}_k$ can be obtained as follows:

$$\boldsymbol{\Phi}_k = \mathrm{diag}\Big(\sqrt{\alpha_{k,1}}e^{j\theta_{k,1}}, \sqrt{\alpha_{k,2}}e^{j\theta_{k,2}}, \cdots, \sqrt{\alpha_{k,N}}e^{j\theta_{k,N}}\Big), \tag{2}$$

where $\sqrt{\alpha_{k,n}} \in [0,1]$ and $\theta_{k,n} \in [0,2\pi)$ are, respectively, the amplitude and phase shifting response of the reflection element n in RIS k. Assuming lossless reflection power at all the RISs, we consider that $\alpha_{k,n} = 1, \forall k, n$. In (1), the signals directly transmitted from the satellite to the UEs are neglected because of severe path loss and shadowing. For a practical air-to-ground (A2G) channel, $\boldsymbol{f}_{i,k}$ is generated by a correlated Rician fading channel composed of a deterministic LoS path and a fading non-LoS (NLoS) path, as follows:

$$\boldsymbol{f}_{i,k} = \bar{\beta}_{i,k}\bar{\boldsymbol{f}}_{i,k} + \tilde{\beta}_{i,k}\tilde{\boldsymbol{f}}_{i,k}, \tag{3}$$

where

$$\bar{\beta}_{i,k} = \sqrt{\frac{\kappa_{i,k}\zeta_{i,k}}{\kappa_{i,k}+1}}, \tag{4}$$

$$\tilde{\beta}_{i,k} = \sqrt{\frac{\zeta_{i,k}}{\kappa_{i,k}+1}}. \tag{5}$$

Here, $\kappa_{i,k}$ and $\zeta_{i,k}$ are, respectively, the Rician factor and the path loss between RIS i and UE k, and $\bar{\boldsymbol{f}}_{i,k} \in \mathbb{C}^{N\times 1}$ comprises the deterministic LoS components whose entries are determined depending on the locations of RIS i and UE k. Defining $\bar{f}_{i,k,n}$ as the nth element of $\bar{\boldsymbol{f}}_{i,k}$, we have $\bar{f}_{i,k,n} = \exp(-j2\pi d_{i,k,n}/\lambda)$, where $d_{i,k,n}$ denotes the distance between reflecting element n of RIS i and UE k, and λ denotes the wavelength of a signal [19]. Also, $\tilde{\boldsymbol{f}}_{i,k} = \boldsymbol{G}_i^{1/2}\hat{\boldsymbol{f}}_{i,k} \in \mathbb{C}^{N\times 1}$ consists of the correlated NLoS components between RIS i and UE k, where $\boldsymbol{G}_i \in \mathbb{C}^{N\times N}$ is the spatial correlation matrix at RIS i, and $\hat{\boldsymbol{f}}_{i,k}$ is the independent NLoS component. In (1), $\boldsymbol{D}_k$ is a rain fading matrix, and $\boldsymbol{B}_k \in \mathbb{C}^{N\times K}$ is the coefficient matrix whose entries represent the gains of the beam radiation pattern with path loss between the satellite and RIS k, respectively, given as [18]:

$$\boldsymbol{D}_k = |c_k|\exp(j\psi_k)\boldsymbol{I}_N, \tag{6}$$

$$\boldsymbol{B}_k = \sqrt{\xi_k}\boldsymbol{a}_k^{\mathrm{T}} \otimes \mathbf{1}_{N\times 1}. \tag{7}$$

In (7), ξ_k and the entries of $\boldsymbol{a}_k \in \mathbb{R}^{K\times 1}$ are the coefficients representing, respectively, the effects of the path loss and beam pattern as given by:

$$\xi_k = \left(\frac{\lambda}{4\pi d_k}\right)^2, \tag{8}$$

$$a_{k,m} = \sqrt{G_{\max}}\left(\frac{J_1(u_{k,m})}{2u_{k,m}} + 36\frac{J_3(u_{k,m})}{u_{k,m}^3}\right), \tag{9}$$

where $a_{k,m}$ is the mth element of $\boldsymbol{a}_k$, d_k is the distance from the satellite to RIS k, $G_{\max}$ is the maximum gain of the satellite beam, $J_n(\cdot)$ is the first-kind Bessel function of order n, and $u_{k,m} = 2.07123\frac{\sin\phi_{k,m}}{\sin\phi_{3\mathrm{dB}}}$. Also, $\phi_{3\mathrm{dB}}$ and $\phi_{k,m}$ are the 3 dB angle of the beam and the angle between the center of beam m and RIS k, respectively. In (6), $|c_k|$ is the rain fading channel gain that follows a double-log-normal distribution such as $|c_k| \sim \mathcal{L}^2\mathcal{N}(\eta_k, \sigma_k^2)$, and ψ_k is the phase of the fading channel at RIS k that is distributed uniformly over $[0, 2\pi)$, as in [18]. Hence, the elements in $\boldsymbol{a}_k$ and $\boldsymbol{B}_k$ are deterministic values depending on the locations of the satellite and UEs.

2.2. ZFBF and Co-Phasing Technique

In (1), the term $\sqrt{\rho_i}\boldsymbol{f}_{k,k}^{\mathrm{H}}\boldsymbol{\Phi}_k\boldsymbol{D}_k\boldsymbol{B}_k\boldsymbol{w}_i x_i$ represents IBI caused by the ith on-board feed $\boldsymbol{w}_i x_i$, while the term $\sqrt{\rho_i}\boldsymbol{f}_{i,k}^{\mathrm{H}}\boldsymbol{\Phi}_i\boldsymbol{D}_i\boldsymbol{B}_i\boldsymbol{w}_i x_i$ represents IRI resulting from the ith on-board feed reflected by the ith RIS. Note that the acquisition of full CSI for RIS-related interference channels, such as $\boldsymbol{f}_{i,k}$ and $\boldsymbol{D}_i$, is challenging in practice [15]. Also, ZFBF and the co-phasing technique can be designed solely based on information about the desired channels, without requiring any information about the RIS-related interference channels [18,20]. Hence, we consider ZFBF at the ES to eliminate IBI while adopting the co-phasing technique at the RIS as a suboptimal strategy to mitigate IRI.

The effective channel between the satellite and UE k via RIS k and the aggregate effective channel for K UEs are, respectively, given as:

$$\boldsymbol{h}_{k,k} = \boldsymbol{f}_{k,k}^{\mathrm{H}}\boldsymbol{\Phi}_k\boldsymbol{D}_k\boldsymbol{B}_k, \tag{10}$$

$$\boldsymbol{H} = \left[\boldsymbol{h}_{1,1}^{\mathrm{T}}, \boldsymbol{h}_{2,2}^{\mathrm{T}}, \cdots, \boldsymbol{h}_{K,K}^{\mathrm{T}}\right]^{\mathrm{T}}. \tag{11}$$

Consider that $\boldsymbol{H}$ is perfectly known at the ES. Then, the corresponding ZFBF matrix is obtained as $\boldsymbol{T} = \boldsymbol{H}^{-1}$. Then, the ZFBF vector for UE k can be expressed as $\boldsymbol{w}_k = \boldsymbol{t}_k/(K\|\boldsymbol{t}_k\|)$, where $\boldsymbol{T} = [\boldsymbol{t}_1, \boldsymbol{t}_2, \cdots, \boldsymbol{t}_K]$. Here, the equal power allocation is assumed because, in high transmit power systems such as SatCom systems, the equal power allocation can achieve the performance resulting from the water-filling solution [21]. Also, to maximize the desired signal power, the phase shifts at the RIS can be designed based on the co-phasing technique such that the reflected N signals from the RIS are aligned in phase as follows: $\theta_{k,n} = -\angle f_{k,k,n}^*, \forall k, n$, where $f_{k,k,n}$ is the nth element of $\boldsymbol{f}_{k,k}$. Then, the received signal-to-interference-plus-noise ratio (SINR) at UE k can be obtained as follows:

$$\gamma_k = \frac{\rho_k}{K^2\|\boldsymbol{t}_k\|^2\left(1 + \sum_{i\neq k}^{K}\rho_i|\boldsymbol{g}_{i,k}^{\mathrm{T}}\boldsymbol{D}_i\boldsymbol{B}_i\boldsymbol{w}_i|^2\right)}, \tag{12}$$

where we define

$$\boldsymbol{g}_{i,k} = \left[|f_{i,k,1}|e^{j\vartheta_{i,k,1}}, |f_{i,k,2}|e^{j\vartheta_{i,k,2}}, \cdots, |f_{i,k,N}|e^{j\vartheta_{i,k,N}}\right]$$

and $\vartheta_{i,k,n} = \angle f_{i,k,n}^* - \angle f_{i,i,n}^*$. We can observe from (12) that the IBI is perfectly eliminated, while the IRI still remains. However, we will demonstrate that the co-phasing technique at the RIS can asymptotically mitigate the IRI as well.

3. Active Beamforming and Phase Shift Design

We provide an asymptotic analysis of γ_k in the following two scenarios: (i) full CSI is available at the ES and (ii) only partial CSI composed of the deterministic information, such as the locations of satellites and UEs, is available at the ES.

3.1. ZFBF and Co-Phasing Technique Using Full CSI

Considering the ZFBF at the ES and the co-phasing technique at the RIS, the effective channel between the satellite and UE k via RIS k and the corresponding ZFBF vector for UE k can be rewritten, respectively, as follows:

$$\boldsymbol{h}_{k,k} = \sqrt{\xi_k}|c_k|\exp(j\psi_k)\boldsymbol{a}_k^{\mathrm{T}}\textstyle\sum_n|f_{k,k,n}|, \tag{13}$$

$$\boldsymbol{w}_k = \frac{\bar{\boldsymbol{t}}_k}{K\exp(j\psi_k)\|\bar{\boldsymbol{t}}_k\|}. \tag{14}$$

Here, $\bar{\boldsymbol{t}}_k$ is defined as the kth column vector of $\bar{\boldsymbol{T}}$, where $\bar{\boldsymbol{T}} = \bar{\boldsymbol{H}}^{-1}$ and $\bar{\boldsymbol{H}} = \left[\boldsymbol{a}_1^{\mathrm{T}}, \boldsymbol{a}_2^{\mathrm{T}}, \cdots, \boldsymbol{a}_K^{\mathrm{T}}\right]^{\mathrm{T}}$. Then, the received signal at UE k can be obtained as:

$$\begin{aligned} y_k = & \frac{\sqrt{\rho_k \xi_k}|c_k| \sum_n |f_{k,k,n}|}{K\|\bar{\boldsymbol{t}}_k\|} x_k \\ & + \sum_{i \neq k}^{K} \frac{\sqrt{\rho_i \xi_i}|c_i| \sum_n |f_{i,k,n}| e^{j\vartheta_{i,k,n}}}{K\|\bar{\boldsymbol{t}}_i\|} x_i + n_k, \end{aligned} \tag{15}$$

where $\vartheta_{i,k,n} = \angle f_{i,k,n}^* - \angle f_{i,i,n}^*$. The second term in the right-hand side of (15) represents the residual IRI generated by the reflected signal from other RISs. As N increases, the residual IRI and the noise become negligible compared to the desired signal as follows:

$$y_k \xrightarrow[(a)]{\text{a.s.}} \frac{\sqrt{\rho_k \xi_k}|c_k| \sum_n |f_{k,k,n}|}{K\|\bar{\boldsymbol{t}}_k\|} x_k. \tag{16}$$

In (16), (a) results from the following two observations: (i) as N increases, $\sum_n |f_{k,k,n}|$ increases with $\mathcal{O}(N)$, while (ii) $\sum_n |f_{i,k,n}| e^{j\vartheta_{i,k,n}}$ and n_k do not increase because the values of $\vartheta_{i,k,n}, \forall n$ are different from each other. Hence, as N increases, we can determine the asymptotic SINR at UE k for ZFBF with the co-phasing approach as follows:

$$\begin{aligned} \gamma_k^{\text{full}} = & \frac{N^2 \rho_k \xi_k |c_k|^2 \left(\sum_n |f_{k,k,n}|/N\right)^2}{K^2 \|\bar{\boldsymbol{t}}_k\|^2} \\ & \xrightarrow{\text{a.s.}} \frac{N^2 \rho_k \xi_k |c_k|^2 \mathrm{E}\left[|f_{k,k,n}|\right]^2}{K^2 \|\bar{\boldsymbol{t}}_k\|^2}. \end{aligned} \tag{17}$$

In (17), $\mathrm{E}\left[|f_{k,k,n}|\right]$ represents the expectation of the random variable $|f_{k,k,n}|$ for n, where n ranges from 1 to N. From (17), it can be observed that, as N increases, the received SINR increases with $\mathcal{O}(N^2)$, and this is equivalent to the asymptotic SNR of an ideal RIS-assisted cellular network, as demonstrated in [22]. While the implementation of the ZFBF and the co-phasing technique may not be optimal, they have the potential to achieve optimal performance in the asymptotic regime. However, a general ZFBF requires complete CSI of $\boldsymbol{h}_{k,k}$, and the co-phasing technique requires complete CSI of $\boldsymbol{f}_{k,k}$. Acquiring complete CSI at the ES leads to a considerable increase in signaling overhead and introduces significant latency due to extended round-trip delays, particularly in SatCom systems. Therefore, in the next section, we propose a deterministic ZFBF and co-phasing technique that utilizes partial CSI.

3.2. ZFBF and Co-Phasing Technique Using Partial CSI

Acquiring complete CSI at the ES results in substantial signaling overhead, leading to significant latency. Additionally, due to the fully passive nature of the RIS, the ES cannot accurately estimate RIS-related channels such as $\boldsymbol{D}_k$ and $\boldsymbol{f}_{k,k}$. In this regard, we propose a deterministic ZFBF and co-phasing technique that utilizes partial CSI.

Based on (13) and (14), we can observe that $\boldsymbol{h}_{k,k}\bar{\boldsymbol{t}}_i = 0, \forall i \neq k$. Motivated by this observation, we propose a ZFBF vector as $\boldsymbol{w}_k = \bar{\boldsymbol{t}}_k/(K\|\bar{\boldsymbol{t}}_k\|)$, which can perfectly eliminate the IBI as follows:

$$\begin{aligned} & \sum_{i \neq k}^{K} \sqrt{\rho_i} \boldsymbol{h}_{k,k} \boldsymbol{w}_i x_i \\ & = \sum_{i \neq k}^{K} \sqrt{\rho_i \xi_k}|c_k| \exp(j\psi_k) \sum_n |f_{k,k,n}| \frac{\boldsymbol{a}_k^{\mathrm{T}} \bar{\boldsymbol{t}}_i}{K\|\bar{\boldsymbol{t}}_i\|} x_i = 0. \end{aligned}$$

To design the proposed ZFBF vector $\boldsymbol{w}_k = \bar{\boldsymbol{t}}_k/(K\|\bar{\boldsymbol{t}}_k\|)$, the ES requires access to $\boldsymbol{a}_k$ for all k, which consists of the deterministic beam pattern coefficients. These coefficients are deterministic information and can be accurately estimated using location-aware channel estimation based on global navigation satellite systems (GNSSs) [18]. Also, to maximize the desired signal power using partial CSI, the phase shifts at the RIS can be designed such that the reflected N signals from N LoS paths are aligned in phase as follows: $\theta_{k,n} = -\angle \bar{f}^*_{k,k,n}, \forall k, n$. Similarly, in (16), as N increases, both the IRI and the noise become negligible compared to the desired signal as follows:

$$\begin{aligned} y_k =& \frac{\sqrt{\rho_k \xi_k}|c_k|e^{j\psi_k} \sum_n \left(\bar{\beta}_{k,k} + \tilde{\beta}_{k,k}\tilde{f}^*_{k,k,n}e^{-j\angle \bar{f}^*_{k,k,n}}\right)}{K\|\bar{\boldsymbol{t}}_k\|} x_k \\ &+ \sum_{i\neq k}^{K} \frac{\sqrt{\rho_i \xi_i}|c_i|e^{j\psi_i} \sum_n |f_{i,k,n}|e^{j\tilde{\vartheta}_{i,k,n}}}{K\|\bar{\boldsymbol{t}}_i\|} x_i + n_k \\ \xrightarrow{\text{a.s.}}& \frac{N\sqrt{\rho_k \xi_k}|c_k|e^{j\psi_k}\left(\bar{\beta}_{k,k} + \tilde{\beta}_{k,k}\mathrm{E}\left[\tilde{f}^*_{k,k,n}e^{-j\angle \bar{f}^*_{k,k,n}}\right]\right)}{K\|\bar{\boldsymbol{t}}_k\|} x_k \\ \underset{(b)}{=}& \frac{N\sqrt{\rho_k \xi_k}|c_k|e^{j\psi_k}\bar{\beta}_{k,k}}{K\|\bar{\boldsymbol{t}}_k\|} x_k, \end{aligned} \tag{18}$$

where $\tilde{f}_{k,k,n}$ is the nth element of $\tilde{\boldsymbol{f}}_{k,k}$ and $\tilde{\vartheta}_{i,k,n} = \angle f^*_{i,k,n} - \angle \bar{f}^*_{i,i,n}$. In (18), (b) arises from the fact that $\mathrm{E}\left[\tilde{f}^*_{k,k,n}e^{-j\angle \bar{f}^*_{k,k,n}}\right] = \mathrm{E}\left[\tilde{f}^*_{k,k,n}\right] \cdot \mathrm{E}\left[e^{-j\angle \bar{f}^*_{k,k,n}}\right] = 0$. Thus, to design the proposed RIS phase shift matrix, the ES only requires $\bar{\boldsymbol{f}}_{k,k}$, which consists of the deterministic LoS components, which can also be accurately estimated using location-aware channel estimation based on GNSSs. Using the proposed deterministic ZFBF and co-phasing technique, we can obtain the asymptotic SINR at UE k as follows:

$$\gamma_k^{\text{part}} = \frac{N^2 \rho_k \xi_k |c_k|^2 \bar{\beta}^2_{k,k}}{K^2\|\bar{\boldsymbol{t}}_k\|^2}. \tag{19}$$

From (19), we can observe that as N increases, the received SINR also increases with $\mathcal{O}(N^2)$, as seen in the case of the ZFBF with co-phasing approach using full CSI. Furthermore, in practical A2G channel models, the average and maximum values of the Rician factor measured in [23] are 10 dB and 20 dB, respectively. As a result, $\mathrm{E}\left[|f_{k,k,n}|\right]$ is approximately similar to $\bar{\beta}_{k,k}|\exp(-j2\pi d_{k,k,n}/\lambda)| = \bar{\beta}_{k,k}$, leading to $\gamma_k^{\text{part}} \approx \gamma_k^{\text{full}}$. In summary, the proposed beamforming scheme comprises two stages. First, to perfectly eliminate the IBI, the ZFBF vector is designed as follows: $\boldsymbol{w}_k = \bar{\boldsymbol{t}}_k/(K\|\bar{\boldsymbol{t}}_k\|)$. Next, to maximize the designed signal power based on the partial CSI, the phase shifting of the RIS is designed as follows: $\theta_{k,n} = -\angle \bar{f}^*_{k,k,n}, \forall k, n$.

3.3. *Asymptotic Performance Analysis*

Next, we conduct an asymptotic analysis of the mean of the capacity (i.e., ergodic capacity) for the proposed deterministic beamforming. From (19), the ergodic capacity can be derived asymptotically for a large N as follows:

$$\begin{aligned} \mu_{R_k} &= \mathrm{E}\left[\log_2\left(1 + \frac{N^2 \rho_k \xi_k |c_k|^2 \bar{\beta}^2_{k,k}}{K^2\|\bar{\boldsymbol{t}}_k\|^2}\right)\right] \\ &\approx \log_2\left(\frac{N^2 \rho_k \xi_k \bar{\beta}^2_{k,k}}{K^2\|\bar{\boldsymbol{t}}_k\|^2}\right) + 2\mathrm{E}[\log_2|c_k|] \\ &\underset{(c)}{=} \log_2\left(\frac{N^2 \rho_k \xi_k \bar{\beta}^2_{k,k}}{K^2\|\bar{\boldsymbol{t}}_k\|^2}\right) - \frac{\log_2 10}{10} e^{\mu_k + \frac{\sigma_k^2}{2}}, \end{aligned} \tag{20}$$

where $\mu_k = \eta_k - \log(\log 10) + \log 20$ [18]. In (20), (c) arises from the characteristic of the double-log-normal distribution of $|c_k| \sim \mathcal{L}^2\mathcal{N}\left(\eta_k, \sigma_k^2\right)$ and a large N. Similarly, the variance of R_k can be obtained by the following:

$$\begin{aligned}\sigma_{R_k}^2 &= \mathrm{Var}\left[\log_2\left(1+\frac{N^2\rho_k\xi_k|c_k|^2\bar{\beta}_{k,k}^2}{K^2\|\bar{\boldsymbol{t}}_k\|^2}\right)\right] \\ &\underset{(d)}{\approx} 4\mathrm{Var}[\log_2|c_k|] \\ &= \left(\frac{\log_2 10}{10}\right)^2\left(e^{\sigma_k^2}-1\right)e^{2\mu_k+\sigma_k^2},\end{aligned} \tag{21}$$

where (d) results from a large N approximation. Given that the values of ρ_k, ξ_k, $\bar{\beta}_{k,k}$, μ_k, σ_k, and $\|\bar{\boldsymbol{t}}_k\|$ are deterministic, it is possible to obtain μ_{R_k} and $\sigma_{R_k}^2$ deterministically. Consequently, we can assess the performance of an RIS-assisted SatCom system in terms of ergodic capacity without the need for extensive simulations.

Based on the scaling law for N, we can observe from (20) that the ergodic capacity increases with $\mathcal{O}(\log N)$, while (21) shows that $\sigma_{R_k}^2$ remains constant regardless of N. Therefore, we can characterize the outage probability, defined as the probability that R_k falls below a target value R_0, as follows:

$$P_o = \Pr(R_k < R_0) = \Pr\left(R_k - \mu_{R_k} < R_0 - \mu_{R_k}\right). \tag{22}$$

Consider a target value $R_0 = \epsilon \cdot \mu_{R_k}$ where $0 < \epsilon \leq 1$. Then, we can obtain the outage probability as follows:

$$P_o = \Pr\left(R_k - \mu_{R_k} < (\epsilon - 1)\mu_{R_k}\right). \tag{23}$$

From (20), we can observe that

$$R_k - \mu_{R_k} \approx 2\log_2|c_k| + \frac{\log_2 10}{10}e^{\mu_k+\frac{\sigma_k^2}{2}}, \tag{24}$$

and thus, a random variable $R_k - \mu_{R_k}$ is unrelated to N. Then, the probability density function (PDF) of the random variable $R_k - \mu_{R_k}$ is not affected by the increase in N, while μ_{R_k} increases with $\mathcal{O}(\log N)$. Hence, as N increases, the PDF of $R_k - \mu_{R_k}$ remains unchanged, and $(\epsilon - 1)\mu_{R_k}$ decreases, resulting in a decreasing P_o. Hence, with the increase in N, the outage probability decreases, thereby verifying the ability of an RIS to offer reliable SatCom services for a large N regime.

4. Simulation Results and Analysis

We conduct extensive simulations to evaluate the performance of a practical RIS-assisted SatCom system, considering a limited number of reflecting elements N. It is assumed that a low equatorial orbit (LEO) satellite serves a cluster of 7 UEs through 7 RISs (i.e., $K = 7$). The noise power at each UE is given by $\kappa_B T B$, where κ_B, T, and B are the Boltzmann constant, noise temperature at each UE, and bandwidth per beam, respectively. To ensure statistical reliability, all simulations are averaged over a large number of independent runs, and the simulation parameters used are provided in Table 1 based on [17].

Table 1. Simulation parameters.

Parameter	Value
Orbit	LEO (3.6 × 700 km)
Distance between UAV and UE	1 km
Carrier frequency	2 GHz (L-band)
K, $G_{\max}$, $\phi_{3\mathrm{dB}}$	7, 49.5 dB, 3.2 °
Rician factor ($\kappa_{i,k}, \forall i,k$)	10 dB

Table 1. *Cont.*

Parameter	Value
UE antenna gain	38.16 dB
Transmit power	50 dBW
Boltzmann's constant (κ_B)	1.38×10^{-23} J/m
Bandwidth per beam (B)	20 MHz
Noise temperature (T)	300 K
RIS element layout	2D square lattice
RIS element spacing	0.01 m
Average rain attenuation ($\forall k$)	3 dB ($\eta_k = -1.5617, \sigma_k = 1$)

In Figure 2, the asymptotic ergodic capacity of the proposed scheme analyzed in (20) is verified. In this figure, "MRT without IBI and IRI" shows an ideal performance representing the case where the ES employs the maximum ratio transmission (MRT) to generate beamforming weights $w_k = h_{k,k}^{\mathrm{H}} / (K \| h_{k,k} \|)$ that align the transmitted signals towards the intended UE. On the other hand, "random phase shift" represents the scenario of random passive beamforming, where the phase shifts $\theta_{k,n}, \forall k$, and n are randomly determined. The figure demonstrates that the asymptotic ergodic capacity derived from (20) aligns well with the simulation results as the value of N increases. The proposed deterministic ZFBF combined with the co-phasing technique utilizes partial CSI but achieves a performance close to that obtained using full CSI. Moreover, when $N = 64$, the proposed deterministic beamforming achieves approximately 92.8% of the ideal performance, and it increases as N increases, further confirming the asymptotic optimality of the proposed scheme.

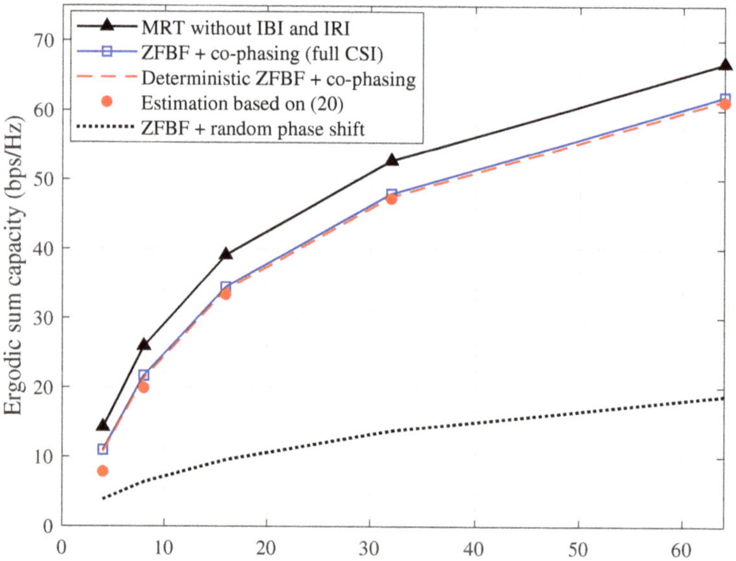

Figure 2. Performance comparison of the ergodic sum capacities for N.

In Figure 3, the outage probability analyzed in (23) is verified. The figure compares the outage probabilities obtained from the simulations for different values of ϵ, specifically for $\epsilon = 0.90, 0.85$, and 0.80. As shown in Figure 3, the outage probabilities tend to decrease towards zero as N increases. This decreasing trend in the outage probability indicates improved system performance and enhanced reliability as the RIS is able to mitigate potential communication outages. Furthermore, Figure 3 demonstrates that the proposed deterministic ZFBF combined with the co-phasing technique utilizes partial CSI while achieving outage performance close to that obtained using full CSI. In conclusion, this result provides empirical evidence that supports the notion of the RIS as a viable solution for ensuring reliable SatCom in the presence of a large number of RIS elements.

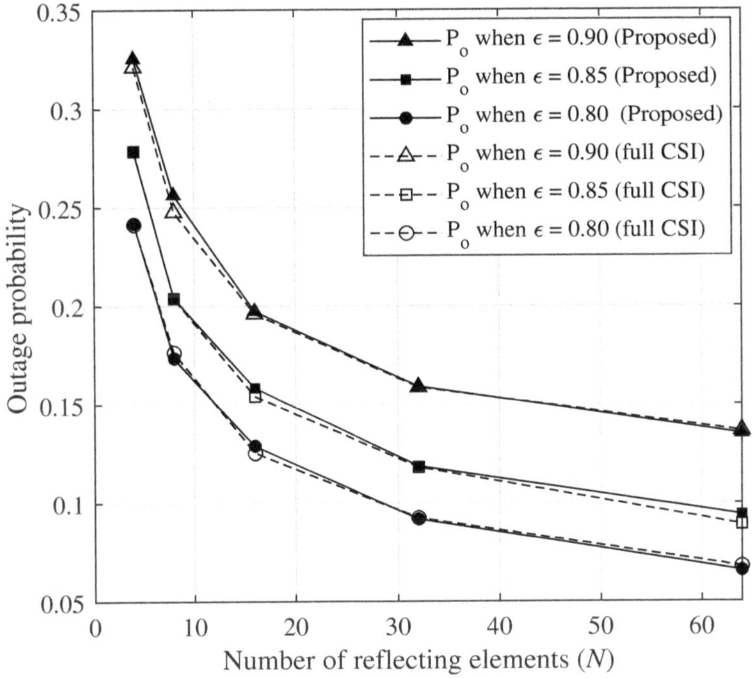

Figure 3. Performance comparison of the outage probabilities for N.

5. Conclusions

In this paper, we conducted an asymptotic analysis of a ZFBF combined with the co-phasing technique for mitigating IBI and IRI in RIS-assisted SatCom systems. In particular, we analyzed the performance of the ZFBF combined with co-phasing technique based on full CSI and demonstrated that IBI and IRI are asymptotically eliminated as the number of RIS elements increases to infinity. To make the analysis more practical, we proposed a novel deterministic ZFBF and co-phasing technique that only requires partial CSI, while still achieving asymptotic elimination of IBI and IRI. Furthermore, we derived a closed-form expression for the asymptotic ergodic capacity of the proposed scheme, allowing us to accurately determine the system's performance without extensive simulations. Additionally, we proved that the outage probability decreases as N increases, confirming the RIS's ability to provide reliable SatCom services in a large N regime. The simulation results validated the accuracy of our asymptotic analyses, and we observed consistent results between the approximated ergodic capacity and actual simulation results. Moreover, the proposed scheme demonstrated asymptotic optimality as it closely achieved the ideal performance, and this verifies the effectiveness of the proposed scheme in achieving optimal performance. Additionally, the outage probability demonstrated a potential convergence to zero as the number of RIS elements increases to infinity, further supporting the reliability and efficiency of the proposed scheme.

Author Contributions: Conceptualization, M.J.; methodology, T.K. and H.S.; software, M.J.; validation, M.J., T.K. and H.S.; formal analysis, M.J.; investigation, M.J.; resources, M.J.; data curation, M.J.; writing—original draft preparation, M.J.; writing—review and editing, T.K. and H.S.; visualization, M.J.; supervision, T.K. and H.S.; project administration, M.J., T.K. and H.S.; funding acquisition, M.J., T.K. and H.S. All authors have read and agreed to the published version of the manuscript.

Funding: This work was supported in part by Institute of Information & Communications Technology Planning & Evaluation (IITP) under the metaverse support program to nurture the best talents (IITP-2023-RS-2023-00254529) grant funded by the Korean government (MSIT) and by the National Research Foundation of Korea (NRF) grants funded by the Korean government (MSIT) (No. 2022R1F1A1064106, No. 2021R1C1C1012950, and No. 2021R1F1A1051075).

Data Availability Statement: Data are contained within the article.

Conflicts of Interest: The authors declare no conflicts of interest.

References

1. Saad, W.; Bennis, M.; Chen, M. A Vision of 6G Wireless Systems: Applications, Trends, Technologies, and Open Research Problems. *IEEE Netw.* **2020**, *34*, 134–142. [CrossRef]
2. Wang, S.; Zhang, L.; Fan, H. An Enhanced Multi-Constraint Optimization Algorithm for Efficient Network Topology Generation. *Mathematics* **2023**, *11*, 3456. [CrossRef]
3. Xu, Y.; Deng, F.; Zhang, J. UDCO-SAGiMEC: Joint UAV Deployment and Computation Offloading for Space–Air–Ground Integrated Mobile Edge Computing. *Mathematics* **2023**, *11*, 4014. [CrossRef]
4. Son, H.; Jung, M. Phase Shift Design for RIS-Assisted Satellite-Aerial-Terrestrial Integrated Network. *IEEE Trans. Aerosp. Electron. Syst.* **2023**, *59*, 9799–9806; *early access*. [CrossRef]
5. Zhang, J.; Qi, C. Channel Estimation for mmWave Satellite Communications with Reconfigurable Intelligent Surface. In Proceedings of the 2021 IEEE Global Communications Conference (GLOBECOM), Madrid, Spain, 7–11 December 2021; pp. 1–6. [CrossRef]
6. Lin, Z.; Niu, H.; An, K.; Wang, Y.; Zheng, G.; Chatzinotas, S.; Hu, Y. Refracting RIS-Aided Hybrid Satellite-Terrestrial Relay Networks: Joint Beamforming Design and Optimization. *IEEE Trans. Aerosp. Electron. Syst.* **2022**, *58*, 3717–3724. [CrossRef]
7. Khan, W.U.; Lagunas, E.; Mahmood, A.; ElHalawany, B.M.; Chatzinotas, S.; Ottersten, B. When RIS Meets GEO Satellite Communications: A New Sustainable Optimization Framework in 6G. In Proceedings of the IEEE Vehicular Technology Conference, Helsinki, Finland, 19–22 June 2022; pp. 1–6. [CrossRef]
8. Rahman, M.H.; Sejan, M.A.S.; Aziz, M.A.; Kim, D.S.; You, Y.H.; Song, H.K. Deep Convolutional and Recurrent Neural-Network-Based Optimal Decoding for RIS-Assisted MIMO Communication. *Mathematics* **2023**, *11*, 3397. [CrossRef]
9. Rahman, M.H.; Sejan, M.A.S.; Aziz, M.A.; Kim, D.S.; You, Y.H.; Song, H.K. Spectral Efficiency Analysis for IRS-Assisted MISO Wireless Communication: A Metaverse Scenario Proposal. *Mathematics* **2023**, *11*, 3181. [CrossRef]
10. Yu, Y.; Wang, J.; Zhou, X.; Wang, C.; Bai, Z.; Ye, Z. Review on Channel Estimation for Reconfigurable Intelligent Surface Assisted Wireless Communication System. *Mathematics* **2023**, *11*, 3235. [CrossRef]
11. Singya, P.K.; Alouini, M.S. Performance of UAV-Assisted Multiuser Terrestrial-Satellite Communication System Over Mixed FSO/RF Channels. *IEEE Trans. Aerosp. Electron. Syst.* **2022**, *58*, 781–796. [CrossRef]
12. Zhao, J.; Gao, F.; Wu, Q.; Jin, S.; Wu, Y.; Jia, W. Beam Tracking for UAV Mounted SatCom on-the-Move With Massive Antenna Array. *IEEE J. Sel. Areas Commun.* **2018**, *36*, 363–375. [CrossRef]
13. Joo, C.; Choi, J. Low-delay broadband satellite communications with high-altitude unmanned aerial vehicles. *J. Commun. Netw.* **2018**, *20*, 102–108. [CrossRef]
14. Perez-Neira, A.I.; Vazquez, M.A.; Shankar, M.B.; Maleki, S.; Chatzinotas, S. Signal Processing for High-Throughput Satellites: Challenges in New Interference-Limited Scenarios. *IEEE Signal Process. Mag.* **2019**, *36*, 112–131. [CrossRef]
15. Hashida, H.; Kawamoto, Y.; Kato, N. Selective Reflection Control: Distributed IRS-Aided Communication With Partial Channel State Information. *IEEE Trans. Veh. Technol.* **2022**, *71*, 11949–11958. [CrossRef]
16. Al-Jarrah, M.; Al-Dweik, A.; Alsusa, E.; Iraqi, Y.; Alouini, M.S. On the Performance of IRS-Assisted Multi-Layer UAV Communications With Imperfect Phase Compensation. *IEEE Trans. Commun.* **2021**, *69*, 8551–8568. [CrossRef]
17. Kim, T.; Jung, M.; Son, H. Joint User Scheduling and Phase Shift Optimization for STAR-RIS-Assisted Multicast Satellite Communications. *IEEE Trans. Aerosp. Electron. Syst.* **2024**, 1–10. [CrossRef]
18. Ahmad, I.; Nguyen, K.D.; Letzepis, N.; Lechner, G.; Joroughi, V. Zero-Forcing Precoding With Partial CSI in Multibeam High Throughput Satellite Systems. *IEEE Trans. Veh. Technol.* **2021**, *70*, 1410–1420. [CrossRef]
19. Tse, D.; Viswanath, P. *Fundamentals of Wireless Communication*; Cambridge University Press: Cambridge, UK, 2005.
20. Basar, E.; Renzo, M.D.; de Rosny, J.; Debbah, M.; Alouini, M.S.; Zhang, R. Wireless Communications through Reconfigurable Intelligent Surfaces. *IEEE Access* **2019**, *7*, 116753–117773. [CrossRef]
21. Yu, W.; Cioffi, J. Constant-power waterfilling: Performance bound and low-complexity implementation. *IEEE Trans. Commun.* **2006**, *54*, 23–28. [CrossRef]

22. Wu, Q.; Zhang, R. Intelligent Reflecting Surface Enhanced Wireless Network via Joint Active and Passive Beamforming. *IEEE Trans. Wirel. Commun.* **2019**, *18*, 5394–5409. [CrossRef]
23. Cai, X.; Song, J.; Rodríguez-Piñeiro, J.; E. Mogensen, P.; Tufvesson, F. Characterizing the Small-Scale Fading for Low Altitude UAV Channels. In Proceedings of the International Conference on Wireless and Mobile Communications, Nice, France, 18–22 July 2021; pp. 16–19.

mathematics

MDPI

Article

HCEL: Hybrid Clustering Approach for Extending WBAN Lifetime

Heba Helal [1], Farag Sallabi [2,*], Mohamed A. Sharaf [1], Saad Harous [3] and Mohammad Hayajneh [2] and Heba Khater [2]

[1] Department of Computer Science and Software, College of Information Technology, United Arab Emirates University, Al Ain 15551, United Arab Emirates; 202090034@uaeu.ac.ae (H.H.); msharaf@uaeu.ac.ae (M.A.S.)
[2] Department of Computer and Network Engineering, College of Information Technology, United Arab Emirates University, Al Ain 15551, United Arab Emirates; mhayajneh@uaeu.ac.ae (M.H.); 202190177@uaeu.ac.ae (H.K.)
[3] Department of Computer Science, College of Computing and Informatics, University of Sharjah, Sharjah 27272, United Arab Emirates; harous@sharjah.ac.ae
* Correspondence: f.sallabi@uaeu.ac.ae

Abstract: Wireless body area networks (WBANs) have emerged as a promising solution for addressing challenges faced by elderly individuals, limited medical facilities, and various chronic medical conditions. WBANs consist of wearable sensing and computing devices interconnected through wireless communication channels, enabling the collection and transmission of vital physiological data. However, the energy constraints of the battery-powered sensor nodes in WBANs pose a significant challenge to ensuring long-term operational efficiency. Two-hop routing protocols have been suggested to extend the stability period and maximize the network's lifetime. These protocols select appropriate parent nodes or forwarders with a maximum of two hops to relay data from sensor nodes to the sink. While numerous energy-efficient routing solutions have been proposed for WBANs, reliability has often been overlooked. Our paper introduces an energy-efficient routing protocol called a Hybrid Clustering Approach for Extending WBAN Lifetime (HCEL) to address these limitations. HCEL leverages a utility function to select parent nodes based on residual energy (RE), proximity to the sink node, and the received signal strength indicator (RSSI). The parent node selection process also incorporates an energy threshold value and a constrained number of serving nodes. The main goal is to extend the overall lifetime of all nodes within the network. Through extensive simulations, the study shows that HCEL outperforms both Stable Increased Throughput Multihop Protocol for Link Efficiency (SIMPLE) and Energy-Efficient Reliable Routing Scheme (ERRS) protocols in several key performance metrics. The specific findings of our article highlight the superior performance of HCEL in terms of increased network stability, extended network lifetime, reduced energy consumption, improved data throughput, minimized delays, and improved link reliability.

Keywords: WBAN; energy-efficient; routing; healthcare; clustering

MSC: 05C85; 68R10

1. Introduction

Healthcare spending has shown a substantial global increase recently, particularly in the United Arab Emirates (UAE). In 2020, healthcare expenditure in the UAE increased by 5.67% [1] and is projected to grow by nearly one-third in the coming years [2]. Several factors contribute to this upward trajectory, including the aging population and the prevalence of lifestyle diseases, highlighting the increasing demand for accessible and effective healthcare services [1,2]. Consequently, there is a pressing need to explore innovative approaches that can meet the needs of patients, especially in remote areas or situations where physical hospital visits pose challenges. Finding cost-effective and convenient ways to deliver healthcare services is essential to ensure that individuals receive the care they require [3–5].

Technological advancements have played an essential role in addressing these challenges. A notable innovation is the integration of remote care providers or hospitals, which allow the delivery of healthcare services from a distance [5,6]. In this context, WBANs have emerged as a significant trend that offers promising opportunities for remote healthcare monitoring [5–9]. WBANs are considered a subset of wireless sensor networks (WSNs) and also represent the next generation of personal area networks (PANs) [5,10–15]. Additionally, WBANs may contain various types of physiological, kinematic, and ambient sensors. These sensors capture and measure different parameters related to the human body and the environment, providing valuable insights into an individual's health status [7,16–18]. In WBANs, small sensor nodes (edge nodes) are strategically placed within a range of 1–2 m [12,13] from the central node, known as the sink node or coordinator node. The sink node has enhanced resource capabilities to handle data from all sensor nodes and transmit them to the cloud for remote analysis.

WBANs may have implanted or wearable sensors [11–13,15,19]. Implanted sensors are designed to be placed inside the human body, while wearable sensors are externally attached to the body, allowing for non-invasive monitoring of physiological and environmental data. The applications of WBANs span medical and non-medical domains [11–13,15,19]. In the medical field, WBANs facilitate real-time health monitoring, remote patient management, and early detection of medical emergencies. For instance, WBANs can monitor vital signs like heart rate, blood pressure, and body temperature, allowing health professionals to remotely assess a patient's health status and provide timely interventions. In non-medical applications, WBANs contribute to various sectors such as monitoring circadian rhythms, supporting battlefield healthcare for military personnel, and enhancing interactive gaming experiences by capturing real-time physiological responses [10,15,16,18,20,21].

Implementing WBANs poses several challenges, with a primary concern of energy efficiency [11–13,15,19]. To address these challenges, various routing protocols have been proposed; however, most existing protocols designed for WSNs are not directly applicable to WBANs due to unique constraints in WBANs. WBANs operate with limitations on the number of sensor nodes and the distances between nodes placed on the human body. Consequently, for instance, finding multiple sensors capable of performing the same task, which is applicable in WSNs, becomes constrained in WBANs. In addition, the routing protocols in WBANs only transmit the packets with a maximum of two hops. One of the key requirements for WBANs is also efficient energy utilization, considering the limited power resources of sensor nodes. The existing limitations in WBANs underscore the need for designing new routing protocols tailored to these networks.

Among the different types of routing protocols, clustering-based protocols have shown promise for WBANs [3–6,22–25]. These protocols group edge nodes into clusters, with a designated parent node for each cluster responsible for data aggregation and transmission to the sink node for remote management and analysis. These clustering-based protocols optimize energy usage by selecting an intermediate node (a parent node) that shortens the distance between the sensor nodes and the sink node. However, the selection criteria for the parent (forwarder) node on a round-by-round basis are based on factors that may lead to an uneven distribution of traffic among sensor nodes, potentially minimizing the network lifespan. Hence, enhancing the selection criteria for parent nodes becomes crucial for improving the overall efficiency of energy utilization in WBANs.

To address the limitations in existing cluster-based routing protocols, this paper introduces a novel routing protocol, called the HCEL (Hybrid Clustering Approach for Extending WBAN Lifetime). The HCEL routing protocol combines the benefits of static clustering (i.e., fixed parent node until it depletes its energy) and dynamic clustering (i.e., flexible parent node selection) to optimize energy consumption and enhance network lifetime. The proposed HCEL routing protocol encompasses three key phases: network deployment and initialization, parent node selection, and data transmission. In the network deployment and initialization phase, the WBAN is established, and initial operations are performed to ensure the smooth functioning of the deployed nodes. Sensor nodes are strategically placed during this phase, and the network is initialized for subsequent operations. The parent node selection phase plays a key role in HCEL. It involves selecting parent nodes within the network that serve as intermediaries for data transmission from sensor nodes to the sink. The selection process is guided by an energy-efficient algorithm considering factors such as (1) node energy levels, (2) distance to the sink, and (3) the received signal strength indicator (RSSI). Moreover, the selected node must adhere to constraints, including the number of serving nodes (τ) and the threshold value of the candidate parent nodes' energy (Th). By carefully choosing parent nodes, the HCEL approach aims to achieve balanced energy consumption and prevent the nodes from prematurely depleting their energy resources.

Once the parent nodes are selected, the data transmission phase begins. Sensor nodes within the WBAN collect data from various physiological sensors and transmit them to the sink node. Critical data are transmitted directly (one hop) to the sink, while regular data are transmitted through the selected parent nodes (two hops) to the sink. The sink node acts as the central hub, receiving and processing the data for further analysis or storage. Throughout this phase, the sink node continuously monitors the energy levels of the sensor nodes. This monitoring mechanism helps to prevent node failures and prolong the overall lifetime of sensor nodes. The proposed HCEL approach offers several advantages over traditional WBAN systems as follows: (1) Combining static and dynamic clustering leverages clustering benefits, resulting in improved energy efficiency and network stability, and (2) the parent node selection process further enhances these benefits by ensuring load balancing among sensor nodes. As a result, the HCEL protocol extends the lifetime of sensor nodes, reducing energy consumption and enhancing the overall performance of the WBAN system.

The paper follows a well-structured organization to present the research findings. Section 2 provides an overview of related work conducted in the context of this article, giving readers a comprehensive understanding of existing research works. Section 3 introduces key preliminaries for comprehending the proposed scheme and the experiments conducted. In Section 4, the paper delves into the heart of the study by presenting the proposed routing protocol, HCEL, in detail. Moving forward, Section 5 presents the simulation results and analysis, shedding light on the performance of the proposed HCEL protocol. Finally, Section 6 closes the article by summarizing the research's key findings, contributions, and implications.

2. Related Work

WBANs encounter many challenges due to the nature of the sensors and the environments in which they operate. One of the challenges in WBANs is the limited data rate, which ranges from a few kilobits per second (kbps) to several megabits per second (Mbps) [12,13]. This limitation poses a hurdle for transmitting sensor data to the sink node. Furthermore, the movement and dynamic nature of the human body [26,27] alter the radio frequency environment, resulting in interference and signal attenuation [11]. The security and privacy of the data transmitted in a WBAN can also be another challenge. Special precautions must be taken to maintain data privacy and integrity due to the sensitive nature of physiological data acquired in WBANs [28–31]. Some methods can be used to guarantee data confidentiality and integrity, such as access control systems or encryption algorithms. However, using these techniques incurs overhead for data transmission.

One of the most significant challenges in designing WBANs is the restricted energy source for wearable sensors. A constrained network lifetime is the result of these sensors often being battery-powered and difficult to recharge or replace. In other words, the energy constraints are mainly due to the limited battery capacity, the small size of the sensors, and the need for the sensors to be continuously operational [7,13]. Therefore, optimizing energy consumption is essential to extending the WBAN's lifespan. The challenge of energy efficiency in WBANs has been addressed using several techniques. One technique is replenishing the sensors' battery using energy harvesting techniques [16–18,32], such as solar or kinetic energy. This approach can increase the lifetime of the WBAN by reducing the need for battery replacement or recharging. Energy harvesting techniques, however, require appropriate hardware and are not ideal for all kinds of sensors. Additionally, sleep scheduling and duty cycling techniques can be used to conserve energy by turning off the sensors when not needed [12,33,34].

In conjunction with the previous techniques, one of the most common techniques is the development of energy-efficient routing protocols [19,35,36]. When designing WBANs, routing is paramount to ensuring effective and reliable data transmission [8,21]. A routing protocol determines the optimal path for data transmission from the sensors to the sink node to minimize the sensors' energy consumption during data transmission. The lifetime and performance of the WBAN are increased by reducing the energy consumption of the sensors. Routing protocols in WBANs can be classified into several categories according to their functionality and mode of operation, including cross-layer, postural, QoS-aware, thermal-aware, and cluster-based [8,11–13,15,19,21].

In cross-layer routing protocols, the routing decision is made based on information gathered from different layers of the protocol stack, such as physical, MAC, and network layers, to optimize network performance [35,37]. In traditional networking, each layer operates independently, and information flows only vertically between adjacent layers. However, in cross-layer design, information can be exchanged horizontally between different layers, allowing for more efficient use of network resources and better performance. It is particularly beneficial for optimizing energy consumption and network lifetime [37,38]. By leveraging information from different protocol stack layers, cross-layer routing protocols can make more informed decisions about routing paths and power management. For example, using information from the MAC layer, such as the status of the medium access control protocol and the buffer status of the nodes, cross-layer routing protocols can make more efficient use of the available bandwidth and reduce collisions, leading to improved network performance. Despite the advantages mentioned earlier, some challenges of cross-layer design include the increased complexity of the protocol stack and conflicts between different layers due to the exchange of information. Examples of cross-layer routing protocols in WBANs include WASP (Wireless Autonomous Spanning Tree Protocol) [39], CICADA (Cascading Information of Controlling Access and Distributed Slot Assignment) [40], TICOSS (Time-Zone Coordinated Sleeping Scheme) [34], Biocomm (a communication protocol for biomedical sensor networks) [41], and AMR (Adaptive Multihop Tree-Based Routing) [42].

Postural routing protocols leverage the specific structure and posture of the human body, such as the relative positions and orientations of the sensors on the body, to improve the efficiency of data transmission [26,27,43,44]. It uses this information to make routing decisions that minimize the energy consumption of the sensor nodes and improve the reliability of the network. For example, postural routing protocols can improve the network's overall reliability and energy efficiency by selecting routes that avoid interference caused by body movement or selecting nodes with stronger signals due to their position on the body. However, postural routing protocols may face some limitations, such as the need for accurate posture sensing and interference from body movement. Examples of a postural routing protocol in WBANs are Opportunistic (opportunistic routing based on posture prediction) [45], DVRPLC (Distance-Vector Routing Protocol with Postural Link Costs) [46], and OBSFR (On-body Store and Flood Routing Protocol) [14].

Quality of Service (QoS) is important in WBANs since some applications require real-time data transmission and reliable communication. QoS-aware routing protocols consider the QoS requirements of different applications, such as delay, reliability, and energy efficiency, and prioritize them for routing decisions accordingly [47]. For example, for applications that require low latency and high reliability, QoS-aware routing protocols can prioritize routes with low delay and high link quality. Conversely, for applications that are more tolerant to delay but require low energy consumption, QoS-aware routing protocols can prioritize routes with low energy consumption and choose one irrespective of how long it takes. QoS-aware routing protocols can improve overall performance by providing differentiated services to different types of traffic and by meeting the QoS requirements of each application. However, it may increase the complexity of the routing protocol and the QoS conflicts between different applications. Examples of QoS-aware routing protocols in WBANs include DMQoS (Data-Centric Multi-Objective QoS-based routing protocol) [48], QPRD (delay-sensitive data QoS-aware routing protocol) [49], EPR (Energy-Aware Peering routing protocol) [50], TMQoS (Thermal-Aware Multi-Constrained Intrabody QoS routing protocol) [51], and DARE (Distance-Aware Relaying Energy-Efficient System) [52].

The thermal-aware routing protocol uses the temperature information from sensor nodes to make routing decisions. It aims to balance the trade-off between energy and temperature control to ensure the reliable operation of the sensors. Temperature information is used to estimate energy consumption and predict the lifetime of sensor nodes [53,54]. Thermal-aware routing protocols can reduce energy consumption and prevent sensor nodes from overheating by avoiding routes that pass through nodes with high temperatures. Furthermore, these protocols can increase the lifetime of the sensor nodes by distributing the traffic load across the network. However, there are some challenges, such as the need for accurate temperature sensing and the complexity of the routing algorithm. Examples of a temperature-based routing protocol in WBANs are M^2E^2 (Multimode Energy-Efficient Multi-Hop protocol) [55], RE-ATTEMPT (Reliability-Enhanced Adaptive Threshold-Based Thermal-Unaware Energy-Efficient Multi-Hop protocol) [56], MATTEMPT (Mobility-Supporting Adaptive Threshold-Based Thermal-Aware Energy-Efficient Multi-Hop protocol) [57], TMQoS (Thermal-Aware Multi-Constrained Intrabody QoS routing protocol) [51], ETPA (Energy-Efficient, Thermal, and Power-Aware routing protocol) [58], ALTR (Adaptive Less Temperature Rise) [59], LTRT (Least Total Route Temperature) [60], and TARA (Thermal-Aware Routing Algorithm) [61].

The cluster-based routing protocol reduces the amount of data transmission and the distance between the sensors and the sink node, leading to reduced energy consumption and improved network lifetime. However, the choice of cluster heads may affect the network's overall performance. Examples of clustered routing protocols in WBANs include SIMPLE (Stable Increased Throughput Multihop Protocol for Link Efficiency) [24], iM-SIMPLE (Improved-SIMPLE) [23], FEEL (Forwarding Energy-Efficient Data with Load Balancing) [4], ERRS (Energy-Efficient Reliable Routing Scheme) [25], LAEEBA (Linked-Aware Energy-Efficient routing protocol) [22], CO-LAEEBA (Cooperative-LAEEBA) [6], and DSCB (Dual Sink Approach Using Clustering in Body Area Network) [5]. On the basis of the above discussions, the cluster-based routing protocol is the most suitable for WBAN applications. This paper aims to improve the energy efficiency of the network and extend the lifetime of the network by clustering the sensor nodes into groups [35]. Each cluster has a cluster head, which serves as an intermediary node and is responsible for collecting data from the sensor nodes and forwarding it to a sink node.

Table 1 summarizes the different cluster-based routing protocols employed in WBANs and compares them with our proposed routing protocol (HCEL). The table highlights essential information, such as the criteria used for parent node selection, advantages, limitations, clustering approach, and the number of sink nodes.

Most cluster-based routing protocols are built on a static clustering strategy. This static approach keeps the parent node (e.g., nodes close to the sink) unchanged until it depletes its energy (i.e., dies), resulting in decreased network efficiency. In contrast, dynamic clustering involves a flexible selection and adjustment of the parent node based on the current state of the network. The choice of the parent node is dynamically modified to adapt to changes in node energy levels, connectivity, or other network conditions. Although dynamic clustering improves network performance, it often incurs additional overhead and complexity, as the cluster structure needs to be frequently updated and modified.

Table 1. Summary of cluster-based routing protocols in WBANs.

Protocol	Parent Selection Criteria	Advantages	Limitations	Type of Clustering	Number of Sink Nodes
SIMPLE [24]	Minimum CF [2] $CF = \frac{dist^{1}}{RE^{5}}$	• Maximize throughput, • Balance residual energy across sensors.	• Unequal distribution of traffic load among sensors, • Low performance of the protocol in crucial data due to frequent data transfer, • Large overhead in communication cost, • Assume stationary nodes.	Static	Single
iM-SIMPLE [23]	Same as SIMPLE	• Handle mobility, • Maximize throughput.	• Picking closer nodes frequently, • Ignoring link quality and path loss, • Large number of dropped packets.	Static	Single
FEEL [4]	Maximum RE [5]	• Balance energy consumption of the network.	• Ignoring path failure, • Less intention to retransmit data packets, • Assuming stationary nodes.	Static	Single
ERRS [25]	Minimum WF [3] $WF = \frac{dist^{1}}{RE^{5}}$	• Load balance by selecting new parent node after several data transmission rounds.	• Ignoring scalability, • Assuming stationary nodes.	Static + Dynamic	Single
LAEEBA [22]	Minimum CF [2] $CF = \frac{\sqrt{dist}^{1}}{RE^{5}}$	• Reliable, • Minimum path loss, • High network lifetime.	• More overhead of data packets during data communication, • Assuming stationary nodes.	Static	Single
CO-LAEEBA [6]	Maximum RE [5] for parent node	• Consider data priority, • Maximum throughput, • High network lifetime.	• Limited number of parent nodes used, • Require many links between sensors and sink for packet delivery, • Assuming stationary nodes.	Static + Dynamic	Single
BEC [3]	Minimum CF [2] $CF = \frac{1}{RE^{5}}$	• Works well in terms of network lifetime, • Uniform load distribution.	• Parent node consumes more energy due to the demanding task of forwarding, • Assuming stationary nodes.	Static	Single
DSCB [5]	Minimum CF [2] $CF = \frac{dist^{1}}{RE^{5} \times TP^{4}}$	• Balance network load due to the double sinks used, • Improve negative consequences of the shadow effect causes.	• Disregarding energy efficiency with node cooperation and energy harvesting for efficient routing, • Assuming stationary nodes.	Static	Double

Table 1. *Cont.*

Protocol	Parent Selection Criteria	Advantages	Limitations	Type of Clustering	Number of Sink Nodes
HCEL	Maximum CF [2] $CF = \frac{dist^{1} \times RSSI}{RE^{5}}$	• Equal distribution of traffic load by combining static and dynamic clustering, • Extend sensor nodes' lifetimes, • Improve overall network performance and reliability.	• Assuming stationary nodes.	Static + Dynamic	Single

[1] *dist* refers to the distance between the sensor node and the sink node, [2] *CF* represents a cost function, [3] *WF* is a weight factor, [4] *TP* is a transmission power, and [5] *RE* represents residual energy.

To overcome the limitations of static and dynamic clustering, we propose an energy-efficient routing strategy called HCEL. In our routing protocol, we prioritize the selection of parent nodes based on factors like distance, residual energy, and RSSI. Unlike static clustering approaches where the same node is chosen until its energy is depleted, our protocol selects a different parent node after several data transmission rounds. Our routing protocol considers other parameters, such as the number of the nodes that the parent node can serve (τ) and the threshold energy for all parent candidates (Th). By incorporating these factors, the selection of the parent node is adjusted based on the current state of the network, effectively combining the benefits of both static and dynamic clustering. This adjustment ensures a balanced distribution of the load across all nodes, preventing any individual node from being overwhelmed with data forwarding tasks. By combining both the static and dynamic clustering, we enhance the network lifetime, stability period, energy consumption, and overall performance.

3. Preliminaries

This section introduces our system, assumptions, energy, and path loss models. Moreover, it presents the various performance measures used in our simulation analysis.

3.1. System Model and Assumptions

Our model is based on graph theory [62], represented as an undirected graph $G = (V, E)$. The graph consists of a set of vertices V representing the network's nodes $v_i \in V$ and a set of edges E that connect nodes together. Any edge $e_{ij} \in E$ is formed between any two nodes, v_i and v_j, when the two nodes are in the transmission range of each other.

This paper relies on assumptions to establish, model, and analyze our network. The first assumption is that all network nodes are stationary, meaning their positions and network topology remain fixed throughout the analysis. Except for the sink node, all nodes are assumed to have the same capabilities, equal importance, and initial energy levels. Data transmission is categorized into one-hop for critical data and two-hop for normal data. Moreover, the nodes, excluding the sink node, are equipped with routing functionality but are constrained regarding resources, encompassing energy, and processing capabilities.

3.2. Energy Model

We use the first-order radio energy model as presented in Ref. [63]. This model provides mathematical equations to characterize the energy consumption during data transmission and reception processes in WBANs. The energy consumption of the transmitter and receiver is given by Equations (1) and (2), respectively:

$$E_{Tx}(k, d) = E_{Tx-elec} \times k + E_{amp} \times \eta \times k \times d^{\eta} \quad (1)$$

$$E_{Rx}(k) = E_{Rx-elec} \times k. \tag{2}$$

In Equation (1), $E_{Tx}(k,d)$ represents the total energy consumption (in joules) required to transmit k bits between two nodes (e.g., v_i and v_j) with a separation distance of d. Here, $E_{Tx-elec}$ is the energy (in joule/bit) needed in the transmitter's circuits, while E_{amp} is the energy (in joule/bit/m$^{\eta}$) consumed by the amplifier circuit in the transmitter. η is the path loss coefficient. Moving on to Equation (2), $E_{Rx}(k)$ is the amount of energy consumed (in joules) by the receiver. Within this equation, $E_{Rx-elec}$ accounts for the energy (in joule/bit) utilized in the receiver's circuits. The values for the path loss coefficient η in Equation (1) are provided as follows [4]:

$$\eta = \begin{cases} 2 & \text{free space;} \\ 3\text{–}4 & \text{line of sight (LOS);} \\ 5\text{–}7.4 & \text{non-line of sight (NLOS).} \end{cases}$$

In our work, we account for the impact of signal degradation caused by various factors in a WBAN. These factors include obstacles like the body, walls, furniture, and indoor objects that can affect the line-of-sight transmission between sensor nodes on different body parts. To incorporate this attenuation into our analysis, we utilize an attenuation factor of $\eta = 3.38$ [23]. This factor quantifies the influence of these obstacles on the strength of the transmitted signal.

The energy parameters used in our analysis, such as $E_{tx-elec}$, $E_{Rx-elec}$, and E_{amp}, are hardware-dependent and specific to the system under consideration. Although several transceivers are commonly used in WBANs, we specifically employ the Chipcon CC2420 transceiver in our research. Detailed energy parameters for this transceiver are provided in Table 2 [4,23,24,57]. By considering the impact of signal attenuation due to obstacles and utilizing the energy parameters specific to the Chipcon CC2420 transceiver, we ensure that our analysis accurately reflects the real-world performance of the WBAN system under investigation.

Table 2. Energy parameters for the Chipcon CC2420 transceiver.

Parameters	Value	Units
$E_{tx-elec}$	96.9	nJ/bit
$E_{Rx-elec}$	172.8	nJ/bit
E_{amp}	2.71×10^{-7}	J/bit/m$^{\eta}$

To calculate the energy consumption in our proposed routing strategy, we consider the residual energy (RE) of each node in the network. This is calculated by subtracting the consumed energy from the initial energy. The energy consumed during data transmission by a sensor node is represented by Equation (1). However, for the parent node, the energy consumption includes the terms of Equation (1) and the energy consumed in the reception process, as expressed by Equation (2). This distinction arises because the parent node not only receives data from the sensor node but also forwards it to the sink.

3.3. Path Loss

Path loss in WBAN refers to the attenuation or reduction in signal power as it travels through the human body or the surrounding environment. It can be captured by different models based on the application and the environment in which sensors are placed [64]. In this work, we use a log-distance path loss model, which considers the effect of obstacles and objects as well as various environmental factors such as attenuation, reflection, diffraction, and scattering. The model is proportional to the logarithm of the distance between the transmitter (e.g., sensor node) and the receiver (e.g., parent or sink node).

In Equation (3), the power loss $PL(d)$ is modeled as a linear function of the distance (d) between the transmitter and receiver [65]:

$$PL(d) = PL_0 - 10\eta \log_{10} \frac{d}{d_0} + X_\sigma, \tag{3}$$

where $d_0 < d$, PL_0 is the power loss at the reference distance d_0 (defined in Equation (4)), and X_σ is the Gaussian random variable with zero mean and standard deviation σ.

In practice, it is difficult to predict the strength of the signal between the transmitter and the receiver [24]. Thus, the random variable (X_σ) is used to model the effect of shadowing and multipath fading. In addition, PL_0 is used as a reference for comparing the path loss at other distances in the case where there is no shadowing or multipath fading:

$$PL_0 = 10 \log_{10}(\frac{4 * \pi * d}{\lambda})^2 \tag{4}$$

where λ is the electromagnetic wavelength, $\lambda = \frac{c}{f}$, c is the speed of light ($3 * 10^8$ m/s), and f is the frequency.

3.4. Performance Metrics

In this section, we present different performance metrics that will be used in our experiment analysis (see Section 5) to assess our proposed routing protocol.

3.4.1. Network Lifetime and Stability Period

The network lifetime and stability period are closely related metrics in WBANs, both vital for ensuring the overall performance and reliability of the network. The network lifetime refers to the duration for which the network can operate before the energy of the nodes is completely depleted. At the same time, the stability period represents the time from network initiation until the first node becomes non-operational due to energy exhaustion. Maximizing the network lifetime is a fundamental objective in WBAN design, and it directly impacts the stability period. Prolonging the stability period, in turn, contributes to extending the overall lifetime of the network, allowing continuous monitoring and data transmission, which is particularly necessary in applications such as healthcare systems.

Several factors influence both the network lifetime and the stability period in WBANs. One significant factor is the energy consumption of the sensor nodes, which occurs during data transmission, reception, processing, and sensing activities. Efficient routing protocols play a vital role in balancing the energy consumption among nodes and optimizing the utilization of energy resources. Uneven energy consumption can lead to premature depletion of certain nodes, creating energy holes and reducing the overall network lifetime [25,47]. To address this, load balancing techniques can distribute the workload evenly across the nodes, preventing energy exhaustion in specific nodes, particularly those near the sink. Furthermore, selecting appropriate parent nodes for data forwarding is paramount for maximizing the network lifetime.

3.4.2. Network Throughput

Network throughput is an important performance metric in WBANs, as it measures the successful reception of packets at the sink, particularly when dealing with vital patient data. The goal is to minimize the packet drop ratio and maximize data reception at the sink to ensure reliable transmission of important healthcare information [4,25]. The choice of routing protocol is essential for improving the network throughput, as it should be designed to optimize throughput in low-power and resource-constrained WBAN environments. By designing efficient and reliable routing protocols, congestion can be avoided, packet loss can be minimized, and data reception at the sink can be maximized, ultimately enhancing network throughput.

3.4.3. Energy Consumption

Energy consumption is the most important parameter that needs to be considered during the design and development of WBANs. The analysis of residual energy is essential because the energy consumption of a single node directly affects the stability period and reliability of the network. Due to the limited energy resources of the sensor nodes in WBANs, it is vital to optimize energy consumption to prolong the operational lifetime of the network. The first-order radio model [63] is often used to analyze energy consumption in WBANs, which is explained in Section 3.2.

3.4.4. End-to-End Delay

End-to-end delay (EED) is a metric that measures the time it takes for a packet to travel from a source (i.e., sensor node) to the destination node (i.e., sink). It is particularly important in WBANs to monitor physiological data, where timely information delivery is necessary for medical applications. EED metrics in WBANs ensure the timely delivery of critical data from sensor nodes to the sink node. The delay needs to be examined and controlled to ensure that the monitored data reach the sink in a timely manner. Delays in delivering data beyond acceptable thresholds may result in inaccurate or outdated information, which can be detrimental to medical diagnosis, treatment, or patient monitoring.

To calculate EED, the time at which the sensor node transmits the first packet is noted, and the time at which that first data packet arrives at the sink is recorded. The difference between these two times provides EED for that specific packet [66]. This calculation is performed for multiple packets, and the overall average EED is determined. Mathematically, the overall average EED in a WBAN can be calculated by dividing the total time required to send all packets (T_{sp}) by the total number of packets received at the sink node (T_{rp}), as in Equation (5) [25,67]. This calculation provides an average value that represents the average delay experienced by packets in reaching their destinations.

$$EED = \frac{T_{sp}}{T_{rp}} \tag{5}$$

3.4.5. Path Loss

Path loss refers to the attenuation or reduction in signal strength as it propagates over a distance through the wireless medium between the transmitter (i.e., the source node) and the receiver (i.e., the sink node). It is an important factor to consider in WBANs, as it directly impacts overall performance and network lifetime. Path loss increases bit error rate and packet loss, and it degrades the communication system throughput. The selection of an appropriate path loss model is crucial to accurately depicting the signal attenuation behavior in WBANs. A commonly used model is the log-distance path loss model [64], explained previously in Section 3.3.

The impact of path loss on the overall performance of WBANs can be significant, as explained above. Higher path loss results in weaker received signals, requiring higher transmission power to maintain reliable communication. This leads to an increase in the energy consumption of the sensor nodes, which can significantly reduce the lifetime of the network. Several techniques can be employed to mitigate the effects of path loss and improve network performance. These include optimizing the placement of sensor nodes to reduce the distance between the nodes and the sink, using higher transmission power or directional antennas to compensate for the signal attenuation and deploying parent nodes to improve signal coverage.

3.4.6. Communication Costs

Communication costs encompass both overhead and payload costs. Payload represents the data transmitted between the sensors and the sink or among different nodes. When the payload size increases, the associated payload cost also rises, demanding more energy for transmission than smaller packets. The packet size is typically predetermined during network initialization, making it challenging to exert direct control over it. Although compression techniques can be used to reduce the payload size, their practicality is limited due to the resource-constrained nature of sensor nodes. Overhead, which includes control information and protocol-related data, can consume a significant portion of the available energy, leaving less energy for actual data transmission and sensing tasks. Minimizing overhead is essential when designing efficient WBAN communication protocols, as it does not directly contribute to the end application.

3.4.7. Expected Transmission Count (ETX)

In the context of wireless networks, the Expected Transmission Count (ETX) is a metric that refers to the number of transmissions required to successfully deliver a packet from a source node (i.e., sensor node) to a destination node (i.e., sink node). It is commonly used to evaluate and optimize the reliability and efficiency of communication links in WBANs. The ETX metric can be calculated based on the packet delivery ratio (PDR) [68], which provides information on the reliability of a link. The intuition behind ETX is that a higher PDR indicates a more reliable link, while a lower PDR suggests a less reliable link. The formula for calculating ETX is as follows:

$$ETX = \frac{1}{PDR}. \tag{6}$$

In WBANs, where energy efficiency is crucial due to the limited battery resources of sensor nodes, ETX serves as an important metric to evaluate link quality and optimize communication performance. It helps to select more reliable links to transmit critical data, minimize the number of dropped packets and retransmissions, and conserve energy. By monitoring and evaluating ETX values for different links in a WBAN, network administrators or routing protocols can make informed decisions about selecting routes or next-hop nodes with lower ETX values, indicating more reliable and efficient communication links. This ultimately leads to improved packet delivery, reduced latency, and better overall network performance.

4. Materials and Methods

In this section, we introduce the rationale behind our proposed routing protocol and elucidate the various phases integral to its implementation.

4.1. Routing Protocol Motivation

One of the main challenges in WBANs is the excessive energy consumption during communication and packet transfer between sensor nodes. This indicates that the significant limitation of the sensor nodes' energy resources affects the overall WBAN performance, including network lifetime and reliability. Various energy-efficient clustering-based routing protocols have been proposed to mitigate this challenge to prolong the network's lifetime. These routing protocols are based on multi-hop communication (i.e., two-hop) to reduce the distance between the sensor nodes and the sink, decreasing the consumed energy. Some of these protocols are highlighted in Table 1. Despite their goal of energy efficiency, existing cluster-based protocols suffer from inappropriate selection criteria for parent nodes and an uneven distribution of traffic load among the deployed sensor nodes. A common problem of these existing protocols is the overloading of nodes near the sink node, as they are frequently selected to forward packets, leading to rapid battery depletion and creating energy holes within the network. Furthermore, the selection criteria of parent nodes for forwarding data introduces inefficiencies in energy consumption and network reliability.

Given these challenges, our proposed routing protocol aims to tackle these issues by implementing a more reliable and efficient approach. Specifically, our protocol focuses on refining the selection criteria for parent nodes and ensuring the traffic burden is evenly distributed among sensor nodes. By improving the criteria for selecting forwarders, the protocol aims to optimize energy usage and avoid situations where certain nodes are overloaded, while others remain underutilized. This helps lessen the energy hole and promotes a more balanced utilization of network resources. Additionally, our proposed protocol also emphasizes the importance of reliability in WBANs. In this context, timely delivery (i.e., delay) and successful packet transfer are important factors, especially in scenarios such as ongoing patient monitoring, where undetected life-threatening events can have severe repercussions. By addressing these reliability concerns, our protocol aims to enhance the quality of patient monitoring by ensuring that data are delivered accurately and on time.

4.2. HCEL Routing Protocol

In this subsection, we introduce three different phases of our proposed routing protocol (HCEL), namely, network deployment and initialization, parent node selection, and data transmission.

4.2.1. Network Deployment and Initialization

The sequence diagram for the network deployment and initialization phase is shown in Figure 1. Initially, the sink node broadcasts its location information to all nodes in the network. This serves as a reference point for the other nodes to determine their positions relative to the sink node. Additionally, it allows the nodes to establish a communication link with the sink node. The two-hop nodes transmit essential information to the sink, including the unique identifier of the node, location coordinates, RSSI value, and residual energy. Simultaneously, each two-hop node shares its identifiers and location coordinates with other nodes. This exchange of information helps to establish a network topology and enables the nodes to identify their neighboring nodes. The one-hop sensor nodes send only their unique identifiers and locations to the sink, as these nodes have not participated in the data forwarding process (i.e., routing).

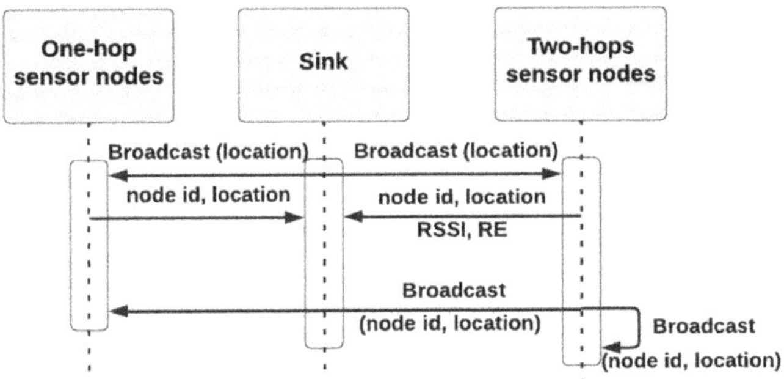

Figure 1. Sequence diagram of the initialization phase.

Following these steps in the network deployment and initialization phase, the WBAN achieves an organized and coordinated setup. The information shared between the sink node and sensor nodes, along with the exchange of information among the two-hop nodes, establishes a foundation for effective communication and data transfer. This initialization phase ensures that the network is properly configured and ready for the subsequent data transmission and routing stages. Additionally, this phase in WBAN is designed to minimize the overhead in communication cost, further discussed in Section 5. In this case, data forwarding and routing decisions can be performed by giving the sink node essential information from the two-hop sensor nodes (i.e., location, RSSI, and residual energy). Through efficient assignment of parent nodes, the network reduces unnecessary communication overhead. This minimization of overhead conserves energy resources and enhances the overall efficiency and reliability of data transmission in WBAN. By striving to minimize the communication cost overhead, the network deployment and initialization phase lays the foundation for an optimized and energy-efficient WBAN system.

4.2.2. Parent Node Selection

After the sink node receives the necessary information from the sensor nodes, it employs a cost function to select the parent nodes. The cost function considers factors such as the distance between a sensor node and the sink, the current residual energy level of the node, and the RSSI value from a sensor node to the sink. The distance between the sink and the sensor node is calculated using the Euclidean formula, while the residual energy is determined as discussed in Section 3.2. The RSSI value, which follows a log-normal distribution, further discussed in Section 5.1, is updated for each transmission. Hence, the cost function, denoted as $CF(i)$ for the sensor node i, is calculated as Equation (7), where d represents the distance between node i and the sink, $RE(i)$ is the residual energy, and $RSSI(i)$ indicates the RSSI value of node i.

$$CF(i) = \frac{d \times RSSI(i)}{RE(i)} \tag{7}$$

By incorporating distance, the cost function considers the physical proximity of each sensor node to the sink node. Including residual energy in the cost function is significant for energy efficiency. Nodes with higher residual energy levels are given high scores, indicating their ability to handle data-forwarding tasks. Additionally, the cost function incorporates RSSI values, which reflect the signal quality between the nodes and the sink. The algorithm prioritizes nodes with stronger and more reliable communication links by considering RSSI. This balanced distribution of responsibilities helps prevent energy imbalance among nodes and contributes to the network's longevity.

In particular, the sink node selects the parent node based on several criteria, including shorter distance, higher RSSI value, and greater residual energy. Once the cost function is computed for all two-hop sensor nodes, the sink node selects the parent node with the maximum cost function value. However, there is a constraint on the number of nodes that a parent node can serve (τ). This constraint ensures optimal utilization of resources within the network.

The selection of parent nodes is an ongoing process that adapts to the changing energy levels of the sensor nodes. When the energy of a selected parent node reaches a threshold value (Th), the sink stops using this node for forwarding packets. At that time, another node has been selected as a parent node, as previously explained. Once the residual energy of all two-hop sensor nodes reaches the threshold (Th), a new round of parent node selection is initiated. In this case, all two-hop sensor nodes can serve as parent nodes. This approach ensures that the flow of traffic and data routing within the WBAN remains efficient and well-balanced. After selecting the parent nodes, their identifiers are broadcast to all two-hop sensor nodes, enabling them to establish a communication link with their designated parent node. It is important to note that reaching the energy threshold value (Th) is an alarm, signaling the network management system to replace or recharge the node.

By selecting the parent nodes based on the maximum cost function value, the threshold energy value (Th), and the limitation in the number of nodes that the parent node can serve (τ), the algorithm redistributes the data traffic among the sensor nodes. This strategy prevents any single node from becoming overwhelmed with excessive data forwarding responsibilities, promoting a greater utilization of network resources. In this scenario, the lifetime of all sensor nodes is extended, and the energy consumption is more evenly distributed throughout the network. The pseudo-code of the parent node selection process is depicted in Algorithm 1, while the flowchart for this process is shown in Figure 2.

Algorithm 1 Parent Node Selection Process

Input: N (Number of nodes), $d[1..N]$ (distances from sensor nodes to the sink), $RE[1...N]$ (residual energy levels of sensor nodes), $RSSI[1..N]$ (RSSI values of sensor nodes), τ (number of serving nodes for parent node), Th (energy threshold), one-hop nodes' id

Output: PN (Parent nodes' id)

```
for i = 1 to N do
    if i not in one-hop nodes' id then
        CF[i] = (d[i] × RSSI[i]) / RE[i]                          ▷ Cost Function
    end if
end for
serve_count[1..N] = 0
SortedCF = Sort(CF)                                               ▷ Ascending Sorting
indices[1..N] = 0
for i = 1 to N do                       ▷ Store indices of nodes after sorting cost function
    indices[i] = IndexOf(CF, SortedCF[i])            ▷ Get indices of nodes after sorting
end for
for j = 1 to N do                                                 ▷ Static Clustering
    for i in indices do
        if serve_count[i] < τ and RE[i] > 0 and RE[i] > Th then
            if i != j then
                serve_count[i] += 1
            end if
            index = i
        end if
    end for
    PN[j] = index
end for
if RE[1..N] <= Th then                                            ▷ Dynamic Clustering
    for j = 1 to N do
        for i in indices do
            if serve_count[i] < τ and RE[i] > 0 and RE[i] <= Th then
                if i != j then
                    serve_count[i] += 1
                end if
                index = i
            end if
        end for
        PN[j] = index
    end for
end if
```

4.2.3. Data Transmission

After the parent node selection and network organization, the WBAN enters the data transmission phase. This phase considers the specific requirements of one-hop sensor nodes, which transmit their critical data directly to the sink. Meanwhile, two-hop sensor nodes transmit their packets to the chosen parent node before reaching the sink node.

After each round of data transmission, the two-hop sensor nodes send their updated residual energy (RE) and RSSI values to the sink. This information is used to update the cost function calculations and, consequently, to select the parent nodes. During this process, the sink actively monitors the residual energy level of the parent node. If the parent node's residual energy reaches a threshold (Th), the sink temporarily suspends its involvement in data forwarding until the residual energy of all two-hop sensor nodes reaches the threshold (Th). At that moment, the sink node selects another two-hop sensor node as a parent node. This ensures that no individual parent node becomes excessively drained of energy, maintaining a balanced traffic load distribution and prolonging the overall network stability.

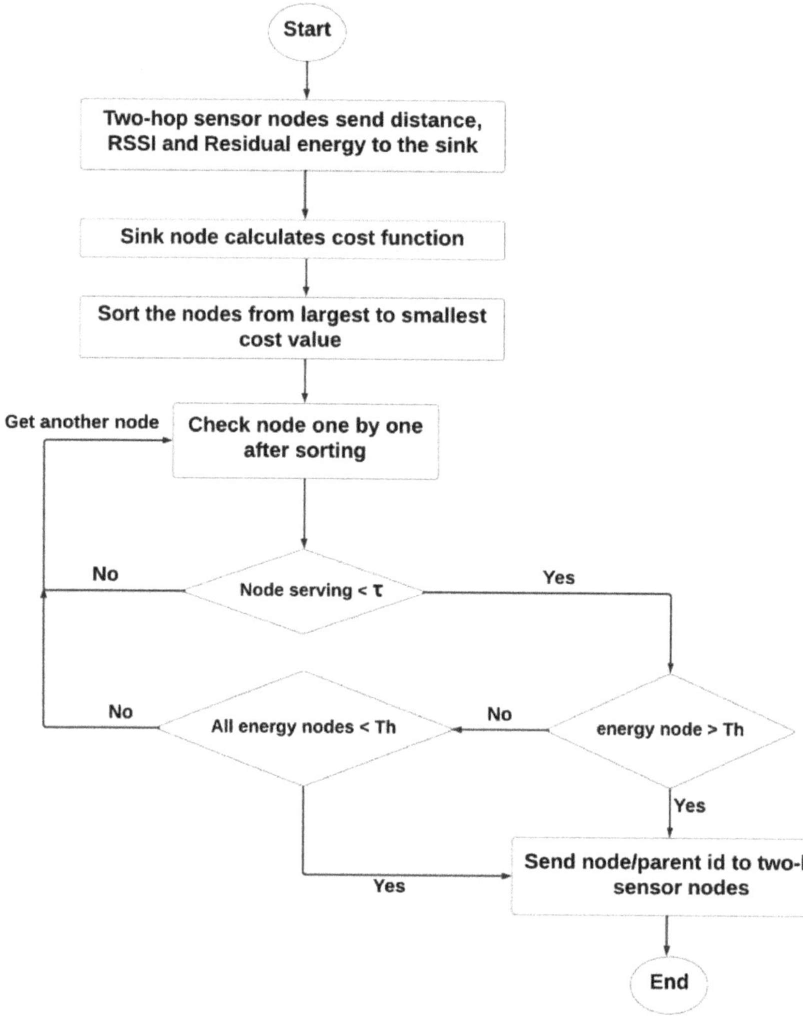

Figure 2. Flowchart of the parent node selection process.

Upon reaching the threshold energy level for all two-hop sensor nodes, a new round of parent node selection is initiated. At this stage, all two-hop sensor nodes are eligible to serve as parent nodes and participate in data forwarding. This approach promotes fairness by giving each two-hop sensor node an opportunity to contribute as a parent node, which in turn extends the lifetime of the network as a whole. The sequential steps and the pseudo-code of the data transmission phase are shown in Figure 3 and Algorithm 2, respectively.

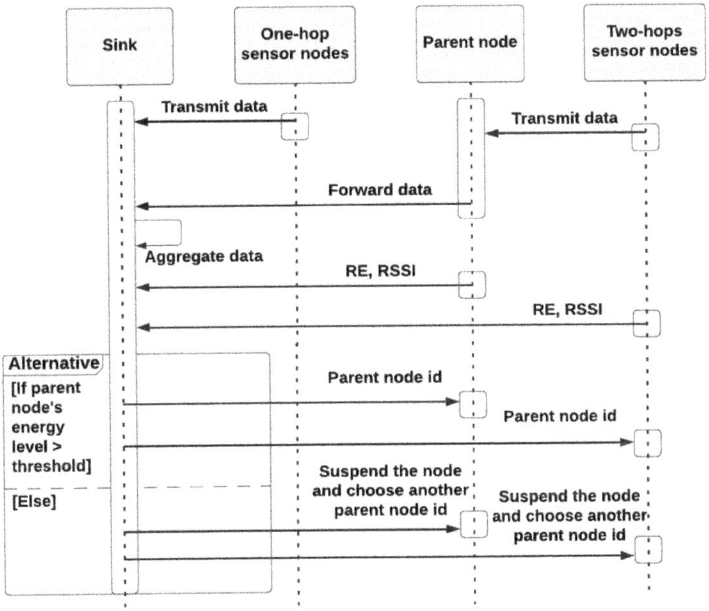

Figure 3. Sequence diagram of the data transmission phase.

Algorithm 2 Data Transmission

Input: N (Number of nodes), R (Number of rounds), $PN[1...N]$ (parent node), $E[1...N]$ (initial node energy), E_{tx} (transmitted energy consumption as in Equation (1), E_{rx} (received energy consumption as in Equation (2), one-hop nodes' id.
Output: Transmission of all node packets to the sink (single-hop, two-hop)

for *round* $= 1$ to R **do**
    for $i = 1$ to N **do**
        if i in one-hop nodes' id **then** ▷ Critical Packet
            if $E[i] > 0$ **then** ▷ Enough Energy
                Send packet directly to the sink
                $E[i] = E[i] - E_{tx}$
            else ▷ Not Enough Energy
                Cannot send the packet
            end if
        else ▷ Normal Packet
            if $E[i] > 0$ **then** ▷ Enough Energy
                Send packet to $PN[i]$
                $E[i] = E[i] - E_{tx}$
                $E[PN[i]] = E[PN[i]] - E_{rx}$
                Send packet from $PN[i]$ to the sink
                $E[PN[i]] = E[PN[i]] - E_{tx}$
            else ▷ Not Enough Energy
                Cannot send the packet
            end if
        end if
    end for
end for

4.3. Time Complexity

We first analyze the time complexity of Algorithm 1 (i.e., the parent node selection process) and Algorithm 2 (i.e., data transmission), and then we combine the time complexities of both algorithms to obtain the overall time complexity for our proposed protocol.

1. Parent node selection process: This algorithm aims to select appropriate parent nodes for transmitting the data from the two-hop sensor node to the selected parent node and then relaying it to the sink. This algorithm runs on the sink node and involves the following steps:
   (a) Calculating the cost function for all two-hop sensor nodes at the sink node. This step has a time complexity of $O(N)$, where N represents the number of nodes in the network (lines 1 to 5 in Algorithm 1);
   (b) Sorting the nodes based on their cost function values. The sorting operation has a time complexity of $O(N\log(N))$ (line 7 in Algorithm 1), using merge-sorting;
   (c) Storing the indices of the sorted nodes requires $O(N)$ time complexity (lines 9 to 11 in Algorithm 1);
   (d) Iterating through the sorted nodes to select those serving a limited number of nodes and having sufficient energy for packet transmission. This process includes both static clustering (lines 12 to 22 in Algorithm 1) and dynamic clustering (lines 23 to 35 in Algorithm 1) parts. This step is applied to each two-hop sensor node. Thus, the time complexity of these iterations has a maximum of $O(N^2)$.

   By summing up the time complexities of these steps, the overall time complexity of the parent node selection algorithm is $O(N) + O(N\log(N)) + O(N) + O(N^2)$, which can be simplified to $O(N^2)$.
2. Data transmission: This algorithm transmits data from the sensor node to the sink node. If a node sends a critical packet with sufficient energy, it transmits it directly to the sink (lines 3 to 6 in Algorithm 2). However, when a node sends a normal packet with enough energy, it sends the packet to its parent node, which the sink node chooses. The parent node then forwards the packet to the sink (lines 10 to 16 in Algorithm 2). The steps include:
   (a) Iterating through all N nodes in the network for varying rounds R. This operation has a time complexity of $O(RN)$.

By combining the time complexities of both algorithms, the total time complexity of our proposed protocol is $O(N^2) + O(RN)$. Since the dominant term is $O(N^2)$, we can express the overall time complexity as $O(N^2)$.

Regarding space complexity, the algorithm mainly requires memory for storing network information, such as node indices, locations, energy levels, RSSI, and any additional data structures used for sorting and calculations. The space complexity depends on the number of nodes and the size of the data structures, but it is typically within reasonable bounds for most WBAN applications.

5. Simulation Results and Analysis

We carried out a series of experiments to evaluate the performance of our proposed protocol, HCEL. Details regarding the performance metrics can be found in Section 3.4. The evaluation involved comparing our proposed protocol (HCEL) with both SIMPLE and ERRS protocols [24,25] across these performance metrics. Our implementation utilized the Networkx package [69] with different experiment parameters outlined in Table 3. To account for the inherent randomness in the RSSI values generated by our protocol, we conducted 15 simulation runs following a log-normal distribution, each lasting 35,000 time units.

Table 3. Experimental parameters.

Parameters	Value
Initial energy of the sensors	4 joules
Number of sink nodes	1
Number of sensor nodes	10
Packet size (k)	127 bytes
Reference distance (d_0)	10 cm
Frequency (f)	2.4 GHz
Path loss coefficient (η)	3.38
Standard deviation (σ)	4.1 dB
Mean (μ)	0
Energy threshold (Th)	2 joules
Number of serving nodes for parent node (τ)	4
Total number of rounds (R)	35,000

5.1. Experimental Testbed Setup

Our experimental testbed was established within an academic unit's laboratory. The network topology in our testbed comprises 10 sensors, denoted as node 1 through node 10, and a central sink node identified as node 0. The system structure of our network is illustrated in Figure 4. To collect data from sensor nodes, we employed Crossbow MICAz motes running the TinyOS embedded operating system in our testbed, each equipped with a 1.2-inch-long monopole antenna with a wavelength of one-quarter inch. These motes were programmed to transmit sensing (i.e., physiological) data to the sink node at 3-min intervals. Communication among the nodes was established using the Chipcon CC2420 (Chipcon, Oslo, Norway) transceiver operating at 2.4 GHz, which facilitates data exchange among sensors.

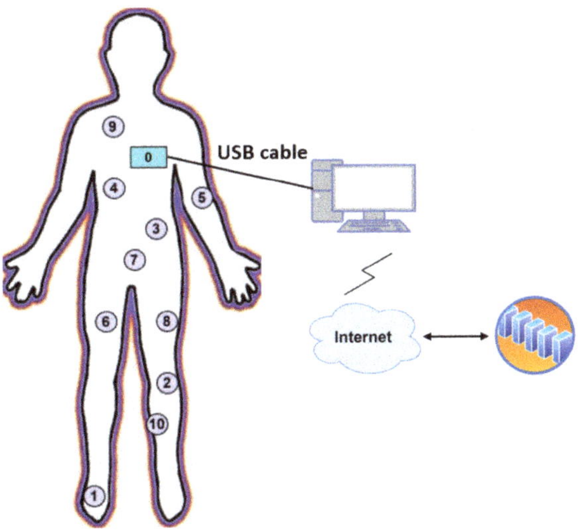

Figure 4. System architecture.

In our network, the sink node (node 0) serves as the gateway, receiving data from the sensors and relaying it to the backend servers. This sink node utilizes USB connectivity to establish communication with the cloud, facilitate configuration, and supply power. Therefore, it could be connected to any PC using a USB cable, as shown in Figure 4. All sensor nodes, except the sink node, were powered by AA batteries and had sufficient power (i.e., RSSI) to communicate with each other and the sink node. The transmission power in our testbed was configured within the range of 0 dBm (1 mW) to −25 dBm

(0.003 mW) to minimize interference and conserve radio power consumption. The primary objective of this testbed was to assess the performance of the WBAN and gather additional data for future investigations.

In wireless networks, high RSSI values indicate reliable data transfer capabilities. Thus, our work leveraged the data collected from the sensor nodes in our testbed to model RSSI using a log-normal distribution. The choice of log-normal distribution was validated through a goodness-of-fit assessment using a QQ-plot, confirming its suitability for our data as shown in Figure 5. The log-normal distribution, widely applied in this domain, offers a probability density function (PDF) that effectively characterizes the distribution of RSSI values. By adopting this approach, we gain a deeper understanding of RSSI behavior and its implications within our work.

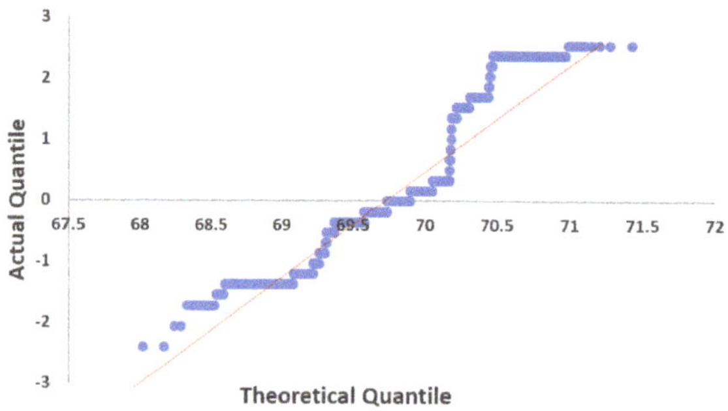

Figure 5. QQ–plot.

5.2. Node Deployment

In our simulation setup, we strategically positioned sensors on the human body, as shown in Figure 6. Among these sensors, one-hop nodes, such as sensor 3 and sensor 7, are crucial in monitoring tasks that require immediate action (e.g., ECG and glucose). Due to their critical nature and time-sensitive requirements, these sensors transmit their data to the sink (node 0) using single-hop communication, as illustrated in Figure 6 (e.g., direct packet transmission from node 3 to node 0). These critical sensors are positioned near the sink node to ensure efficiency and prompt data delivery.

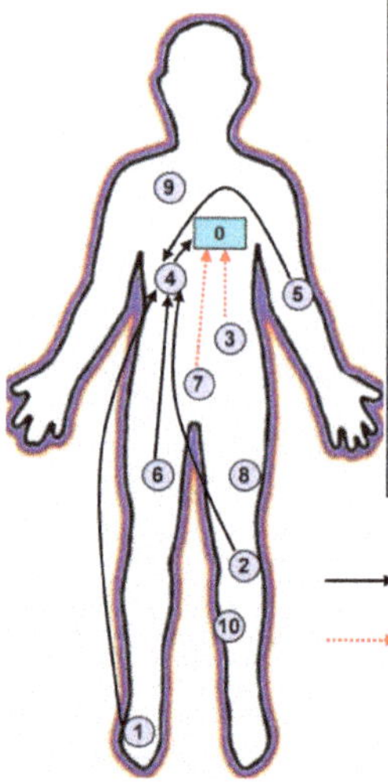

Node ID	X	Y
0	0.4	1
1	0.3	0.1
2	0.5	0.3
3	0.45	0.65
4	0.35	0.9
5	0.7	0.8
6	0.3	0.55
7	0.37	0.6
8	0.5	0.54
9	0.35	1.1
10	0.4	0.25

Figure 6. Node deployment.

On the other hand, the two-hop sensor nodes are responsible for normal monitoring functions, encompassing all other deployed sensor nodes shown in Figure 6. These nodes adopt a two-hop communication approach, transmitting their data through the intermediate parent node to reach the sink. The transmission path of packets from some two-hop sensor nodes to the sink is showcased in Figure 6 (e.g., packet transmission from sensor node 1 to node 4, the selected parent node, and then to node 0, the sink node). Using a two-hop communication extends the network lifetime by distributing the energy consumption among nodes, reducing the distance to the sink, and mitigating the energy drain on any single sensor. This design choice ensures a balance between critical monitoring requirements and the efficient utilization of network resources for normal monitoring tasks.

In the experiment depicted in Figures 7 and 8, we compared the network lifetime and stability period of our proposed routing protocol (HCEL) with both SIMPLE and ERRS protocols. The results demonstrate the superiority of the HCEL protocol in extending both the stability period and the overall network lifetime. The effectiveness of our proposed protocol stems from selecting parent nodes based on residual energy (i.e., not exceeding the threshold value Th), ensuring that the energy load is evenly distributed across the network and providing the chance for every sensor node to become a parent node. In contrast, the low performance of the SIMPLE protocol arises from the repeated selection of nodes closer to the sink as parent nodes, leading to their quick energy depletion and subsequent death of other sensor nodes. While the ERRS protocol does improve network lifetime compared to SIMPLE, it still falls short of the performance achieved by our HCEL protocol. ERRS achieves this improvement by balancing the load and selecting a new parent node after a set number of transmission rounds. Moreover, ERRS considers both distance and residual energy when selecting parent nodes. However, our HCEL protocol takes it a step further by considering an additional parameter, RSSI from the node to the sink, to make better parent node selections. This additional refinement helps optimize energy usage and contributes to the high performance of our protocol.

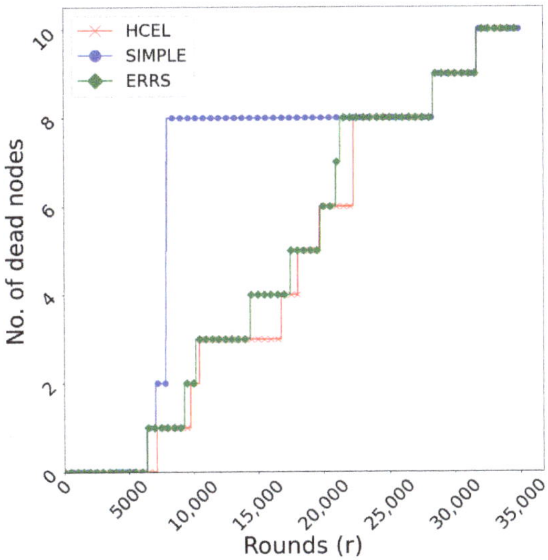

Figure 7. Analysis of network lifetime.

We observed the number of dead nodes over different rounds to evaluate the network lifetime, as shown in Figure 7. The results indicate the higher performance of our proposed routing protocols compared to both the SIMPLE and ERRS protocols. Our protocol increases the stability period, with the HCEL protocol achieving approximately 7300 rounds, surpassing the 6383 rounds attained by the SIMPLE protocol and the 6425 rounds achieved by the ERRS protocol. Additionally, in both HCEL and ERRS protocols, two-hop sensor nodes exhibit extended lifetimes due to the balanced distribution of traffic load. This achievement is made possible by combining static and dynamic clustering during the parent node selection process; thus, the stability period and network lifetime increase. Notably, nodes 3 and 7 in all protocols, which do not participate in data forwarding or routing, exhibit longer lifetimes, as their communication is direct to the sink (i.e., the last two dead nodes).

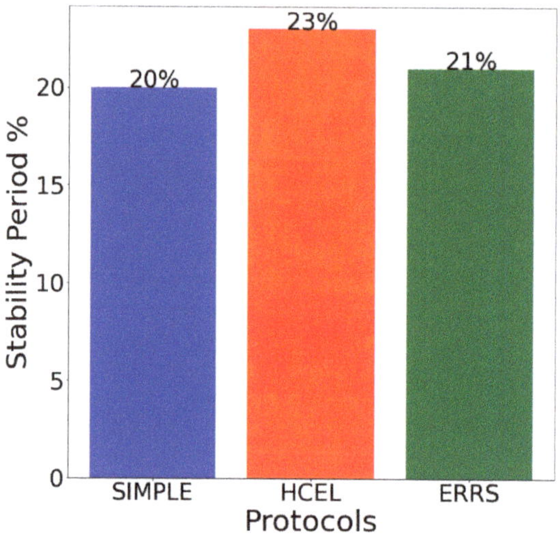

Figure 8. Comparing the stability period among the HCEL, SIMPLE, and ERRS protocols.

In the stability period experiment, we focus on the duration from network initiation until the first node's depletion for our proposed protocol (HCEL), along with the SIMPLE and ERRS protocols. From Figure 8, our proposed protocol achieves a stability period of 23% of the total operational duration, slightly surpassing both the ERRS protocol at 21% and the SIMPLE protocol at 20%. These results highlight the effectiveness of our proposed protocol in providing a longer stability period, indicating an extension in the lifetimes of nodes beyond the first node's depletion.

The next experiment, shown in Figures 9–11, evaluates the network throughput based on the number of packets successfully received at the sink, the packet delivery ratio (PDR), and the number of dropped packets. To demonstrate the superior performance of our proposed routing protocol (HCEL), we compared the results with both the SIMPLE and ERRS protocols, aiming to establish its maximum throughput advantage.

Figure 9 depicts the number of packets successfully delivered to the sink node in different rounds. When there are no dead nodes in all protocols (i.e., from round 0 until round 6383), the sink node receives the same number of packets sent. However, as nodes in the network become inactive or die, the number of received packets at the sink decreases. The delivery of packets to the sink depends on the number of active or alive nodes in the network. Therefore, more alive nodes lead to more data packets to the sink, ultimately maximizing network throughput. Due to the extended lifetimes of sensor nodes in our protocol (HCEL), the number of packets received at the sink is larger than that of the SIMPLE and ERRS protocols. The gap between the HCEL and SIMPLE protocols increases as the number of rounds increases. This occurs because the SIMPLE protocol quickly loses its nodes after the first node's depletion, causing a decrease in the number of received packets at the sink. However, the difference between the HCEL and ERRS is less pronounced due to the extended network lifetime observed in the ERRS protocol.

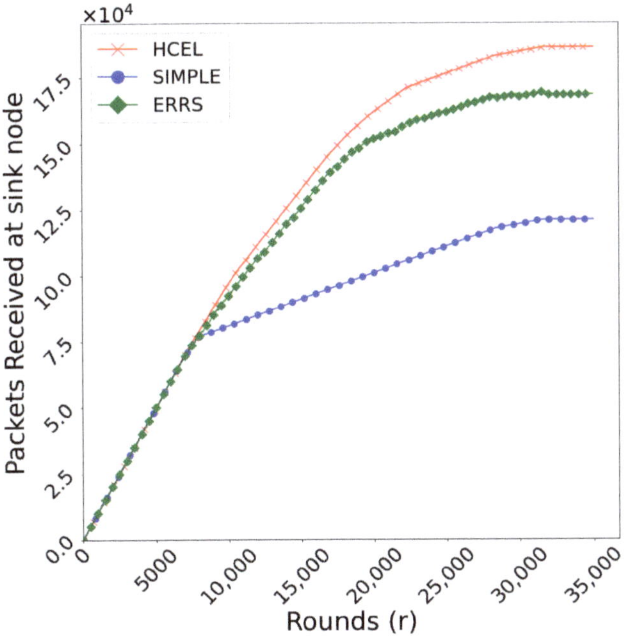

Figure 9. Analysis of total number of packets successfully received at the sink.

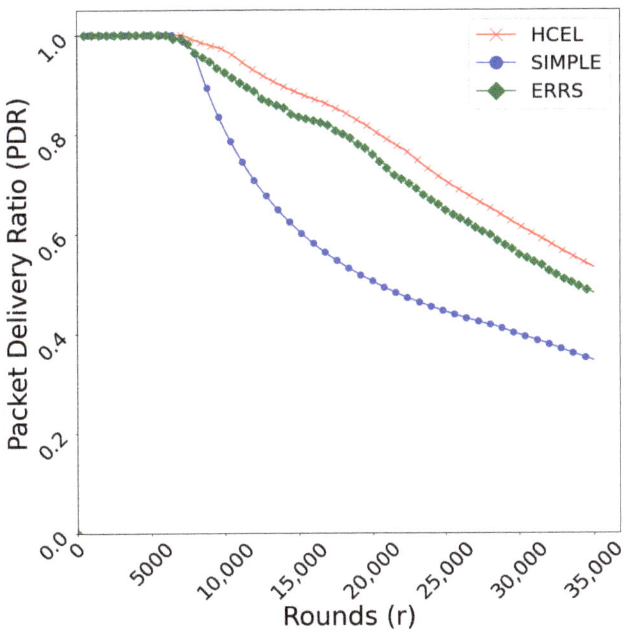

Figure 10. Analysis of packet delivery ratio (PDR).

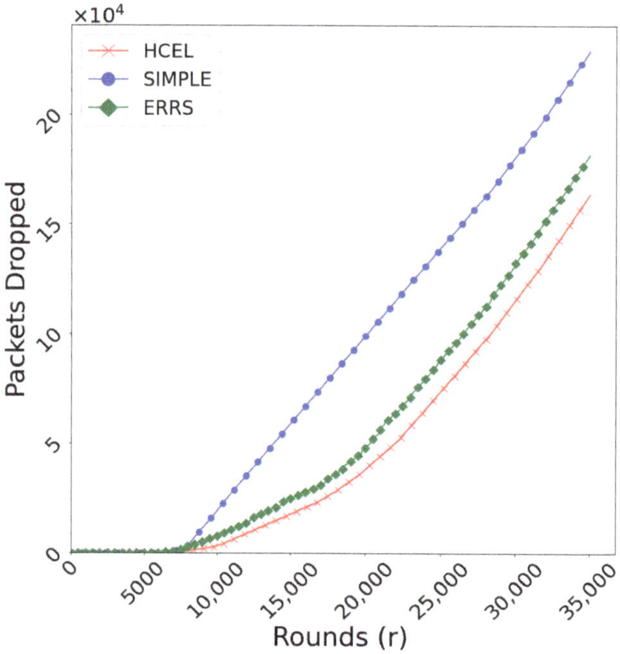

Figure 11. Analysis of total number of dropped packets.

Figure 10 shows the packet delivery ratio (PDR) across various rounds. PDR represents the ratio of packets successfully delivered to the sink to the total number of transmitted packets. As our protocol (HCEL) successfully delivers a larger number of packets to the sink, it exhibits higher throughput regarding PDR compared to both the SIMPLE and ERRS protocols. The superiority of our protocol and ERRS protocol can be attributed to its longer stability period resulting from uniform load distribution and increased lifetime of sensor nodes. However, our protocol's performance surpasses that of the ERRS protocol, due to its consideration of the RSSI between the node and sink during parent node selection. When $PDR = 1$, all packets were received successfully to the sink. However, as PDR decreases, the number of packets dropped increases.

Figure 11 illustrates that the number of dropped packets in the HCEL is lower than in the ERRS and SIMPLE protocols. This is due to the longer stability period of our protocol and to the consideration of different parameters when selecting the parent node. In particular, during the stability period, when all nodes are alive, no packets are dropped. The increase in the number of dead nodes and the speed of depletion of nodes influence the total number of packets dropped in different rounds.

We measure the average residual energy of 10 nodes across different rounds in our next experiment, shown in Figure 12. Since the initial energy of each node is 4 joules, the average energy in round 0 is 4 joules. The simulation runs until all deployed sensor nodes exhaust their energy (i.e., RE = 0). Figure 12 depicts the decrease in the residual energy of all deployed nodes as the rounds progress. In the SIMPLE protocol, the residual energy of sensor nodes is drastically reduced after the first node dies (at round 6383). This rapid reduction occurs due to the speed at which all other nodes deplete their energy after the death of the first node. However, our proposed protocol (HCEL) and the ERRS protocol outperform the SIMPLE protocol, especially after the depletion of the first node. The better performance of our proposed and ERRS protocols is due to the use of a static and dynamic clustering-based routing mechanism. By allowing every sensor node to become a parent node, the load is evenly distributed among all nodes, extending their lifetime.

Furthermore, the efficient selection of parent nodes and appropriate route selection in our proposed scheme (HCEL) play a significant role in the energy conservation of the deployed sensor nodes, thus extending the network lifetime compared to both the ERRS and SIMPLE protocols.

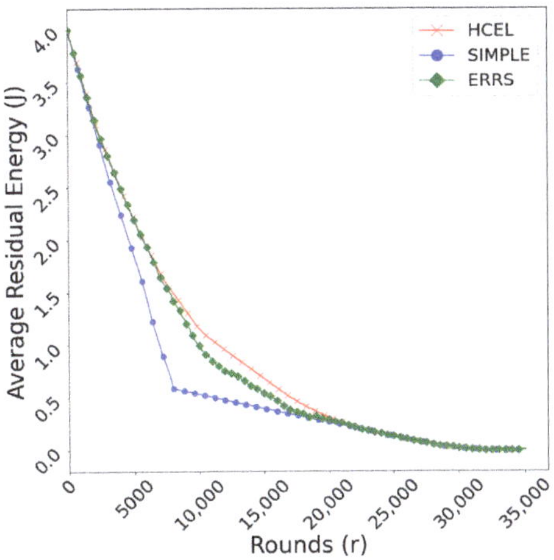

Figure 12. Analysis of residual energy.

The experiment in Figure 13 calculates an average EED for our proposed protocol (HCEL), as well as the SIMPLE and ERRS protocols. The results, shown in Figure 13, indicate that our proposed protocol achieves a lower delay than the SIMPLE and ERRS protocols. The key factor contributing to the improved EED in our protocol lies in the use of an adaptive static cluster routing mechanism, coupled with the careful selection of the parameters during parent node selection. That is, unlike the frequent selection of nodes closest to the sink as parent nodes in the SIMPLE protocol, our approach ensures a more balanced distribution of the traffic load among the nodes. This measure effectively mitigates rapid energy depletion at specific nodes, thereby enhancing the successful reception of packets at the sink and reducing the overall delay. As a result, our protocol shows an impressive improvement of approximately 56% compared to the SIMPLE protocol and approximately 40% compared to the ERRS protocol.

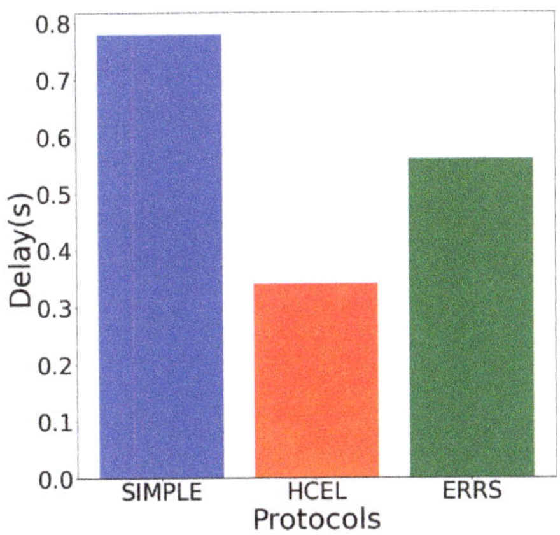

Figure 13. Comparing the end-to-end delay among the HCEL, SIMPLE, and ERRS protocols.

The following experiment, shown in Figure 14, analyzes the network cumulative path loss for our proposed protocol (HCEL), as well as for the SIMPLE and ERRS protocols. Using two-hop transmission in the routing protocol helps minimize the data transmission distance from the sensor nodes to the sink, resulting in lower path loss. Initially, our proposed protocol (HCEL) and ERRS protocol perform well compared to the SIMPLE protocol. However, starting from round 7981, all two-hop sensor nodes in the SIMPLE protocol deplete their energy, leading to a significant decrease in path loss. Consequently, the SIMPLE protocol exhibits a greater reduction in path loss compared to our proposed protocol (HCEL) and to the ERRS protocol from round 7981 to round 22,268. It is worth noting that the cumulative path loss is minimal when the number of alive nodes is minimal. Additionally, the increase in path loss observed in our proposed protocol (HCEL) and ERRS protocol during this period (rounds 6383 to 22,268) is due to its longer stability period and a higher number of alive nodes, contributing to the increased cumulative path loss.

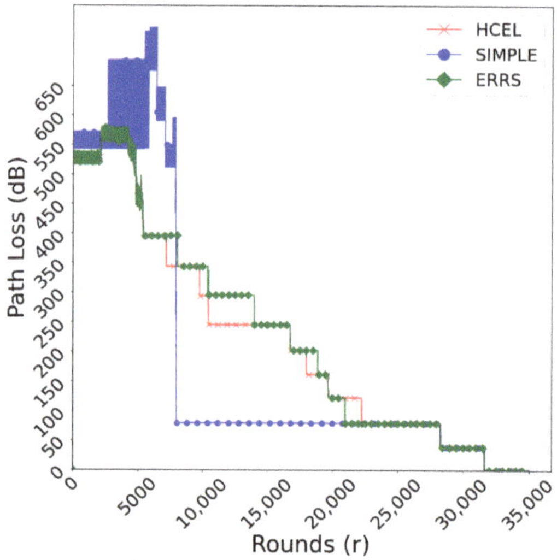

Figure 14. Analysis of network path loss.

To address the overhead issue, efficient routing protocols aim to reduce the size of control packets. In our experiment, depicted in Figure 15, we analyze the communication cost in terms of bits for various numbers of rounds for our proposed protocol (HCEL), as well as for the SIMPLE and ERRS protocols. While the payload size remains the same (127 bytes) in all protocols, the difference lies in the overhead calculation. In the HCEL protocol, the overhead during the initialization phase in round 0 (63,424 bits) is lower compared to the SIMPLE protocol (94,080 bits). This is because one-hop sensor nodes do not participate in data forwarding or routing, eliminating the need to send their energy levels and RSSI values to other nodes or the sink in our protocol. After each round of data transmission, our protocol only sends the unique identifier of the respective parent node to two-hop sensor nodes, without requiring the transmission of the cost function to all nodes for parent node selection, as done in the SIMPLE protocol. Furthermore, one-hop sensor nodes in our protocol do not need to send their updated residual energy to the sink for cost function calculation, as required in the SIMPLE protocol. However, in the ERRS protocol, the overhead during the initialization phase is 53,281 bits, which is less than in our HCEL protocol. This discrepancy occurs because nodes in the ERRS protocol transmit only their residual energy to the sink node during initialization, whereas in the HCEL protocol, they transmit both residual energy and RSSI values. This leads to an increase in the number of bits during the initialization phase in the HCEL protocol. Additionally, after each data transmission round in the HCEL protocol, all nodes must send their updated residual and

RSSI to the sink node to calculate the cost value and then choose the appropriate parent node. Conversely, in the ERRS protocol, nodes only send updated residual energy to select the parent node, resulting in a decrease in communication cost across different rounds compared to our HCEL protocol.

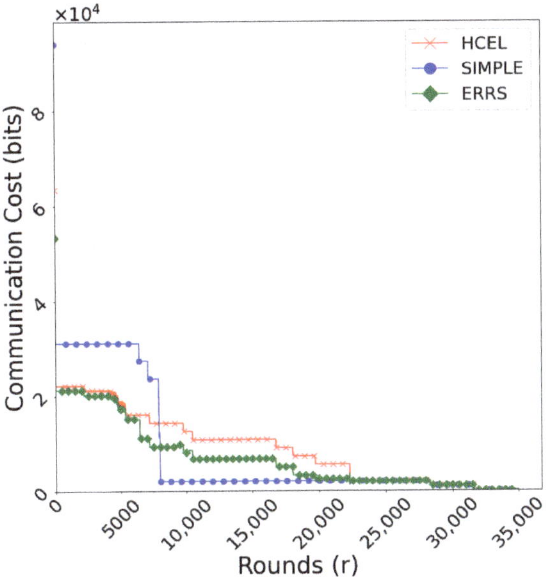

Figure 15. Analysis of network communication cost.

It is observable that the SIMPLE protocol generates a smaller number of bits for communication from rounds 7981 to 22,268 compared to the HCEL and ERRS protocols. That is because no two-hop sensor nodes are still alive for any communication. However, when all protocols have the same number of alive nodes, ERRS and our proposed protocol (HCEL) require fewer bits for communication between sensor nodes and the sink, as in the case from rounds 0 to 6383. This reduction in communication cost is attributed to the optimized overhead calculation and efficient exchange mechanisms employed in our proposed protocol. By optimizing these aspects, our protocol reduces communication costs between sensor nodes and the sink compared to the SIMPLE protocol.

ETX is calculated, as discussed in Section 3.4.7, to measure the link quality of our proposed protocol (HCEL), as well as the SIMPLE and ERRS protocols in different rounds. In Figure 16, it can be seen that, when all sensor nodes are operational for all protocols, the ETX value stands at 1. This implies that no packets are lost and that every packet is successfully transmitted to the sink. However, as the packet loss between nodes increases, the ETX value also increases. The ETX values in the SIMPLE protocol are larger than those of our proposed protocol (HCEL) and ERRS protocol, especially after round 7300 for HCEL and after round 6425 for ERRS. This indicates that the link quality in the SIMPLE protocol is poorer compared to both the HCEL and ERRS protocols. In general, as the number of rounds increases, the gap in ETX values between our protocol and SIMPLE widens, referring to a larger number of successfully received packets in the HCEL protocol compared to the SIMPLE protocol. However, this gap narrows between the HCEL and ERRS protocols due to the improved network lifetime observed in the ERRS protocol compared to the SIMPLE protocol.

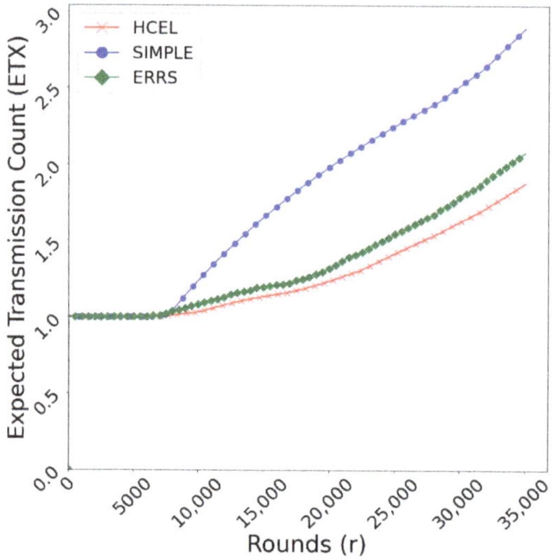

Figure 16. Analysis of expected transmission count (ETX).

6. Conclusions

In this paper, we present HCEL, a novel, energy-efficient, and reliable routing protocol specifically designed to enhance the overall performance of WBANs. By integrating static and dynamic clustering routing techniques in the parent node selection process, HCEL significantly improves the stability period and extends network lifetimes, thereby maximizing the reliability of WBANs. By conducting thorough simulations using Networkx, we provide empirical evidence showcasing the superiority of HCEL over both the SIMPLE and ERRS protocols across various performance metrics. HCEL substantially enhances network stability periods, network lifetimes, throughput, residual energy, path loss, and expected transmission counts. Particularly noteworthy is the remarkable 56% and 40% reduction in end-to-end delay achieved by HCEL compared to the SIMPLE and ERRS protocols, respectively, highlighting its efficiency and reliability as a routing solution for WBANs. To further advance the capabilities and applications of WBANs in healthcare and beyond, our future work will focus on incorporating mobility and security considerations. By addressing these key issues, we can further enhance the versatility and robustness of WBANs, enabling their effective deployment in a wide range of real-world scenarios.

Author Contributions: Conceptualization, H.H. and F.S.; Methodology, H.H. and S.H.; Software, H.H.; Validation, F.S., M.A.S., S.H., M.H. and H.K.; Formal analysis, H.H. and S.H.; Data curation, F.S.; Writing—original draft, H.H.; Writing—review & editing, F.S., M.A.S. and M.H.; Funding acquisition, M.A.S. All authors have read and agreed to the published version of the manuscript.

Funding: This research received no external funding.

Data Availability Statement: Data sharing is not applicable to this article.

Conflicts of Interest: The authors declare no conflicts of interest.

References

1. Available online: https://data.worldbank.org/ (accessed on 8 June 2023).
2. Available online: https://www.zawya.com/en/uae (accessed on 8 June 2023).
3. Sahndhu, M.M.; Javaid, N.; Imran, M.; Guizani, M.; Khan, Z.A.; Qasim, U. BEC: A novel routing protocol for balanced energy consumption in wireless body area networks. In Proceedings of the 2015 International Wireless Communications and Mobile Computing Conference (IWCMC), Dubrovnik, Croatia, 24–28 August 2016; pp. 653–658.

4. Shu, M.M.; Javaid, N.; Akbar, M.; Najeeb, F.; Qasim, U.; Khan, Z.A. FEEL: forwarding data energy efficiently with load balancing in wireless body area networks. In Proceedings of the 2014 IEEE 28th International Conference on Advanced Information Networking and Applications, Victoria, BC, Canada, 13–16 May 2014; pp. 783–789.
5. Ullah, Z.; Ahmed, I.; Razzaq, K.; Naseer, M.K.; Ahmed, N. DSCB: Dual sink approach using clustering in body area network. *Peer Peer Netw. Appl.* **2019**, *12*, 357–370. [CrossRef]
6. Ahmed, S.; Javaid, N.; Yousaf, S.; Ahmad, A.; Shu, M.M.; Imran, M.; Khan, Z.A.; Alrajeh, N. Co-LAEEBA: Cooperative link aware and energy efficient protocol for wireless body area networks. *Comput. Hum. Behav.* **2015**, *51*, 1205–1215. [CrossRef]
7. Cavallari, R.; Martelli, F.; Rosini, R.; Buratti, C.; Verdone, R. A survey on wireless body area networks: Technologies and design challenges. *IEEE Commun. Surv. Tutor.* **2014**, *16*, 1635–1657. [CrossRef]
8. Goyal, R.; Mittal, N.; Gupta, L.; Surana, A. Routing protocols in wireless body area networks: Architecture, challenges, and classification. *Wirel. Commun. Mob. Comput.* **2023**, *2023*, 9229297. [CrossRef]
9. Khan, J.Y.; Yuce, M.R.; Bulger, G.; Harding, B. Wireless body area network (WBAN) design techniques and performance evaluation. *J. Med. Syst.* **2012**, *36*, 1441–1457. [CrossRef]
10. Chen, M.; Gonzalez, S.; Vasilakos, A.; Cao, H.; Leung, V.C. Body area networks: A survey. *Mob. Networks Appl.* **2011**, *16*, 171–193. [CrossRef]
11. Ghamari, M.; Janko, B.; Sherratt, R.S.; Harwin, W.; Piechockic, R.; Soltanpur, C. A survey on wireless body area networks for ehealthcare systems in residential environments. *Sensors* **2016**, *16*, 831. [CrossRef] [PubMed]
12. Latré, B.; Braem, B.; Moerman, I.; Blondia, C.; Demeester, P. A survey on wireless body area networks. *Wirel. Netw.* **2011**, *17*, 1–18. [CrossRef]
13. Movassaghi, S.; Abolhasan, M.; Lipman, J.; Smith, D.; Jamalipour, A. Wireless body area networks: A survey. *IEEE Commun. Surv. Tutor.* **2014**, *16*, 1658–1686. [CrossRef]
14. Muhannad, Q.; Biswas, S. On-body packet routing algorithms for body sensor networks. In Proceedings of the 2009 First International Conference on Networks & Communications NetCoM (2009), Chennai, India, 27–29 December 2009; pp. 171–177.
15. Ullah, S.; Khan, P.; Ullah, N.; Saleem, S.; Higgins, H.; Kwak, K.S. A review of wireless body area networks for medical applications. *arXiv* **2010**, arXiv:1001.0831.
16. Cornet, B.; Fang, H.; Ngo, H.; Boyer, E.W.; Wang, H. An overview of wireless body area networks for mobile health applications. *IEEE Netw.* **2022**, *36*, 76–82. [CrossRef]
17. Hasan, K.; Biswas, K.; Ahmed, K.; Nafi, N.S.; Islam, M.S. A comprehensive review of wireless body area network. *J. Netw. Comput. Appl.* **2019**, *143*, 178–198. [CrossRef]
18. Wu, T.; Wu, F.; Redoute, J.M.; Yuce, M.R. An autonomous wireless body area network implementation towards IoT connected healthcare applications. *IEEE Access* **2017**, *5*, 11413–11422. [CrossRef]
19. Qu, Y.; Zheng, G.; Ma, H.; Wang, X.; Ji, B.; Wu, H. A survey of routing protocols in WBAN for healthcare applications. *Sensors* **2019**, *19*, 1638. [CrossRef] [PubMed]
20. Crosby, G.V.; Ghosh, T.; Murimi, R.; Chin, C.A. Wireless body area networks for healthcare: A survey. *Int. J. Hoc, Sens. Ubiquitous Comput.* **2012**, *3*, 1–26. [CrossRef]
21. Elhadj, H.B.; Chaari, L.; Kamoun, L. A survey of routing protocols in wireless body area networks for healthcare applications. *Int. J. Health Med. Commun.* **2012**, *3*, 1–18. [CrossRef]
22. hmed, S.; Javaid, N.; Akbar, M.; Iqbal, A.; Khan, Z.A.; Qasim, U. LAEEBA: Link aware and energy efficient scheme for body area networks. In Proceedings of the 2014 IEEE 28th International Conference on Advanced Information Networking and Applications AINA (2014), Victoria, BC, Canada, 13–16 May 2014; pp. 435–440.
23. Javaid, N.; Ahmad, A.; Nadeem, Q.; Imran, M.; Haider, N. iM-SIMPLE: iMproved stable increased-throughput multi-hop link efficient routing protocol for Wireless Body Area Networks. *Comput. Hum. Behav.* **2015**, *51*, 1003–1011. [CrossRef]
24. Nadeem, Q.; Javaid, N.; Mohammad, S.N.; Khan, M.Y.; Sarfraz, S.; Gull, M. Simple: Stable increased-throughput multi-hop protocol for link efficiency in wireless body area networks. In Proceedings of the 2013 Eighth International Conference on Broadband and Wireless Computing, Communication and Applications, Compiegne, France, 28–30 October 2013; pp. 221–226.
25. Ullah, F.; Khan, M.Z.; Faisal, M.; Rehman, H.U.; Abbas, S.; Mubarek, F.S. An energy efficient and reliable routing scheme to enhance the stability period in wireless body area networks. *Comput. Commun.* **2021**, *165*, 20–32. [CrossRef]
26. Wang, F.; Hu, F.; Wang, L.; Du, Y.; Liu, X.; Guo, G. Energy-efficient medium access approach for wireless body area network based on body posture. *China Commun.* **2015**, *12*, 122–132. [CrossRef]
27. Wang, F.; Hu, F.; Wang, L.; Du, Y.; Liu, X. Posture-aware medium access approach under walking scenery for wireless body area network. In Proceedings of the 2014 IEEE International Conference on Communication Systems, Macau, China, 19–21 November 2014; pp. 457–461.
28. Gaikwad, V.D.; Ananthakumaran, S. A Review: Security and Privacy for Health Care Application in Wireless Body Area Networks. *Wirel. Pers. Commun.* **2023**, *130*, 673–691. [CrossRef]
29. Saba, T.; Haseeb, K.; Ahmed, I.; Rehman, A. Secure and energy-efficient framework using Internet of Medical Things for e-healthcare. *J. Infect. Public Health* **2020**, *13*, 1567–1575. [CrossRef] [PubMed]
30. Wu, L.; Zhang, Y.; Li, L.; Shen, J. Efficient and anonymous authentication scheme for wireless body area networks. *J. Med. Syst.* **2016**, *40*. [CrossRef] [PubMed]

31. He, Z.G.; Chung, C.; Poon, Y.; Zhang, Y.T. A review on body area networks security for healthcare. *Int. Sch. Res. Not.* **2011** *2011*, 692592.
32. Fayaz, A.; Rehmani, M.H. Energy harvesting for self-sustainable wireless body area networks. *IT Prof.* **2017**, *19*, 32–40.
33. Lu, G.; Sadagopan, N.; Krishnamachari, B.; Goel, A. Delay efficient sleep scheduling in wireless sensor networks. In Proceedings of the IEEE 24th Annual Joint Conference of the IEEE Computer and Communications Societies, Miami, FL, USA, 13–17 March 2005; pp. 2470–2481.
34. Ruzzelli, A.G.; Jurdak, R.; O'Hare, G.M.; Van Der Stok, P. Energy-efficient multi-hop medical sensor networking. In Proceedings of the 1st ACM SIGMOBILE International Workshop on Systems and Networking Support for Healthcare and Assisted Living Environments, San Juan, Puerto Rico, 11 June 2007; pp. 37–42.
35. Mehdi, E.; Dehghan, M.; Rahmani, A.M. A comprehensive survey of energy-aware routing protocols in wireless body area sensor networks. *J. Med. Syst.* **2016**, *40*. [CrossRef]
36. Zhang, Yu; Zhang, B.; Zhang, S. A lifetime maximization relay selection scheme in wireless body area networks. *Sensors* **2017**, *17*, 1267. [CrossRef] [PubMed]
37. Laurie, H.; Wang, X.; Chen, T. A review of protocol implementations and energy efficient cross-layer design for wireless body area networks. *Sensors* **2012**, *12*, 14730–14773. [CrossRef]
38. Weilian, S.; Lim, T.L. Cross-layer design and optimisation for wireless sensor networks. *Int. J. Sens. Netw.* **2009**, *6*, 3–12.
39. Braem, B.; Latre, B.; Moerman, I.; Blondia, C.; Demeester, P. The wireless autonomous spanning tree protocol for multihop wireless body area networks. In Proceedings of the 2006 Third Annual International Conference on Mobile and Ubiquitous Systems: Networking & Services, San Jose, CA, USA, 17–21 July 2006; pp. 1–8.
40. Braem, B.; Latré, B.; Moerman, I.; Blondia, C.; Reusens, E.; Joseph, W.; Demeester, P. A low-delay protocol for multishop wireless body area networks. In Proceedings of the 4th International Conference on Mobile and Ubiquitous Systems: Networking and Services, Philadelphia, PA, USA, 6–10 August 2007; pp. 38–45.
41. Anirban, B.; Bassiouni, M.A. Biocomm—A cross-layer medium access control (MAC) and routing protocol co-design for biomedical sensor networks. *Int. J. Parallel Emergent Distrib. Syst.* **2009**, *24*, 85–103.
42. Ortiz, A.M.; Ababneh, N.; Timmons, N.; Morrison, J. Adaptive routing for multihop IEEE 802.15. 6 wireless body area networks In Proceedings of the SoftCOM 2012, 20th International Conference on Software, Telecommunications and Computer Networks, Split, Croatia, 11–13 November 2012; pp. 1–5.
43. Hu, F.; Wang, L.; Wang, S.; Liu, X.; He, G. A human body posture recognition algorithm based on BP neural network for wireless body area networks. *China Commun.* **2016**, *13*, 198–208. [CrossRef]
44. Talpur, A.; Baloch, N.; Bohra, N.; Shaikh, F.K.; Felemban, E. Analyzing the impact of body postures and power on communication in WBAN. *Procedia Comput. Sci.* **2014**, *32*, 894–899. [CrossRef]
45. Maskooki, A.; Soh, C.B.; Gunawan, E.; Low, K.S. Opportunistic routing for body area network. In Proceedings of the 2011 IEEE Consumer Communications and Networking Conference (CCNC), Las Vegas, NV, USA, 9–12 January 2011; pp. 237–241.
46. Muhannad, Q.; Biswas, S. DTN routing in body sensor networks with dynamic postural partitioning. *Ad Hoc Netw.* **2010**, *8*, 824–841.
47. Ullah, F.; Khan, M.Z.; Mehmood, G.; Qureshi, M.S.; Fayaz, M. Energy efficiency and reliability considerations in wireless body area networks: A survey. *Comput. Math. Methods Med.* **2022**, *2022*, 1090131. [CrossRef] [PubMed]
48. Razzaque, M.A.; Hong, C.S.; Lee, S. Data-centric multiobjective QoS-aware routing protocol for body sensor networks. *Sensors* **2011**, *11*, 917–937. [CrossRef]
49. Khan, Z.A.; Sivakumar, S.; Phillips, W.; Robertson, B. A QoS-aware routing protocol for reliability sensitive data in hospital body area networks. *Procedia Comput. Sci.* **2013**, *19*, 171–179. [CrossRef]
50. Khan, Z.; Aslam, N.; Sivakumar, S.; Phillips, W. Energy-aware peering routing protocol for indoor hospital body area network communication. *Procedia Comput. Sci.* **2012**, *10*, 188–196. [CrossRef]
51. Monowar, M.M.; Mehedi, Hassan, M.; Bajaber, F.; Hamid, M.A.; Alamri, A. Thermal-aware multiconstrained intrabody QoS routing for wireless body area networks. *Int. J. Distrib. Sens. Netw.* **2014**, *10*, 676312. [CrossRef]
52. Tauqir, A.; Javaid, N.; Akram, S.; Rao, A.; Mohammad, S.N. Distance aware relaying energy-efficient: Dare to monitor patients in multi-hop body area sensor networks. In Proceedings of the 2013 Eighth International Conference on Broadband and Wireless Computing, Communication and Applications, Compiegne, France, 28–30 October 2013; pp. 206–213.
53. Samir, B. Thermal Aware Routing Protocols for Wireless Body Area Networks: Review and Open Research Issues. *Karbala Int. J. Mod. Sci.* **2022**, *8*, 356–374.
54. Jamil, F.; Iqbal, M.A.; Amin, R.; Kim, D. Adaptive thermal-aware routing protocol for wireless body area network. *Electronics* **2019**, *8*, 47. [CrossRef]
55. Omid, R.; Lighvan, M.Z. M2E2: A novel multi-hop routing protocol for wireless body sensor networks. *Int. J. Comput. Netw. Commun. Secur.* **2014**, *2*, 260–267.
56. Ahmad, A.; Javaid, N.; Qasim, U.; Ishfaq, M.; Khan, Z.A. RE-ATTEMPT: A new energy-efficient routing protocol for wireless body area sensor networks. *Int. J. Distrib. Sens. Netw.* **2014**, *10*, 464010. [CrossRef]
57. Javaid, N.; Abbas, Z.; Fareed, M.S.; Khan, Z.A.; Alrajeh, N. M-ATTEMPT: A new energy-efficient routing protocol for wireless body area sensor networks. *Procedia Comput. Sci.* **2013**, *19*, 224–231. [CrossRef]

58. Samaneh, M.; Abolhasan, M.; Lipman, J. Energy efficient thermal and power aware (ETPA) routing in body area networks. In Proceedings of the 2012 IEEE 23rd International Symposium on Personal, Indoor and Mobile Radio Communications (PIMRC), Sydney, NSW, Australia, 9–12 September 2012; pp. 1108–1113.
59. Anirban, B.; Bassiouni, M.A. Energy efficient thermal aware routing algorithms for embedded biomedical sensor networks. In Proceedings of the 2006 IEEE International Conference on Mobile Ad Hoc and Sensor Systems, Vancouver, BC, Canada, 9–12 October 2006; pp. 604–609.
60. Daisuke, T.; Xiao, Y.; Hu, F. LTRT: Least total-route temperature routing for embedded biomedical sensor networks. In Proceedings of the IEEE GLOBECOM 2007—IEEE Global Telecommunications Conference, Washington, DC, USA, 26–30 November 2007; pp. 641–645.
61. Tang, Q.; Tummala, N.; Gupta, S.K.; Schwiebert, L. TARA: thermal-aware routing algorithm for implanted sensor networks. In Proceedings of the First IEEE International Conference DCOSS 2005, Marina del Rey, CA, USA, 30 June–1 July 2005; pp. 206–217.
62. Diestel, R. *Graph Theory*, 5th ed.; Graduate Texts in Mathematics 173; Springer: Berlin/Heidelberg, Germany, 2018.
63. Heinzelman, W.R.; Chandrakasan, A.; Balakrishnan, H. Energy-efficient communication protocol for wireless microsensor networks. In Proceedings of the Proceedings of the 33rd Annual Hawaii International Conference on System Sciences, Maui, HI, USA, 7 January 2000.
64. Rahman, C.H.; Hassan, M.S.; Ahmed, A. Analysis of path loss characteristics in body area network for different physical structures. In Proceedings of the 2016 9th International Conference on Electrical and Computer Engineering (ICECE), Dhaka, Bangladesh, 20–22 December 2016; pp. 299–302.
65. Benaissa, S.; Plets, D.; Tanghe, E.; Vermeeren, G.; Martens, L.; Sonck, B.; Tuyttens, F.A.; Vaele, L.; Hoebeke, J.; Stevens, N.; et al. Characterization of the on-body path loss at 2.45 GHz and energy efficient WBAN design for dairy cows. *IEEE Trans. Antennas Propag.* **2016**, *64*, 4848–4858. [CrossRef]
66. Balaji, R.G.; Raghavendran, P.S.; Ashokan, R. Performances enhancement in wireless body area network (WBAN). *Int. J. Eng. Innov. Technol.* **2013**, *3*, 2008–2011.
67. Pratheema, S.; Srinivasagan, K.G.; Naskath, J. Minimizing end-to-end delay using multipath routing in wireless sensor networks. *Int. J. Comput. Appl.* **2011**, *21*, 20–26.
68. Javaid, N.; Javaid, A.; Khan, I.A.; Djouani, K. Performance study of ETX based wireless routing metrics. In Proceedings of the 2009 2nd International Conference on Computer, Control and Communication, Karachi, Pakistan, 17–18 February 2009; pp. 1–7.
69. Available online: https://networkx.org/ (accessed on 24 May 2023).

mathematics

MDPI

Article

Optimal Design of Group Orthogonal Phase-Coded Waveforms for MIMO Radar

Tianqu Liu [1], Jinping Sun [1], Guohua Wang [2], Xianxun Yao [1,*] and Yaqiong Qiao [1]

1 School of Electronic and Information Engineering, Beihang University, Beijing 100191, China; tqliu@buaa.edu.cn (T.L.); sunjinping@buaa.edu.cn (J.S.); qiaoyaqiong@crscd.com.cn (Y.Q.)
2 Hertzwell Pte Ltd., Singapore 138565, Singapore; wghwood@hotmail.com
* Correspondence: xianxun.yao@buaa.edu.cn; Tel.: +86-010-8231-7240

Abstract: Digital radio frequency memory (DRFM) has emerged as an advanced technique to achieve a range of jamming signals, due to its capability to intercept waveforms within a short time. multiple-input multiple-output (MIMO) radars can transmit agile orthogonal waveform sets for different pulses to combat DRFM-based jamming, where any two groups of waveform sets are also orthogonal. In this article, a group orthogonal waveform optimal design model is formulated in order to combat DRFM-based jamming by flexibly designing waveforms for MIMO radars. Aiming at balancing the intra- and intergroup orthogonal performances, the objective function is defined as the weighted sum of the intra- and intergroup orthogonal performance metrics. To solve the formulated model, in this article, a group orthogonal waveform design algorithm is proposed. Based on a primal-dual-type method and proper relaxations, the proposed algorithm transforms the original problem into a series of simple subproblems. Numerical results demonstrate that the obtained group orthogonal waveforms have the ability to flexibly suppress DRFM-based deceptive jamming, which is not achievable using p-majorization–minimization (p-MM) and primal-dual, two of the most advanced orthogonal waveform design algorithms.

Keywords: waveform design; optimization; MIMO radar; group orthogonal; phase-coded; radar countermeasures

MSC: 90C26

1. Introduction

Waveform diversity and multiantenna technologies are utilized in multiple-input multiple-output (MIMO) radars to improve their angular resolution, antijamming ability, and other target detection abilities [1–4]. By transmitting orthogonal waveform sets, an MIMO radar can separate the received waveforms transmitted by different antennas. Generally, an MIMO radar uses a matched filter bank to process the echoes. Thus, the cross-correlation functions among the orthogonal waveforms should be as low as possible. Meanwhile, in order to achieve a good pulse compression performance, the autocorrelation functions' side-lobes should also be as low as possible.

Designing multiple different phase-coding sequences with a low cross-correlation is one of the most common ways to realize an orthogonal waveform set [5–7]. With a large number of transmitted waveforms, a phase-coded waveform set is difficult to intercept by traditional jammers [8]. Integrated side-lobe level (ISL) and peak side-lobe level (PSL) are two commonly used metrics for orthogonal MIMO radar phase-coded waveform sets [9–11]. Aiming to minimize the ISL, multicyclic-algorithm-new (multi-CAN [12]), majorization–minimization correlation (MM-Corr [13]), ISL-New [14], alternating-direction-method-of-multipliers (ADMM [15]), etc., algorithms have been proposed, although they cannot minimize the PSL. PSL minimization is more complex and harder than ISL minimization. Thus far, some researchers have proposed effective PSL optimization algorithms [16–20], with the method based on primal-dual achieving the best performance [18]. Particularly, the p-MM algorithm can obtain almost the same PSL metrics as primal-dual, with much lower ISL metrics [19]. All the abovementioned ISL and PSL optimization algorithms lead to a single set of orthogonal waveforms. With a well-designed orthogonal phase-coded waveform set, the MIMO radar achieves a high waveform diversity gain and a low probability of intercept.

With the advent of digital radio frequency memory (DRFM) [21,22], more advanced jamming technologies have been rapidly developed. DRFM-based jammers pose a great threat to MIMO radar, and can perform read and copy functions and diverse parameter modulations (like delay, Doppler, etc.) within a short time. Therefore, DRFM-based jammers may seriously affect the operational capabilities of MIMO radars in the future.

The transmitting agile waveform is an effective way to combat modern DRFM-based jamming [23,24]. Although DRFM-based jammers have a strong ability to intercept waveforms, they must delay at least one pulse repetition time (PRT) to complete jamming signal generation. If the correlation between adjacent pulses is low, it is difficult for the DRFM-based jammers to interfere with the waveforms. From the perspective of orthogonality, MIMO radars should transmit an orthogonal waveform set within each pulse. In different pulse intervals, the waveform sets should also be orthogonal to each other. Thus, even though the jammers have intercepted the waveforms in the previous pulse, they cannot interfere with the subsequent pulses. Therefore, in this article, the MIMO radar antijamming agile waveform is modeled as multiple groups of orthogonal waveform sets. The abovementioned PSL and ISL optimization algorithms can be used to design all groups of waveforms directly. However, they cannot finely control the intra- and intergroup orthogonal performance.

In order to balance the intra- and intergroup orthogonal performances, in this article, an optimization model for designing group orthogonal waveforms is proposed, which is able to maximize the waveform diversity gain when diverse DRFM-based jamming signals are encountered. The proposed model can overcome the shortcomings of existing methods. Considering that target detection is based on correlation peak detection, in the proposed model, the objective function is formulated considering the following two aspects. The first is to minimize the cross-correlation peak and autocorrelation side-lobe peaks within each group of orthogonal waveforms, which can be evaluated by the traditional PSL metric. The second is to minimize the cross-correlation peak among different groups of orthogonal waveforms, which can be evaluated by the peak cross-correlation level (PCL) metric. The objective function of the proposed model is the weighted sum of the PSL and PCL metrics. These metrics evaluate the MIMO waveform diversity gain and jamming suppression performance, respectively.

In order to solve the proposed optimization model, in this article, a group orthogonal waveform optimal design algorithm is proposed. Minimizing the correlation function peak in the proposed model is difficult. Based on some relaxations, ref. [18] divided the origin PSL minimization problem into a series of convex subproblems. Inspired by the relaxation approach in [18], in this article, the intra- and intergroup correlation functions are processed separately and the proposed optimization model is transformed into a series of solvable subproblems. Finally, the proposed algorithm generates solutions to each subproblem. The numerical results show that the proposed algorithm can effectively design multiple groups of orthogonal waveform sets. The cross-correlation functions between the waveform sets are also low. The antijamming simulation results demonstrate that the designed group orthogonal waveforms can balance the DRFM-based deceptive jamming suppression and range compression performances of MIMO radar by adjusting the weighting factors.

2. Problem Formulation

The optimization variables in group orthogonal waveform design are G groups of waveforms, with each group consisting of M waveforms. Considering phase-coded waveform design, group orthogonal waveforms with intra- and intergroup orthogonality can be realized by assigning different waveforms with different phase-coded sequences. Since the pulse length, chip length, carrier frequency, and other radar system parameters have little effect on the correlation function side-lobe level, in this article, the group orthogonal waveforms $\{\mathbf{x}_i\}_{i=1}^{GM}$ are modeled as

$$x_i[n] = \exp(\mathrm{j}\varphi_{i,n}), i = 1,2,\ldots,GM, n = 1,2,\ldots,N \tag{1}$$

where N is the sequence length. The phase value continuously satisfies $\varphi_{i,n} \in [-\pi,\pi)$. The correlation functions of $\{\mathbf{x}_i\}_{i=1}^{GM}$ are defined as

$$r_{ij}[k] = \sum_{n=1}^{N} x_i[n+k]\overline{x}_j[n], k = -N+1,\ldots,N-1 \tag{2}$$

where $r_{ij}[k]$ is the auto- or cross-correlation function and $\overline{(\cdot)}$ represents the complex conjugate. The set of the indexes of intra- and intergroup correlation functions is defined as $\{i,j,k|i,j = 1,2,\ldots,GM, k = -N+1,\ldots,N-1\}$.

To evaluate the group orthogonal waveforms' correlation properties, we consider the intra- and intergroup correlation functions separately. We use the traditional peak side-lobe level (PSL) metric [9–11] to evaluate the waveforms within each group. The PSL metric for the g-th group is defined as

$$\mathrm{PSL}_g \triangleq \frac{1}{N^2} \max_{(i,j,k)\in\mathcal{K}_g} \left|r_{ij}[k]\right|^2, g = 1,2,\ldots,G \tag{3}$$

$$\mathcal{K}_g = \{i,j,k|i,j \in [M(g-1)+1,\ Mg], i \neq j \text{ or } k \neq 0\} \tag{4}$$

where set $\mathcal{K}_g$ contains the correlation function indexes within the g-th group of waveforms, except for the indexes of the autocorrelation peak values that are equal to N. Because there are only cross-correlation functions between different groups of waveforms, the following peak cross-correlation level (PCL) is defined to evaluate the intergroup cross-correlation peak.

$$\mathrm{PCL} \triangleq \frac{1}{N^2} \max_{(i,j,k)\in\mathcal{G}} \left|r_{ij}[k]\right|^2 \tag{5}$$

$$\mathcal{G} = \left\{i,j,k\middle|(i,j,k) \notin \cup_{g=1}^{G}\mathcal{K}_g, i \neq j\right\} \tag{6}$$

The set $\mathcal{G}$ in Equation (6) contains all the indexes of intergroup cross-correlation functions.

The object of group orthogonal waveform design is to minimize the PSL_1, PSL_2,…, PSL_G and PCL metrics, which are functions of the optimization variables $\{\mathbf{x}_i\}_{i=1}^{GM}$. Thus, the optimization model can be expressed as

$$\begin{array}{ll} \min & \mathbf{f}(\{\mathbf{x}_i\}_{i=1}^{GM}) = (\mathrm{PSL}_1, \mathrm{PSL}_2, \cdots, \mathrm{PSL}_G, \mathrm{PCL}) \\ \text{s.t.} & |x_i[n]| = 1, \quad i = 1,2,\ldots,GM,\ n = 1,2,\ldots,N \end{array} \tag{7}$$

where $\mathbf{f}$ represents the objective function vector, which contains the convolution and max($\cdot$) operations. In addition, the waveforms $\{\mathbf{x}_i\}_{i=1}^{GM}$ have constant modulus constraints. The optimization model (7) is a complex minimax problem with nonconvex constraints. In order to simplify problem (7), we introduce the variables ε_1, ε_2,…, ε_G and γ to constrain the PSL_1, PSL_2,…, PSL_G and PCL metrics values, respectively. Meanwhile, in order to balance the intra- and intergroup orthogonal performances, we introduce a weighting factor w, thereby transforming the original problem into the following single-objective problem.

$$\begin{aligned}
&\min w \cdot \left(\sum_{g=1}^{G} \varepsilon_g^2/G\right) + (1-w)\cdot\gamma^2 \\
&\text{s.t. } \varepsilon_g \geq \left|r_{ij}[k]\right|^2/N^2, \forall (i,j,k) \in \mathcal{K}_g, g = 1,2,\ldots,G \\
&\quad \gamma \geq \left|r_{ij}[k]\right|^2/N^2, \forall (i,j,k) \in \mathcal{G} \\
&\quad r_{ij}(k) = \sum_n x_i[n+k]\overline{x}_j[n], \forall (i,j,k) \\
&\quad |x_i[n]| = 1, \forall (i,n)
\end{aligned} \tag{8}$$

Note that the convolution operations and constant modulus constraints in problem (8) are not conducive to solving this problem. Inspired by ref. [18], we introduced the auxiliary variables $\{\mathbf{h}_i\}_{i=1}^{GM}$ and $\rho_{ij}(k)$ to decompose the complex nonlinear convolutions. Then, the correlation function constraints are replaced by the following linear equality constraints:

$$\begin{cases} \rho_{ij}(k) = \sum_n x_i[n+k]\overline{h}_j[n], \forall (i,j,k) \\ \mathbf{h}_i^H \mathbf{x}_i = N, \|\mathbf{h}_i\|_2^2 \leq N, \forall i \end{cases} \tag{9}$$

In Equation (9), if $\mathbf{h}_i = \mathbf{x}_i$ holds for all i = 1, 2,. . ., GM, then $\rho_{ij}(k)$ is equal to the correlation functions $r_{ij}(k)$. This equivalent constraint condition in Equation (9) is correct according to the proposition below.

Proposition 1. *The constraints* $\mathbf{h}_i^H \mathbf{x}_i = N$, $\|\mathbf{h}_i\|_2^2 \leq N$ *are equivalent to the constraint* $\mathbf{h}_i = \mathbf{x}_i$ *for all* i = 1, 2,. . ., GM.

Proof of Proposition 1. According to the Cauchy–Schwartz inequality,

$$\mathbf{h}_i^H \mathbf{x}_i = N = \left|\mathbf{h}_i^H \mathbf{x}_i\right| \leq \|\mathbf{h}_i\|_2 \|\mathbf{x}_i\|_2 = \sqrt{N}\|\mathbf{h}_i\|_2 \tag{10}$$

If $\|\mathbf{h}_i\|_2 \leq \sqrt{N}$ holds, then inequality (10) implies $\|\mathbf{h}_i\|_2 = \sqrt{N}$. Using the equality condition of the Cauchy–Schwartz inequality, we obtain

$$\mathbf{x}_i = \kappa \mathbf{h}_i, \kappa \in \mathbb{C} \tag{11}$$

According to Equation (11) and $\mathbf{h}_i^H \mathbf{x}_i = N$, it is not difficult to find $\kappa = 1$. Thus, $\mathbf{h}_i = \mathbf{x}_i$ is obtained and the proof is complete. □

By introducing the variables $\{\mathbf{h}_i\}_{i=1}^{GM}$ and $\rho_{ij}(k)$, the nonlinear constraints are relaxed to linear convex constraints. However, the constant modulus constraints of $\{\mathbf{x}_i\}_{i=1}^{GM}$ in problem (8) are still coupled with its other constraints, which leads to difficulties in solving for $\{\mathbf{x}_i\}_{i=1}^{GM}$. Therefore, we introduce the variables $\{\mathbf{y}_i\}_{i=1}^{GM}$ to simplify the subproblem regarding $\{\mathbf{x}_i\}_{i=1}^{GM}$. The following conditions are added to ensure the obtained waveforms have a constant modulus.

$$x_i[n] = y_i[n], \ |y_i[n]| = 1, \ \forall (i,n) \tag{12}$$

Then, the constant modulus constraints are transferred to the subproblem regarding $\{\mathbf{y}_i\}_{i=1}^{GM}$, which is easy to solve, even with constant modulus constraints. The subproblem regarding $\{\mathbf{x}_i\}_{i=1}^{GM}$ becomes an unconstrained convex problem.

After introducing series auxiliary variables into the optimization model (8), we formulate the following group orthogonal waveform optimal design model:

$$\begin{aligned}
&\min \ w \cdot \left(\sum_{g=1}^{G} \varepsilon_g^2/G\right) + (1-w)\cdot\gamma^2 \\
&\text{s.t.} \ \left|\rho_{ij}(k)\right| \leq \varepsilon_g, \forall (i,j,k) \in \mathcal{K}_g, g = 1,2,\ldots,G \\
&\quad \left|\rho_{ij}(k)\right| \leq \gamma, \forall (i,j,k) \in \mathcal{G} \\
&\quad \rho_{ij}(k) = \sum_n x_i[n+k]\overline{h}_j[n], \forall (i,j,k) \\
&\quad \mathbf{h}_i^H \mathbf{x}_i = N, \|\mathbf{h}_i\|_2^2 \leq N, \forall i \\
&\quad x_i[n] = y_i[n], |y_i[n]| = 1, \forall (i,n)
\end{aligned} \tag{13}$$

The weighting factor w can balance the intra- and intergroup correlation peak values. The optimization variables include $\{\mathbf{x}_i\}_{i=1}^{GM}$, $\{\mathbf{y}_i\}_{i=1}^{GM}$, $\{\mathbf{h}_i\}_{i=1}^{GM}$, $\rho_{ij}(k)$, γ, and ε_g, $g = 1, 2, \ldots, G$. Although the dimensions of the optimization variables increase, the intractable nonlinear and nonconvex constraints in the optimization model (8) are relaxed into a series of linear and convex constraints.

3. Proposed Group Orthogonal Waveform Design Algorithm

Based on the characteristics of the objective function and the constraints in the optimization model (13), we formulated an augmented Lagrange function [25] and transformed problem (13) into the following constrained minimization problem:

$$\begin{aligned} \min\ & L\left(\varepsilon_1, \ldots, \varepsilon_G, \gamma, \{\rho_{ij}(k)\}, \{\mathbf{x}_i\}, \{\mathbf{y}_i\}, \{\mathbf{h}_i\}, \{\lambda_i\}, \{\alpha_{ijk}\}, \{\beta_{in}\}\right) \\ \text{s.t. } & |\rho_{ij}(k)| \leq \varepsilon_g, \forall (i,j,k) \in \mathcal{K}_g, g = 1, 2, \ldots, G \\ & |\rho_{ij}(k)| \leq \gamma, \forall (i,j,k) \in \mathcal{G} \\ & \|\mathbf{h}_i\|_2^2 \leq N, \forall i \\ & |y_i[n]| = 1, \forall (i,n) \end{aligned} \tag{14}$$

The augmented Lagrange function L can be expressed as follows:

$$\begin{aligned} L = & w \cdot \left(\sum_{g=1}^{G} \varepsilon_g^2 / G\right) + (1-w) \cdot \gamma^2 + \mathrm{Re}\left[\sum_i \overline{\lambda}_i \left(\mathbf{h}_i^H \mathbf{x}_i - N\right)\right] \\ & + \mathrm{Re}\left[\sum_{(i,j,k) \in (\cup_{g=1}^{G} \mathcal{K}_g) \cup \mathcal{G}} \overline{\alpha}_{ijk} \left(\rho_{ij}(k) - \sum_n x_i[n+k] \overline{h}_j[n]\right)\right] \\ & + \mathrm{Re}\left[\sum_{i,n} \overline{\beta}_{in} (x_i[n] - y_i[n])\right] + \tfrac{\theta_1}{2}\left(\sum_i \left|\mathbf{h}_i^H \mathbf{x}_i - N\right|^2\right) \\ & + \tfrac{\theta_2}{2}\left(\sum_{(i,j,k) \in (\cup_{g=1}^{G} \mathcal{K}_g) \cup \mathcal{G}} \left|\rho_{ij}(k) - \sum_n x_i[n+k] \overline{h}_j[n]\right|^2\right) \\ & + \tfrac{\theta_3}{2}\left(\sum_{i,n} |x_i[n] - y_i[n]|^2\right) \end{aligned} \tag{15}$$

where $\mathrm{Re}[\cdot]$ represents the real part of a complex number. θ_1, θ_2, and θ_3 are penalty parameters. $\{\lambda_i\}$, $\{\alpha_{ijk}\}$, and $\{\beta_{in}\}$ are Lagrangian multipliers, also dual variables. To solve problem (14), we propose a group orthogonal waveform design algorithm based on a primal-dual type method. The proposed algorithm decomposes problem (14) into a series of simple subproblems. By sequentially updating the optimization variables and dual variables, the proposed algorithm minimizes the augmented Lagrange function L in Equation (15) after several iterations. According to optimization model (13), the variables $\left\{\mathbf{x}_i^{(l)}\right\}_{i=1}^{GM}$, $\{\mathbf{y}_i\}_{i=1}^{GM}$, and $\left\{\mathbf{h}_i^{(l)}\right\}_{i=1}^{GM}$ converge to the same point. The convergence condition is $\sum_i \|\mathbf{h}_i^{(l)} - \mathbf{x}_i^{(l)}\|_2^2 / \sum_i \|\mathbf{x}_i^{(l)}\|_2^2 < \eta$. Algorithm 1 summarizes the proposed algorithm. The subproblems of the proposed group orthogonal waveform design algorithm can be expressed as follows:

$$\mathcal{P}_1: \left\{ \varepsilon_1, \cdots, \varepsilon_G, \gamma, \ \rho_{ij}(k)\right\}^{(l+1)} = \mathop{\arg\min}_{\substack{|\rho_{ij}(k)| \leq \varepsilon_g,\ \forall (i,j,k) \in \mathcal{K}_g,\ g=1,2,\ldots,G \\ |\rho_{ij}(k)| \leq \gamma,\ \forall (i,j,k) \in \mathcal{G}}} L\left(\begin{array}{c} \varepsilon_1, \cdots, \varepsilon_G, \gamma, \rho_{ij}(k) \\ \left\{\mathbf{x}_i, \mathbf{y}_i, \mathbf{h}_i, \lambda_i, \alpha_{ijk}, \beta_{in}\right\}^{(l)} \end{array} \right) \tag{16}$$

$$\mathcal{P}_2: \left\{ \{\mathbf{h}_i\}^{(l+1)} = \mathop{\arg\min}_{\|\mathbf{h}_i\|_2^2 \leq N, \forall i} L\left(\begin{array}{c} \{\varepsilon_1, \cdots, \varepsilon_G, \gamma, \rho_{ij}(k)\}^{(l+1)}, \\ \{\mathbf{h}_i\}, \left\{\mathbf{x}_i, \mathbf{y}_i, \lambda_i, \alpha_{ijk}, \beta_{in}\right\}^{(l)} \end{array} \right) \right. \tag{17}$$

$$\mathcal{P}_3: \left\{ \{\mathbf{x}_i\}^{(l+1)} = \mathrm{argmin} L(\ldots, \{\mathbf{x}_i\}, \ldots) \right. \tag{18}$$

$$\mathcal{P}_4 : \left\{ \{\mathbf{y}_i\}^{(l+1)} = \underset{|y_i[n]|=1,\forall(i,n)}{\operatorname{argmin}} \; L(\ldots,\{\mathbf{y}_i\},\ldots) \right. \tag{19}$$

$$\left\{ \begin{array}{l} \lambda_i^{(l+1)} = \lambda_i^{(l)} + c\theta_1\left(\left(\mathbf{h}_i^{(l+1)}\right)^H \mathbf{x}_i^{(l+1)} - N \right) \\ \alpha_{ijk}^{(l+1)} = \alpha_{ijk}^{(l)} + c\theta_2\left(\rho_{ij}^{(l+1)}(k) - \sum_n x_i^{(l+1)}[n+k]\overline{h}_j^{(l+1)}[n] \right) \\ \beta_{in}^{(l+1)} = \beta_{in}^{(l)} + c\theta_3\left(x_i^{(l+1)}[n] - y_i^{(l+1)}[n] \right) \end{array} \right. \tag{20}$$

Parameter c in Equation (20) is the step length. Superscript (l) represents the values of the variables at the l-th iteration. Except for subproblem $\mathcal{P}_1$, subproblems $\mathcal{P}_2$, $\mathcal{P}_3$, and $\mathcal{P}_4$ are actually the same as the primal-dual algorithm [18], and can be solved using similar methods to primal-dual. In the remainder of this section, methods to solve these subproblems are introduced.

Algorithm 1 Group Orthogonal Waveform Design Algorithm

Initialization

Randomly select $\left\{\mathbf{x}_i^{(0)}\right\}_{i=1}^{GM}$ (constant modulus is not required).

Set constant modulus $\left\{\mathbf{y}_i^{(0)}\right\}_{i=1}^{GM}$ using the phases of $\left\{\mathbf{x}_i^{(0)}\right\}_{i=1}^{GM}$.

Randomly select $\{\lambda_i\}$, $\{\alpha_{ijk}\}$, $\{\beta_{in}\}$ and $\left\{\mathbf{h}_i^{(0)}\right\}_{i=1}^{GM}$, set $l = 0$.

Repeat

Compute $\left\{\varepsilon_1,\cdots,\varepsilon_G,\gamma,\rho_{ij}(k)\right\}^{(l+1)}$ by solving subproblem (16).

Compute $\{\mathbf{h}_i\}^{(l+1)}$ by solving subproblem (17).

Compute $\{\mathbf{x}_i\}^{(l+1)}$ by solving subproblem (18).

Compute $\{\mathbf{y}_i\}^{(l+1)}$ by solving subproblem (19).

Compute $\lambda_i^{(l+1)}, \alpha_{ijk}^{(l+1)}, \beta_{in}^{(l+1)}$ using (20), $l = l+1$.

Until $\sum_i \|\mathbf{h}_i^{(l)} - \mathbf{x}_i^{(l)}\|_2^2 / \sum_i \|\mathbf{x}_i^{(l)}\|_2^2 < \eta$.

3.1. Solving Subproblem $\mathcal{P}_1$

According to the augmented Lagrange function (15), subproblem (16) can be separated into two independent parts as follows:

$$\begin{array}{c} \min\limits_{|\rho_{ij}(k)|\le\varepsilon_g} w\varepsilon_g^2/G + \sum\limits_{(i,j,k)\in\mathcal{K}_g} \left[a_{ijk}|\rho_{ij}(k)|^2 + \operatorname{Re}\left(\overline{b}_{ijk}\rho_{ij}(k)\right) \right] \\ g = 1,2,\ldots,G \end{array} \tag{21}$$

$$\min_{|\rho_{ij}(k)|\le\gamma} (1-w)\gamma^2 + \sum_{(i,j,k)\in\mathcal{G}} \left[a_{ijk}|\rho_{ij}(k)|^2 + \operatorname{Re}\left(\overline{b}_{ijk}\rho_{ij}(k)\right) \right] \tag{22}$$

where $a_{ijk} = \theta_2/2$, $b_{ijk} = \alpha_{ijk}^{(l)} - \theta_2 \cdot \sum_n x_i^{(l)}[n+k]\overline{h}_j^{(l)}[n]$. Considering problem (21), when ε_g is fixed, the optimal solution of $\rho_{ij}(k)$ for every index (i,j,k) can be expressed as

$$\rho_{ijk}^*(\varepsilon_g) = \left\{ \begin{array}{l} -b_{ijk}/\left(2a_{ijk}\right), \left|b_{ijk}/\left(2a_{ijk}\right)\right| \le \varepsilon_g \\ -\left(b_{ijk}/\left|b_{ijk}\right|\right)\varepsilon_g, \text{ others} \end{array} \right. \tag{23}$$

where $\rho_{ijk}^*(\varepsilon_g)$ is a function of ε_g. According to Equation (23), problem (21) is equivalent to

$$\min_{\varepsilon_g\ge0} f(\varepsilon_g) \triangleq w\varepsilon_g^2/G + \sum_{(i,j,k)\in\mathcal{K}_g} f_{ijk}^*(\varepsilon_g),\; g = 1,2,\ldots,G \tag{24}$$

where

$$f_{ijk}^*(\varepsilon_g) = a_{ijk}\left|\rho_{ijk}^*(\varepsilon_g)\right|^2 + \mathrm{Re}\left(\overline{b}_{ijk}\rho_{ijk}^*(\varepsilon_g)\right) = \begin{cases} -\left|b_{ijk}\right|^2/\left(4a_{ijk}\right), \left|b_{ijk}/\left(2a_{ijk}\right)\right| \le \varepsilon_g \\ a_{ijk}\varepsilon_g^2 - \left|b_{ijk}\right|\varepsilon_g, \text{ others} \end{cases} \tag{25}$$

Because $f(\varepsilon_g) \le f(-\varepsilon_g)$, the condition $\varepsilon_g \ge 0$ can be ignored. The optimal ε_g^* value can be determined by solving the following equation:

$$\nabla f(\varepsilon_g) = 2w\varepsilon_g + \sum_{(i,j,k)\in\mathcal{K}_g} \min\left\{2a_{ijk}\varepsilon_g - \left|b_{ijk}\right|, 0\right\} = 0 \tag{26}$$

Because $a_{ijk} = \theta_2/2 > 0$, $\nabla f(\varepsilon_g)$ is a monotonically increasing function of ε_g. The optimal $\varepsilon_g^{(l+1)} = \varepsilon_g^*$ value can be obtained efficiently via the bisection method. Similarly, the optimal solution $\gamma^{(l+1)} = \gamma^*$ of problem (22) can be obtained by solving the following equation:

$$2(1-w)\gamma + \sum_{(i,j,k)\in\mathcal{G}} \min\left\{2a_{ijk}\gamma - \left|b_{ijk}\right|, 0\right\} = 0 \tag{27}$$

3.2. Solving Subproblem $\mathcal{P}_2$

According to Equation (15), subproblem (17) is separable for each $\mathbf{h}_i$, i = 1, 2,..., GM. The minimization problem can be expressed as follows:

$$\min_{\|\mathbf{h}_i\|_2^2<N} f(\mathbf{h}_i) \triangleq \mathbf{h}_i^H \mathbf{A}_i \mathbf{h}_i + \mathrm{Re}\left(\mathbf{t}_i^H \mathbf{h}_i\right) \tag{28}$$

where

$$\mathbf{A}_i = (\theta_2/2) \cdot \sum\nolimits_{j,k} \mathbf{x}_{j,(-k)}^{(l)} \mathbf{x}_{j,(-k)}^{(l)H} \tag{29}$$

$$\mathbf{t}_i = \overline{\lambda}_i^{(l)} \mathbf{x}_i^{(l)} - \theta_1 N \mathbf{x}_i^{(l)} - \sum\nolimits_{j,k} \left(\overline{\alpha}_{jik}^{(l)} + \theta_2 \overline{\rho}_{ji}^{(l+1)}(k)\right) \mathbf{x}_{j,(-k)}^{(l)} \tag{30}$$

where $\mathbf{x}_{j,k}$ represents the aperiodic delayed copy of the discrete signal $\mathbf{x}_j$. For $k \ge 0$, $\mathbf{x}_{j,k} = (x_j[k+1], x_j[k+2], \cdots, x_j[N], \mathbf{0}_{1\times k})^{\mathrm{T}}$; for $k \le 0$, $\mathbf{x}_{j,k} = (\mathbf{0}_{1\times|k|}, x_j[1], x_j[2], \cdots, x_j[N-|k|])^{\mathrm{T}}$. The augmented Lagrange function of problem (28) can be expressed as follows:

$$L_i(\mathbf{h}_i, \lambda) = \mathbf{h}_i^H \mathbf{A}_i \mathbf{h}_i + \mathrm{Re}\left(\mathbf{t}_i^H \mathbf{h}_i\right) + \lambda\left(\|\mathbf{h}_i\|_2^2 - N\right) \tag{31}$$

The Karush–Kuhn–Tucker (KKT) conditions for problem (28) are as follows:

$$\begin{cases} \|\mathbf{h}_i^*\|_2^2 \le N, & 2\mathbf{A}_i\mathbf{h}_i^* + \mathbf{t}_i + 2\lambda^*\mathbf{h}_i^* = 0 \\ \lambda^* \ge 0, & \lambda^*\left(\|\mathbf{h}_i^*\|_2^2 - N\right) = 0 \end{cases} \tag{32}$$

Obviously, when $\lambda^* = 0$, KKT condition (32) is equivalent to

$$\|\mathbf{h}_i^*\|_2^2 \le N,\ \mathbf{h}_i^* = -\mathbf{A}_i^{-1}\mathbf{t}_i/2 \tag{33}$$

If condition (33) is true, then $\mathbf{h}_i^{(l+1)} = \mathbf{h}_i^*$; otherwise, the optimal solution should found under the condition of $\lambda^* > 0$.

When $\lambda^* > 0$, condition (32) is equivalent to

$$\|\mathbf{h}_i^*\|_2^2 = N, \mathbf{h}_i^* = -(\mathbf{A}_i + 2\lambda^*\mathbf{I})^{-1}\mathbf{t}_i/2 \tag{34}$$

In order to solve Equation (34), the value of λ^* should be determined. According to Equation (31), for any fixed value of $\lambda > 0$, the optimal $\mathbf{v}(\lambda)$ with minimal $L_i(\mathbf{v}(\lambda), \lambda)$ is below.

$$\mathbf{v}(\lambda) = -(\mathbf{A}_i + 2\lambda\mathbf{I})^{-1}\mathbf{t}_i/2 \tag{35}$$

If $0 < \lambda' < \lambda$, then

$$\begin{cases} L_i(\mathbf{v}(\lambda), \lambda) \leq L_i(\mathbf{v}(\lambda'), \lambda) \\ L_i(\mathbf{v}(\lambda'), \lambda') \leq L_i(\mathbf{v}(\lambda), \lambda') \end{cases} \tag{36}$$

According to function (31), and after summing the two inequalities in (36), then

$$(\lambda' - \lambda)\|\mathbf{v}(\lambda')\|_2^2 \leq (\lambda' - \lambda)\|\mathbf{v}(\lambda)\|_2^2 \tag{37}$$

Therefore, $\|\mathbf{v}(\lambda)\|_2^2$ is a monotone decreasing function of λ. Solving the KKT conditions when $\lambda^* > 0$ is equivalent to finding the zero of function $\|\mathbf{v}(\lambda)\|_2^2 - N$, which can be achieved via the bisection method.

3.3. Solving Subproblem $\mathcal{P}_3$

According to Equation (15), subproblem (18) can be separated into the following unconstrained convex problem for each $\mathbf{x}_i$, $i = 1, 2, \ldots, GM$:

$$\min_{\mathbf{x}_i} f(\mathbf{h}_i) \triangleq \mathbf{x}_i^H \mathbf{C}_i \mathbf{x}_i + \mathrm{Re}\left(\mathbf{d}_i^H \mathbf{x}_i\right) \tag{38}$$

where

$$\mathbf{C}_i = (\theta_2/2) \cdot \sum\nolimits_{j,k} \mathbf{h}_{j,k}^{(l+1)} \mathbf{h}_{j,k}^{(l+1)H} + (\theta_3/2)\mathbf{I} \tag{39}$$

$$\mathbf{d}_i = \lambda_i^{(l)} \mathbf{h}_i^{(l+1)} + \boldsymbol{\beta}_i^{(l)} - \theta_1 N \mathbf{h}_i^{(l+1)} - \sum\nolimits_{j,k} \left(\alpha_{ijk}^{(l)} + \theta_2 \rho_{ij}^{(l+1)}(k)\right) \mathbf{h}_{j,k}^{(l+1)} - \theta_3 \mathbf{y}_i^{(l)} \tag{40}$$

where $\mathbf{h}_{j,k}$ represents the periodic delayed copy of $\mathbf{h}_j$, similar to $\mathbf{x}_{j,k}$ in Equations (29) and (30). It is easy to determine that $\mathbf{C}_i$ in Equation (39) is an $N \times N$ positive definite matrix. Therefore, subproblem (38) is an unconstrained quadratic optimization problem, the optimal solution of which is $\mathbf{x}_i^{(l+1)} = \mathbf{x}_i^* = -\mathbf{C}_i^{-1}\mathbf{d}_i/2$, which can be found efficiently via conjugate gradient methods.

3.4. Solving Subproblem $\mathcal{P}_4$

According to Equation (15), subproblem (19) can be separated for each $y_i[n]$, $i = 1, 2, \ldots, GM$, $n = 1, 2, \ldots, N$, as follows:

$$\min_{|y_i[n]|=1} \mathrm{Re}[\overline{u}_{in} y_i[n]] \tag{41}$$

where

$$u_{in} = -\beta_{in}^{(l)} - \theta_3 x_i^{(l+1)}[n] \tag{42}$$

Define $\overline{u}_{in} = |\overline{u}_{in}| \cdot \exp(\mathrm{j}\phi_{in})$. Then, the solution of subproblem (41) can be expressed as $y_i^{(l+1)}[n] = y_i^*[n] = \exp(\mathrm{j} \cdot (\pi - \phi_{in}))$.

4. Numerical Results

In order to demonstrate the effectiveness of Algorithm 1, in this section, a series of numerical simulations is shown. The parameters are initialized as $c = 0.5$, $\theta_1 = \theta_2 = 10$. The parameter is set as $\theta_3 = \max\{10, \min\{2(l+1), 10^5\}\}$, where l is the current iteration. The parameter of the convergence condition is $\eta = 0.5 \times 10^{-3}$. All experiments were implemented in MATLAB 2020a on a PC with one Intel Core i7-7700 CPU and 16 GB RAM. All the obtained correlation function values and PSL and PCL metrics in dB were calculated as $10\log_{10}(|r_{ij}(k)|^2/N^2)$.

4.1. Effect of Weighting Factor w

Figure 1 shows the convergence curves of the variables γ and ε_g, $g = 1, 2, \ldots, G$ with different parameters. The computational complexity per iteration of the proposed algorithm is $\mathcal{O}(MGN^2 + M^2G^2N\log N)$. The algorithm running time with $M = 2$, $G = 2$, and $N = 256$ is around 700 s on average. The convergence curves of the variables γ and ε_g are not monotonically decreasing because the proposed algorithm minimizes the augmented Lagrange function L in Equation (15). When the convergence condition is satisfied, the variables $\left\{\mathbf{x}_i^{(l)}\right\}_{i=1}^{GM}$ and $\left\{\mathbf{h}_i^{(l)}\right\}_{i=1}^{GM}$ are almost the same, and the correlation function $r_{ij}(k)$ satisfies $r_{ij}(k) = \rho_{ij}(k)$. Thus, PCL, PSL_1, $\text{PSL}_2, \ldots$, PSL_G are minimized effectively with the help of the auxiliary variables.

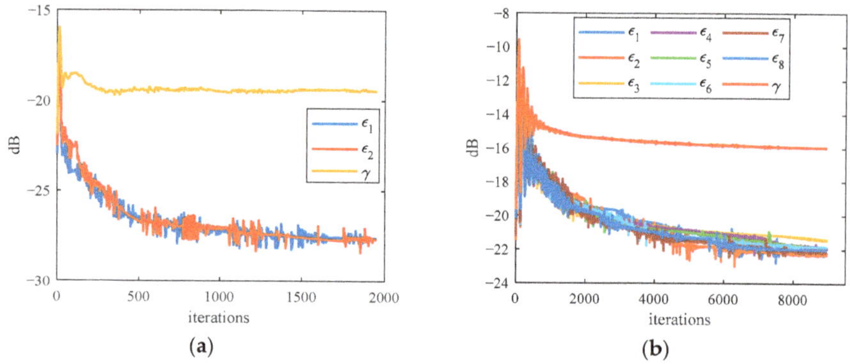

Figure 1. The convergence curves of γ and ε_g, $g = 1,2,\ldots,G$. (**a**) $w = 0.1$, $M = 2$, $G = 2$, $N = 256$; (**b**) $w = 0.7$, $M = 8$, $G = 8$, $N = 256$.

Figure 2a shows the PSL_1, PSL_2, and PCL values obtained with different values of w, with fixed parameters of $M = 2$, $G = 2$, and $N = 256$. Figure 2b shows similar results with $M = 8$, $G = 8$, and $N = 256$. The maximum and minimum values of PSL_1, $\text{PSL}_2, \ldots$, PSL_G are shown, denoted by PSL_{max} and PSL_{min}, respectively. It can be seen that the smaller the value of w, the lower the obtained intragroup PSL values and the higher the obtained intergroup PCL value. The results in Figure 2 demonstrate that the proposed algorithm is able to adjust the weighting factor w to balance the intra- and intergroup correlation performances, which is suitable for different MIMO radar applications.

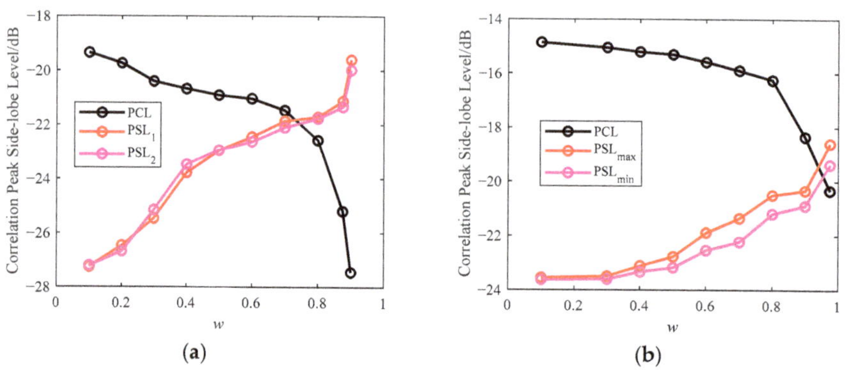

Figure 2. The convergence curves of γ and ε_g, $g = 1,2,\ldots,G$. (**a**) $M = 2$, $G = 2$, $N = 256$; (**b**) $M = 8$, $G = 8$, $N = 256$.

4.2. Obtained Correlation Functions

In his subsection, plots of the intra- and intergroup correlation function curves are detailed. The waveforms obtained by the proposed group orthogonal waveform algorithm are compared to other advanced orthogonal waveforms under different parameters. In Figure 3, random set represents the waveforms generated directly by random numbers. Multi-CAN is one of the best ISL minimization algorithms, and the primal-dual and p-MM algorithms are the current best PSL optimization algorithms. The up- and down-chirp signals have the lowest cross-correlation function when $M = 2$. Therefore, by setting the same time-bandwidth product equal to $N = 256$, the up- and down-chirp signals can be compared to the intragroup orthogonal performance of the waveforms when $M = 2$.

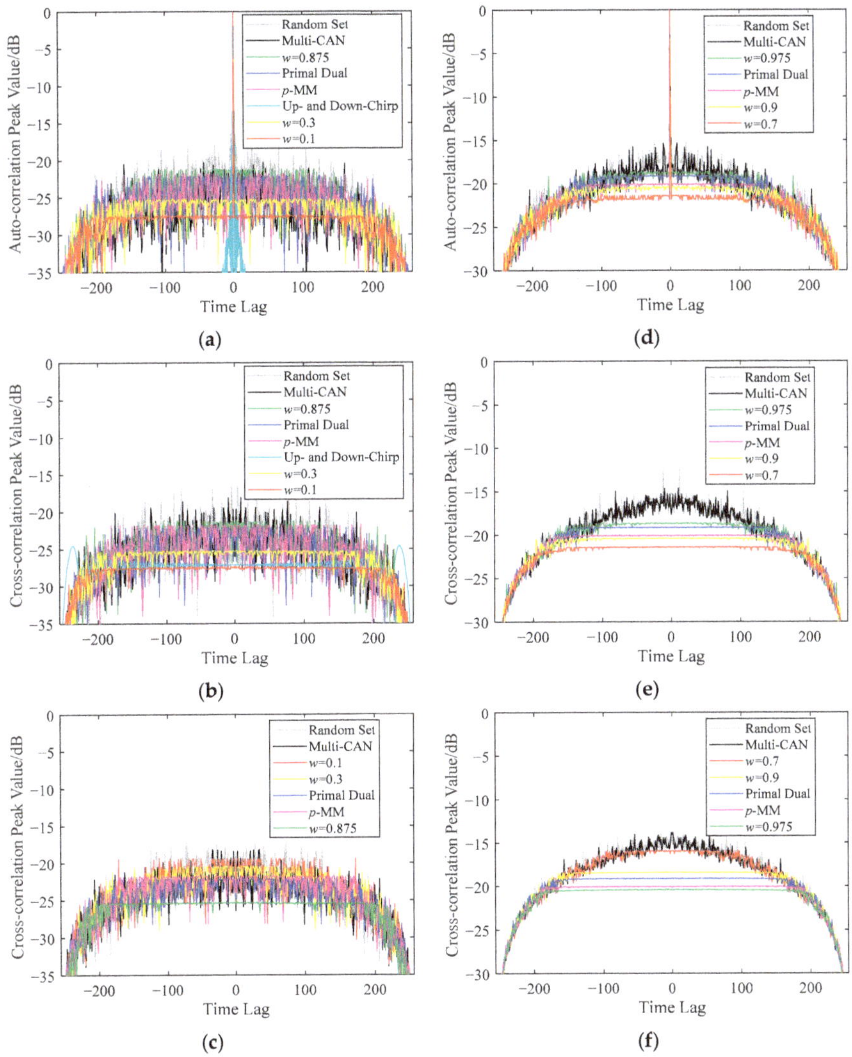

Figure 3. The correlation peak value obtained by the proposed algorithm when $M = 2$, $G = 2$, and $N = 256$: (**a**) autocorrelation; (**b**) intragroup cross-correlation; (**c**) intergroup cross-correlation, and when $M = 8$, $G = 8$, and $N = 256$: (**d**) autocorrelation; (**e**) intragroup cross-correlation; (**f**) intergroup cross-correlation.

Figure 3a shows the autocorrelation function peak value of the *GM* waveforms when $M = 2$, $G = 2$, and $N = 256$. Figure 3b shows the intragroup cross-correlation function peak value, while Figure 3c shows the intergroup cross-correlation function peak value. When $w = 0.1$, the intragroup correlation side-lobe peak value obtained by the proposed algorithm is the lowest in comparison to other waveforms with relatively higher intergroup cross-correlation function peak values. When $w = 0.875$, although the intragroup auto- and cross-correlation functions of the waveforms obtained by the proposed algorithm are relatively high, the intergroup cross-correlation functions are the lowest. The results when $w = 0.3$ lie between those when $w = 0.1$ and $w = 0.875$. Figure 3d–f show the results when $M = 8$, $G = 8$, and $N = 256$.

The results in Figure 3 demonstrate that, with typical parameter values, the proposed group orthogonal waveform design algorithm is able to effectively balance the intra- and intergroup correlation function performances. To briefly summarize, compared with designing all the *GM* waveforms directly, as performed in the primal-dual algorithm, the proposed group orthogonal waveform design algorithm is able to obtain lower intragroup PSL metrics by sacrificing a small portion of the intergroup cross-correlation performance and vice versa. Therefore, for antijamming agile waveform applications, the proposed algorithm is more flexible. The proposed group orthogonal waveform design algorithm is able to design multiple groups of orthogonal waveform sets at the same time.

4.3. Effect of M, N, G Parameters

In Figure 4a,b, the effect of the phase coding sequence length N on the results is analyzed. In the simulations, the weighting factor w and the waveform parameters $M = 2$, $G = 2$ were unchanged. The waveforms were obtained by the proposed algorithm with $N = 32, 64, 96, 128, 160, 192, 224$, and 256. After calculating the PCL and intragroup PSL metrics (the PSL metrics are denoted as PSL_1 and PSL_2 because $G = 2$), Figure 4a,b show the obtained metric values when $w = 0.3$ and $w = 0.875$, respectively. The results demonstrate that the larger the sequence length, the lower the obtained PCL and PSL values.

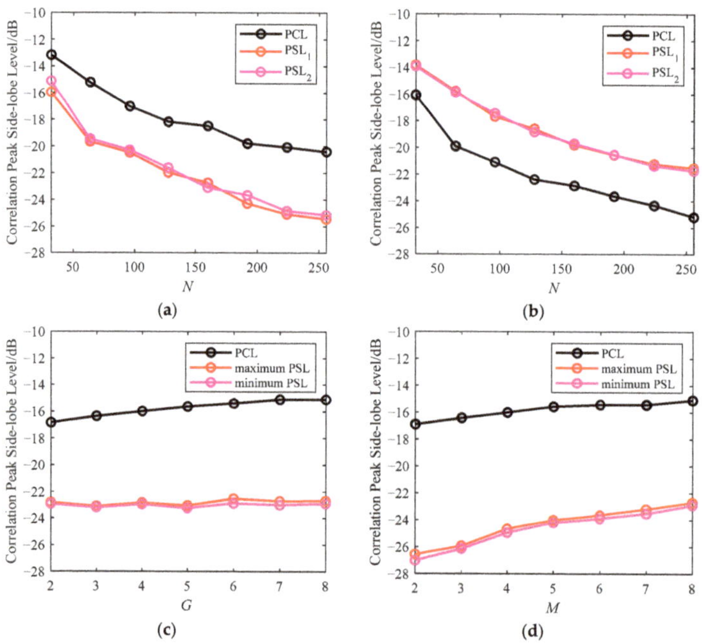

Figure 4. The effect of parameters M, N, and G on the obtained metric values. (**a**) $w = 0.3$, $M = 2$, $G = 2$; (**b**) $w = 0.875$, $M = 2$, $G = 2$; (**c**) $w = 0.5$, $M = 8$, $N = 256$; (**d**) $w = 0.5$, $G = 8$, $N = 256$.

In Figure 4c,d, the effects of the number of groups G and the number of intragroup waveforms M on the results are analyzed. Firstly, the weighting factor was set to $w = 0.5$ and the parameters were set to $M = 8$, $N = 256$. Then, the proposed algorithm was used to obtain the waveforms at $G = 2, 3, 4, 5, 6, 7$, and 8. Finally, the PCL metric value was calculated as well as the maximum and minimum values of $\text{PSL}_1, \text{PSL}_2, \ldots, \text{PSL}_G$. The results are shown in Figure 4c. It can be seen that the PCL increases with an increase in G, because the total number of waveforms, GM, increases. Meanwhile, the obtained intragroup PSL values change little with an increase in G, because the number of intragroup waveforms $M = 8$ remains unchanged. In Figure 4d, the parameter $G = 8$ is unchanged and the proposed algorithm is initiated with $M = 2, 3, 4, 5, 6, 7$, and 8. It can be seen that the intragroup PSL metric values increase with an increase in M. Overall, the PCL metric value mainly depends on the total number of waveforms GM. The intragroup PSL metric values mainly depend on the parameter M.

4.4. Antijamming Simulation

In this section, the jamming suppression performances of MIMO radar using the different waveforms in Table 1 are analyzed. The transmitted phase-coded pulse signal generated using the group orthogonal waveforms $\{\mathbf{x}_i\}_{i=1}^{GM}$ can be expressed as follows:

$$x_i(t) = \exp(\mathrm{j}2\pi\frac{\mathrm{c}}{\lambda}t)\cdot\sum_{n=0}^{N-1} p_\tau(t-n\tau)x_i[n],\ i = 1,2,\ldots,GM,\ p_\tau(t) = \begin{cases} 1, 0 < t \le \tau \\ 0,\ \text{others} \end{cases} \tag{43}$$

where c is the speed of light, τ is the chip length, and λ is the carrier wavelength. There are N_t transmitting elements and N_r receiving elements; the digital signal after matched filtering can be expressed as $\mathbf{y}[n] = (y_1[n], \ldots, y_{N_t N_r}[n])^{\mathrm{T}}$. Then, the digital beam forming output is equal to

$$\mathbf{I}(\theta, n) = \left[\mathbf{a}_r(\theta) \otimes \mathbf{a}_t^{\mathrm{H}}(\theta)\right]^{\mathrm{H}} \mathbf{y}[n] \tag{44}$$

where $\mathbf{I}$ is the angle–range image. $\mathbf{a}_t(\theta) = (1, \exp(-\mathrm{j}\alpha_t), \exp(-\mathrm{j}2\alpha_t), \ldots, \exp(-\mathrm{j}(N_t-1)\alpha_t))^{\mathrm{T}}$ is the transmit steering vector and $\mathbf{a}_r(\theta) = (1, \exp(-\mathrm{j}\alpha_r), \exp(-\mathrm{j}2\alpha_r), \ldots, \exp(-\mathrm{j}(N_r-1)\alpha_r))^{\mathrm{T}}$ is the receive steering vector, with $\alpha_t = 2\pi\sin\theta d_t/\lambda$ and $\alpha_r = 2\pi\sin\theta d_r/\lambda$. The uniform linear array consists of $N_t = 3$ transmit elements spaced at $d_t = 16\lambda$ apart, and $N_r = 16$ receive elements spaced at $d_r = 0.5\lambda$ apart. $\otimes$ represents the Kronecker product and θ represents the direction of arrival. Figure 5 shows a block diagram of the processing in the simulation. The size of range bins is 100 m. The range corresponding to the first bin is 50 km. The angles, ranges, and signal-to-noise ratios of the four true targets are $(-15^\circ, 400, 3\ \text{dB})$, $(0^\circ, 400, 3\ \text{dB})$, $(20^\circ, 400, 3\ \text{dB})$, and $(20^\circ, 50, 5\ \text{dB})$, respectively, where the range parameters' units are range bins. DRFM-based deceptive jamming causes two false targets, whose angles, ranges, and jamming-to-noise ratios are $(-15^\circ, 50, 5\ \text{dB})$ and $(0^\circ, 50, 5\ \text{dB})$, respectively. The noise is Gaussian random white noise.

Table 1. Parameters and performance metrics of simulated waveforms.

Waveform Design Method	Title 2	Title 3
Fixed waveform	$M = 3$, $N = 256$	PSL = −23.22 dB
Multi-CAN [12]	$MG = 6$, $N = 256$	PSL = −17.02 dB
Primal-dual [18]	$MG = 6$, $N = 256$	PSL = −20.18 dB
p-MM [19]	$MG = 6$, $N = 256$	PSL = −20.42 dB
Proposed method when $w = 0.9$	$M = 3$, $G = 2$, $N = 256$	PSL_1 = −19.18 dB
		PSL_2 = −19.39 dB
		PCL = −25.56 dB
Proposed method when $w = 0.1$	$M = 3$, $G = 2$, $N = 256$	PSL_1 = −25.64 dB
		PSL_2 = −25.88 dB
		PCL = −18.51 dB

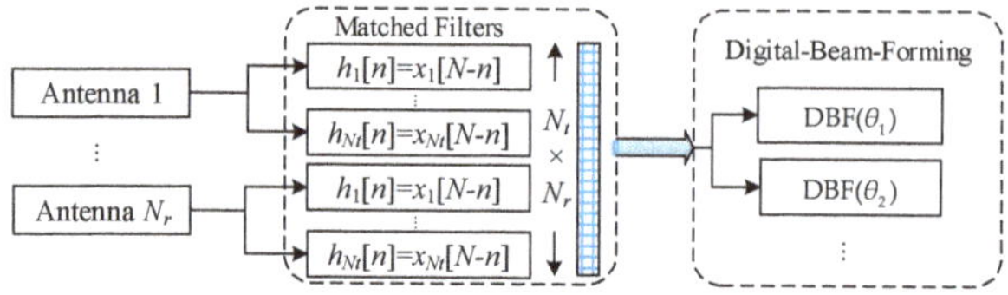

Figure 5. The block diagram of the matched filtering and digital beam forming processes.

Assuming that one pulse of the waveform has been intercepted, Figure 6 shows the angle–range images formed with MIMO radar using adjacent pulse echoes. The results show that the proposed algorithm is able to balance the antijamming and range compression performances by adjusting the weighing factor. On this basis, the waveform diversity gain of MIMO radar can be maximized by selecting a proper waveform set as the best response to flexible adaptive deceptive jamming [26].

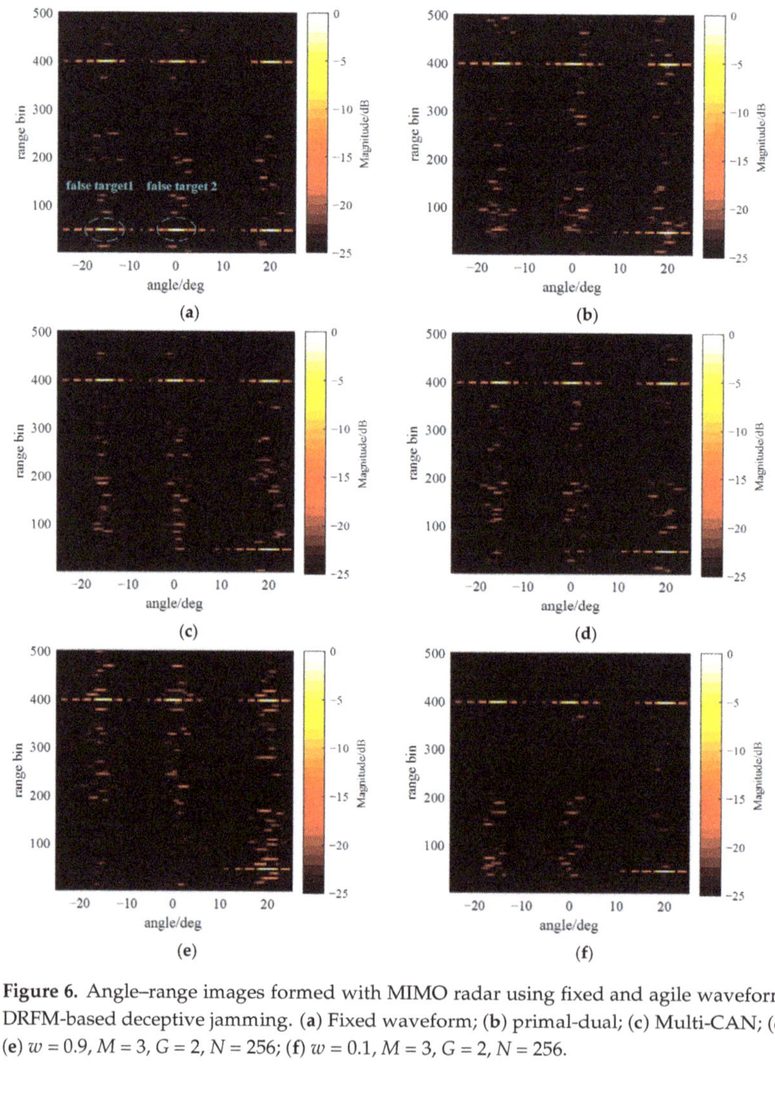

Figure 6. Angle–range images formed with MIMO radar using fixed and agile waveforms under DRFM-based deceptive jamming. (**a**) Fixed waveform; (**b**) primal-dual; (**c**) Multi-CAN; (**d**) p-MM; (**e**) $w = 0.9$, $M = 3$, $G = 2$, $N = 256$; (**f**) $w = 0.1$, $M = 3$, $G = 2$, $N = 256$.

5. Conclusions

Transmitting agile group orthogonal waveforms is an effective way for MIMO radar to combat DRFM jamming. Aiming at balancing the waveform diversity gain and jamming suppression performance for an MIMO radar system, in this article, a group orthogonal waveform optimal design model was proposed. The objective function separates the waveform performances into two parts. The first part is the peak value of intragroup auto- and cross-correlation functions. The other part is the peak value of the intergroup cross-correlation functions. To solve the proposed optimization problem, after proper relaxations, in this article, a group orthogonal waveform design algorithm was proposed, transforming the minimization of the augmented Lagrange function into a series of subproblems. The numerical results showed that the proposed algorithm can minimize the intra- and inter-group correlation functions effectively. The antijamming simulation results showed that the proposed algorithm is able to achieve a balance between the DRFM-based deceptive jamming suppression and range compression performances. The proposed algorithm is flexible and has potential in adaptive antijamming applications for MIMO radars.

Author Contributions: Conceptualization, T.L. and J.S.; methodology, T.L.; software, T.L.; validation, J.S., G.W., X.Y. and Y.Q.; formal analysis, T.L.; investigation, T.L.; resources, J.S.; data curation, T.L.; writing—original draft preparation, T.L.; writing—review and editing, J.S., G.W., X.Y. and Y.Q.; visualization, T.L.; supervision, J.S.; project administration, X.Y.; funding acquisition, J.S. All authors have read and agreed to the published version of the manuscript.

Funding: This work was supported in part by National Natural Science Foundation of China under grant 62131001 and grant 62171029, Key R&D Plan of Zhejiang Province (2023C01148) and the Hangzhou Leading Innovation and Entrepreneurship Team (TD2022006).

Data Availability Statement: The data will be made available by the authors on request.

Conflicts of Interest: Guohua Wang was employed by Hertzwell Pte Ltd. The remaining authors declare that the research was conducted in the absence of any commercial or financial relationships that could be construed as a potential conflict of interest. Hertzwell Pte Ltd. had no role in the design of the study; in the collection, analyses, or interpretation of data; in the writing of the manuscript, or in the decision to publish the results.

References

1. Li, J.; Stoica, P. MIMO Radar with Colocated Antennas. *IEEE Signal Process. Mag.* **2007**, *24*, 106–114. [CrossRef]
2. Fishler, E.; Haimovich, A.; Blum, R.S.; Cimini, L.J.; Chizhik, D.; Valenzuela, R.A. Spatial Diversity in Radars—Models and Detection Performance. *IEEE Trans. Signal Process.* **2006**, *54*, 823–838. [CrossRef]
3. Haimovich, A.M.; Blum, R.S.; Cimini, L.J. MIMO Radar with Widely Separated Antennas. *IEEE Signal Process. Mag.* **2008**, *25*, 116–129. [CrossRef]
4. Bergin, J.; Guerci, J.R. *MIMO Radar: Theory and Application*; Artech House: Boston, MA, USA, 2018; pp. 45–49.
5. Friedlander, B. Waveform Design for MIMO Radars. *IEEE Trans. Aerosp. Electron. Syst.* **2007**, *43*, 1227–1238. [CrossRef]
6. Kajenski, P.J. Design of Low-Sidelobe Phase-Coded Waveforms. *IEEE Trans. Aerosp. Electron. Syst.* **2019**, *55*, 2891–2898. [CrossRef]
7. Hassan, A.K.; Al-Saggaf, U.M.; Moinuddin, M.; Alshoubaki, M.K. Statistical Method Based Waveform Optimization in Collocated MIMO Radar Systems. *Mathematics* **2023**, *11*, 680. [CrossRef]
8. Pace, P. *Detecting and Classifying Low Probability of Intercept Radar*, 2nd ed.; Artech House: Boston, MA, USA, 2008; pp. 358–377.
9. Keel, B.M.; Baden, J.M.; Heath, T.H. A Comprehensive Review of Quasi-Orthogonal Waveforms. In Proceedings of the 2007 IEEE Radar Conference, Waltham, MA, USA, 17–20 April 2007.
10. Liu, H.; Zhou, S.; Zang, H.; Cao, Y. Two Waveform Design Criteria for Colocated MIMO Radar. In Proceedings of the 2014 International Radar Conference, Lille, France, 13–17 October 2014.
11. Gini, F.; Maio, A.D.; Patton, L. *Waveform Design and Diversity for Advanced Radar Systems*; IET: London, UK, 2012; pp. 89–117.
12. He, H.; Stoica, P.; Li, J. Designing Unimodular Sequence Sets with Good Correlations—Including an Application to MIMO Radar. *IEEE Trans. Signal Process.* **2009**, *57*, 4391–4405. [CrossRef]
13. Song, J.; Babu, P.; Palomar, D.P. Sequence Set Design with Good Correlation Properties via Majorization-Minimization. *IEEE Trans. Signal Process.* **2016**, *64*, 2866–2879. [CrossRef]
14. Li, Y.; Vorobyov, S.A. Fast Algorithms for Designing Unimodular Waveform(s) with Good Correlation Properties. *IEEE Trans. Signal Process.* **2018**, *66*, 1197–1212. [CrossRef]
15. Wang, J.; Wang, Y. Designing Unimodular Sequences with Optimized Auto/Cross-Correlation Properties via Consensus-ADMM/PDMM Approaches. *IEEE Trans. Signal Process.* **2021**, *69*, 2987–2999. [CrossRef]

16. Esmaeili-Najafabadi, H.; Ataei, M.; Sabahi, M.F. Designing Sequence with Minimum PSL Using Chebyshev Distance and its Application for Chaotic MIMO Radar Waveform Design. *IEEE Trans. Signal Process.* **2017**, *65*, 690–704. [CrossRef]
17. Alaee-Kerahroodi, M.; Modarres-Hashemi, M.; Naghsh, M.M. Designing Sets of Binary Sequences for MIMO Radar Systems. *IEEE Trans. Signal Process.* **2019**, *67*, 3347–3360. [CrossRef]
18. Lin, Z.; Pu, W.; Luo, Z.Q. Minimax Design of Constant Modulus MIMO Waveforms for Active Sensing. *IEEE Signal Process. Lett.* **2019**, *26*, 1531–1535. [CrossRef]
19. Liu, T.; Sun, J.; Wang, G.; Du, X.; Hu, W. Designing Low Side-Lobe Level-Phase Coded Waveforms for MIMO Radar Using p-Norm Optimization. *IEEE Trans. Aerosp. Electron. Syst.* **2023**, *59*, 3797–3810. [CrossRef]
20. Hu, J.; Wei, Z.; Li, Y.; Li, H.; Wu, J. Designing Unimodular Waveform(s) for MIMO Radar by Deep Learning Method. *IEEE Trans. Aerosp. Electron. Syst.* **2021**, *57*, 1184–1196. [CrossRef]
21. Feng, D.; Xu, L.; Pan, X.; Wang, X. Jamming Wideband Radar Using Interrupted-Sampling Repeater. *IEEE Trans. Aerosp. Electron. Syst.* **2017**, *53*, 1341–1354. [CrossRef]
22. Pace, P. *Developing Digital RF Memories and Transceiver Technologies for Electromagnetic Warfare*; Artech House: Boston, MA, USA, 2022.
23. Akhtar, J. Orthogonal Block Coded ECCM Schemes against Repeat Radar Jammers. *IEEE Trans. Aerosp. Electron. Syst.* **2009**, *45*, 1218–1226. [CrossRef]
24. Zhang, J.; Xu, N. Discrete Phase Coded Sequence Set Design for Waveform-Agile Radar Based on Alternating Direction Method of Multipliers. *IEEE Trans. Aerosp. Electron. Syst.* **2020**, *56*, 4238–4252. [CrossRef]
25. Boyd, S.; Vandenberghe, L. *Convex Optimization*; Cambridge University Press: Cambridge, UK, 2009; pp. 215–271.
26. Panoui, A.; Lambotharan, S.; Chambers, J.A. Game Theoretic Distributed Waveform Design for Multistatic Radar Networks. *IEEE Trans. Aerosp. Electron. Syst.* **2016**, *52*, 1855–1865. [CrossRef]

mathematics

MDPI

Article

UDCO-SAGiMEC: Joint UAV Deployment and Computation Offloading for Space–Air–Ground Integrated Mobile Edge Computing

Yinghao Xu [1,†], Fukang Deng [2,3,†] and Jianshan Zhang [3,4,*]

1 Maynooth University International Engineering College, Fuzhou University, Fuzhou 350108, China; 832103303@fzu.edu.cn
2 College of Computer and Data Science, Fuzhou University, Fuzhou 350108, China; 231020058@fzu.edu.cn
3 Fujian Key Laboratory of Network Computing and Intelligent Information Processing, Fuzhou 350108, China
4 College of Computer and Control Engineering, Minjiang University, Fuzhou 350108, China
* Correspondence: jszhang@mju.edu.cn
† These authors contributed equally to this work.

Abstract: Computation-intensive applications offloading is challenging, especially in the designated regions where communication infrastructure is absent or compromised. In this paper, we present a Space–Air–Ground integrated Mobile Edge Computing (SAGiMEC) system for these regions to provide quality computational services, where the unmanned aerial vehicles (UAVs) act as in-fight edge servers to provide low-latency edge computing and the satellite provides resident cloud computing. A joint optimization problem is formulated considering UAV deployment, ground device (GD) access, and computation offloading to minimize the system average response latency. To cope with the problem's complexity, we propose a Particle Swarm Optimization (PSO) and Greedy Strategy (GS)-based algorithm (PSO&GS) to obtain an approximate optimal solution. Extensive simulations validate the convergence of the proposed algorithm. Numerical results show that the proposed approach has excellent convergence, and the system average response latency is about 0.65x–0.85x that of the benchmark algorithm.

Keywords: space–air–ground integrated; mobile edge computing; UAV deployment; device access; computation offloading

MSC: 68M10

1. Introduction

With the rapid development of the Internet-of-Things (IoT) and 5th-generation (5G) networks, a myriad of practical and promising applications and services have emerged, such as smart homes, high definition (HD) live streaming, and autonomous driving. These applications and services bring great convenience to people's learning, living, and working, benefiting from the ultra-high data rate, low latency, high reliability, and massive connectivity provided by 5G networks [1–3]. However, efficient and reliable communication and a wide range of applications require a lot of computational capabilities [4]. For example, HD live streaming requires real-time processing for video streams, and autonomous driving requires fast handling for Artificial Intelligence (AI) models. These computation-intensive applications severely challenge resource-constrained IoT devices' energy storage and computational capabilities. Mobile Cloud Computing (MCC) can significantly reduce the strain on IoT devices by offloading computation-intensive applications to cloud servers. Still, the excessive transmission distance potentially causes it to fail to meet the latency requirements of these applications. Mobile Edge Computing (MEC) has been widely studied as a prospective approach to address these issues. In the MEC paradigm, ultra-dense network edge devices such as macro/small cell base stations and WiFi access points will be deployed, which can provide exponentially growing edge computational resources for wireless networks [1]. Further, IoT devices can offload computation-intensive applications to nearby edge servers, overcoming the network congestion caused by centralized processing in MCC, thus effectively improving the Quality-of-Service (QoS) for users [5].

However, 5G networks may not provide ubiquitous coverage in regions where communication infrastructure is absent or compromised. The IoT devices deployed in these areas cannot access computational services through the typical edge and cloud computing paradigms. The Space–Air–Ground Integrated Network (SAGIN) architecture enables the offloading of IoT applications in the above scenarios. The SAGIN combines space, air, and ground networks to provide reliable network coverage and flexible computational services to a designated region. It can be applied to many promising fields, such as intelligent transportation systems, remote area monitoring, disaster relief, and large-scale high-speed mobile Internet access [6]. The SAGIN consists of a space segment, an air segment, and a ground segment, which are affected by different constraints. On the one hand, the air network nodes can act as in-flight edge servers to provide low-latency edge computational services for IoT devices. On the other hand, satellite communications, although they may have lower communication rates and higher transmission latency, can provide resident cloud computational services through seamless coverage and satellite backbone networks [7]. However, computation offloading in SAGIN will encounter many problems that must be solved. First, the deployment positions of different in-flight edge servers will result in different wireless channel conditions and edge computational server coverage. The deployment positions of these in-flight edge servers should be carefully considered before performing computation offloading. Second, device access and computation offloading strategy determine the computational resource utilization in the SAGIN. An efficient access scheme and offloading strategy should be designed to improve computational resource utilization.

In this paper, we propose a Space–Air–Ground integrated Mobile Edge Computing (SAGiMEC) system to provide quality computational services for ground devices (GDs). We design an efficient joint optimization approach for unmanned aerial vehicle (UAV) deployment, GD access, and computation offloading that minimizes the system average response latency while considering constraints such as UAV coverage capability and computation task latency requirements. First, we develop a position model to describe the position of each computational node in the proposed SAGiMEC system and define the quality and characteristics of the wireless channel between all computational nodes. Second, we construct a task model to characterize various computation tasks and describe three computing models. Based on the above position, task, and computing model, we formulate the joint optimization problem of UAV deployment, GD access, and computation offloading. Finally, we propose an approach based on particle swarm optimization (PSO) and greedy strategy and perform a simulation. Numerical results show the proposed approach has excellent convergence and system average response latency optimization performance.

The rest of this paper is organized as follows. In Section 2, we present the related work. In Section 3, the system model is introduced and the formulated joint optimization problem of UAV deployment and computation offloading is presented. In Section 4, we propose an approach to solve the formulated problem for the SAGiMEC system. Performance evaluation results are presented in Section 5, followed by the conclusions in Section 6.

2. Related Work

2.1. Mobile Edge Computing

MEC was initially proposed by the European Telecommunications Standards Institute (ETSI) in [8], and it is capable of providing cloud computational services close to users in wireless networks [9]. Computation offloading is a crucial technology for MEC to enhance user experience. The end devices can reduce latency and energy consumption by offloading computation-intensive tasks to edge servers for processing. Therefore, performing efficient computation offloading is vital for MEC systems [10].

According to the performance metrics, the existing works about MEC computation offloading can be broadly classified into three types, i.e., energy-efficient computation offloading, latency-efficient computation offloading, and computational offloading that optimizes the weighted sum of latency and energy consumption [11–13]. In [11], the authors proposed a MEC system that utilizes Non-Orthogonal Multiple Access (NOMA) for computation offloading and minimizes the total energy consumption by jointly optimizing the transmit power, transmission time allocation, and task offloading. However, the system neglects to optimize the latency. Yang et al. [12] studied a MEC system consisting of mobile devices and heterogeneous edge servers supporting various wireless access technologies. They formally defined the optimal offloading node selection strategy as a Markov Decision Process (MDP) based on the heterogeneous edge servers' available bandwidth and the mobile devices' location. The authors also proposed a value iterative algorithm for solving the problem to achieve latency minimization. In [13], the authors proposed an energy-aware computation offloading that can jointly optimize communication and computational resource allocation with limited energy and higher latency sensitivity, achieving a trade-off between energy consumption and latency.

In addition, the task offloading mode is also a key direction for MEC computation offloading. The computation offloading in MEC is divided into two main modes: full and partial offloading. Full offloading refers to the complete offloading of computing tasks from the end device to the server for processing. Full offloading is used in many works as a traditional computation offloading mode. In [14], the authors proposed a multi-user MEC network driven by wireless power transmission in which each energy harvesting terminal follows a full offloading mode. They proposed an optimization algorithm based on the coordinate descent method and an alternating direction multiplier method to solve the proposed optimization problem. Similarly, Huang et al. [15] studied a wireless-powered MEC network using a full offloading strategy. They proposed an online offloading framework based on deep reinforcement learning (DRL) [16] capable of learning full offloading decisions from experience. Partial offloading employs techniques such as code decomposition to divide computation tasks and offload some to remote servers for processing. Partial offloading enables fuller utilization of computational resources in the network than full offloading [17]. In [18], the authors studied the partial offloading problem for a single user. They proposed a multi-user computation offloading framework that supported partial offloading and designed a heuristic algorithm to make offloading decisions dynamically. In [19], You et al. studied the resource allocation for the multi-user MEC partial offloading problem considering the TDMA and OFDMA cases. In [20], Wu et al. studied a joint optimization approach for partial computation offloading and radio resource allocation.

2.2. Space–Air–Ground Integrated Network

The SAGIN architecture can provide ubiquitous network coverage for end devices in communication networks and has received extensive academic attention in recent years. The different SAGIN architectures have been discussed in many works [21–23]. In [21], the authors proposed a Space–Air–Ground–Sea Integrated Network architecture supporting satellites, UAVs, terrestrial base stations, and maritime speedboats for the complexity of marine communication services, which significantly compensates for the lack of maritime communication resources. In [22], the authors proposed an onboard network architecture supporting the SAGIN. They addressed communication security, efficiency, and reliability between the SAGIN and the onboard network. In [23], the authors studied a civil aircraft-enhanced SAGIN and proposed a fair optimization strategy, including resource allocation and auction.

The combination of MEC and SAGIN is expected to break through the various limitations of traditional ground communication and further improve the flexibility of computational and communication resources. It has received continuous attention in recent years. In [24], the authors studied the computation offloading problem in the SAGIN, where multiple end devices collaborate to utilize computational resources. The authors formulated an optimization problem for minimizing the total system latency considering the dynamic nature of tasks, the mobility of UAVs, and the variability of end devices, and expressed it as an MDP. However, only one UAV is used as an edge server in this work, and the proposed model may not apply to scenarios with larger user sizes, considering the limited computational resources of the UAV. In [25,26], the authors studied an IoT system supporting a single satellite, a single UAV, and multiple GDs, where the UAV acts as an edge server capable of collecting computation tasks from GDs during flight and offloading them to a base station or satellite for processing. In addition, the authors adopted a full offloading strategy. At the same time, they ignored the computational capabilities of the end devices, thus failing to fully utilize the computational resources of various end devices within the system. Further, the joint optimization of UAV deployment and computation offloading is one of the research priorities of the SAGIN. Tang et al. [27] proposed a reinforcement learning-based traffic offloading strategy by considering the high mobility of nodes and the frequently changing network traffic and link states. However, such optimization algorithms must provide extensive data for training the model. In [28], the authors proposed a joint optimization algorithm based on the block coordinate descent method that can quickly solve the problem of UAV deployment, task offloading, and resource allocation in a SAGIN. However, the complexity of such algorithms based on convex optimization theory grows exponentially with the size of the problem and is, therefore, more suitable for small-scale scenarios.

The comparative analysis of the previous work has been illustrated in Table 1. In summary, although some work has been carried out to investigate the critical technologies of the SAGiMEC, most of it only considers supporting a single UAV. They thus cannot be applied to practical and complex scenarios. In addition, to simplify the model, most of the work adopts a full offloading strategy, resulting in low resource utilization for all types of computational nodes in the system.

Table 1. The comparative analysis of different work ("+": involved; "-": not involved).

Reference	Off-Mode		UAV		Constraint		Object		
	All	Partial	Single	Multiple	Energy	Deadline	Time	Energy	Other
Our work	-	+	-	+	-	+	+	-	-
[24]	-	+	+	-	+	-	-	+	-
[25]	-	+	+	-	+	-	-	+	-
[26]	-	+	+	-	-	+	-	-	+
[27]	-	+	+	-	+	-	+	-	+
[28]	+	-	+	-	-	+	-	+	+

3. System Model and Problem Formulation

In this section, we describe the SAGiMEC system and analyze the system average response latency. We also formulated the joint optimization problem of UAV deployment and computation offloading to minimize the system average response latency. For ease of reference, the key notations used throughout this paper are listed in Table 2.

Table 2. Summary of Key Notations.

Notation	Definition
$\mathcal{M} = \{1, 2, \ldots, M\}$	The set of the GDs in the ground layer
$\mathcal{N} = \{1, 2, \ldots, N\}$	The set of the UAVs in the air layer
$\boldsymbol{u}_m^{\mathrm{GD}} = (x_m^{\mathrm{GD}}, y_m^{\mathrm{GD}}, 0), \forall m \in \mathcal{M}$	The coordinate of GD $m \in \mathcal{M}$
$\boldsymbol{u}_n^{\mathrm{UAV}} = (x_n^{\mathrm{UAV}}, y_n^{\mathrm{UAV}}, H), \forall n \in \mathcal{N}$	The coordinate of UAV $n \in \mathcal{N}$
$I_m, \forall m \in \mathcal{M}$	The computation task of GD $m \in \mathcal{M}$
$D_m, \forall m \in \mathcal{M}$	The computational task-input data size of I_m

Table 2. *Cont.*

Notation	Definition
$S_m, \forall m \in \mathcal{M}$	The computational complexity of I_m
$O_m, \forall m \in \mathcal{M}$	The computation result size of I_m
$E_m, \forall m \in \mathcal{M}$	The maximum tolerable latency of I_m
$\alpha_m^{\mathrm{GD}}, \forall m \in \mathcal{M}$	The local processing ratio of D_m
$\alpha_m^{\mathrm{UAV}}, \forall m \in \mathcal{M}$	The ratio of D_m offloaded to the UAVs
$\alpha_m^{\mathrm{LEO}}, \forall m \in \mathcal{M}$	The ratio of D_m offloaded to the LEO satellite
$f_m^{\mathrm{GD}}, \forall m \in \mathcal{M}$	The CPU-cycle frequency of GD $m \in \mathcal{M}$
$T_m^{\mathrm{GD}}, \forall m \in \mathcal{M}$	The local processing latency of I_m
$h_{m,n}, \forall m \in \mathcal{M}, n \in \mathcal{N}$	The wireless channel gain between GD m and UAV n
$d_{m,n}, \forall m \in \mathcal{M}, n \in \mathcal{N}$	The spatial distance between GD $m \in \mathcal{M}$ and UAV $n \in \mathcal{N}$
h_0	The wireless channel gain at a reference distance
$R_{m,n}, \forall m \in \mathcal{M}, n \in \mathcal{N}$	The wireless channel transmission rate between GD $m \in \mathcal{M}$ and UAV $n \in \mathcal{N}$
B	The bandwidth between the GDs and the UAVs
$P_m, \forall m \in \mathcal{M}$	The transmission power of GD $m \in \mathcal{M}$
σ^2	The additive white Gaussian noise
$\boldsymbol{\beta} = \{\beta_{m,n} \| m \in \mathcal{M}, n \in \mathcal{N}\}$	The access between the GDs and the UAVs
$\Lambda_{\max}$	The maximum parallel tasks number of UAVs
$L_{\max}$	The maximum service distance of the UAVs
$T_m^{\mathrm{UAV}}, \forall m \in \mathcal{M}$	The edge processing latency of I_m
$f_{m,n}^{\mathrm{UAV}}, \forall m \in \mathcal{M}, n \in \mathcal{N}$	The CPU-cycle frequency of UAV $n \in \mathcal{N}$ when it computes the input data of I_m
R_s	The transmission rate between the GDs and the LEO satellite
$f_m^{\mathrm{LEO}}, \forall m \in \mathcal{M}$	The CPU-cycle frequency of the LEO satellite when it computes the input data of I_m
$T_m^{\mathrm{LEO}}, \forall m \in \mathcal{M}$	The cloud processing latency of I_m
$T_m, \forall m \in \mathcal{M}$	The processing delay of I_m
$T(\boldsymbol{u}^{\mathrm{UAV}}, \boldsymbol{\alpha}, \boldsymbol{\beta})$	The system average response latency

3.1. System Model

As shown in Figure 1, a SAGiMEC system consists of three layers: the space layer with the LEO satellite, the air layer with N UAVs, and the ground layer with M GDs. We denote the set of the GDs and the UAVs as $\mathcal{M} = \{1, \ldots, M\}$ and $\mathcal{N} = \{1, \ldots, N\}$, respectively. In the ground layer, GDs with limited computational capabilities are distributed in areas where communication equipment is lacking or damaged, such as remote mountainous and disaster areas, and process computation-intensive tasks. To enable the GDs to process computation tasks satisfactorily, the UAVs in the air layer act as edge computing servers to provide computational services to GDs. In addition, the LEO satellite in the space layer can provide centralized cloud computational services to GDs within its coverage area. Due to the limited computational capabilities of the GDs, they can offload some of their computation tasks to UAVs or LEO satellites to ease the pressure on computational capabilities.

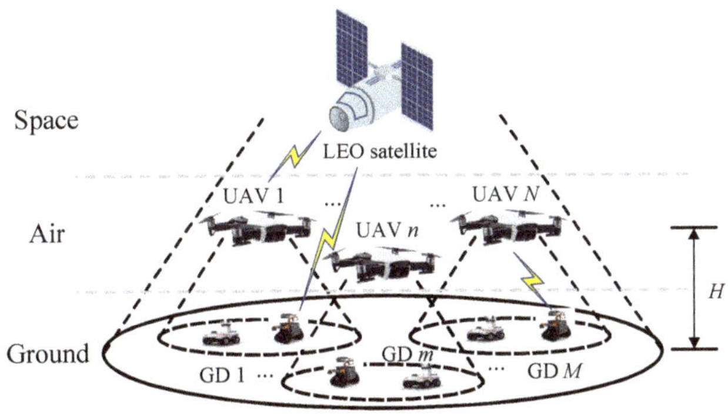

Figure 1. The SAGiMEC system.

3.1.1. Position Model

A three-dimensional (3D) Euclidean coordinate system is adopted without loss of generality, whose coordinates are measured in meters. Specifically, the coordinates of GD $m \in \mathcal{M}$ and UAV $n \in \mathcal{N}$ can be presented as $\boldsymbol{u}_m^{\mathrm{GD}} = (x_m^{\mathrm{GD}}, y_m^{\mathrm{GD}}, 0)$ and $\boldsymbol{u}_n^{\mathrm{UAV}} = (x_n^{\mathrm{UAV}}, y_n^{\mathrm{UAV}}, H)$, respectively, where x_m^{GD} and y_m^{GD} are the horizontal positions of GD $m \in \mathcal{M}$. x_n^{UAV} and y_n^{UAV} are the horizontal positions of UAV $n \in \mathcal{N}$. H is the flying altitude of the UAVs. We assume that the UAVs hover at a fixed height $H > 0$ during the computation tasks processing. $H > 0$ is assumed to be the minimum height appropriate to the work terrain and can avoid obstructions without frequent descending and ascending.

3.1.2. Task Model

In the considered time slot, the computation task of GD $m \in \mathcal{M}$ is denoted as a positive tuple $< D_m, S_m, O_m, E_m, >$, where D_m represents the size of the computational task-input data in bits (e.g., the length of the encoded program codes and input parameters); S_m is the amount of required computational resource for computing 1-bit of input data (i.e., the required number of CPU cycles); O_m is the size of the computation result; E_m is the maximum tolerable latency for I_m in seconds. The characteristics of computation tasks, i.e., the size of the computational task-input data, the computational complexity, the size of the computation result, and the maximum tolerable delay, are determined by the GD. Moreover, to ensure the efficiency of the GDs, a computation task processing latency exceeding its maximum tolerable delay is not allowed.

Due to the limited computational capabilities of the GDs, the computation-intensive tasks may not be processed before the maximum tolerable latency and, therefore, need to be offloaded to the UAVs or the LEO satellite. We adopt a partial offloading strategy to fully utilize the computational capabilities in the ground, air, and space layers. In the offloading strategy, the GDs can process part of the tasks locally and also offload some tasks to the UAV or LEO satellite. Let α_m^{GD}, α_m^{UAV}, and α_m^{LEO} denote the percentage of computation task I_m locally processed, offloaded to the UAV, and offloaded to the LEO satellite, respectively. We assume that computation task partitioning does not generate additional computational loads [29], i.e., it is constrained by

$$\alpha_m^{\mathrm{GD}} + \alpha_m^{\mathrm{UAV}} + \alpha_m^{\mathrm{LEO}} = 1, \forall m \in \mathcal{M}. \tag{1}$$

3.1.3. Computing Model

Depending on the processing location of the computation tasks, we define three different computing modes, i.e., local computing mode for local processing, edge computing mode for UAV processing, and cloud computing mode for LEO satellite processing. These three computing modes use orthogonal computational resources to proceed in parallel.

Local Computing Mode

In the local computing mode, based on the computational task-input data size and the local processing ratio, the input data size for local processing can be quantified as $\alpha_m^{\mathrm{GD}} D_m, \forall m \in \mathcal{M}$. We define f_m^{GD} as the CPU-cycle frequency of GD $m \in \mathcal{M}$. The local processing latency of computation task I_m is quantified as [26,27,30]

$$T_m^{\mathrm{GD}} = \frac{\alpha_m^{\mathrm{GD}} D_m S_m}{f_m^{\mathrm{GD}}}, \forall m \in \mathcal{M}. \tag{2}$$

Edge Computing Mode

In the edge computing mode, the latency for processing the computation task consists of three parts: (1) the communication latency for transmitting the computation task-input data from the GD to the UAV; (2) the processing latency for the computation task-input data at the UAV; (3) the communication delay for transmitting the computation result from the UAV to the GD.

The wireless channel between GD $m \in \mathcal{M}$ and UAV $n \in \mathcal{N}$ is assumed to be dominated by Line-of-Sight (LoS) links regarding recent field experiments by Qualcomm [31]. Therefore, the wireless channel gain between GD $m \in \mathcal{M}$ and UAV $n \in \mathcal{N}$ is quantified as

$$h_{m,n} = d_{m,n}^{-2} h_0, \forall m \in \mathcal{M}, n \in \mathcal{N}, \tag{3}$$

where $d_{m,n} = \sqrt{\|\boldsymbol{u}_m^{\text{GD}} - \boldsymbol{u}_n^{\text{UAV}}\|^2}, \forall m \in \mathcal{M}, n \in \mathcal{N}$ is the spatial distance between GD $m \in \mathcal{M}$ and UAV $n \in \mathcal{N}$; h_0 is the channel power gain at a reference distance $d_0 = 1$ m. The wireless channel transmission rate between GD $m \in \mathcal{M}$ and UAV $n \in \mathcal{N}$ can be quantified as [26,27,30]

$$R_{m,n} = B \log\left(1 + \frac{P_m h_{m,n}}{\sigma^2}\right), \forall m \in \mathcal{M}, n \in \mathcal{N}, \tag{4}$$

where P_m is the transmission power of GD $m \in \mathcal{M}$, σ^2 is additive white Gaussian noise, and B is the wireless channel bandwidth. Similar to [1], we assume that the UAVs are allocated equal bandwidth for the GDs by frequency division multiple access (FDMA).

We define $\beta = \{\beta_{m,n} | m \in \mathcal{M}, n \in \mathcal{N}\}$ as the access between the GDs and the UAVs, where $\beta_{m,n}$ denotes the access between GD $m \in \mathcal{M}$ and UAV $n \in \mathcal{N}$, when GD $m \in \mathcal{M}$ offloads the computation task to UAV $n \in \mathcal{N}$, $\beta_{m,n} = 1$, otherwise $\beta_{m,n} = 0$. We assume that each GD only offloads computation task to one UAV for processing, then we have the following constraint:

$$\sum_{n=1}^{N} \beta_{m,n} \leq 1, \forall m \in \mathcal{M}. \tag{5}$$

In addition, due to the UAVs' limited parallel computational capabilities, we assume that the number of computation tasks received by each UAV in the considered time slot cannot exceed the maximum number of parallel tasks $\Lambda_{\max}$. Thus, we have another constraint:

$$\sum_{m=1}^{M} \beta_{m,n} \leq \Lambda_{\max}, \forall n \in \mathcal{N}. \tag{6}$$

To ensure QoS, we limit the UAV to providing edge computing services only for the GDs within a certain spatial distance, i.e.,

$$\beta_{m,n} d_{m,n} \leq L_{\max}, \forall m \in \mathcal{M}, n \in \mathcal{N}, \tag{7}$$

where $L_{\max}$ is the maximum service distance of the UAVs.

The computation result size tends to be much smaller than the computational task-input data size, and the communication latency generated by its transmission from the UAV to the GD is small, so we ignore the communication latency generated by this process. Ignoring the back-propagation process of the computation result can significantly reduce the model complexity, which has been adopted in many works [30,32,33]. Therefore, the edge processing latency of computation task I_m is quantified as

$$T_m^{\text{UAV}} = \sum_{n=1}^{N} \beta_{m,n} \left(\frac{\alpha_m^{\text{UAV}} D_m}{R_{m,n}} + \frac{\alpha_m^{\text{UAV}} D_m S_m}{f_{m,n}^{\text{UAV}}} \right), \forall m \in \mathcal{M}, \tag{8}$$

where $f_{m,n}^{\text{UAV}}$ is the CPU-cycle frequency of UAV $n \in \mathcal{N}$ when it computes the input data of I_m.

Cloud Computing Mode

In the cloud computing model, the GDs adopt ground-space transmission technology [1] to offload some computation tasks to the LEO satellite for processing. Referring to [1], we denote the transmission rate between the GDs and the LEO satellite by R_s. Due to the long spatial distance between the GDs and the LEO satellite, R_s is usually smaller than the transmission rate between the GDs and the UAVs, i.e., $R_s < R_{m,n}, \forall m \in \mathcal{M}, n \in \mathcal{N}$. We define $f_{m,n}^{\text{LEO}}$ as the CPU-cycle frequency of the LEO satellite when it computes the input data of I_m. The cloud processing latency of computation task I_m is quantified as

$$T_m^{\text{LEO}} = \frac{\alpha_m^{\text{LEO}} D_m}{R_s} + \frac{\alpha_m^{\text{LEO}} D_m S_m}{f_{m,n}^{\text{LEO}}}, \forall m \in \mathcal{M}. \tag{9}$$

Since the local, edge, and cloud computing modes can be performed in parallel, the processing latency of I_m is quantified as

$$T_m = \max(T_m^{\text{GD}}, T_m^{\text{UAV}}, T_m^{\text{LEO}}), \forall m \in \mathcal{M}. \tag{10}$$

The system average response latency is defined as the average processing delay of all computation tasks, as described in the following:

$$T(\boldsymbol{u}^{\text{UAV}}, \boldsymbol{\alpha}, \boldsymbol{\beta}) = \frac{1}{M} \sum_{m=1}^{M} T_m. \tag{11}$$

3.2. Problem Formulation

We expect to minimize the system average response latency of the SAGiMEC system by jointly optimizing the UAV deployment, the computation offloading, and the GD access for a given set of system parameters. The above joint optimization problem can be formulated as

$$\mathbf{P}: \min_{\boldsymbol{u}^{\text{UAV}}, \boldsymbol{\alpha}, \boldsymbol{\beta}} \quad T(\boldsymbol{u}^{\text{UAV}}, \boldsymbol{\alpha}, \boldsymbol{\beta}) \tag{12}$$

$$\text{s.t.} \quad T_m \le E_m, \forall m \in \mathcal{M} \tag{12a}$$

$$\alpha_m^{\text{GD}}, \alpha_m^{\text{UAV}}, \alpha_m^{\text{LEO}} \in [0,1], \forall m \in \mathcal{M} \tag{12b}$$

$$\alpha_m^{\text{GD}} + \alpha_m^{\text{UAV}} + \alpha_m^{\text{LEO}} = 1, \forall m \in \mathcal{M} \tag{12c}$$

$$\beta_{m,n} \in [0,1], \forall m \in \mathcal{M}, n \in \mathcal{N} \tag{12d}$$

$$\sum\nolimits_{n=1}^{N} \beta_{m,n} \le 1, \forall m \in \mathcal{M} \tag{12e}$$

$$\sum\nolimits_{m=1}^{M} \beta_{m,n} \le \Lambda_{\max}, \forall n \in \mathcal{N} \tag{12f}$$

$$\beta_{m,n} d_{m,n} \le L_{\max}, \forall m \in \mathcal{M}, n \in \mathcal{N} \tag{12g}$$

where (12a) states that the computation task must be completed before its maximum tolerable latency; (12b) and (12c) define the range of computation task offloading ratio; (12d)–(12f) denote the constraints for the access indicator between the GDs and the UAVs; and (12g) states the each UAV provides edge computing services only for the GDs within a certain spatial distance.

4. Approach

In this section, we first analyze the characteristics of the joint optimization problem, and then describe the proposed approach.

4.1. Problem Analysis

The joint optimization problem is a complicated mixed integer non-linear programming (MINLP) because of the non-convexity of the objective function, the non-linearity of the optimization variables, and the non-linear couplings exist among the optimization variables. To address these issues, we propose a Particle Swarm Optimization (PSO) and Greedy Strategy (GS)-based algorithm (PSO&GS) to obtain an approximate optimal solution for the joint optimization problem. In solving the optimization problem, traditional PSO does not require the gradient information of the objective function and has no constraints on the continuity and derivability of the objective function. It has the potential of solving the joint optimization problem. However, traditional PSO has the following issues [34,35]:

- The joint optimization problem involves both continuous variables $(\boldsymbol{u}^{\text{UAV}}, \boldsymbol{\alpha})$ and discrete variables $\boldsymbol{\beta}$. It is a typical mixed decision variable optimization problem which is difficult to be solved directly by traditional PSO.
- In the joint optimization problem, there is the non-linear couplings among the optimization variables. On the one hand, since the UAVs can only provide edge computing services to the GDs within their respective coverage areas, the access between the GDs and the UAVs depends on the UAV deployment. On the other hand, the UAV deployment needs to be adjusted according to the computation task offloading to get the best system performance.
- Traditional PSO is prone to fall into local optimality and cannot obtain optimal or near-optimal solutions to the joint optimization problem.

To sum up, it is inefficient to use traditional PSO to solve the joint optimization problem. Therefore, we improved the traditional PSO by introducing the update operator of the genetic algorithm (GA) and incorporated the idea of GS to make it have the ability to solve the joint optimization problem efficiently.

4.2. Algorithm Design

We propose an inner and outer double-layer nested joint optimization approach by integrating the traditional PSO, GA, and greedy strategy to solve the issues arising from the traditional PSO in solving the joint optimization problem XX. During each iteration of the proposed joint optimization approach, the outer layer achieves UAV deployment optimization by the PSO-GA, while the inner layer realizes GD access and computation offloading optimization by the greedy strategy. The details of the algorithm are as follows.

4.2.1. Outer: UAVs Deployment Optimization

The first sub-problem of problem $\mathbf{P}$, denoted $\mathbf{P}_1$, is to solve for the optimal UAV deployment $\boldsymbol{u}^{\text{UAV}*}$ with given GD access $\boldsymbol{\alpha}^*$ and computation offloading $\boldsymbol{\beta}^*$. $\mathbf{P}_1$ can be formulated as follows:

$$\mathbf{P}_1 : \min_{\boldsymbol{u}^{\text{UAV}}} \quad T(\boldsymbol{u}^{\text{UAV}}, \boldsymbol{\alpha}^*, \boldsymbol{\beta}^*) \tag{13}$$

$$\text{s.t.} \quad T_m \leq E_m, \forall m \in \mathcal{M} \tag{13a}$$

To solve $\mathbf{P}_1$, we propose PSO-GA, which is based on traditional PSO and introduces the crossover and mutation operators of GA in the particle update process. While inheriting the advantages of easy implementation and rapid convergence of PSO, PSO-GA incorporates the characteristics of GA with excellent global search capability, which can effectively overcome the shortcomings of traditional PSO, which is prone to fall into local optimum. Then, PSO-GA is described in detail.

Problem Encoding

As PSO-GA is an approach based on the swarm intelligence algorithm, the particle encoding method will directly affect the algorithm's efficiency and the quality of the solution. Obviously, traditional particle encoding methods, such as binary and integer encoding, are difficult to represent feasible solutions for problem $\mathbf{P}_1$. Therefore, we adopt the following particle encoding method. Let $\mathcal{K} = \{1, \ldots, K\}$ denote K particles contained in the population, where each particle denotes a UAV deployment scheme. Each particle is encoded as a set containing N two-dimensional vectors, where each two-dimensional vector represents a UAV's deployment coordinates at the current time slot. Further, the T

iterations undergone by the PSO-GA are denoted by $\mathcal{T} = \{1, \ldots, T\}$. Specifically, particle $k \in \mathcal{K}$ at $t(\in \mathcal{T})$-th iteration can be denoted as

$$U_k^t = (u_{k,1}^t, u_{k,2}^t, \ldots, u_{k,N}^t), \forall t \in \mathcal{T}, k \in \mathcal{K}, \tag{14}$$

where $u_{k,n}^t = (x_{k,n}^t, y_{k,n}^t)$ denotes the deployment coordinate of UAV $n \in \mathcal{N}$ in the UAV deployment corresponding to particle $k \in \mathcal{K}$.

Fitness Function

The fitness function judges the superiority or inferiority of the two particles participating in the comparison. We utilize the system average response latency function defined in Equation (11) as the fitness function of PSO-GA. Therefore, the fitness value of a particle can be obtained by the UAV deployment corresponding to the particle and GD access and computation offloading obtained by the greedy strategy, which will be described in Section 4.2.2. Since the objective of problem $\mathbf{P}_1$ is to minimize the system average response latency, the particle with the smaller fitness value has the better quality.

Population Update

For traditional PSO, each particle moves in a specific direction and velocity in the solution space. During the iteration process, each particle is updated by its personal best state and the current global best state. The particle's velocity and position are updated by

$$V_k^{t+1} = wV_k^t + c_1 r_1 (pBest_k - U_k^t) + c_2 r_2 (gBest - U_k^t), \forall t \in \mathcal{T}, k \in \mathcal{K}, \tag{15}$$

and

$$U_k^{t+1} = V_k^{t+1} + U_k^t, \forall t \in \mathcal{T}, k \in \mathcal{K}, \tag{16}$$

respectively. V_k^t and U_k^t are the velocity and position for particle k in the t-th iteration. $pBest_k$ and $gBest$ are the personal best for particle k and the global best for the population after t iterations. w is the inertia factor, which determines the effect of the current velocity on the velocity during the next iteration. c_1 and c_2 are the individual and social learning factors, respectively, reflecting the ability of the particles to learn the personal and global best. r_1 and r_2 are two random numbers that add randomness to the iterative search process.

PSO-GA improves the particle update process of traditional PSO by introducing crossover and mutation operators of GA, and the update of particle k is modified to

$$U_k^{t+1} = c_2 \oplus C_g(c_1 \oplus C_p(w \oplus M_u(U_k^t), pBest_k), gBest), \forall t \in \mathcal{T}, k \in \mathcal{K}, \tag{17}$$

where $M_u()$ is the mutation operator; $C_p()$ and $C_g()$ are the crossover operator.

In PSO-GA, the particle inertia part follows

$$\begin{aligned} A_k^{t+1} &= w \oplus M_u(U_k^t) \\ &= \begin{cases} M_u(U_k^t), & r_1 < w \\ U_k^t, & r_1 \geq w \end{cases}, \forall t \in \mathcal{T}, k \in \mathcal{K}, \end{aligned} \tag{18}$$

where $r_1 \in [0, 1]$. $M_u(U_k^t)$ means a randomly selected encoding position on particle k for random mutation. Figure 2 shows the mutation operator in the inertial part. In this example, the encoding position mp_1 is selected, and $(11, 162)$ replaces $(72, 27)$ to get the new particle.

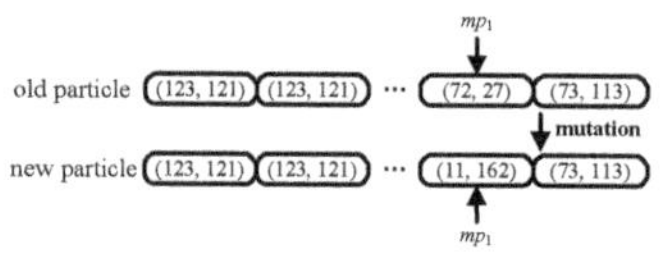

Figure 2. Mutation operator.

Further, the individual and social learning parts of PSO-GA are combined with the crossover operator of GA. They are updated by

$$B_k^{t+1} = w \oplus C_p\left(A_k^{t+1}, pBest_k\right) = \begin{cases} C_p\left(A_k^{t+1}, pBest_k\right), & r_2 < c_1 \\ A_k^{t+1}, & r_2 \geq c_1 \end{cases}, \forall t \in \mathcal{T}, k \in \mathcal{K}, \quad (19)$$

and

$$C_k^{t+1} = w \oplus C_g\left(B_k^{t+1}, gBest\right) = \begin{cases} C_g\left(B_k^{t+1}, gBest\right), & r_3 < c_2 \\ B_k^{t+1}, & r_3 \geq c_2 \end{cases}, \forall t \in \mathcal{T}, k \in \mathcal{K}, \quad (20)$$

respectively, where $r_2, r_3 \in [0,1]$. The crossover operator randomly picks the encoding fragment between two encoding positions and replaces it with the encoding fragment at the same position in $pBest_k$ or $gBest$. Figure 3 illustrates the crossover operator in the individual or social learning part. In this example, the crossover operator picks the encoding fragment between cp_1 and cp_2 and replaces it with the encoding fragment at the same position of $pBest_k$ or $gBest$. Finally, the new particle is obtained.

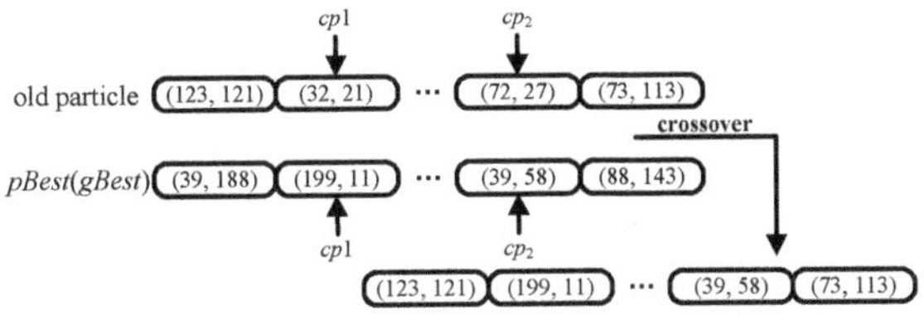

Figure 3. Crossover operator.

Parameter Settings

The inertia weight factor for traditional PSO can affect the convergence velocity and search capability. The algorithm has a strong global search capability when the inertia weight factor is large, and the probability of particle mutation is higher. Conversely, the probability of particle mutation is smaller, and the algorithm has a powerful local search capability. In the early and late stages of algorithm execution, we need to focus more on search diversity and local search capability. Therefore, we adopt a linear adjustment strategy for the inertia weight factor. In this strategy, the inertia weight factor decreases linearly with the increase in the number of iterations. The adjustment strategy can be defined as

$$w = w_{\max} - iters_{\text{cur}} \times \frac{w_{\max} - w_{\min}}{T}, \quad (21)$$

where $w_{\max}$ and $w_{\min}$ are the maximum and minimum values of the inertia weight factor, respectively, and $iters_{\text{cur}}$ is the current number of iterations.

Similarly, we apply a linear adjustment strategy to the individual and social learning factors in Equation (17), as follows

$$c_1 = c_1^{\text{start}} - iters_{\text{cur}} \times \frac{c_1^{\text{start}} - c_1^{\text{end}}}{T}, \quad (22)$$

and

$$c_2 = c_2^{\text{start}} + iters_{\text{cur}} \times \frac{c_2^{\text{end}} - c_2^{\text{start}}}{T}, \quad (23)$$

where c_1^{start} and c_2^{start} indicate the initial values of c_1 and c_2, and c_1^{end} and c_2^{end} indicate the final values of c_1 and c_2.

Algorithm Flowchart

The main steps of the PSO-GA are given in Algorithm 1.

Algorithm 1 PSO-GA

Require: $\boldsymbol{u}^{\mathrm{GD}}, \boldsymbol{I}, \boldsymbol{\alpha}, \boldsymbol{\beta}$
Ensure: $\boldsymbol{u}^{\mathrm{UAV}}$
1: Initialize the parameter settings of PSO-GA and generate the population;
2: **for** each $k \in \mathcal{K}$ **do**
3: Call Algorithm 2 to obtain the computation offloading for particle k;
4: Calculate the fitness of particle k according to Equation (11);
5: Initialize the personal best $pBest_k$ of particle k;
6: **end for**
7: Initialize the global best particle of the population;
8: $t = 1$;
9: **while** $t < T$ **do**
10: **for** each $k \in \mathcal{K}$ **do**
11: Update particle k according to Equation (17);
12: Call Algorithm 2 to obtain the computation offloading for particle k;
13: Calculate the fitness of particle k according to Equation (11);
14: **if** $Fitness(k) < Fitness(pBest_k)$ **then**
15: Update $pBest_k$;
16: **if** $Fitness(k) < Fitness(gBest)$ **then**
17: Update $gBest$;
18: **end if**
19: **end if**
20: **end for**
21: **end while**
22: Mapping $gBest$ to UAV deployment;
23: **return** $\boldsymbol{u}^{\mathrm{UAV}}$

The algorithm initializes the values of relevant parameters such as population size K, maximum number of iterations T, inertia weight factor w, and the maximum and minimum values of the cognitive factors. It then randomly generates the initial population (Line 1). For each particle in the initialized population, Algorithm 2 is called to get the computation offloading for particle k (Line 3) and calculate the fitness of each particle according to Equation (11) (Line 4). The initial state of each particle is set to its personal best (Line 5), while the optimal particle in the initial particle population is taken to the global best (Line 7). Then, the algorithm begins to iterate. In each iteration, each particle in the population is updated one by one according to Equation (17) (Line 11), and Algorithm 2 is recalled to obtain the computation offloading and calculate the fitness for each particle (Lines 12–13). If the fitness of the updated particle is less than the fitness of its personal best state, its updated state is set to its best state (Lines 14–15). In addition, if the fitness of the updated particle is smaller than the fitness of the global best particle, its updated state is set as the global best particle (Lines 16–17). After completing T iterations, map $gBest$ to the UAV deployment $\boldsymbol{u}^{\mathrm{UAV}}$ and return (Lines 22–23).

4.2.2. Inner: GDs Access and Computation Offloading Optimization

The second sub-problem of problem $\mathbf{P}$, denoted $\mathbf{P}_2$, is to solve for the optimal GD access $\boldsymbol{\alpha}^*$ and computation offloading $\boldsymbol{\beta}^*$ with given UAV deployment $\boldsymbol{u}^{\mathrm{UAV}*}$. $\mathbf{P}_2$ can be formulated as

$$\begin{aligned}
\mathbf{P}_2 : \min_{\boldsymbol{\alpha},\boldsymbol{\beta}} \quad & T(\boldsymbol{u}^{\mathrm{UAV}*}, \boldsymbol{\alpha}, \boldsymbol{\beta}) && (24)\\
\text{s.t.} \quad & T_m \le E_m, \forall m \in \mathcal{M} && (24a)\\
& \alpha_m^{\mathrm{GD}}, \alpha_m^{\mathrm{UAV}}, \alpha_m^{\mathrm{LEO}} \in [0,1], \forall m \in \mathcal{M} && (24b)\\
& \alpha_m^{\mathrm{GD}} + \alpha_m^{\mathrm{UAV}} + \alpha_m^{\mathrm{LEO}} = 1, \forall m \in \mathcal{M} && (24c)\\
& \beta_{m,n} \in [0,1], \forall m \in \mathcal{M}, n \in \mathcal{N} && (24d)\\
& \sum\nolimits_{n=1}^{N} \beta_{m,n} \le 1, \forall m \in \mathcal{M} && (24e)\\
& \sum\nolimits_{m=1}^{M} \beta_{m,n} \le \Lambda_{\max}, \forall n \in \mathcal{N} && (24f)\\
& \beta_{m,n} d_{m,n} \le L_{\max}, \forall m \in \mathcal{M}, n \in \mathcal{N} && (24g)
\end{aligned}$$

As the computation offloading problem formulated in Equation (24) is NP-Hard [36], the traditional algorithm is inefficient for this problem. Therefore, we tackle the problem with a greedy strategy-based algorithm. Greedy is a low-complexity approximation algorithm that always makes the current optimal decision in solving the problem. Therefore, it can quickly obtain a near-optimal solution. According to Equation (4), it can be concluded that when the GD is closer to the accessed UAV, the greater the channel gain between them. Therefore, accessing the GDs to their nearest UAVs can reduce the transmission latency during computation offloading. However, since the service coverage of the UAV is finite, the GDs outside the coverage area cannot offload the tasks to the UAV. In addition, the parallel processing capability of UAVs is constrained and cannot process excessive computation tasks simultaneously. Therefore, we classify the GDs into three types according to their distribution:

1. The GDs not within the coverage area of any one UAV and the set of such GDs is defined by $\mathcal{M}_1$;
2. The GDs within the coverage area of only one UAV and the set of such GDs is defined by $\mathcal{M}_2$;
3. The GDs within the coverage area of multiple UAVs and the set of such GDs is denoted by $\mathcal{M}_3$.

The first type of GD has the highest priority because they cannot be accessed by UAVs and do not occupy the computational resources of UAVs, so they do not affect the computation offloading policies of the other two types of ground devices. The second type of GD can only be accessed to the nearest UAV and has the next highest priority. The third type of GD can be accessed to the most appropriate UAV among multiple UAVs, and these GDs have the lowest priority.

Further, we decide on the computation offloading of the GDs. For the first type of GD, the computation tasks are divided into local and LEO satellite processing parts since they cannot be accessed by the UAV, i.e., $\alpha_m^{\mathrm{UAV}} = 0, \forall m \in \mathcal{M}_1$. For the second and third types of GD, the computation tasks are split into local, UAV, and LEO satellite processing parts. As the UAV deployment and the GD access are already determined, the computation offloading optimization problem is a standard linear programming problem that can be solved by employing existing toolkits (e.g., CVXPY).

We propose the following greedy strategy based on the above analysis.

1. For $\mathcal{M}_1$, they do not have access to UAVs.
2. For $\mathcal{M}_2$, they are accessed to the nearest UAV. If there are currently $\Lambda_{\max}$ GD accessed by a UAV, the access of this UAV with its accessed GD with the highest transmission latency is canceled, and the computation offloading for this GD is adjusted.
3. For $\mathcal{M}_3$, they always try to access the nearest UAV. If $\Lambda_{\max}$ GDs are currently accessed by this UAV, select the GD in $\mathcal{M}_3$ that is already accessed by this UAV and has the highest transmission latency, and access it to other UAVs that are nearest. If the original access cannot be changed, the GD in $\mathcal{M}_3$ relinquishes access to the UAV and adjusts the computation offloading.

The main steps of the Greedy are given in Algorithm 2.

Algorithm 2 Greedy

Require: $u^{\text{GD}}, u^{\text{UAV}}, I$
Ensure: α, β
1: Classify the GDs to generate $\mathcal{M}_1, \mathcal{M}_2, \mathcal{M}_3$;
2: **for** each $m \in \mathcal{M}_1$ **do**
3:     **for** each $n \in \mathcal{N}$ **do**
4:         $\beta_{m,n} = 0$;
5:     **end for**
6:     Calculate the computation offloading ratio for GD m;
7: **end for**
8: **for** each $m \in \mathcal{M}_2$ **do**
9:     **for** each $n \in \mathcal{N}$ **do**
10:         **if** n is the closest UAV to GD m **then**
11:             $\beta_{m,n} = 1$;
12:         **end if**
13:     **end for**
14:     **if** the number of GD accessed to UAV n is greater than $\Lambda_{\max}$ **then**
15:         Calculate the transmission latency of all GDs in $\mathcal{M}_2$ accessed to UAV n;
16:         Select GD m' with the largest transmission latency so that $\beta_{m',n}$;
17:         Calculate the computation offloading ratio of GD m';
18:     **end if**
19:     Calculate the computation offloading ratio of GD m;
20: **end for**
21: **for** each $m \in \mathcal{M}_3$ **do**
22:     **for** each $n \in \mathcal{N}$ **do**
23:         **if** n is the closest UAV to GD m **then**
24:             $\beta_{m,n} = 1$;
25:         **end if**
26:     **end for**
27:     **if** the number of GD accessed to UAV n is greater than $\Lambda_{\max}$ **then**
28:         Calculate the transmission latency of all GDs in $\mathcal{M}_3$ accessed to UAV n;
29:         Select GD m'' with the largest transmission latency, and let it access the second closest UAV;
30:         Calculate the computation offloading ratio of GD m'';
31:     **end if**
32: **end for**
33: **return** α, β

The algorithm first classifies the GDs to obtain $\mathcal{M}_1$, $\mathcal{M}_2$, and $\mathcal{M}_3$ (Line 1). computation offloading for the GDs is obtained according to the proposed greedy strategy (Lines 2–32). Finally, α and β are returned (Line 33).

5. Performance Evaluation

In this section, extensive simulations are conducted to investigate and validate the effectiveness of the proposed PSO&GS.

5.1. Simulation Settings

In the simulation, we consider a rectangular area of 200×200 m^2, where the CPU-cycle frequencies of the GDs are 0.8–1.2 GHz. The computational task-input data size follows a random distribution of [10, 15] Mbit. The computational complexities for computation tasks are set to 80–120 cycles/bit, and the maximum tolerable latency is 1 s. To provide computational services to GDs, we assumed that $N = 4$ UAVs are deployed as edge servers in the considered area, and their CPU-cycle frequencies are set to 2 GHz. In addition, an LEO satellite is used as a cloud server, which is assigned a CPU-cycle frequency of 3 GHz for each GD. The transmission rate between the GDs and the LEO satellite is 20 Mbit/s. The basic simulation settings are listed in Table 3 [28].

Table 3. Simulation Settings.

Parameter	Value
H	40 m
B	10 MHz
P	1 W
σ^2	−130 dBm
h_0	−30 dB
M	20, 30, 40
$\Lambda_{\max}$	8
$L_{\max}$	{40, 80} m

All the simulations in this section are carried out on a PC equipped with an i5-8500 CPU and 32 GB RAM. The operating system version is Windows 10-2004. PSO&GS and all benchmark algorithms are implemented in Python 3.7. The relevant parameters of PSO&GS refer to [29] and are set as $T = 1000$, $K = 100$, $w_{\max} = 0.8$,$w_{\min} = 0.2$, $c_1^{\text{start}} = 0.9$, $c_1^{\text{end}} = 0.2$, $c_2^{\text{start}} = 0.4$, and $c_2^{\text{end}} = 0.9$.

5.2. Numerical Results Analysis

(Result#1) Discussion of the convergence of PSO&GS

First, we analyze the convergence of the proposed PSO&GS. Figure 4 shows the variation of the system average response latency obtained by PSO&GS with the increase in the number of iterations. As shown in Figure 4, PSO&GS always converges to a stable result within 500 iterations, no matter how M and $L_{\max}$ vary. Specifically, the convergence of PSO&GS is slower when $L_{\max} = 40$ m compared to $L_{\max} = 80$ m. It is because more GDs will be outside the coverage of the UAV when its coverage is small. In such cases, more stringent requirements for UAV deployment are needed to provide computational service to more GDs. Therefore it takes more time for UAV deployment employing PSO&GS. With a large maximum service distance of the UAVs, almost all GDs are within their coverage, so PSO&GS takes less time searching for UAV deployment. Further, the convergence curves of PSO&GS under different conditions show that the increase in the number of GDs will make the problem solution space larger and cause the convergence of PSO&GS to slow down. In summary, the convergence performance of the proposed PSO&GS is excellent in various scenarios.

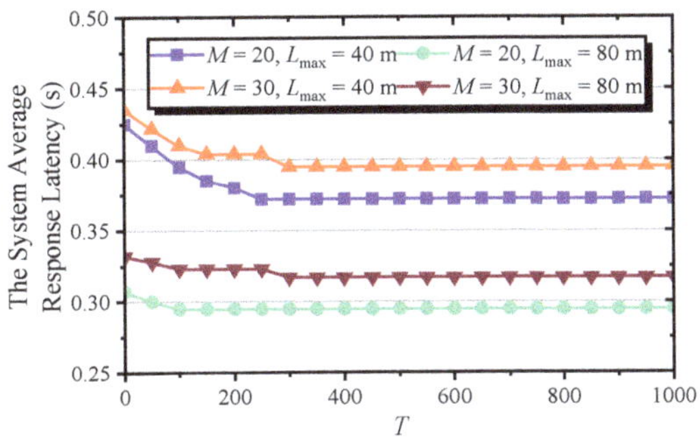

Figure 4. (Result#1) The system average response latency obtained by PSO&GS.

(Result#2) Analysis of the effectiveness of PSO&GS outer PSO-GA

To better evaluate the performance of PSO&GS outer PSO-GA, we conduct the comparison by introducing benchmark algorithms as follows.

- **RanDep**: In this algorithm, a random algorithm is used to replace PSO&GS outer PSO-GA as the UAV deployment optimization approach, and the proposed Greedy is employed to optimize computation offloading. The average result of 1000 repetitions was used as the final result.
- **DeDep**: In this algorithm, the differential evolutionary algorithm is employed as the UAV deployment optimization approach, and the inner adopts the proposed Greedy to optimize computation offloading.

Figure 5 shows the system average response latency obtained by PSO&GS and the benchmark algorithms for different M and L_{max}. As shown in Figure 5, regardless of the variation of M and L_{max}, the proposed PSO&GS has a significant performance advantage over RanDep. Specifically, the performance advantage of PSO&GS is particularly substantial for larger L_{max}. The reason for the above phenomenon is that when the maximum service distance of the UAV is large, high-quality UAV deployment enables the UAV to cover more GDs, so that more GDs offload their computation tasks to the UAV for processing, thus making full use of the UAVs' computational resources. When low-quality UAV deployment is employed, even if the maximum service distance of the UAV is large enough, it results in some GDs being unable to access the UAV, thus limiting the utilization of computational resources. In addition, PSO&GS are better than DeDep in optimizing the system average response latency except for $M = 30$ and $L_{max} = 80$ m. In summary, we can demonstrate the effectiveness and superiority of PSO&GS outer PSO-GA in achieving optimization of UAV deployment.

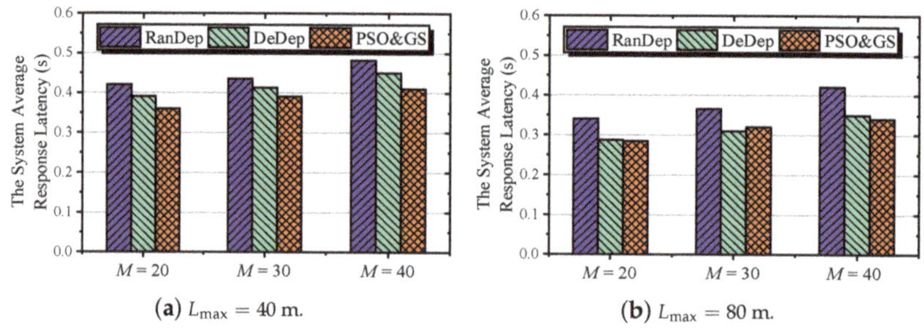

Figure 5. (Result#2) The system average response latency obtained by RanDep, DeDep, and PSO&GS.

(Result#3) Analysis of the effectiveness of PSO&GS inner Greedy

Similarly, we compare by introducing benchmark algorithms to evaluate the performance of PSO&GS inner Greedy as follows.

- **RanOff**: In this algorithm, the outer employs an approach consistent with PSO&GS for UAV deployment, while the inner adopts a random approach for computation offloading.
- **ProAve**: In this algorithm, the outer applies the same approach as PSO&GS for UAV deployment. At the same time, the inner uses the proximity principle to access GDs to UAVs and distributes the computation tasks on average to each end.

Figure 6 shows the system average response latency obtained by PSO&GS, RanOff, and ProAve for different M and L_{max}. As shown in Figure 6, the proposed PSO&GS performs best in various scenarios, followed by ProAve and the worst by RanOff. It is because the proposed PSO&GS is optimized in both device access and computation offloading compared to random-based computation offloading and thus can significantly improve the performance. For ProAve, the proposed PSO&GS is significantly better than the average distribution in computation offloading optimization. Therefore, the Greedy algorithm proposed in this paper can achieve better optimization results when the maximum service distance of the UAV is larger.

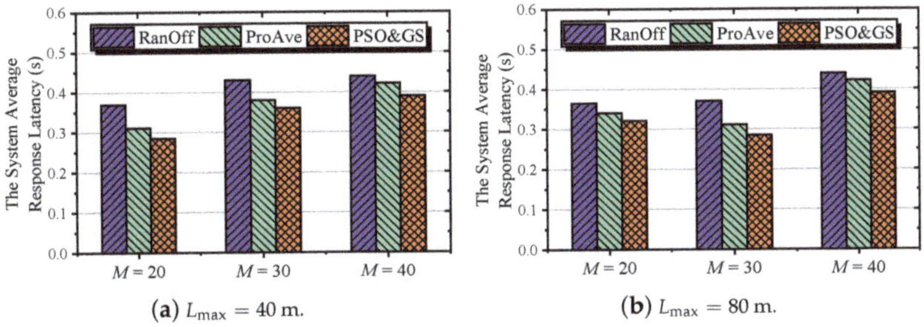

Figure 6. (Result#3) The system average response latency obtained by RanOff, ProAve, and PSO&GS.

6. Conclusions

In this paper, we proposed a joint optimization of UAV deployment, GD access, and computation offloading in the SAGiMEC system to minimize the system average response latency. We formulated the problem as a MINLP and designed an inner and outer double-layer optimization algorithm (i.e., PSO&GS). Numerical results show that the proposed PSO&GS converges well and achieves better system average response latency than other benchmark algorithms. In our future work, we aim to investigate the impact of the NOMA and the successive interference cancellation (SIC) decoding order on the wireless channel performance.

Author Contributions: Y.X. and F.D. drafted the original manuscript and designed the experiments. J.Z. provided ideas and suggestions. All authors have read and agreed to the published version of the manuscript.

Funding: This work was partly supported by the National Natural Science Foundation of China under Grant No. 62072108, Fuzhou-Xiamen-Quanzhou National Independent Innovation Demonstration Zone Collaborative Innovation Platform under grant No. 2022FX5, and the Funds for Scientific Research of Fujian Provincial Department of Finance under Grant No. 83021094.

Data Availability Statement: Data in this paper are available from the corresponding authors upon request.

Conflicts of Interest: The authors declare no conflict of interest.

References

1. Cheng, N.; Lyu, F.; Quan, W.; Zhou, C.; He, H.; Shi, W.; Shen, X. Space/Aerial-Assisted Computing Offloading for IoT Applications: A Learning-Based Approach. *IEEE J. Sel. Areas Commun.* **2019**, *37*, 1117–1129. [CrossRef]
2. Shafi, M.; Molisch, A.F.; Smith, P.J.; Haustein, T.; Zhu, P.; De Silva, P.; Tufvesson, F.; Benjebbour, A.; Wunder, G. 5G: A Tutorial Overview of Standards, Trials, Challenges, Deployment, and Practice. *IEEE J. Sel. Areas Commun.* **2017**, *35*, 1201–1221. [CrossRef]
3. Xiao, H.; Hu, Z.; Yang, K.; Du, Y.; Chen, D. An Energy-Aware Joint Routing and Task Allocation Algorithm in MEC Systems Assisted by Multiple UAVs. In Proceedings of the International Wireless Communications and Mobile Computing (IWCMC), Limassol, Cyprus, 15–19 June 2020.
4. Chen, X.; Huang, Q.; Wang, P.; Liu, H.; Chen, Y.; Zhang, D.; Zhou, H.; Wu, C. MTP: Avoiding Control Plane Overload with Measurement Task Placement. In Proceedings of the IEEE Conference on Computer Communications (INFOCOM), Vancouver, BC, Canada, 10–13 May 2021.
5. Chen, X.; Huang, Q.; Wang, P.; Meng, Z.; Liu, H.; Chen, Y.; Zhang, D.; Zhou, H.; Zhou, B.; Wu, C. LightNF: Simplifying Network Function Offloading in Programmable Networks. In Proceedings of the IEEE/ACM 29th International Symposium on Quality of Service (IWQOS), Tokyo, Japan, 25–28 June 2021.
6. Liu, J.; Shi, Y.; Fadlullah, Z.M.; Kato, N. Space–Air–Ground Integrated Network: A Survey. *IEEE Commun. Surv. Tutor.* **2018**, *20*, 2714–2741. [CrossRef]
7. Hu, Y.; Li, V. Satellite-based Internet: a tutorial. *IEEE Commun. Mag.* **2001**, *39*, 154–162.

8. Patel, M.; Naughton, B.; Chan, C.; Sprecher, N.; Abeta, S.; Neal, A. Mobile-edge computing introductory technical white paper. *White Pap. Mob.-Edge Comput. (MEC) Ind. Initiat.* **2014**, *29*, 854–864.
9. Mach, P.; Becvar, Z. Mobile Edge Computing: A Survey on Architecture and Computation Offloading. *IEEE Commun. Surv. Tutor.* **2017**, *19*, 1628–1656. [CrossRef]
10. Flores, H.; Hui, P.; Tarkoma, S.; Li, Y.; Srirama, S.; Buyya, R. Mobile code offloading: From concept to practice and beyond. *IEEE Commun. Mag.* **2015**, *53*, 80–88. [CrossRef]
11. Pan, Y.; Chen, M.; Yang, Z.; Huang, N.; Shikh-Bahaei, M. Energy-Efficient NOMA-Based Mobile Edge Computing Offloading. *IEEE Commun. Lett.* **2019**, *23*, 310–313. [CrossRef]
12. Yang, G.; Hou, L.; He, X.; He, D.; Chan, S.; Guizani, M. Offloading Time Optimization via Markov Decision Process in Mobile-Edge Computing. *IEEE Internet Things J.* **2021**, *8*, 2483–2493. [CrossRef]
13. Zhang, J.; Hu, X.; Ning, Z.; Ngai, E.C.H.; Zhou, L.; Wei, J.; Cheng, J.; Hu, B. Energy-Latency Tradeoff for Energy-Aware Offloading in Mobile Edge Computing Networks. *IEEE Internet Things J.* **2018**, *5*, 2633–2645. [CrossRef]
14. Bi, S.; Zhang, Y.J. Computation Rate Maximization for Wireless Powered Mobile-Edge Computing With Binary Computation Offloading. *IEEE Trans. Wirel. Commun.* **2018**, *17*, 4177–4190. [CrossRef]
15. Huang, L.; Bi, S.; Zhang, Y.J.A. Deep Reinforcement Learning for Online Computation Offloading in Wireless Powered Mobile-Edge Computing Networks. *IEEE Trans. Mob. Comput.* **2020**, *19*, 2581–2593. [CrossRef]
16. Gheisari, M.; Ebrahimzadeh, F.; Rahimi, M.; Moazzamigodarzi, M.; Liu, Y.; Dutta Pramanik, P.K.; Heravi, M.A.; Mehbodniya, A.; Ghaderzadeh, M.; Feylizadeh, M.R.; et al. Deep learning: Applications, architectures, models, tools, and frameworks: A comprehensive survey. *CAAI Trans. Intell. Technol.* **2023**, *8*, 581–606. [CrossRef]
17. Chouhan, S. Energy Optimal Partial Computation Offloading Framework for Mobile Devices in Multi-access Edge Computing. In Proceedings of the International Conference on Software, Telecommunications and Computer Networks (SoftCOM), Split, Croatia, 19–21 September 2019.
18. Ning, Z.; Dong, P.; Kong, X.; Xia, F. A Cooperative Partial Computation Offloading Scheme for Mobile Edge Computing Enabled Internet of Things. *IEEE Internet Things J.* **2019**, *6*, 4804–4814. [CrossRef]
19. You, C.; Huang, K.; Chae, H.; Kim, B.H. Energy-Efficient Resource Allocation for Mobile-Edge Computation Offloading. *IEEE Trans. Wirel. Commun.* **2017**, *16*, 1397–1411. [CrossRef]
20. Wu, Y.; Qian, L.P.; Mao, H.; Yang, X.; Zhou, H.; Tan, X.; Tsang, D.H.K. Secrecy-Driven Resource Management for Vehicular Computation Offloading Networks. *IEEE Netw.* **2018**, *32*, 84–91. [CrossRef]
21. Pang, Y.; Wang, D.; Wang, D.; Guan, L.; Zhang, C.; Zhang, M. A Space–Air–Ground Integrated Network Assisted Maritime Communication Network Based on Mobile Edge Computing. In Proceedings of the IEEE World Congress on Services (SERVICES), Beijing, China, 18–24 October 2020.
22. Niu, Z.; Shen, X.S.; Zhang, Q.; Tang, Y. Space-air-ground integrated vehicular network for connected and automated vehicles: Challenges and solutions. *Intell. Converg. Netw.* **2020**, *1*, 142–169. [CrossRef]
23. Chen, Q.; Meng, W.; Han, S.; Li, C. Service-Oriented Fair Resource Allocation and Auction for Civil Aircrafts Augmented Space–Air–Ground Integrated Networks. *IEEE Trans. Veh. Technol.* **2020**, *69*, 13658–13672. [CrossRef]
24. Wang, Z.; Yu, H.; Zhu, S.; Yang, B. Curriculum Reinforcement Learning-Based Computation Offloading Approach in Space–Air–Ground Integrated Network. In Proceedings of the International Conference on Wireless Communications and Signal Processing (WCSP), Hunan, China, 20–22 October 2021.
25. Zhou, C.; Wu, W.; He, H.; Yang, P.; Lyu, F.; Cheng, N.; Shen, X. Delay-Aware IoT Task Scheduling in Space–Air–Ground Integrated Network. In Proceedings of the IEEE Global Communications Conference (GLOBECOM), Waikoloa, HI, USA, 9–13 December 2019.
26. Zhou, C.; Wu, W.; He, H.; Yang, P.; Lyu, F.; Cheng, N.; Shen, X. Deep Reinforcement Learning for Delay-Oriented IoT Task Scheduling in SAGIN. *IEEE Trans. Wirel. Commun.* **2021**, *20*, 911–925. [CrossRef]
27. Tang, F.; Hofner, H.; Kato, N.; Kaneko, K.; Yamashita, Y.; Hangai, M. A Deep Reinforcement Learning-Based Dynamic Traffic Offloading in Space–Air–Ground Integrated Networks (SAGIN). *IEEE J. Sel. Areas Commun.* **2022**, *40*, 276–289. [CrossRef]
28. Mao, S.; He, S.; Wu, J. Joint UAV Position Optimization and Resource Scheduling in Space–Air–Ground Integrated Networks With Mixed Cloud-Edge Computing. *IEEE Syst. J.* **2021**, *15*, 3992–4002. [CrossRef]
29. Chen, X.; Zhang, J.; Lin, B.; Chen, Z.; Wolter, K.; Min, G. Energy-Efficient Offloading for DNN-Based Smart IoT Systems in Cloud-Edge Environments. *IEEE Trans. Parallel Distrib. Syst.* **2022**, *33*, 683–697. [CrossRef]
30. Liu, Y.; Jiang, L.; Qi, Q.; Xie, S. Energy-Efficient Space–Air–Ground Integrated Edge Computing for Internet of Remote Things: A Federated DRL Approach. *IEEE Internet Things J.* **2023**, *10*, 4845–4856. [CrossRef]
31. Qualcomm, L. Unmanned Aircraft Systems—Trial Report, 2017. Available online: https://www.qualcomm.com/news/onq/2017/05/qualcomm-technologies-releases-lte-drone-trial-results (accessed on 19 September 2023).
32. He, L.; Li, J.; Wang, Y.; Zheng, J.; He, L. Balancing Total Energy Consumption and Mean Makespan in Data Offloading for Space–Air–Ground Integrated Networks. *IEEE Trans. Mob. Comput.* **2022**, 1–14. [CrossRef]
33. Gong, Y.; Yao, H.; Wu, D.; Yuan, W.; Dong, T.; Yu, F.R. Computation Offloading for Rechargeable Users in Space–Air–Ground Networks. *IEEE Trans. Veh. Technol.* **2023**, *72*, 3805–3818. [CrossRef]
34. Sadoughi, F.; Ghaderzadeh, M. A hybrid particle swarm and neural network approach for detection of prostate cancer from benign hyperplasia of prostate. In *e-Health–For Continuity of Care*; IOS Press: Amsterdam, The Netherland, 2014; pp. 481–485.

35. Sadoughi, F.; Ghaderzadeh, M.; Solimany, M.; Fein, R. An intelligent system based on back propagation neural network and particle swarm optimization for detection of prostate cancer from benign hyperplasia of prostate. *J. Health Med. Inform.* **2014**, *5*, 3.
36. Zhang, J.; Zheng, H.; Chen, Z.; Chen, X.; Min, G. Device Access, Sub-Channel Division, and Transmission Power Allocation for NOMA-Enabled IoT Systems. *IEEE Internet Things J.* **2023**. [CrossRef]

mathematics

MDPI

Article

Deep Convolutional and Recurrent Neural-Network-Based Optimal Decoding for RIS-Assisted MIMO Communication

Md Habibur Rahman [1,2], Mohammad Abrar Shakil Sejan [1,2], Md Abdul Aziz [1,2], Dong-Sun Kim [3], Young-Hwan You [2,4] and Hyoung-Kyu Song [1,2,*]

1 Department of Information and Communication Engineering, Sejong University, Seoul 05006, Republic of Korea; habibur@sju.ac.kr (M.H.R.); sejan@sejong.ac.kr (M.A.S.S.); aziz@sju.ac.kr (M.A.A.)

2 Department of Convergence Engineering for Intelligent Drone, Sejong University, Seoul 05006, Republic of Korea; yhyou@sejong.ac.kr

3 Department of Semiconductor Systems Engineering, Sejong University, Seoul 05006, Republic of Korea; dongsun.kim07@sejong.ac.kr

4 Department of Computer Engineering, Sejong University, Seoul 05006, Republic of Korea

* Correspondence: songhk@sejong.ac.kr

Abstract: The reconfigurable intelligent surface (RIS) is one of the most innovative and revolutionary technologies for increasing the effectiveness of wireless systems. Deep learning (DL) is a promising method that can enhance system efficacy using powerful tools in RIS-based environments. However, the lack of extensive training of the DL model results in the reduced prediction of feature information and performance failure. Hence, to address the issues, in this paper, a combined DL-based optimal decoding model is proposed to improve the transmission error rate and enhance the overall efficiency of the RIS-assisted multiple-input multiple-output communication system. The proposed DL model is comprised of a 1-dimensional convolutional neural network (1-D CNN) and a gated recurrent unit (GRU) module where the 1-D CNN model is employed for the extraction of features from the received signal with further process over the configuration of different layers. Thereafter, the processed data are used by the GRU module for successively retrieving the transmission signal with a minimal error rate and accelerating the convergence rate. It is initially trained offline using created OFDM data sets, after which it is used online to track the channel and extract the transmitted data. The simulation results show that the proposed network performs better than the other technique that was previously used in terms of bit error rate and symbol error rate. The outcomes of the model demonstrate the suitability of the proposed model for use with the next-generation wireless communication system.

Keywords: reconfigurable intelligent surface (RIS); 1-dimensional convolutional neural network (1-D CNN); gated recurrent unit (GRU); bit-error rate (BER); symbol-error rate (SER)

MSC: 94A14

1. Introduction

The growth of the fifth-generation (5G) mobile communication has caused a significant bandwidth demand. In wireless networks, greater frequency bands are allotted and used to enhance higher data rates and system capacity. The 5G New Radio (5G NR) is given to mmWave frequencies between 24 GHz and 50 GHz. This technical development is expected to lead to the employment of the sub-THz bands of 114–300 GHz in wireless communication systems beyond the (B5G) or 6G systems [1]. For the next B5G or 6G wireless networks, the reconfigurable intelligent surface (RIS) is being studied as a viable new technology [2]. The real-time programmable controller in RIS is controlled by a number of passive devices that may independently alter the phase of electromagnetic wave propagation [3,4]. The RIS can be implemented in the walls of the inside and outside environment of the buildings and windows glasses [5–7]. The usage of RIS might be used to enhance wireless communications by removing channel blockages, boosting non-line-of-sight connections, expanding the coverage, and lowering inter-user interference (IUI). As each of its components has the capacity to alter the phase and amplitude of the incoming signal, RIS may offer passive beam-forming (BF) [8] by modifying the phase shift and amplitude reflection coefficients of its constituent parts. In contrast to the conventional multiple-input multiple-output (MIMO) systems, which place a greater emphasis on the base station (BS) and user BF, the RIS-assisted communication system requires the joint design of active BF for BS and user, as well as passive BF for RIS, in order to achieve the passive BF gain. Additionally, RIS-assisted systems enhance spectral efficiency (SE) and energy efficiency (EE) while using less money and little power. This is implemented by employing large-scale passive reflecting components [9–11]. The literature has recently been devoted to several research activities covering various aspects, such as channel modeling [12], channel (CE) estimation [13], modulation and encoding [14], SE analysis [6,15], outage probability [16], symbol error probability (SER) [17], energy efficiency [18], weighted sum rate [19,20], and the evaluation of performance in RIS-supported wireless networks [21]. This is because RIS-enabled settings provide significant potential advantages.

Deep learning (DL) approaches have the potential to significantly improve wireless communication performance [22–26]. Many works were performed in RIS-based wireless systems to handle problems related to channel state information (CSI) [27], CE [28], and performance maximization [29,30] using the DL approaches. To avoid performance loss using a DL model, in [29], the authors proposed an end-to-end training-based RIS-assisted MIMO system. The proposed system optimized the signal processing operations at the access point, RIS, and user at the same time, with active BF for the access point and user and passive BF for RIS. For estimating and identifying symbols in signals sent by RIS, the authors of [30] designed a DL method that employs fully connected layers for prediction when estimating channels and phase angles from a reflected signal received through a RIS. In [31], the authors proposed an effective RIS-assisted MIMO system based on a DNN multi-stage training strategy model for perfect CSI acquisition. The authors in [32] proposed a deep denoising neural-network-assisted compressive sensing broadband for the mmWave RIS system. An indoor RIS-based communication environment was considered in [33], where a trained DNN was used to determine the optimal phase shift by using the target user position. A deep reinforcement learning framework was proposed in [34] for predicting the RIS reflection matrix with minimal beam training overhead.

For natural language processing and machine vision, the convolutional neural network (CNN) model is basically used. It can extract the high level of features from the input and also can be used for the time-series data processing [35]. The CNN model is composed of an input layer, convolutional layer, pooling layer, fully connected layer, and output layer. The structure of the 1-D CNN is illustrated in Figure 1a.

In [36], the authors proposed a CNN-based RIS-enhanced multiple-input single-output system for the goal of sum-rate maximization. To archive the accurate CSI and improve performance, an ordinary differential equation-based CNN structured in the RIS-orthogonal frequency division multiplexing (OFDM) single antenna receiver was proposed in [37]. In [38], a CNN-based RIS-assisted OFDM with a single-input single-output system was proposed for achieving the maximization of the achievable rate. In [39], the authors presented a long short-term memory (LSTM)-based energy-efficient RIS wireless system where they evaluated the system EE and achievable rate. The study in [40] presented a CNN-based demodulation method for a multi-user RIS wireless system where, for channel modeling, an OFDM system based on MIMO was taken into consideration. In [40], the authors compared the achievement of the proposed demodulation (Demod)-CNN with the conventional method in terms of bit-error rate (BER) and SER. In addition, the temporal neural network (TCN)-based RIS-assisted MIMO wireless communication was presented in [41], where the proposed TCN analyzed the achievable rate in terms of different modulation schemes.

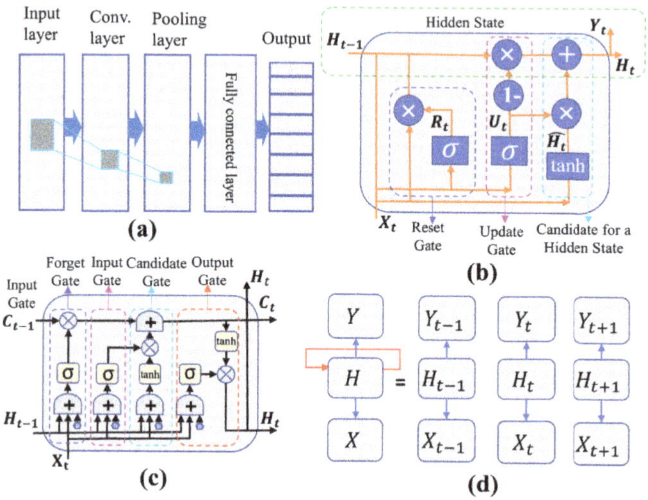

Figure 1. (**a**) The architecture of the 1-D CNN model; (**b**) the architecture of the GRU model; (**c**) the architecture of the LSTM model; (**d**) the architecture of the RNN model.

LSTM and gated recurrent unit (GRU) are examples of recurrent neural networks (RNNs) that are frequently employed to address sequence difficulties. In comparison to traditional RNN, GRU is designed to store past state information more effectively. The traditional RNN training process's gradient vanishing and gradient explosion problems can be addressed [42]. The hidden unit in the typical RNN structure, which may selectively recall relevant information and discard unnecessary information, is replaced by a gate structure in the GRU and LSTM. The update gate and the reset gate are used in the GRU instead of the input gate, forget gate, and output gate of the LSTM [43]. The training setting can be decreased to produce a quicker convergence speed, provided that the prediction accuracy of the GRU is not less than that of the LSTM. Figure 1b shows the internal cell structure of the GRU network. Figure 1c,d show the structure of LSTM and typical RNN network, respectively. For a system with a one-bit ADC and massive MIMO, in [44], the authors proposed the LSTM-GRU to evaluate the channel matrix and showed better performance. For the purpose of CE of data subcarriers, in [45], the authors proposed a DL-GRU neural network-based OFDM system in which the computing demands may be greatly decreased and achieved a more satisfactory gain than other methods.

The performance in the aforementioned studies has a major influence on the SER and BER by using CNN- and RNN-based models. Thus, the combination of 1-D CNN with GRU can be a promising method to achieve high accuracy performance in RIS-assisted communication systems. Motivated by the previous work and enhanced efficacy, herein, a combined DL model, comprising a 1-D CNN and GRU network, is proposed. In terms of SER and BER with various signal-to-noise (SNR) levels, the suggested model outperforms the [40].

The key contributions of the proposed system can be summarized as follows:

- DL-cascaded model-based effective decoding is proposed for the RIS-assisted MIMO system. The proposed DL model is structured with a 1-D CNN model, which extracts the features in the form of complex numbers of the received channel matrix, and a GRU network, which uses extracted data for obtaining the optimal prediction.
- Inter-carrier interference and inter-symbol interference (ISI) are taken into consideration while segmenting the data of the received signals using a 1-D CNN-based feature extractor. For handling the considerable ISI and accelerating the training speed of the model, the GRU layers are utilized. Thus, the combined model enhances demodulation accuracy.

- The different layers of the proposed model are configured for improved accuracy and effectiveness and are examined by an Adam optimizer.
- The efficiency of the proposed model is compared with a conventional system, the LSTM model implemented in our experiment, TCN [41], and (Demod)-CNN-based RIS model [40], with respect to the different SNR values where the proposed model outperforms other methods at different SNRs with the measurement error of BER and SER.

The remaining parts of the paper are arranged as follows. The system model for RIS-based communication, which includes the channel model and signal transmission, is described in Section 2. The training data preparation, training, and inference of the proposed deep learning model are covered in Section 2.4. The simulation outcomes and complexity analysis are reported in Section 3, and Section 4 presents the conclusion.

Notations: A vector and matrix are denoted by the lowercase and uppercase bold letters $\mathbf{h}$ and $\mathbf{H}$, respectively. $\mathbf{H}^H$ stands for the conjugate transpose matrix of $\mathbf{H}$, and $\text{diag}(\mathbf{x})$ stands for the diagonal matrix with vector $\mathbf{x}$ on its diagonal.

2. RIS-Assisted System and Channel Model

In this section, the RIS-assisted system, channel modeling, and selected precoder structure are described.

2.1. System Model

Figure 2 depicts a RIS communication system where I user equipment (UE) contains a single antenna, and the base station (BS) contains M uniform linear array (ULA) antennas. The total number of RIS is taken to be uniform planner array (UPA) N elements. For the i-th UE ($i = 1, 2, ..., I$), the received signal through RIS can be stated as follows [6]:

$$\mathbf{y}_\text{i} = \boldsymbol{H}_{2,i}^H \boldsymbol{\Omega} \boldsymbol{H}_{br} \boldsymbol{x} + \mathbf{n}_\text{i}, \tag{1}$$

where $\mathbf{x} \in \mathbb{C}^{M\times 1}$ represents the precoded transmit signal with Q subcarriers in each OFDM symbol. Inverse discrete Fourier transform (IDFT) is used to process the OFDM data symbols after they are transformed to the time domain. Then, the CP data are inserted as a guard interval. Discrete Fourier transform (DFT) is carried out at the UE after the CP data are removed. Thereafter, $\mathbf{y}_\text{i}$ is the signal that is received at the i-th UE. $\boldsymbol{H}_{br} \in \mathbb{C}^{N\times M}$ is the channel matrix from BS to RIS, the channel matrix between RIS and the i-th UE is $\boldsymbol{H}_{2,i} \in \mathbb{C}^{N\times 1}$, and at the i-th user, the additive white Gaussian noise is represented $n_i \sim \mathcal{CN}(0, \sigma^2)$. The phase shift values of RIS elements are represented by $\boldsymbol{\Omega} = \text{diag}(\boldsymbol{\Upsilon}) \in \mathbb{C}^{N\times N}$, which is the diagonal matrix of $\boldsymbol{\Omega}$. Each element can be characterized as $\boldsymbol{\Upsilon} = [\gamma_1 e^{j\zeta_1}, \gamma_2 e^{j\zeta_2}, \ldots, \gamma_N e^{j\zeta_N}]^T \in \mathbb{C}^{N\times 1}$, where the phase shift coefficient and the amplitude for the n-th reflective elements are $\zeta_n \in [0, 2\pi]$ and $\gamma_n \in [0, 1]$. Consequently, the symbol $\boldsymbol{\Omega}$ can be written as follows:

$$\boldsymbol{\Omega} = \begin{bmatrix} \gamma_1 e^{j\zeta_1} & \cdots & \cdots & \cdots \\ \cdots & \gamma_2 e^{j\zeta_2} & \cdots & \cdots \\ \vdots & \vdots & \cdots & \vdots \\ \cdots & \cdots & \cdots & \gamma_N e^{j\zeta_N} \end{bmatrix}. \tag{2}$$

The constant amplitude coefficient is assumed to be $\zeta_n = 1$ for convenience in calculations [9]. In this study, the direct communication channel $\boldsymbol{H}_d \in \mathbb{C}^{M\times 1}$ between BS to UEs is also considered. The BS to the i-th UEs channel $\boldsymbol{H}_{2,i}^H \boldsymbol{\Omega} \boldsymbol{H}_{br} \in \mathbb{C}^{M\times 1}$ via RIS is assumed. In particular, it is important to remember that the matrix $\boldsymbol{\Omega} = \text{diag}(\boldsymbol{\Upsilon})$ is a diagonal matrix. Thereafter, the previous whole channel matrix $\boldsymbol{H}_{2,i}^H \boldsymbol{\Omega} \boldsymbol{H}_{br}$ can be reformed as follows:

$$\mathbf{H}_{2,i}^{H}\Omega\mathbf{H}_{br} = \mathbf{H}_{2,i}^{H}\text{diag}(\Psi)\mathbf{H}_{br} = \Psi^{\mathrm{T}}\text{diag}(\mathbf{H}_{2,i}^{H})\mathbf{H}_{br}, \tag{3}$$

$$\mathbf{H}_{cas} = \mathbf{H}_{2,i}^{H}\Omega\mathbf{H}_{br}, \tag{4}$$

where the cascaded channel for the i-th UEs is $\mathbf{H}_{cas} \in \mathbb{C}^{N\times M}$ and only relies on the downlink CSI [46]. Therefore, the total received signal $\mathbf{y_{rit}}$ at the i-th UEs via the cascaded channel and the direct communication link can be written as follows [47]:

$$\mathbf{y_{rit}} = H_{2,i}^{H}\Omega H_{br}x + H_{d}x + \mathbf{n_i}, \tag{5}$$

where H_d is the direct communication link from BS to the i-th UEs.

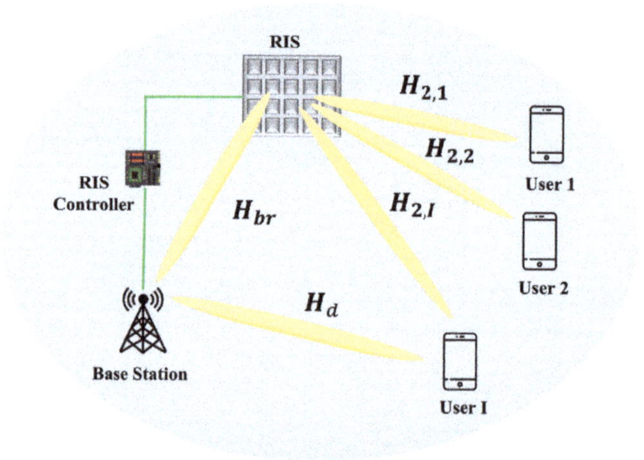

Figure 2. The architecture of RIS-assisted downlink MIMO communication.

2.2. Channel Model

The three-dimensional (3D) Saleh-Valenzuela channel model is used in this research for the mmWave propagation [48], which is a statistical channel model in a multipath propagation scenario. The prevalent theory of the channel model for mmWave communication can be represented as follows [49]:

$$H = \sqrt{\frac{N}{L}}\sum_{l=1}^{L}\eta_{l}\mathbf{a}(\zeta_{l}^{H},\phi_{l}^{H}), \tag{6}$$

where the channel vector is represented as H, the complex gain of the l-th path is η_l, and the number of total paths is L. The azimuth angle of departure and the elevation angle of departure are ζ_l^H and ϕ_l^H, respectively. The array response vector is defined as $\mathbf{a}(\zeta_l^H, \phi_l^H)$. The array response vector for a standard $N_1 \times N_2$ is expressed as follows [49]:

$$\mathbf{a}(\zeta,\phi) = \frac{1}{\sqrt{N}}[e^{-j2\pi d\sin(\zeta)\cos(\phi)\mathbf{n_1}/\lambda}] \otimes [e^{-j2\pi d\sin(\phi)\mathbf{n_2}/\lambda}], \tag{7}$$

where $\mathbf{n_1} = [0, 1, \ldots, N_1 - 1]$ and $\mathbf{n_2} = [0, 1, \ldots, N_2 - 1]$, the carrier wavelength is λ, and the antenna spacing is d, the antenna inter-space satisfies the condition $d = \lambda/2$. Accordingly, the BS to RIS channel H_{br} can be presented as follows:

$$H_{br} = \sqrt{\frac{MN}{L_1}}\sum_{l_1=1}^{L_1}\eta_{l_1}\mathbf{b}(\zeta_{l_1}^{H_r},\phi_{l_1}^{H_r})\mathbf{a}^{H}(\zeta_{l_1}^{H_t},\phi_{l_1}^{H_t})^{T}, \tag{8}$$

where L_1 stands for the pathway numbers among the BS and RIS, η_l stands for the complex gain of those paths, $\mathbf{b}(\zeta_l^{H_r}\phi_l^{H_r})$ is the steering vector related to the RIS, and $\mathbf{a}(\zeta_l^{H_t},\phi_l^{H_t})$ is the steering vector connected to the BS for the l-th path.

The following definitions can be used to describe the channel $\boldsymbol{H}_{2,i}^{H}$ between the RIS and the UE, as follows:

$$\boldsymbol{H}_{2,i}^{H} = \sqrt{\frac{N}{L_2}} \sum_{l_2=1}^{L_2} \eta_{l_2} \mathbf{a}^{H}(\zeta_{l_2}^{H_t}, \phi_{l_2}^{H_t}), \tag{9}$$

where the complex gain of paths is η_{l_2}, and L_2 is the number of pathways among the RIS and the i-th UE. $\zeta_{l_2}^{H_t}$ and $\phi_{l_2}^{H_t}$ represent the departure angles in azimuth and elevation, respectively. The array response vector is defined as $\mathbf{a}(\zeta_{l_2}^{H_t}, \phi_{l_2}^{H_t})$. In conclusion, according to (4), (8), and (9), the cascaded channel matrix $\boldsymbol{H}_{cas}$ from BS to the i-th UE can be expressed as follows [6]:

$$\begin{aligned} \boldsymbol{H}_{cas} = & \sqrt{\frac{MN}{L_1 L_2}} \sum_{l_1=1}^{L_1} \sum_{l_2=2}^{L_2} \eta_{l_1}\eta_{l_2} \\ & \mathrm{diag}(\mathbf{a}^{H}(\zeta_{l_2}^{H_t}, \phi_{l_2}^{H_t}))\mathbf{b}(\zeta_{l_1}^{H_r}, \phi_{l_1}^{H_r})\mathbf{a}^{H}(\zeta_{l_1}^{H_t}, \phi_{l_1}^{H_t})^{T}. \end{aligned} \tag{10}$$

2.3. Design of Precoder

Dirty paper coding (DPC), successive interference cancellation (SIC), maximum ratio transmission (MRT), zero-forcing (ZF), and minimal mean squared error (MMSE) are examples of linear and non-linear precoders that can be taken into consideration in a MIMO system. Although non-linear precoders exhibit the best performance, implementation is impractical because of the enormous computational complexity. Liner precoders can be implemented in the MIMO system due to the low complexity with sub-optimal performance. In multi-user systems, the linear precoder of ZF performs IUI removal more efficiently than the linear precoder of MRT. Additionally, the ZF precoder performs similarly to the MMSE precoder when the noise value is taken into account [50]. Hence, ZF precoder $\mathbf{Z} = [\mathbf{z}_1, ..., \mathbf{z}_I] \in \mathbb{C}^{M \times I}$ is implemented in this study. For total matrix channel $\mathbf{H}_{all} = [\mathbf{H}_1, ..., \mathbf{H}_I]^T \in \mathbb{C}^{I \times M}$, the precoder Z can be written as follows:

$$\mathbf{Z} = \overline{\mathbf{Z}}\sqrt{\mathbf{P}}, \tag{11}$$

where the ZF precoder matrix stands for $\overline{\mathbf{Z}}$, and the power allocation matrix is represented by $\mathbf{P} = \mathrm{diag}(\delta_i, ..., \delta_I)$. The i-th UE's transmit power at the BS is represented by the δ_i symbol.

2.4. Proposed Deep Learning Model

The dataset generation, structure, and operation of the model are described, and the training and testing procedures are described at the end of this section.

2.4.1. Dataset Preparation

In this paper, 128 OFDM subcarriers are taken into consideration for the transmission of data, and quadrature phase shift-keying (QPSK) modulation is used to map the bits belonging to symbols. The data are transmitted over a RIS-assisted cascaded and direct channel. The BS sends the whole OFDM packet to the UE along with the additional AWGN noise. The generated feature vector $\mathbf{V_{ce}}$ and matching label are used to store the received OFDM packet as training data samples. The number of labels is created by the combination of data from BS antennas where 4 unique symbols can be generated by each antenna; in this way, $4^2 = 16$ labels are generated where 2 is the BS antennas number. The feature vector $\mathbf{V_{ce}}$ holds the complex signal of each symbol in the OFDM packet. To train the suggested model, the complex signal is first divided into real $\mathfrak{R}_\mathfrak{e}$ and imaginary $\mathfrak{I}_\mathrm{m}$ components. The generated transmission and reception data are divided into training ξ and validation τ

data, respectively. A total of 256,000 data samples are generated where 80% are used for training ξ and 20% for validation τ. Finally, the dataset and label samples are stored for the training and validation process. The example datasets of the real and imaginary parts of symbols with corresponding labels are shown in Figure 3.

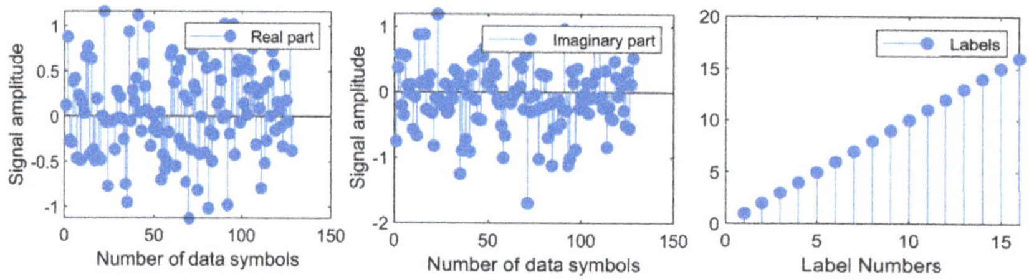

Figure 3. Dataset examples including real and imaginary parts with its labels.

2.4.2. Model Construction and Operation

The proposed model is comprised of two different networks such as 1-D CNN and the GRU network. The proposed CNN-GRU-based network design is shown in Figure 4. The input layer, convolutional layer, pooling layer, GRU layer, and output layer make up the proposed model's structure. The input OFDM demodulated information is extracted by the convolutional layer, and afterward, the pooling layers are used to compensate for the features. Then, the GRU layer is utilized for memorizing features of the previous 1-D CNN output. In the end, the classification is performed by the softmax layer for GRU output. The operation of each layer of the proposed model is described as follows:

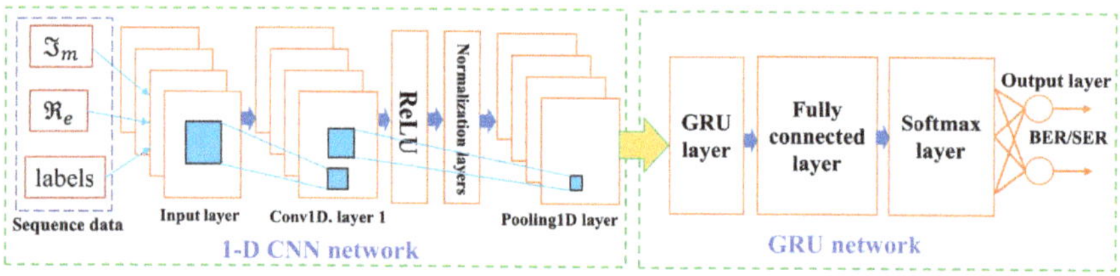

Figure 4. The architecture of the proposed CNN-GRU network: the 1-D CNN network with its layers (**left side**) and the GRU network layers (**right side**).

Input layer: We consider the input sequence features matrix of $Q = [S_1{}^{\mathfrak{R}_e}, S_1{}^{\mathfrak{I}_m}, S_2{}^{\mathfrak{R}_e}, S_2{}^{\mathfrak{I}_m}, S_3{}^{\mathfrak{R}_e}, S_3{}^{\mathfrak{I}_m}, \ldots, S_N{}^{\mathfrak{R}_e}, S_N{}^{\mathfrak{I}_m}]$, which is composed with two numerical values together and the corresponding label classes. The i-th output of the sequence matrix is S_i, so that $S_0 = Q$. Eight real and imaginary values, which correspond to label 1, are contained in each sequence. The dimension of 8×1 in the input layer is assumed to equal the size of the input features. Two numerical values of sequence features with their associated labels are supplied into the CNN's input layer from the created dataset, where the configuration sets the number of features in the input data to be equal to the input size.

Convolutional layer (i = 1): To extract local features from the sequence data, we utilize the convolutional layer. The convolutional layer's parameters ($k \times s$), where k is the kernel size and s is the input data dimension, comprise a collection of learnable filters. In the convolutional layer, a total of 64 filters in 3×3 various sizes are used. The outcome feature matrix S_i may, thus, be represented as follows:

$$S_i = f(S_{i-1} \otimes w_i + b_i), \tag{12}$$

where w_i stands for the weight of the i-th layer, and b_i stands for the layer's bias. The following layers make use of a non-linear or piecewise linear activation function called a rectified linear unit (ReLU). If the output of the ReLU function is positive, it will be directly entered; otherwise, it will be 0. The ReLU function is shown mathematically as follows: $f(x) = max(0, x)$.

Pooling layer (i = 2): The main function of the pooling layer is to compress the features that the convolutional layer has recovered in order to decrease the information dimension and the risk of network overfitting. This layer's function in the study is to cut down on the computing time required to process sequence characteristics from earlier hidden layers using a 1-D global average pooling approach. The output of the pooling layer is the highest of the earlier features matrix. Consequently, the outcome feature matrix S_i may be expressed as follows:

$$S_i = f_p(S_{i-1}), \tag{13}$$

where the pooling function f_p is used. The dimension of S_2 is calculated from the pooling layer as $m/z \times n$, where z represents the scale value of the current layer of the pooling layer, m represents the number of filters, and n represents the number of input data time steps.

GRU layer (i = 3): The GRU layer train is the feature vectors that the CNN extracts where the GRU layer retains the internal workings of many features. According to the structure of the GRU layer, the GRU operating system may be summed up in the next following sentences. The input state information X_t at the present instant and the hidden layer information H_{t-1} that is previously learned are used to compute the update U_t and reset gate R_t in the first step. The reset gate R_t is used in the second phase to count the number of new pieces of information that are stored in the node $\widehat{H_t}$. The third step involves using the update gate U_t to determine the hidden layer output at the present time. The GRU computation procedure is explained using the formulas below:

$$U_t = \sigma(W_U X_t + V_U H_{t-1} + b_U) \tag{14}$$

$$R_t = \sigma(W_R X_t + V_R H_{t-1} + b_R) \tag{15}$$

$$\widehat{H_t} = \tanh(W_H X_t + V_H (R_t \otimes H_{t-1}) + b_H) \tag{16}$$

$$H_t = (1 - U_t) \otimes H_{t-1} + U_t \otimes \widehat{H_t}, \tag{17}$$

where the values for the bias are b_U, b_R, and b_H; The gate activation function is calculated using the sigmoid function, which is denoted by $\sigma(c) = (1 + e^{-c})^{-1}$. The hyperbolic tangent function known as (tanh) is in charge of calculating the state activation function. W_U, W_R, W_H, V_U, V_R, and V_H are weight matrices. While the input state X_t and the output of the hidden layer at the previous instant are merged to generate $\widehat{H_t}$, the output of the hidden layer at the current instant is represented by H_t. The Hadamard product is represented as $\otimes$ for element-wise multiplication.

Fully connected layer (i = 4): The task of activation with the softmax is provided to the fully linked layer, which plays a significant role in classification, and the final classification is performed in this layer. Here, the model calculates the probability that each sample now belongs to one of the classes and then derives the expression of a feature (Y_{pred}), which is stated as follows:

$$Y_{pred}(i) = f(L = l_i | S_3; (W, b)), \tag{18}$$

where the GRU features are S_3 and the activation function for softmax is $f(\cdot)$. The computation output for the i-th classes of the input data is represented by the variables l_i, where W and b stand for the weight and bias values, respectively.

The cross-entropy operation calculates the cross-entropy loss for single-label and multi-label classification tasks between network predictions and target values. Cross-entropy is frequently the best choice when output probability models are dealt with. Furthermore, *L*2 regularization may be viewed as a successful compromise between the identification of tiny weights and the reduction of the cost function [51]. Therefore, overfitting is avoided by using cross-entropy and $L2$ regularization. The following formulation of the loss computation serves as the basis for the goal of model training:

$$Loss(W,b) = -\sum_{i=1}^{n_s}\sum_{t=1}^{c}(Y^{(t)}(i) * \log(Y_{pred}^{(t)}(i)) + \frac{\lambda}{2}\sum_{i=1}^{n_s} W_i^2, \tag{19}$$

where $(Y^{(t)}(i)$ represents the prospect of the known goal, $(Y_{pred}^{(t)}(i))$ is the probability that the i-th sample belongs to the t-th class, the number of samples is n_s, the number of the class is c, and finally, the term "λ" defines the coefficient of the regularization of $L2$. The Adam optimizing strategy is applied to reduce the loss [52].

2.4.3. Overview of Training and Testing Process

In Figure 5, an illustration of the proposed RIS-assisted MIMO wireless communication's training and inference process is presented. After the model is successfully trained, it is validated to check for optimal performance and inference accuracy. Figure 6 shows the flow chart for the testing procedure after training. The training procedure stops and the model for inference is stored when it reaches a model accuracy of 99.95%. The training and validation performance for the 50 epoch according to its loss and accuracy are shown in Figure 7. Figure 7 illustrates that the model performs consistently after 11 epochs, demonstrating that the parameters have learned from the data by the model. During model training, the Adam optimizer is used to analyze the data at a learning rate of 0.01. The training and inference process is shown in Algorithm 1.

Algorithm 1 The training and inference process

Training process:

Load dataset χ.
split χ into a training ξ and a validation τ set at a ratio of 80% and 20%, i.e., 4:1.
for e: 1 to E **do**:
Pass the processed data to build the model and configure the layers.
Configure the model parameters such as learning rate ψ, maximum epochs E_{max}, minibatch size mB, gradient threshold gT, and validation frequency f_v.
Calculate the loss function using (19).
Computing the corrective parameters and obtaining the optimal performance while updating the parameters, using the Adam optimization algorithm.
end for
Save model for inference step.
Output: proposed model.

Inference process:

Load the trained model.
Initialize the parameters.
for s: 1 to SNR **do**:
for i: 1 to Iteration **do**:
Generate data symbol to transmit.
Data transmit by H_{cas} and H_d channel matrix.
Match the label classes with a trained model to classify.
SER and BER performance with different SNR.
end for
end for
Output: SER and BER results.

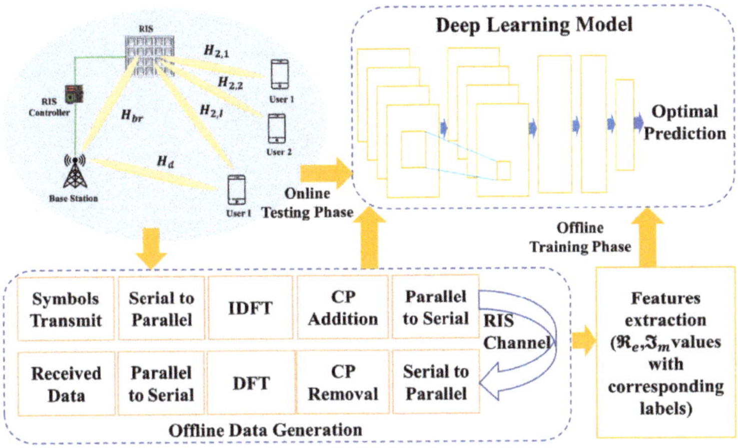

Figure 5. The structure of the proposed CNN-GRU network: training and testing phases.

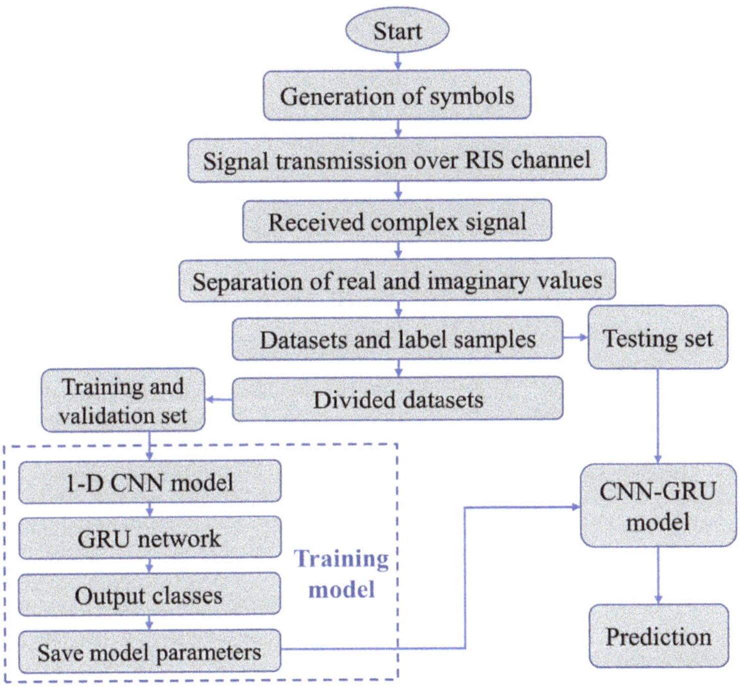

Figure 6. Flowchart for the testing process of the proposed CNN-GRU model.

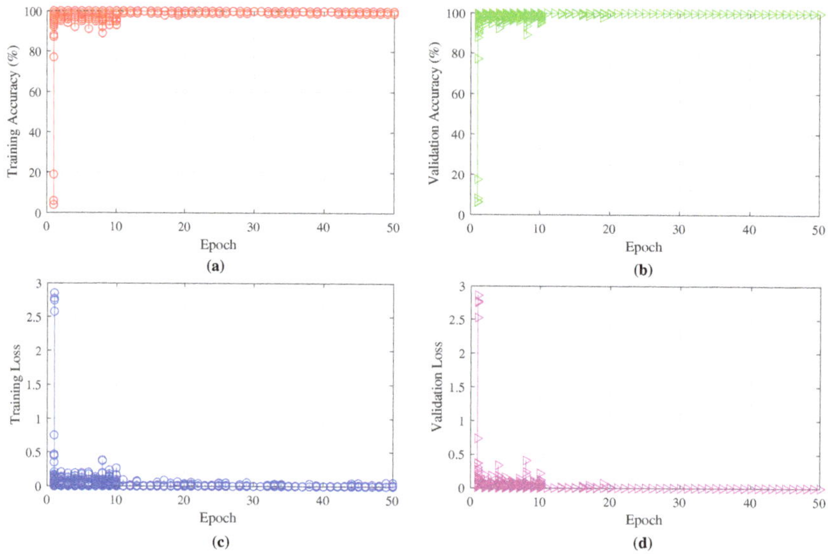

Figure 7. The model training performance analysis: (**a**) the training accuracy; (**b**) the validation accuracy; (**c**) the training loss; (**d**) the validation loss.

3. Simulation Results

The simulation results of the proposed DL-based strategy are provided in this part for wireless MIMO communication systems with RIS assistance. The simulation work is performed with MATLAB and the Windows 10 Pro operating system. The DL layers are connected to create the DNN. The DL Toolbox^{TM} creates the DL model and keeps track of the training procedure. The training performance is improved by using an NVIDIA graphics card. In the mmWave propagation environment, it is indicated that there are $P_1 = 2$ and $P_2 = 4$ channel paths, respectively, between the BS and RIS and between RIS and each UE. The RIS to UEs channel $\boldsymbol{H}_{2,i}^{H}$ is tuned to the non-line-of-sight (LOC), while the BS to RIS channel $\boldsymbol{H}_{br}$ is set to the LOC channel with 15 dB of Rician K-factor. The obstructions prevent access to the BS to UEs direct channel $\mathbf{H}_d$. Additionally, this simulation takes into account a wireless system with RIS service in which transmit antennas are deployed in ULA configurations and RIS components are deployed in UPA configurations, with each transmit antenna and RIS element separated by a half-wavelength $(\lambda/2)$. The dimension of RIS elements is configured as $N = 32 \times 16$. Table 1 presents the simulation settings for the proposed RIS communication system. In Sections 2.4.1 and 2.4.3, the training data generation and training and testing procedure are presented. The model training is implemented by setting the parameters listed in Table 2. The SNR value of 30dB is configured while the training datasets *chi* are being created. When evaluating the simulation's performance during the online testing phase, the [0:4:20] dB SNR range is taken into consideration. To calculate the channel error rate, 128 quadrature phase-shift keying (QPSK) symbols are produced using OFDM. A length of 32 cyclic prefix (CP) as a guard interval is introduced to lessen ISI. In this study, the simulation is run to assess the BER and SER analysis in comparison to the proposed DL model and earlier studies. The data received by the i-th UEs are taken into account at a time for the BER and SER calculation. The average error rate (E_R) for the i-th UEs is represented by the BER and SER. The data from the i-th UEs are first demodulated individually, then $(E_R) = (UE1 + UE2, ..., +UEI)/T$ is used to determine the (E_R), where T is the number of the i-th UEs at the receiver terminal. The inaccurate demodulated bit at the receiver is referred to as the (E_R) for BER. (E_R) for SER refers to the incorrect categorization of the received symbol. The effectiveness of the proposed DL model is compared with the LSTM model, Demod-CNN model [40], and TCN model [41], and traditional method in [40]. The effectiveness of the suggested CNN-GRU is improved by cascading with the GRU module with 1-D CNN and different parameter adjustments.

Table 1. Simulation parameters of the RIS design.

Parameters	Value
The number of RIS elements	32 × 16
BS to RIS paths	2
RIS to user paths	4
BS to user paths	2
Transmit antenna spacing	0.5 λ
RIS elements spacing	0.5 λ
Number of transmitter antenna	2
User number	2
BS antenna number	2
OFDM subcarrier number	128
Number of cyclic prefixes	32
Modulation type	QPSK
Channel noise	AWGN

Table 2. Simulation parameters of the proposed model.

Parameters	Value
Maximum epochs number	50
No. of filters in convolutional layer	64
No. of hidden units in GRU layer	100
Fully connected layer	16
Size of minibatch	100
Learning rate	0.01
Gradient threshold	1
Validation frequency	50
Used optimizer	Adam

3.1. Performance Evaluation

Figure 8 shows the prediction performance of the proposed CNN-GRU according to the predicated and tested data where the examined model is learned with the Adam optimizer with learning rate = 0.01 at the time of training. In Figure 8a, the confusion matrix is analyzed in terms of predicted and true classes for the evaluation of prediction accuracy. There are a total of 16 classes for prediction, and it is seen that the classification error of classes is very marginal. Additionally, the performance of the test sequence prediction between predicated and tested data is shown in Figure 8b. In Figure 8b, the orange and blue color-covered regions are presented with test data and predicted data, respectively. To be more clearly visible, the zoom view of the portion of the output is indicated by the blue arrow. It is evident that the miss prediction rate of class with time step is very low and the proposed model is highly efficient.

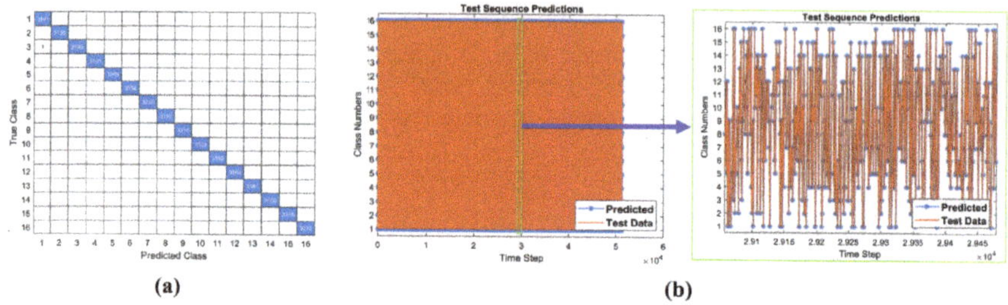

Figure 8. CNN-GRU-based RIS-assisted MIMO communication prediction performance: (**a**) Confusion matrix of predictions and true class; (**b**) Results of test sequence predictions between test and prediction data (blue arrow indicates the zoom view of a portion of the output for more visible).

Figure 9 depicts the comparison of the proposed CNN-GRU decoding performance with the traditional method, Demod-CNN model, TCN, and LSTM model according to the BER versus different SNR values. The graph absolutely exhibits that the proposed CNN-GRU-based demodulation system outperforms the traditional method, the Demod-CNN model, TCN, and LSTM model. Due to the low SNR value at the SNR (0–5) dB range, the proposed conclusions are less precise but not less than those of the other systems. However, the model performance improves as the SNR values rise up to (5–20) dB. Evidently, the proposed model significantly outperforms other systems at SNR ranging from 0 to 20 dB and BER reaches approximately 10^{-5}. In contrast, conventional and Demod-CNN systems cover SNR ranging from 0 to 30 dB and BER reaches around 10^{-4}. Up to 5 dB, the BER for the proposed CNN-GRU model follows a similar pattern to all of the methods. After this threshold, the BER for the proposed CNN-GRU model significantly decreases and ends at 20 dB SNR. Additionally, the CNN-GRU is also compared with the LSTM model and TCN model. It can be shown that the suggested model performs better than the LSTM and TCN models with the SNR (0–20) dB range. Because of the low SNR values, the LSTM model performance is degraded more than all other methods, but after 12 dB SNR, its performance follows the proposed CNN-GRU model. In addition, TCN shows a little better performance with (0–3) dB SNR, and after that, the performance is degraded and follows the LSTM trends. This demonstrates how effectively the proposed CNN-GRU model can demodulate and enhance the BER system performance.

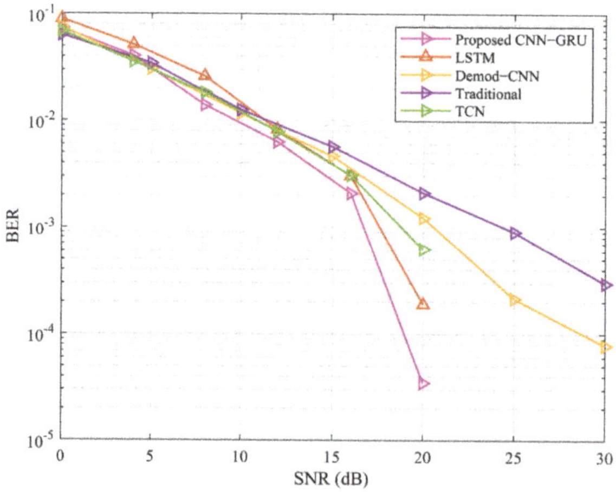

Figure 9. BER simulation results of CNN-GRU-based RIS-assisted MIMO communication.

Figure 10 shows the comparison of the proposed CNN-GRU decoding performance with the traditional demodulation system, Demod-CNN model, and LSTM model according to the SER versus different SNR values. The graph shows that the proposed CNN-GRU-based demodulation system outperforms all other methods. Evidently, the proposed model vastly outperforms other systems at SNR ranging from 0 to 20 dB and SER reaches approximately 10^{-4}. Contrarily, the conventional method and the Demod-CNN model have SNR ranges of 0 to 30 dB and SER of approximately 10^{-3} and 10^{-4}, respectively. Unlike the BER accuracy of the proposed CNN-GRU model, SER performance is also higher than the conventional, Demod-CNN, TCN, and LSTM models and shows significant improvement in decoding capability. However, due to the lower SNR, the LSTM model shows lower performance than the conventional and Demod-CNN models, but with higher SNR, it follows the proposed model performance. Additionally, the TCN model has the same performance with SNR range (0–6) dB, and after that, it follows the LSTM. It is seen from the above simulation outputs that the CNN-GRU gains a noticeable performance difference from the other four schemes. The summary is from the above simulation results; in order to enhance BER and SER for RIS-based systems, the proposed model can be an effective solution.

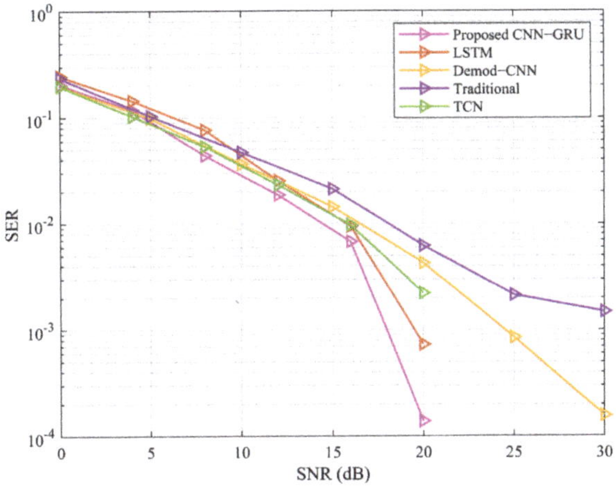

Figure 10. SER simulation results of CNN-GRU-based RIS-assisted MIMO communication.

Figure 11 shows the analysis of SER and BER of the proposed CNN-GRU model by changing different model parameters during the training of the model. The changing parameters are considered three well-known optimization algorithms, namely, Adam, SGDm, and RMSprop. In addition, to conform to the model learning rate (LR), we have configured the training parameters by considering the LR of 0.01 and 0.001, respectively. Figure 11a,b indicate that the simulation results are taken under the LR of 0.01 and three optimization algorithms. In Figure 11a, with the same LR, the Adam optimizer shows better SER performance than the others with different SNR values, and among SDGm and RMSprop, RMSprop achieves a better SER performance than SGDm. A similar performance happens for BER of Figure 11b. On the other hand, the simulation results of SER and BER with LR of 0.001 and three optimizers are shown in Figure 11c,d. Figure 11c,d indicate that with LR of 0.001, the SER and BER performance of the proposed CNN-GRU model is less than LR 0.01. Furthermore, the Adam optimizer shows a better performance compared to the others. It is indicated that in the lower SNR part, the three optimization algorithms show similar performance and follow the same trend, which confirms the proposed model robustness with three optimization algorithms.

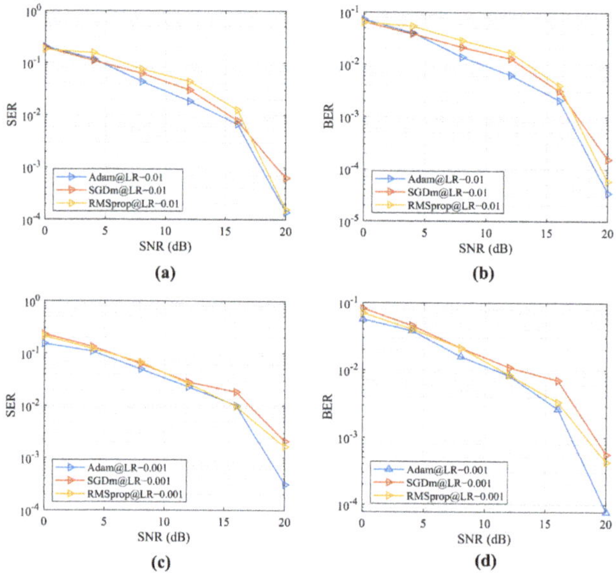

Figure 11. SER and BER simulation results of CNN-GRU model with different learning rate (LR) and optimization algorithms for RIS-assisted MIMO communication: (**a**) SER with LR = 0.01; (**b**) BER with LR = 0.01; (**c**) SER with LR = 0.001; (**d**) BER with LR = 0.001.

Figure 12 shows the throughput analysis comparison of the proposed CNN-GRU with the traditional method, Demod-CNN model, TCN, and LSTM model regarding different SNR values. The graph demonstrates that the suggested model performs almost as well as other approaches in the SNR regions of (0–5) dB. At the initial values of the SNR, the achievable rate of the CNN-GRU is 1.69 bps, which is similar to the other methods and is 1.85 bps with 5 dB SNR. After 5 dB SNR, the proposed CNN-GRU outperforms the other methods, and it achieves the maximum rate at a 20 dB SNR value. Thus, the proposed model exhibits the robustness of handling optimal prediction rates in different SNR ranges.

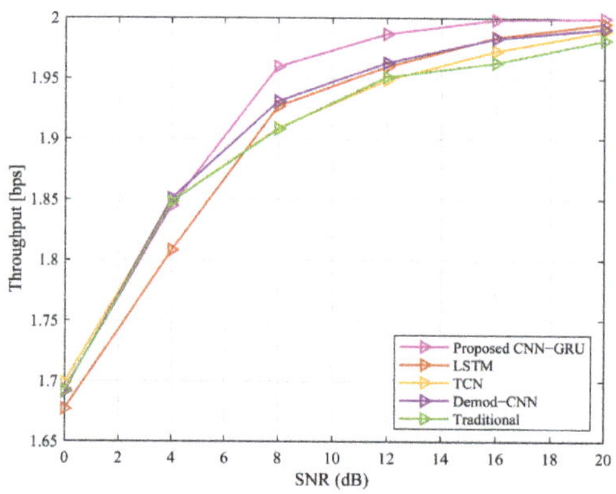

Figure 12. Throughput simulation results of CNN-GRU-based RIS-assisted MIMO communication.

3.2. Complexity Analysis

The suggested combined model's computational complexity is represented by the following formula: $\mathcal{O}(B_s \times E_s(r_c \times p_c \times t_c) + t_s t_h(3t_i + 3t_h + 3)$, where B_s represents the number of input packets obtained, E_s represents the size of an OFDM block, and r_c represents the size of the CNN input; the number of filter size and neuron size of CNN are p_c and t_c, respectively; the number of features in the input vector is t_i; t_s is the input time sequence size; and t_h is the number of hidden units. On the other hand, the computational complexity of Demod-CNN [40] can be defined as $\mathcal{O}(B_s \times E_s(r_c \times p_c \times t_c) \times 2)$. Additionally, the computational complexity of the traditional OFDM system can be expressed as $\mathcal{O}(M_s)$, where the modulation order is M_s [53]. Because only IFFT and FFT are employed, the traditional OFDM is computationally efficient. We consider the parameter values to be $B_s = 100$, $E_s = 128$, $r_c = 8 \times 1$, $p_c = 3 \times 64$, $t_c = 100$, $t_s = 64$, $t_i = 300 \times 64$, and $t_h = 300 \times 100$ for example evaluation. After the consideration of these parameters using the Demod-CNN expression, we require 3.9×10^9 operations. In the case of the proposed model, we need 2.8×10^{11} operations for the given parameters. The research above demonstrates that the time complexity of the suggested model is greater than that of the conventional technique and the Demod-CNN model. Although the proposed model requires more complexity than others, it outperforms them all in terms of performance, and GPU parallel computing can speed up the process.

4. Conclusions

In this paper, a combined DL network to obtain the optimal decoding rate is proposed in the RIS-assisted MIMO wireless system. The proposed model is configured with a 1-D CNN model and GRU module for acquiring the maximum prediction rate. The proposed model is trained with simulated OFDM data over the RIS-assisted channel network in the offline phase. Then, the learned model is deployed in the online phase to retrieve the original data in the UE terminal to test the efficiency of the proposed model. As a result, it is possible to extract the transmitted data and determine the BER and SER performance. The error performance of the proposed model is evaluated with the well-known Adam optimization algorithm. According to the simulation results, the proposed method has a 10^{-5} BER and 10^{-4} SER in 20 dB and is better compared to other benchmarking models. In addition, the proposed CNN-GRU model has the highest throughput of 2 in 20 dB SNR. Thus, it is shown that the performance of the proposed model outperforms the conventional, Demod-CNN, TCN, and LSTM models in terms of BER, SER, and throughput. The proposed system might offer a way to make the 5G communication system better. Future applications of the proposed model could include more advanced, complicated systems for RIS-assisted wireless communication systems.

Author Contributions: Conceptualization, M.H.R. and M.A.S.S.; methodology, M.H.R.; software, M.H.R., M.A.S.S. and M.A.A.; validation, M.H.R. and M.A.S.S.; formal analysis, M.H.R., M.A.S.S. and M.A.A.; investigation, M.H.R. and M.A.S.S.; resources, H.-K.S.; data curation, M.H.R. and M.A.S.S.; writing—original draft preparation, M.H.R. and M.A.S.S.; writing—review and editing, M.H.R., M.A.S.S., M.A.A., D.-S.K. and H.-K.S.; visualization, M.H.R., M.A.S.S. and M.A.A.; supervision, Y.-H.Y., D.-S.K. and H.-K.S.; project administration, Y.-H.Y., D.-S.K. and H.-K.S.; funding acquisition, H.-K.S. All authors have read and agreed to the published version of the manuscript.

Funding: This work was supported in part by the ICT R&D Program of MSIT/IITP [IITP-2022-2021-0-01816, A Research on Core Technology of Autonomous Twins for Metaverse] and in part by the Basic Science Research Program through the National Research Foundation of Korea (NRF) funded by the Ministry of Education (2020R1A6A1A03038540).

Data Availability Statement: Not applicable.

Conflicts of Interest: The authors declare no conflict of interest.

References

1. Saad, W.; Bennis, M.; Chen, M. A vision of 6G wireless systems: Applications, trends, technologies, and open research problems *IEEE Netw.* **2019**, *34*, 134–142. [CrossRef]
2. Wu, Q.; Zhang, R. Towards smart and reconfigurable environment: Intelligent reflecting surface aided wireless network. *IEEE Commun. Mag.* **2019**, *58*, 106–112. [CrossRef]
3. Basar, E.; Di Renzo, M.; De Rosny, J.; Debbah, M.; Alouini, M.S.; Zhang, R. Wireless communications through reconfigurable intelligent surfaces. *IEEE Access* **2019**, *7*, 116753–116773. [CrossRef]
4. Di Renzo, M.; Zappone, A.; Debbah, M.; Alouini, M.S.; Yuen, C.; De Rosny, J.; Tretyakov, S. Smart radio environments empowered by reconfigurable intelligent surfaces: How it works, state of research, and the road ahead. *IEEE J. Sel. Areas Commun.* **2020**, *38*, 2450–2525. [CrossRef]
5. Basharat, S.; Hassan, S.A.; Pervaiz, H.; Mahmood, A.; Ding, Z.; Gidlund, M. Reconfigurable intelligent surfaces: Potentials, applications, and challenges for 6G wireless networks. *IEEE Wirel. Commun.* **2021**, *28*, 184–191. [CrossRef]
6. Shin, B.S.; Oh, J.H.; You, Y.H.; Hwang, D.D.; Song, H.K. Limited Channel Feedback Scheme for Reconfigurable Intelligent Surface Assisted MU-MIMO Wireless Communication Systems. *IEEE Access* **2022**, *10*, 50288–50297. [CrossRef]
7. Do, T.N.; Kaddoum, G.; Nguyen, T.L.; Da Costa, D.B.; Haas, Z.J. Aerial reconfigurable intelligent surface-aided wireless communication systems. In Proceedings of the 2021 IEEE 32nd Annual International Symposium on Personal, Indoor and Mobile Radio Communications (PIMRC), Helsinki, Finland, 13–16 September 2021; IEEE: Piscataway Township, NJ, USA, 2021; pp. 525–530.
8. Huang, C.; Yang, Z.; Alexandropoulos, G.C.; Xiong, K.; Wei, L.; Yuen, C.; Zhang, Z.; Debbah, M. Multi-hop RIS-empowered terahertz communications: A DRL-based hybrid beamforming design. *IEEE J. Sel. Areas Commun.* **2021**, *39*, 1663–1677. [CrossRef]
9. Jung, J.S.; Park, C.Y.; Oh, J.H.; Song, H.K. Intelligent reflecting surface for spectral efficiency maximization in the multi-user MISO communication systems. *IEEE Access* **2021**, *9*, 134695–134702. [CrossRef]
10. Guo, H.; Liang, Y.C.; Chen, J.; Larsson, E.G. Weighted sum-rate maximization for reconfigurable intelligent surface aided wireless networks. *IEEE Trans. Wirel. Commun.* **2020**, *19*, 3064–3076. [CrossRef]
11. Yang, Y.; Zheng, B.; Zhang, S.; Zhang, R. Intelligent reflecting surface meets OFDM: Protocol design and rate maximization. *IEEE Trans. Commun.* **2020**, *68*, 4522–4535. [CrossRef]
12. Özdogan, Ö.; Björnson, E.; Larsson, E.G. Intelligent reflecting surfaces: Physics, propagation, and pathloss modeling. *IEEE Wirel. Commun. Lett.* **2019**, *9*, 581–585. [CrossRef]
13. You, C.; Zheng, B.; Zhang, R. Channel estimation and passive beamforming for intelligent reflecting surface: Discrete phase shift and progressive refinement. *IEEE J. Sel. Areas Commun.* **2020**, *38*, 2604–2620. [CrossRef]
14. Basar, E. Reconfigurable intelligent surface-based index modulation: A new beyond MIMO paradigm for 6G. *IEEE Trans. Commun.* **2020**, *68*, 3187–3196. [CrossRef]
15. Wei, L.; Huang, C.; Alexandropoulos, G.C.; Wei, E.; Zhang, Z.; Debbah, M.; Yuen, C. Multi-user holographic MIMO surfaces: Channel modeling and spectral efficiency analysis. *IEEE J. Sel. Top. Signal Process.* **2022**, *16*, 1112–1124. [CrossRef]
16. Tan, X.; Sun, Z.; Koutsonikolas, D.; Jornet, J.M. Enabling indoor mobile millimeter-wave networks based on smart reflect-arrays. In Proceedings of the IEEE INFOCOM 2018-IEEE Conference on Computer Communications, Honolulu, HI, USA, 16–19 April 2018; IEEE: Piscataway Township, NJ, USA, 2018; pp. 270–278.
17. Ye, J.; Guo, S.; Alouini, M.S. Joint reflecting and precoding designs for SER minimization in reconfigurable intelligent surfaces assisted MIMO systems. *IEEE Trans. Wirel. Commun.* **2020**, *19*, 5561–5574. [CrossRef]
18. Huang, C.; Zappone, A.; Alexandropoulos, G.C.; Debbah, M.; Yuen, C. Reconfigurable intelligent surfaces for energy efficiency in wireless communication. *IEEE Trans. Wirel. Commun.* **2019**, *18*, 4157–4170. [CrossRef]
19. Zhang, Z.; Dai, L. A joint precoding framework for wideband reconfigurable intelligent surface-aided cell-free network. *IEEE Trans. Signal Process.* **2021**, *69*, 4085–4101. [CrossRef]
20. Jiang, T.; Cheng, H.V.; Yu, W. Learning to reflect and to beamform for intelligent reflecting surface with implicit channel estimation. *IEEE J. Sel. Areas Commun.* **2021**, *39*, 1931–1945. [CrossRef]
21. Jung, M.; Saad, W.; Jang, Y.; Kong, G.; Choi, S. Performance analysis of large intelligent surfaces (LISs): Asymptotic data rate and channel hardening effects. *IEEE Trans. Wirel. Commun.* **2020**, *19*, 2052–2065. [CrossRef]
22. Dong, P.; Zhang, H.; Li, G.Y.; Gaspar, I.S.; NaderiAlizadeh, N. Deep CNN-based channel estimation for mmWave massive MIMO systems. *IEEE J. Sel. Top. Signal Process.* **2019**, *13*, 989–1000. [CrossRef]
23. Ye, H.; Li, G.Y.; Juang, B.H. Power of deep learning for channel estimation and signal detection in OFDM systems. *IEEE Wirel. Commun. Lett.* **2017**, *7*, 114–117. [CrossRef]
24. Sejan, M.A.S.; Rahman, M.H.; Shin, B.S.; Oh, J.H.; You, Y.H.; Song, H.K. Machine Learning for Intelligent-Reflecting-Surface-Based Wireless Communication towards 6G: A Review. *Sensors* **2022**, *22*, 5405. [CrossRef] [PubMed]
25. Rahman, M.H.; Sejan, M.A.S.; Yoo, S.G.; Kim, M.A.; You, Y.H.; Song, H.K. Multi-User Joint Detection Using Bi-Directional Deep Neural Network Framework in NOMA-OFDM System. *Sensors* **2022**, *22*, 6994. [CrossRef]
26. Zappone, A.; Di Renzo, M.; Debbah, M. Wireless networks design in the era of deep learning: Model-based, AI-based, or both? *IEEE Trans. Commun.* **2019**, *67*, 7331–7376. [CrossRef]
27. Huang, C.; Mo, R.; Yuen, C. Reconfigurable intelligent surface assisted multiuser MISO systems exploiting deep reinforcement learning. *IEEE J. Sel. Areas Commun.* **2020**, *38*, 1839–1850. [CrossRef]

28. Rahman, M.H.; Sejan, M.A.S.; Aziz, M.A.; Baik, J.I.; Kim, D.S.; Song, H.K. Deep Learning Based Improved Cascaded Channel Estimation and Signal Detection for Reconfigurable Intelligent Surfaces-Assisted MU-MISO Systems. *IEEE Trans. Green Commun. Netw.* **2023**. [CrossRef]
29. Jiang, H.; Dai, L.; Hao, M.; Mackenzie, R. End-to-End Learning for RIS-Aided Communication Systems. *IEEE Trans. Veh. Technol.* **2022**, *71*, 6778–6783. [CrossRef]
30. Khan, S.; Khan, K.S.; Haider, N.; Shin, S.Y. Deep-learning-aided detection for reconfigurable intelligent surfaces. *arXiv* **2019**, arXiv:1910.09136.
31. Gao, S.; Dong, P.; Pan, Z.; Li, G.Y. Deep multi-stage CSI acquisition for reconfigurable intelligent surface aided MIMO systems. *IEEE Commun. Lett.* **2021**, *25*, 2024–2028. [CrossRef]
32. Liu, S.; Gao, Z.; Zhang, J.; Di Renzo, M.; Alouini, M.S. Deep denoising neural network assisted compressive channel estimation for mmWave intelligent reflecting surfaces. *IEEE Trans. Veh. Technol.* **2020**, *69*, 9223–9228. [CrossRef]
33. Huang, C.; Alexandropoulos, G.C.; Yuen, C.; Debbah, M. Indoor signal focusing with deep learning designed reconfigurable intelligent surfaces. In Proceedings of the 2019 IEEE 20th International Workshop on Signal Processing Advances in Wireless Communications (SPAWC), Cannes, France, 2–5 July 2019; IEEE: Piscataway Township, NJ, USA, 2019; pp. 1–5.
34. Taha, A.; Zhang, Y.; Mismar, F.B.; Alkhateeb, A. Deep reinforcement learning for intelligent reflecting surfaces: Towards standalone operation. In Proceedings of the 2020 IEEE 21st International Workshop on Signal Processing Advances in Wireless Communications (SPAWC), Atlanta, GA, USA, 26–29 May 2020; IEEE: Piscataway Township, NJ, USA, 2020; pp. 1–5.
35. Liao, G.P.; Gao, W.; Yang, G.J.; Guo, M.F. Hydroelectric generating unit fault diagnosis using 1-D convolutional neural network and gated recurrent unit in small hydro. *IEEE Sens. J.* **2019**, *19*, 9352–9363. [CrossRef]
36. Yang, B.; Cao, X.; Huang, C.; Yuen, C.; Qian, L.; Di Renzo, M. Intelligent spectrum learning for wireless networks with reconfigurable intelligent surfaces. *IEEE Trans. Veh. Technol.* **2021**, *70*, 3920–3925. [CrossRef]
37. Xu, M.; Zhang, S.; Zhong, C.; Ma, J.; Dobre, O.A. Ordinary differential equation-based CNN for channel extrapolation over RIS-assisted communication. *IEEE Commun. Lett.* **2021**, *25*, 1921–1925. [CrossRef]
38. Zhang, S.; Zhang, S.; Gao, F.; Ma, J.; Dobre, O.A. Deep learning optimized sparse antenna activation for reconfigurable intelligent surface assisted communication. *IEEE Trans. Commun.* **2021**, *69*, 6691–6705. [CrossRef]
39. Gupta, K.D.; Nigam, R.; Sharma, D.K.; Dhurandher, S.K. LSTM-Based Energy-Efficient Wireless Communication With Reconfigurable Intelligent Surfaces. *IEEE Trans. Green Commun. Netw.* **2022**, *6*, 704–712. [CrossRef]
40. Sejan, M.A.S.; Rahman, M.H.; Song, H.K. Demod-CNN: A Robust Deep Learning Approach for Intelligent Reflecting Surface-Assisted Multiuser MIMO Communication. *Sensors* **2022**, *22*, 5971. [CrossRef]
41. Sejan, M.A.S.; Rahman, M.H.; Aziz, M.A.; You, Y.H.; Song, H.K. Temporal Neural Network Framework Adaptation in Reconfigurable Intelligent Surface-Assisted Wireless Communication. *Sensors* **2023**, *23*, 2777. [CrossRef]
42. Zhao, R.; Wang, D.; Yan, R.; Mao, K.; Shen, F.; Wang, J. Machine health monitoring using local feature-based gated recurrent unit networks. *IEEE Trans. Ind. Electron.* **2017**, *65*, 1539–1548. [CrossRef]
43. Cho, K.; Van Merriënboer, B.; Gulcehre, C.; Bahdanau, D.; Bougares, F.; Schwenk, H.; Bengio, Y. Learning phrase representations using RNN encoder-decoder for statistical machine translation. *arXiv* **2014**, arXiv:1406.1078.
44. Helmy, I.; Tarafder, P.; Choi, W. LSTM-GRU Model-Based Channel Prediction for One-Bit Massive MIMO System. *IEEE Trans. Veh. Technol.* **2023**, 1–6. [CrossRef]
45. Ali, M.H.E.; Rabeh, M.L.; Hekal, S.; Abbas, A.N. Deep Learning Gated Recurrent Neural Network-Based Channel State Estimator for OFDM Wireless Communication Systems. *IEEE Access* **2022**, *10*, 69312–69322.
46. Chen, J.; Liang, Y.C.; Cheng, H.V.; Yu, W. Channel estimation for reconfigurable intelligent surface aided multi-user MIMO systems. *arXiv* **2019**, arXiv:1912.03619.
47. Taha, A.; Alrabeiah, M.; Alkhateeb, A. Enabling large intelligent surfaces with compressive sensing and deep learning. *IEEE Access* **2021**, *9*, 44304–44321. [CrossRef]
48. Busari, S.A.; Huq, K.M.S.; Mumtaz, S.; Dai, L.; Rodriguez, J. Millimeter-wave massive MIMO communication for future wireless systems: A survey. *IEEE Commun. Surv. Tutorials* **2017**, *20*, 836–869. [CrossRef]
49. Wei, X.; Shen, D.; Dai, L. Channel estimation for RIS assisted wireless communications—Part II: An improved solution based on double-structured sparsity. *IEEE Commun. Lett.* **2021**, *25*, 1403–1407. [CrossRef]
50. Björnson, E.; Larsson, E.G.; Marzetta, T.L. Massive MIMO: Ten myths and one critical question. *IEEE Commun. Mag.* **2016**, *54*, 114–123. [CrossRef]
51. Van Laarhoven, T. L2 regularization versus batch and weight normalization. *arXiv* **2017**, arXiv:1706.05350.
52. Kingma, D.P.; Ba, J. Adam: A method for stochastic optimization. *arXiv* **2014**, arXiv:1412.6980.
53. Jaradat, A.M.; Hamamreh, J.M.; Arslan, H. Modulation options for OFDM-based waveforms: Classification, comparison, and future directions. *IEEE Access* **2019**, *7*, 17263–17278. [CrossRef]

mathematics

Article

Energy-Efficient Resource Allocation for D2D-V2V Communication with Load Balancing

Jie Bi, Xizhong Qin * and Zhenhong Jia

College of Information Science and Engineering, Xinjiang University, Urumqi 830000, China
* Correspondence: qinxz@xju.edu.cn

Abstract: The significance of vehicle-to everything (V2X) communication in ensuring road safety is undeniable. In addition, real-time vehicle communication requires an ample amount of spectrum resources. However, the existing spectrum resources are seriously scarce, and the utilization rate is not high, leading to high delays in V2X communication and other unfavorable factors in the case of fast-moving vehicles, bringing great safety risks to driving. Load balancing is one of the most effective methods to improve spectrum utilization. However, the existing load balancing schemes merely focus on static conditions, with a lack of joint scheduling schemes, which cannot support the communication framework of dynamic V2X. To address both of these issues, in this paper, a new communication method is proposed. In addition, this paper studies a joint load balancing scheme of mobility vehicle-to-vehicle (V2V) and user association under incomplete channel state information (CSI) and realizes the load balancing management of a cross-cell V2X network. An algorithm combining power control and resource allocation mode selection is proposed. In particular, according to different coverage areas, different allocation algorithms are adopted to maximize the overall system efficiency. The simulation results show that this strategy can maintain low latency and effectively improve the system energy efficiency of vehicle users.

Keywords: vehicle-to-vehicle (V2V) communications; vehicle load balancing; mode selection; energy efficiency

MSC: 94-05

1. Introduction

V2V is expected to provide safety applications for vehicles, such as collision warning, traffic information prompts and intersection warning [1]. V2V communication has attracted wide attention from researchers in recent years. V2V network topologies based on device-to-device (D2D) communication have become a research hotspot in recent years, as they can offer high reliability and less delays for vehicles [2,3], and provide a high-quality service experience for vehicles during their movement. However, the exponential growth of communication has brought great pressure to cellular networks. Cisco's report shows that global mobile data traffic may reach 160 eb per month by 2025. With the increasing demands of mobile data, the existing spectrum resources are seriously insufficient, which leads to a serious overload problem [4]. They cannot meet the requirements of safe V2V communication.

At present, many studies have shown that in the coverage area of adjacent base stations (BSs), there is a large gap in peak traffic [5]. For example, in daily study and work, a large number of users are concentrated in some fixed areas, while a few users are located in other areas. This situation will lead to an uneven traffic distribution among cellular networks [6]. This can lead to the further waste of a large amount of spectrum resources, meaning that safe V2V communication cannot be effectively guaranteed. In addition, for vehicle users, due to the high-speed mobility of vehicles, vehicles are sensitive to the transmission scenarios of cellular networks, resulting in highly dynamic topologies of V2X wireless networks [2]; thus, studying dynamic vehicle load balancing strategies in cellular network vehicle communication is a great challenge.

There are other methods to effectively solve the shortage of spectrum resources in safety-related V2V communication. The 3rd Generation Partnership Project (3GPP) Release 14 proposes two resource allocation methods for V2V communication based on the D2D:mode 3 (resource allocation by the evolved node B (eNB)) and mode 4 (resource selection by user equipment (UE) [7]. In mode 3, the UE requests the transmission resources of the eNB, and then the eNB allocates the granted resource blocks (RBs) to the UE for data transmission. Mode 3 is also divided into the following two sub-modes: the dedicated mode, which will not be interfered with by other users when using dedicated RBs for direct communication, and the reuse mode, where vehicular user equipment (VUE) can communicate by reusing RBs occupied by primary user equipment (PUE), which will cause co-channel interference. In mode 4, VUEs independently select resources from the preconfigured unauthorized resource pool [8]. Moreover, mode 4 can provide V2X services in and out of coverage without the eNB [9].

In [10,11], it is proved that different areas (residential areas, commercial areas, transportation areas, entertainment areas and comprehensive areas) have different transportation modes at different time points and in different peak periods. In order to make rational use of spectrum resources and improve their utilization rate, it is critical for vehicles to achieve load balance among different BSs [4].

With regard to load balancing, [12] considers an ultra-dense small cell network, which balances the cellular traffic among cells through D2D routing and maximizes the sum rate of the system. In [4], through a D2D load balancing mechanism, the D2D communication between cells was used to realize the transfer of traffic between different cells, and explore the relationship between traffic demand and D2D overhead. The authors of [10,11] used intelligent user association to dynamically associate users with different BSs and realize load and flow balance management.

It has been found that D2D load balancing schemes and user association schemes have different effects on different time scales. By using a smart user association scheme for large time scales and using a D2D scheme for small time scales, better results can be achieved. In daily cellular networks, adjacent BSs often produce overlapping coverage areas [13]. When the VUE receiver (VUR) or VUE transmitter (VUT) is located in the overlapping area and the VUT or VUR is located in the non-overlapping area, a D2D scheme is suitable, and when both the VUR and VUT are located in the overlapping area, a user association scheme is suitable, which can achieve better network performance. Researchers seldom consider combining D2D with user association to achieve load balancing and energy efficiency improvement. At the same time, due to the mobility of vehicles, the topology of vehicle wireless networks is highly dynamic, which will introduce the Doppler effect on vehicle links. It is difficult to obtain accurate CSI of vehicle links. Therefore, the above works are not directly applicable to V2X networks since they do not consider the chief factor of the Doppler effect [14–16].

The authors of [17–19] discussed the maximization of V2X system throughput. The authors of [17] studied a resource allocation scheme based on an interference hypergraph; and the authors of [18] put forward a new v2v resource allocation scheme, which improves the reliability and delay of vehicles; and the authors of [19] put forward a resource allocation method based on deep reinforcement learning to maximize throughput. The author of [20,21] proposed a novel Reverse Auction-based Computation Offloading and Resource Allocation Mechanism, which minimizes the cost of the Cloud Service Center. The author of [22–24] improved the reliability and spectral efficiency of V2X communication through deep learning. However, due to the increase in the number of vehicles and the deterioration of traffic conditions, the energy consumption of vehicle networking will increase greatly, and the total fuel consumption may increase by 120% [25], which will aggravate energy consumption and trigger severe environmental issues. The methods used to improve the energy efficiency of vehicle systems have not been well studied.

In addition, the shortcomings of the work discussed above are as follows:

- Only static user load balancing and resource allocation were studied, which cannot meet the requirements of the communication framework of dynamic V2X.
- There is a lack of research on D2D and user association joint scheduling.
- Therefore, in this paper, we considered a joint load balancing scheme of mobility V2V and user association under incomplete CSI, through a new method for power control and overlapping area and non-overlapping area mode selection, to further improve the overall energy efficiency of the system.
- In view of the above problems, the contributions of this paper are as follows:
- This paper mainly studies the load balancing scheme between two adjacent BSs, and a dynamic energy efficiency optimization mechanism of joint load balancing across cells is proposed. In different cellular systems, the spectrum traffic will fluctuate unevenly, and there will be a load balancing problem between two cells (one cell's spectrum resources are not fully utilized, and the other cell's spectrum resources are seriously insufficient). In order to rationally utilize spectrum resources and improve their utilization rate, a dynamic load balancing strategy for vehicles is proposed in combination with D2D-V2V and user association schemes.
- According to the proposed system model, the mathematical expression is an NP-hard problem, and a method of joint power control and resource allocation mode selection is proposed. For the power control problem, a power control method based on Dienkelbach [26] is proposed, which maximizes the energy efficiency of a single VUE. An iterative algorithm for power control optimization is designed, which can significantly reduce the computational complexity. Using the obtained results, the original optimization problem is further simplified to the problem of resource allocation mode selection.
- For the resource allocation problem, according to the coverage situation of different real BSs, two resource allocation mode selection algorithms are proposed. The unoverlapped region resource allocation algorithm (URRAA) is used in non-overlapping areas, and the overlapped region resource allocation algorithm (ORRAA) is used in overlapping areas. These two algorithms optimize the sum of energy efficiency through three different modes of joint resource scheduling.
- The main contents of the remaining chapters of the paper are as follows: Section 2 introduces the system model and the modeling and definition of the formulas; Section 3 focuses on the modeling and definition of energy efficiency under the background of V2X communication; Section 4 describes the power control mode selection algorithm; Section 5 evaluates and analyzes the system performance of the proposed algorithm; finally, Section 6 briefly summarizes our main work and presents some conclusions.

2. System Model

2.1. System Model for V2X Network

The V2V communication network based on D2D is shown in Figure 1. A V2X network based on an urban environment is proposed.

Without losing generality, we considered two cell scenarios, in which the BS is a circular coverage area with radius R; BS1 belongs to a cellular structure with busy user visits, and BS2 belongs to a cellular structure with less user visits (there are unused spectrum resources). We divided the BS1 users into two different parts: first, there are users in the BS overlapping coverage area (including all vehicle user pairs in the coverage area, one vehicle in the non-overlapping area of BS1, and one vehicle in the overlapping area), along with other users (vehicle user pairs in the non-overlapping area of BS1).

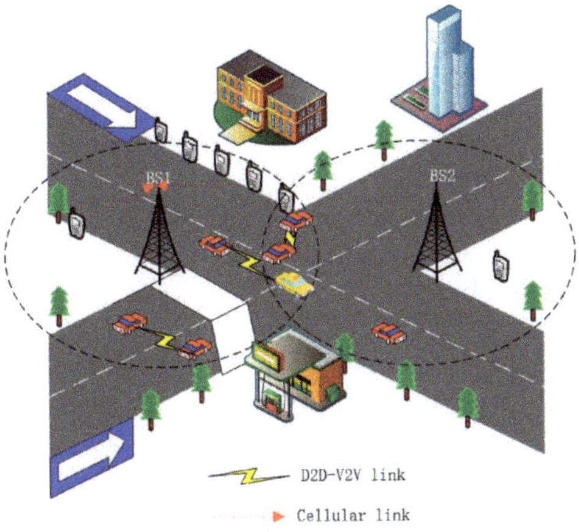

Figure 1. The system model of the D2D-V2V network.

On the premise of orthogonal frequency division multiple access (OFDMA) [27], each PUE is assigned a unique (orthogonal) channel in advance to reduce the interference between PUEs. At the same time, each pair of VUEs multiplexes one and only one subchannel.

The following definitions are used: there are $L1$ vacant licensed RBs and U vacant unlicensed RBs in BS1. $\widetilde{F}_{L1} = (F_{11}, \ldots, F_{l1}, \ldots, F_{L1})$ is the set of vacant licensed RBs in BS1, and $\widetilde{F}_U = (F_1, \ldots, F_u, \ldots, F_U)$ is the set of vacant unlicensed RBs. There are $L2$ vacant licensed RBs in BS2. $\widetilde{F}_{L2} = (F_{12}, \ldots, F_{l2}, \ldots, F_{L2})$ is the set of vacant licensed RBs in BS2. $C_1 = \{P_1UE_1, P_1UE_2, \ldots, P_1UE_{M1}\}$ is the set of PUEs in BS1, and $C_2 = \{P_2UE_1, P_2UE_2, \ldots, P_2UE_{M2}\}$ is the set of PUEs in BS2. $V = \{VUE_1, VUE_2, \ldots, VUE_N\}$ is the set of VUEs in BS1, $V_z = \{V_zUE_1, V_zUE_2, \ldots, V_zUE_{Nz}\}$ is the set of vehicle users in non-overlapping areas, $V_y = V - V_z$ is the set of vehicle users in overlapping areas and $Ny = N - Nz$ represents the number of VUEs in the overlapping coverage area of BS1.

In addition, we assume that in all lanes, the VUE transmitters uniformly fall in different positions in the lanes by adopting the spatial Poisson process. In order to maintain a safe distance between VUE pairs, the number of VUE pairs varies according to different speeds, and the faster the vehicle speed, the lower the number of VUE pairs. The cluster distribution model [28] was considered for each VUE pair, so as to ensure normal communication and connection reliability between the VUR and VUT. Due to road congestion in urban traffic peak hours, we approximately assumed that each VUT and its VUR have the same movement direction and speed, and the vehicle density fitting is as follows [7]:

$$\lambda = \frac{1}{2.5 \times v} \tag{1}$$

where v is the absolute vehicle speed, km/h. In this model, communication between the VUR and VUT is realized through D2D technology. VUEs communicate in different modes (mode 3 or mode 4) according to different situations. Mode 3 can be divided into special mode or reuse mode. In the reuse mode, VUEs reuse the uplink cellular resource RBs [28], where each VUE pair can occupy one RB, and each RB can be occupied by at most one VUE. The meanings of the key numeric symbols are shown in Table 1.

Table 1. Key mathematical symbol.

Symbol	Definition
$M1$	Number of PUEs in BS1
$M2$	Number of PUEs in BS2
N	Number of VUEs pairs in BS1
Nz	Number of VUEs of the non-overlapping coverage area of BS1
Ny	Number of VUEs of the overlapping coverage area of BS1
$\widetilde{F}_L, \widetilde{F}_U$	Number of vacant licensed and unlicensed RBs
C, V	Sets of PUEs and VUEs pairs
v	Absolute vehicle speed
g_i	Channel gain between VUT i and VUR i
g_{jB}	Channel gain between PUE j and eNB1
g_{iB}	Channel gain between VUT i and eNB1
$g_{i,j}$	Channel gain between PUE j and VUR i
σ^2	White Gaussian noise
X	Mode selection matrix
$x^{(1)}$	Indication matrix for VUEs in mode 4
$x^{(2)}$	Indication matrix for VUEs in dedicated mode
$x^{(3)}$	Indication matrix for VUEs in reuse mode
$p^{(1)}$	Transmit power vector for VUEs in mode 4
$p^{(2)}$	Transmit power vector for VUEs in dedicated mode
$p^{(3)}$	Transmit power vector for VUEs in reuse mode
$\zeta_{min}^{C_{min}^{V}}$	Minimum SINR of PUEs and VUEs
$p_{max}^{C_{max}^{V}}$	Maximum transmit power of PUEs and VUEs
$pC*_{i,j}, p_{i,j}^{(3)*}$	Optimal power allocation vector

2.2. Communication Channel Model for V2X Network

In the channel modeling, we considered both large-scale fading and small-scale fading, which are independent block fading channels [29]. For example, $g_i = [g_{iz}, g_{iy}]$ is defined as the channel gain between VUT $\boldsymbol{i}$ and its corresponding receiver, g_{iz} represents the channel gain of the non-overlapping area, and g_{iy} represents the channel gain of the overlapping area, which can be expressed as

$$g_i = |h_i|^2 L_i \tag{2}$$

where $L_i = d_i^{-\alpha}$ represents the large-scale fading effect of VUE i and h_i represents the small-scale fading effect. α represents the path-loss exponent and d_i indicates the distance between the transmitting and receiving ends of the vehicle.

In each TTI, other large-scale fading effects, such as the shadowing effect, are approximately constant, which can be obtained by eNB. However, for small-scale fading, different analyses can be used for different situations. Considering the links directly connected to the eNB shown in Figure 1 and the g_{iz} and g_{iy} of VUE i, we assume that the small-scale fading information of these links can be obtained by the eNB and distributed independently and identically according to $CN(0,1)$ [2]. The reasons are as follows:

- The Doppler effect will significantly affect small-scale fading, but the direction and speed of motion between the VUR and VUT are the same, meaning the influence of the Doppler effect can be ignored.
- In the V2V establishment stage, the channel information of each VUE pair is obtained by the eNB.

The interference link $g_{i,j}$ from the PUE transmitter $\boldsymbol{j}$ to the VUE receiver $\boldsymbol{i}$ is given as [29]

$$g_{i,j} = |h_{i,j}|^2 L_{i,j} \tag{3}$$

Because the PUE transmitter j is relatively stationary or only moves slightly, while the VUE receiver i moves rapidly, there will be a large relative movement speed between them, which will lead to a serious Doppler frequency shift effect and imperfect CSI of $g_{i,j}$, which will have a great impact on the small-scale fading $h_{i,j}$ [30], and the eNB will not be able to obtain accurate knowledge of the small-scale fading. In this case, [30] proves that the first-order Gaussian–Markov process can provide more accurate modeling. The modeling of $h_{i,j}$ is as follows:

$$h_{i,j} = \varepsilon\widetilde{h}_{i,j} + \sqrt{1-\varepsilon^2}e_{i,j} \tag{4}$$

where $\widetilde{h}_{i,j}$ represents the estimated channel gain, $e_{i,j}$ represents the distribution of the small-scale fading and estimation error, $\widetilde{h}_{i,j}$ and $e_{i,j}$ are independently and identically distributed and obey $CN(0,1)$ and the coefficient $\varepsilon(0<\varepsilon<1)$ quantifies the channel correlation of two consecutive time slots, where ε is given as [31]

$$\varepsilon = J_0(2\pi f_D T) \tag{5}$$

where J_0 is the first type of zero-order Bessel function; f_D represents the maximum Doppler frequency shift; $f_D = vf_c/c$, where v indicates the vehicle speed, f_c is the carrier frequency, and $c = 3\times\frac{10^8 m}{s}$; T is the period feedback latency.

2.3. Data Transmission and Energy Consumption for V2V Communication

When VUE i is in dedicated mode or mode 4, it will not be interfered with by the PUE. Tts signal-to-interference-plus-noise -ratio (SINR) ξ_i^V and instantaneous data rate r_i^V can be defined as

$$\xi_i^V = \frac{p_i^V g_i}{\sigma^2} \tag{6}$$

$$r_i^V = log_2(1+\xi_i) \tag{7}$$

where p_i^V is the transmission power of VUE i and σ^2 is additive white Gaussian noise. VUE i multiplexes the resource block of PUE j; its SINR $\xi_{i,j}^{(3)}$ and data rate $r_{i,j}^{(3)}$ can be expressed as

$$\xi_{i,j}^{(3)} = \frac{p_{i,j}^{(3)} g_i}{p_{i,j}^C g_{i,j} + \sigma^2} \tag{8}$$

$$r_{i,j}^{(3)} = log_2(1+\xi_{i,j}^{(3)}) \tag{9}$$

where $p_{i,j}^{(3)}$ indicates the transmission power of VUE i in reuse mode, and $p_{i,j}^C$ indicates the transmission power of PUE j.

In the same way, when PUE j does not share an RB with VUE i, its SINR ξ_j^C and data rate r_j^C can be expressed as

$$\xi_j^C = \frac{p_j^C g_{i,B}}{\sigma^2} \tag{10}$$

$$r_j^C = log_2(1+\xi_j^C) \tag{11}$$

where p_j^C represents the transmission power of PUE j when there is no interference signal signal. However, when the resource block RB of PUE j is reused by VUE i, the SINR $\xi_{i,j}^C$ and data rate $r_{i,j}^C$ of PUE j can be expressed as

$$\xi_{i,j}^C = \frac{p_{i,j}^C g_{j,B}}{p_{i,j}^{(3)} g_{i,B} + \sigma^2} \tag{12}$$

$$r_{i,j}^C = log_2(1+\xi_{i,j}^C) \tag{13}$$

V2X communication requires a much stricter delay and reliability than traditional cellular communication. The minimum SINR requirement of V2X communication needs to be guaranteed. The SINR ξ_i^V needs to be above a certain threshold ξ_{min}^V, as follows:

$$\xi_i^V \geq \xi_{min}^V \tag{14}$$

The energy consumed by the I-th VUE link is expressed as

$$E_i = P_i^d + \eta p_i^V \tag{15}$$

The total energy consumed is

$$E_{total} = \sum_{i=1}^{N} E_i \tag{16}$$

where p_i^d denotes the fixed circuit power of both the transmitter and the receiver of the I-th VUE link, and η is a factor accounting for the transmit amplifier efficiency.

3. Problem Formulation

For different communication modes, we define different binary indicator matrices, in which $X^{(1)}$ is the autonomous resource selection mode, $X^{(2)}$ is the dedicated mode and $X^{(3)}$ is the reuse mode. We assumed that each VUE can only choose one communication mode; here, its binary indicator variable was set to 1, and the other indicator variables were set to 0.

The throughput of communication in the autonomous selection mode is

$$R^{(1)} = \sum_{i=1}^{N} X^{(1)} log_2(1 + \frac{p_i^{(1)} g_i}{\sigma^2}) \tag{17}$$

The system throughput in the dedicated mode is

$$R^{(2)} = \sum_{i=1}^{N} X^{(2)} log_2(1 + \frac{p_i^{(2)} g_i}{\sigma^2}) \tag{18}$$

The throughput of the vehicle system in the reuse mode is

$$R^{(3)} = \sum_{i=1}^{N} \sum_{j=1}^{M} X^{(3)} log_2(1 + \frac{p_{i,j}^{(3)} g_i}{p_{i,j}^C g_{i,j} + \sigma^2}) \tag{19}$$

The total vehicle throughput is

$$R_{total} = R^{(1)} + R^{(2)} + R^{(3)} \tag{20}$$

At the same time, due to the rapid mobility of vehicles and the change in communication environment, the current V2X communication practice also has potential deficiencies in energy efficiency (EE) [6]. Therefore, our goal was to maximize the energy efficiency of vehicle systems. The optimization problem is shown in Formula (21).

$$\max_{\{p_i^V, p_{i,j}^C, X^{(1)}, X^{(2)}, X^{(3)}\}} EE = \frac{R_{total}}{E_{total}} \tag{21}$$

$$subject\ to$$

$$x_i^{(1)}, x_i^{(2)}, x_{i,j}^{(3)} \in \{0,1\}, \forall i,j \tag{21a}$$

$$\sum_{i=1}^{N} x_i^{(1)} \le U, \forall i \tag{21b}$$

$$\sum_{i=1}^{N} x_i^{(2)} \le L - M, \forall i \tag{21c}$$

$$x_i^{(1)} + x_i^{(2)} + \sum_{i=1}^{N} x_{i,j}^{(3)} \le 1, \forall i \tag{21d}$$

$$\sum_{i=1}^{N} x_{i,j}^{(3)} \le 1, \forall j \tag{21e}$$

$$\sum_{i=1}^{N} x_{i,j}^{(3)} \xi_{i,j}^{C} + \left(1 - \sum_{i=1}^{N} x_{i,j}^{(3)}\right) \xi_j^C \ge \xi_{min}^{C}, \forall j \tag{21f}$$

$$x_i^{(1)} \xi_i^{(1)} + x_i^{(2)} \xi_i^{(2)} + \sum_{j=1}^{M} x_{i,j}^{(3)} \xi_{i,j}^{(3)} \ge \xi_{min}^{V}, \forall j \tag{21g}$$

$$\sum_{i=1}^{N} x_{i,j}^{(3)} p_{i,j}^{C} + \left(1 - \sum_{i=1}^{N} x_{i,j}^{(3)}\right) p_j^C \le p_{max}^{C}, \forall j \tag{21h}$$

$$x_i^{(1)} p_i^{(1)} + x_i^{(2)} p_i^{(2)} + \sum_{j=1}^{M} x_{i,j}^{(3)} p_{i,j}^{(3)} \le p_{max}^{V}, \forall j \tag{21i}$$

In the optimization problem (21), $p_i^v = \{p_i^{(1)}, p_i^{(2)}, p_{i,j}^{(3)}\}$, $p_i^{(1)}$ represents the transmission power of the VUE in autonomous resource selection mode, $p_i^{(2)}$ represents the transmission power in dedicated mode, $p_{i,j}^{(3)}$ represents the transmission power in reuse mode and $p_{i,j}^c$ represents the transmission power of the PUE being reused when the VUE is in reuse mode.

Formula (21b) stipulates that the number of VUEs using mode 4 for communication must be less than or equal to the total number of unauthorized RBs. In the same way, Formula (21c) stipulates that the number of VUEs using the dedicated mode for communication must be less than or equal to the total number of authorized RBs. Formula (21d) indicates that a VUE pair can only work in one resource allocation mode at most, and Formula (21e) indicates that each RB can only be shared by one PUE and one VUE pair.

In order to meet the data transmission requirements of the PUE and VUE, we used the constraints (21f) and (21g) to limit the minimum SINR requirements of the PUE and VUE. The maximum transmission power of the PUE and VUE is restricted by (21h) and (21i), respectively.

Because the optimization formula contains binary variables $x^{(1,2,3)}$ and continuous variables $p_i^V, p_{i,j}^C$, the optimization problem is NP-hard, and it is difficult to obtain the optimal solution in polynomial time. Therefore, we solved the optimization problem through a distributed algorithm, i.e., power control and mode selection. We optimized the power in three modes, and then carried out the mode selection algorithm after obtaining the optimal power allocation.

4. Power Control Mode Selection Algorithm

4.1. Power Control Algorithm Based on DIENKEBACH

In this section, we calculate the optimal power allocation value of a single VUE in different modes and propose a power allocation algorithm to reduce the complexity of the problem. When in mode 4 (unlicensed spectrum), the energy efficiency optimization formula of a single VUE can be expressed as

$$\max_{\{p_i^{(1)}\}} EE = \frac{log_2(1 + \frac{p_i^{(1)} g_i}{\sigma^2})}{2p_i^d + \eta p_i^{(1)}}$$

$$subject\ to$$

$$\xi_i^{(1)} \geq \xi_{min}^V \tag{22}$$

$$p_i^{(1)} \leq P_{max}^V$$

According to the principle of non-linear programming, the objective function of the above problem can be expressed as

$$q^* = max\frac{R}{E} \tag{23}$$

The form of the optimization problem is as follows:

$$F(q^*) = max\{R - q^*E\} = 0 \tag{24}$$

Therefore, the above optimization problem is transformed into

$$\underset{\{p_i^{(1)}\}}{max} : log_2(1 + \tfrac{p_i^{(1)}g_i}{\sigma^2}) - q\{2p_i^d + \eta p_i^{(1)}\}$$

$$subject\ to$$

$$log_2(1 + \frac{p_i^{(1)}g_i}{\sigma^2}) \geq \xi_{min}^V \tag{25}$$

$$p_i^{(1)} \leq P_{max}^V$$

This problem is solvable.

We aimed to verify that the above problem is a standard definition of a convex optimization problem.

$$f(p_i^{(1)}) = log_2(1 + \frac{p_i^{(1)}g_i}{\sigma^2}) \tag{26}$$

By finding the second derivative of Formula (26), the second-order judgment condition of the convex function can be obtained. It is easy to prove that f is a concave function, meaning that constraint 1 of the above problem is a concave function, and constraint 2 is a linear function, which is a standard convex optimization problem [32].

When in reuse mode, the optimization formula can be expressed as

$$\underset{\{p_{i,j}^{(3)},p_{i,j}^C\}}{\max}\ EE = \frac{log_2(1 + \frac{p_{i,j}^{(3)}g_i}{p_{i,j}^C g_{i,j} + \sigma^2})}{2p_i^d + \eta p_i^{(3)}}$$

$$subject\ to$$

$$\xi_{i,j}^C = \frac{p_{i,j}^C g_{j,B}}{p_{i,j}^{(3)} g_{j,B} + \sigma^2} \geq \xi_{min}^C \tag{27}$$

$$\xi_{i,j}^{(3)} = \frac{p_{i,j}^{(3)} g_j}{p_{i,j}^C g_{i,j} + \sigma^2} \geq \xi_{min}^V$$

$$p_{i,j}^C \leq p_{max}^C$$

$$p_{i,j}^{(3)} \leq P_{max}^{V}$$

The solution method is the same as the above, and the above problem can be transformed into

$$\max_{\{p_{i,j}^{(3)}, p_{i,j}^{C}\}} : log_2(1 + \frac{p_{i,j}^{(3)} g_i}{p_{i,j}^{C} g_{i,j} + \sigma^2}) - q\{2p_i^d + \eta p_i^{(3)}\}$$

$$subject\ to$$

$$\xi_{i,j}^{C} = \frac{p_{i,j}^{C} g_{j,B}}{p_{i,j}^{(3)} g_{j,B} + \sigma^2} \geq \xi_{min}^{C} \tag{28}$$

$$\xi_{i,j}^{(3)} = \frac{p_{i,j}^{(3)} g_j}{p_{i,j}^{C} g_{i,j} + \sigma^2} \geq \xi_{min}^{V}$$

$$p_{i,j}^{C} \leq p_{max}^{C}$$

$$p_{i,j}^{(3)} \leq P_{max}^{V}$$

Define function $f_1(p_{i,j}^{(3)}, p_{i,j}^{C})$ as

$$f_1(p_{i,j}^{(3)}, p_{i,j}^{C}) = log_2(1 + \frac{p_{i,j}^{(3)} g_{i,B}}{p_{i,j}^{C} g_{i,B} + \sigma^2}) \tag{29}$$

Define function $f_2(p_{i,j}^{(3)}, p_{i,j}^{C})$ as

$$f_2(p_{i,j}^{(3)}, p_{i,j}^{C}) = log_2(1 + \frac{p_{i,j}^{(3)} g_i}{p_{i,j}^{C} g_{i,j} + \sigma^2}) \tag{30}$$

The second derivative of f_1 and f_2 can be found to be less than or equal to 0. The objective function and constraint function of the optimization problem are concave functions, and the other constraints are linear functions, which is a standard convex optimization problem. The MATLAB convex optimization toolkit can be used to find the optimal solution and the optimal power allocation.

The specific process is shown in Algorithm 1.

Algorithm 1: Power control algorithm based on Dienkelbach

1: Initialize q, $p_{i,j}^{C}, p_{i,j}^{(3)}$. And set the precision error as ε;
2: Get $p_{i,j}^{C*}, p_{i,j}^{(3)*}$ according to Formula (27);
3: Get $F(q^*)$ according to Formula (24);
4: Compared $F(q^*)$ with ε; If $F(q^*) \leq \varepsilon$, then through (23), get q^*, jump to step 6, otherwise, skip to step 5;
5: If $F(q^*) \geq \varepsilon$, then get q^* through (23). Go back to step 2;
6: Power control algorithm is finished. Obtain $q^*, pC*_{i,j}, p_{i,j}^{(3)*}$;

4.2. Mode Selection Algorithm

In order to improve the energy efficiency of the whole system, it is necessary to switch as many V2V pairs as possible to unauthorized and dedicated modes. If one V2V pair works in the dedicated mode, it will not produce co-channel interference, and will provide additional orthogonal spectrum resources for other VUE pairs, which will significantly reduce the overall interference of the system and provide more multiplexing opportunities for VUE pairs in the multiplexing mode.

We substituted p_i^v and $p_{i,j}^c$ obtained through the power allocation algorithm in different modes into Formula (21). In this way, Formula (21) can be simplified to Formula (31).

$$\begin{gathered}
\max_{\{X^{(1)},X^{(2)},X^{(3)}\}} EE = \frac{R_{total}}{E_{total}} \\
subject\ to \\
x_i^{(1)}, x_i^{(2)}, x_{i,j}^{(3)} \in \{0,1\}, \forall i,j \\
\sum\nolimits_{i=1}^{N} x_i^{(1)} \le U, \forall i \\
\sum\nolimits_{i=1}^{N} x_i^{(2)} \le L - M, \forall i \\
\sum\nolimits_{i=1}^{N} x_{i,j}^{(3)} \le 1, \forall j
\end{gathered} \tag{31}$$

Formula (31) only contains binary indicator variables, namely $x^{(1)}, x^{(2)},$ and $x^{(3)}$. Therefore, Formula (31) is a 0–1 integer programming problem. In order to obtain the optimal system energy efficiency, we propose a joint resource scheduling method with different communication modes in different coverage areas.

4.2.1. Unoverlapped Region Resource Allocation Algorithm

As shown in Algorithm 2, in step 1, in order to obtain the best energy efficiency value, VUE users in the left non-overlapping area preferentially choose to work in mode 4. The VUE can independently select appropriate resources for communication according to the mapping relationship between the geographical area and resource pool or cognitive radio mode. When the VUE selects suitable unauthorized RBs for communication, the VUE exchanges scheduling information with other VUEs through broadcasting and deletes the unauthorized RBs from the resource pool. This process does not depend on the control of the core network and eNB. Upon detecting that there are no idle unauthorized RBs, the eNB will receive the report information of the VUE, and step 1 ends.

In step 2, the remaining VUEs that are not communicating will switch to the dedicated mode. In the dedicated mode, the eNB is responsible for uniquely allocating authorized RBs to the VUE. Once the VUE occupies an RB, the eNB deletes the used RB from the idle licensed RB set. The algorithm terminates when all idle licensed RBs are allocated or all VUEs can communicate through the obtained free authorized RBs. Otherwise, the remaining VUEs will be able to transfer to reuse mode.

$X^{(1)}$ and $X^{(2)}$ can be obtained by executing the first two steps of the URRAA, and Formula (31) can be further simplified to Formula (32). Formula (32) only contains a binary indicator variable, $x^{(3)}$, which is equivalent to solving the problem concerning which PUE resource block should be reused by the VUE. The matching algorithm was analyzed as follows:

$$\begin{gathered}
\max_{\{X^{(3)}\}} EE = \frac{R_{total}}{E_{total}} \\
subject\ to \\
x_{i,j}^{(3)} \in \{0,1\}, \forall i,j \\
\sum\nolimits_{i=1}^{N} x_{i,j}^{(3)} \le 1, \forall j
\end{gathered} \tag{32}$$

$$\sum_{j=1}^{M} x_{i,j}^{(3)} \leq 1, \forall j$$

For the analysis of the matching algorithm and in order to meet the Qos requirements of VUE users and the information rate requirements of PUE users, the matching problem between the VUE and its reusable PUE can be transformed into a marriage matching problem, because this not only ensures the needs of both parties, but also achieves a better energy efficiency improvement. Both the improved Gale–Shapley (GS) algorithm and Hungarian algorithm can solve the problem of stable matching between two sets, and the number of PUEs and VUEs can be uncertain, but the EE improvement of the improved GS algorithm is not significant. Meanwhile, the Hungarian algorithm can achieve stable matching between the VUE and PUE and obtain a better EE value to meet the demands for minimal delays among vehicles. Therefore, the Hungary algorithm was adopted in this paper.

Algorithm 2: Unoverlapped region resource allocation algorithm

```
Vz represents the set of unmatched VUEs;
Vr represents the set of VUEs sharing same RBs with PUEs;
while Vz ≠ ∅ do;
Step 1: Autonomous Resource Selection Mode Selection
if F̃_U ≠ ∅ then;
  VUE i gets access to vacant unlicensed RB F̂_i by sensing and broadcasting technology;
  x_i^(1) = 1;
  Remove RB F̂_i from vacant unlicensed RBs set F_U;
  Remove VUE i from Vz;
Step 2: Dedicated Mode Selection
 else if F̃_L1 ≠ ∅ then;
        VUE i gets access to vacant licensed RB Fi by the scheduling of eNB;
        x_i^(2) = 1;
        Remove RB Fi from vacant licensed RBs set F̃_L1;
        Remove VUE i from Vz;
     else if F̃_U = ∅&F̃_L1 = ∅ then;
          The Hungarian algorithm obtains the optimal match;
     end if;
 end while;
 URRAA algorithm is finished. Obtain X^(1),X^(2) and X^(3).
```

4.2.2. Overlapped Region Resource Allocation Algorithm

As shown in Algorithm 3, in step 1, because VUE users in overlapping areas can adopt the load balancing scheme of D2D and user joint scheduling proposed in this paper, they can dynamically select the spectrum resources of their own base station and neighboring base stations according to the SINR requirements, so as to obtain the optimal system energy efficiency. This leads to step 2.

In step 2, the remaining uncommunicated VUEs will use the multiplexing mode for communication. Because the first base station is a congested base station, and there are many users, in order to guarantee the service quality of PUE users, all users in the overlapping area reuse the spectrum resources of the second PUE users for communication. In this paper, the Hungarian algorithm was used to match the best PUEs for the vehicles, and one-to-one matching was adopted.

Algorithm 3: Overlapped region resource allocation algorithm

Vy represents the set of unmatched VUEs;
Vr represents the set of VUEs sharing same RBs with PUEs;
while $Vy \neq \varnothing$ do;
Step 1: Base Station Selection
  if $\widetilde{F}_U \neq \varnothing$ then;
    VUE i gets access to vacant unlicensed RB $\hat{F}_i$ by sensing and broadcasting technology;
    $x_i^{(1)} = 1$;
    Remove RB $\hat{F}_i$ from vacant unlicensed RBs set $\widetilde{F}_U$;
    Remove VUE i from Vy;
  else if $\widetilde{F}_{L1}, \widetilde{F}_{L2} \neq \varnothing$ then;
    In dedicated mode, Calculate the best SINR ξ_1 of RBs in BS1 used by VUE $\boldsymbol{i}$;
    Calculate the best SINR ξ_2 of RBs in BS2 used by VUE i;
      if $\xi_1 > \xi_2$;
        Select RBs F_i^1 of BS1;
      else if;
        Select RBs F_i^2 of BS2;
        $x_i^{(2)} = 1$;
        Remove RB F_i^1 (F_i^2) from vacant licensed RBs set $\widetilde{F}_{L1}, \widetilde{F}_{L2}$;
        Remove VUE i from Vy;
    else if $\widetilde{F}_U = \varnothing \& (\widetilde{F}_{L1}, \widetilde{F}_{L2}) = \varnothing$ then;
Step 2: Resource Allocation
  The Hungarian algorithm obtains the optimal match;
  end if;
end while;
ORRAA algorithm is finished. Obtain $X^{(1)}$,$X^{(2)}$ and $X^{(3)}$.

5. Performance Analysis

In this section, we evaluate the performance of the proposed algorithm. We followed the city case simulation in 3GPP TR36.885 [7] (density, speed, vehicle channel, V2V data traffic, etc.). The vehicle position uniformly falls in different positions in the lane through the spatial Poisson process. In order to verify the performance of the mathematical model proposed in this paper, MATLAB 2018a was used to conduct simulation experiments under the configuration of Intel (R) Core (TM) i5-6500CPU@3.20 GHz, RAM8GB and Windows10 Professional.

The main simulation parameters are presented in Table 2. In addition, we simulated and tested the average EE performance more than 1000 times.

Table 2. Simulation parameters.

Simulation Parameter	Value
Cell radius	500 m
Bandwidth of each RB	180 kHz
Noise spectral density	−174 dBm/Hz
Path-loss exponent	4
SINR threshold PUE VUE	8 dB,10 dB
MAX transmit power	23 dBm
Vehicle speed	40 km/h
Number of vacant licensed RBs, L	1
Number of vacant unlicensed RBS, U	3
Circuit power of vehicles	50 mW
Transmit amplifier efficiency	5
Carrier frequency	2 GHz
Channel feedback latency, T	0.5 ms
MAX VUE pair distance	35 m

In order to evaluate the performance of the proposed algorithm, the results of this algorithm were compared with those of an exhaustive algorithm, an algorithm without load balancing (WLB) and the improved GS algorithm (IGS).

Figure 2 shows a comparison of the energy efficiency values of the different algorithm systems. It can be clearly observed that the load balancing strategy proposed in this paper improved the energy efficiency of the vehicles. It was significantly superior to the other algorithms, and its performance was close to that of the optimization algorithm, with an average system performance of 97% of that of the exhaustive algorithm. This is because this paper used spectrum resources that are not fully utilized in neighboring cells, meaning that more vehicle users can communicate in the special mode of the authorized mode without interference from the main users, and the channel condition information is better, enabling the energy efficiency value of the system to increase. Furthermore, with the increase in VUE distance, the energy efficiency of the system showed a downward trend, because with the increase in distance, the path loss increased significantly, and the overall energy efficiency decreased. Energy efficiency increased by an average of 17%. With the increase in the number of resource blocks, the energy efficiency of the system increased accordingly, because the number of vehicles that can be used in special mode also increased, and more vehicles were used in dedicated mode in the overlapping areas, which significantly improved the energy efficiency of the system.

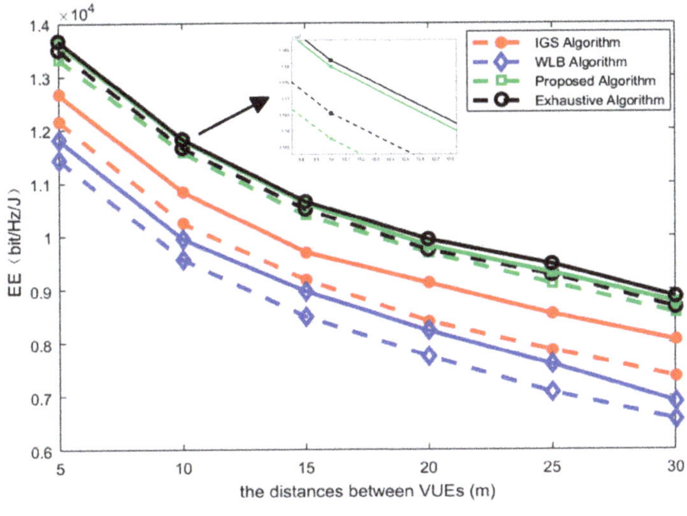

Figure 2. The EE values of VUEs for different distances between the VUR and VUT.

Figure 2 proves that the energy efficiency of this system is close to 97% of the system performance of the exhaustive algorithm, and as shown in Figure 3, the time efficiency of this algorithm was significantly better than that of the exhaustive algorithm, which can better meet the requirements for minimal delays among vehicles. Additionally, with the increase in PUE number, the time of the exhaustive algorithm increased, but the time change of this algorithm was not significant, which shows that the performance of this algorithm is close to that of the exhaustive algorithm. Moreover, the time efficiency was greatly reduced, and the running speed of the algorithm did not slow down with the increase in the number of users.

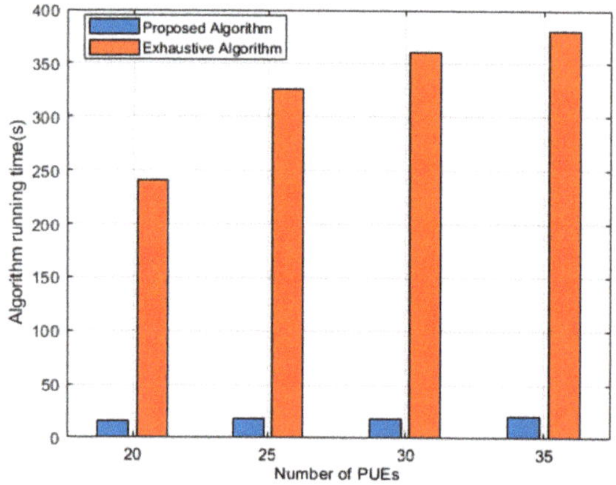

Figure 3. Algorithm running time for different numbers of PUEs.

In this scheme, according to Formula (1), different vehicle speeds and different vehicle densities λ lead to different numbers of VUE pairs, and different VUE pairs will also have a great impact on the system energy efficiency.

It can be observed in Figure 4 that, with the increase in vehicle speed, the energy efficiency of the system showed a downward trend. This is because the vehicle modeling in this paper obeyed the spatial Poisson distribution, the vehicle density was 0 and the number of vehicles was determined by the vehicle speed, that is, with the increase in vehicle speed, the number of vehicles and the energy efficiency of the system will decrease.

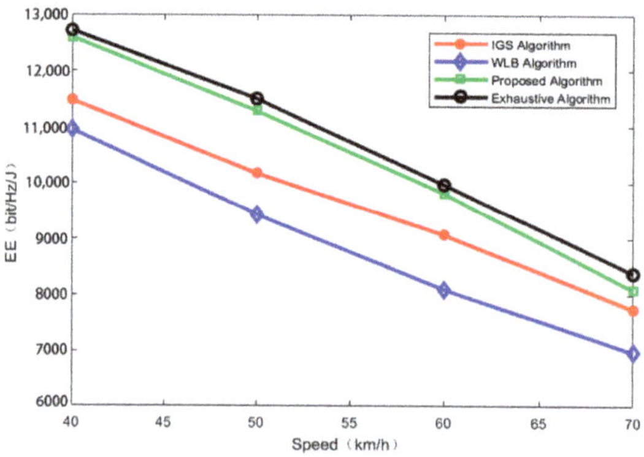

Figure 4. The EE values of VUEs for different vehicle speeds.

As shown in Figure 5, we compared the energy efficiency of the system with the iteration times of the proposed algorithm. It can be observed in the figure that the energy efficiency converged within a small number of iterations and basically reached the convergence state in only two steps. The convergence speed was faster, and by comparing the results of the algorithms with or without load balancing, it can be observed that the energy efficiency of the load balancing algorithm in this paper was obviously better than that of the algorithm without load balancing.

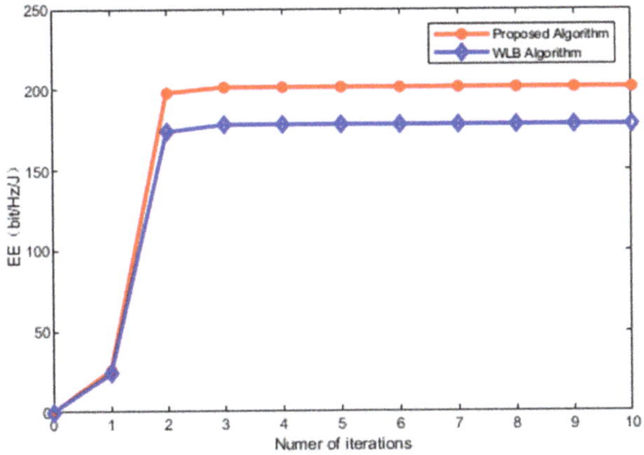

Figure 5. The EE values for different numbers of iterations.

It can be observed in Figure 5 above that the iterative convergence speed of the power control algorithm in this paper was faster, and as shown in Figure 6, the energy efficiency of the algorithm with power control in this paper was significantly improved compared with the algorithm without power control, because the power control algorithm in this paper can iteratively find the power value that makes the energy efficiency of a single VUE system optimal, and can significantly reduce the interference from the main users. Moreover, with the increase in the number of V2V pairs, the accumulated energy efficiency trend will be better than that of the algorithm without power control.

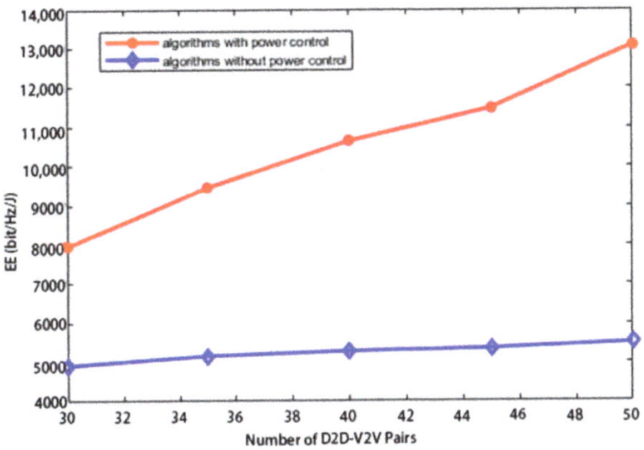

Figure 6. The EE values of VUEs for different numbers of V2V pairs.

In Figure 7, with the increase in the number of primary users, the energy efficiency value of the vehicles decreased. This is because the number of vehicles communicating in dedicated mode decreased, which were interfered with by the primary users, and the energy efficiency value decreased. Moreover, the performance of the algorithm system in this paper was better than that of the case without load balancing.

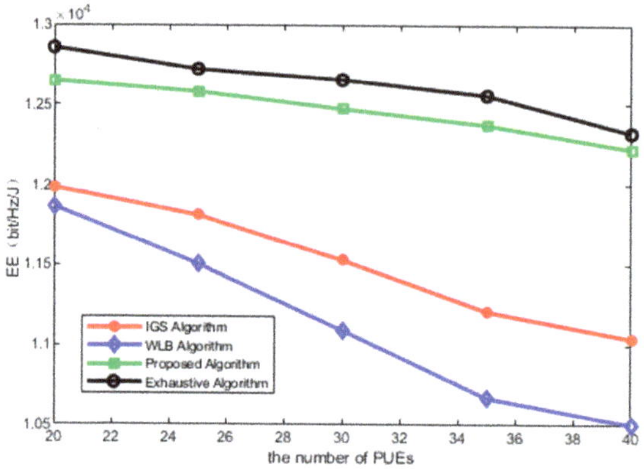

Figure 7. The EE values of VUEs for different numbers of PUEs.

As shown in Figure 8, the energy efficiency of the system with a radius of 500 was obviously higher than that with a radius of 200. This is because when the radius is small, the overlapping area of two base stations will be reduced, the number of vehicles in the overlapping area will be reduced with the same number of vehicles, and the number of vehicles that can reasonably utilize the spectrum resources of BS2 will also be reduced, meaning the system energy efficiency will be reduced. However, regardless of the radius, the system energy efficiency with load balancing was always higher than that without load balancing.

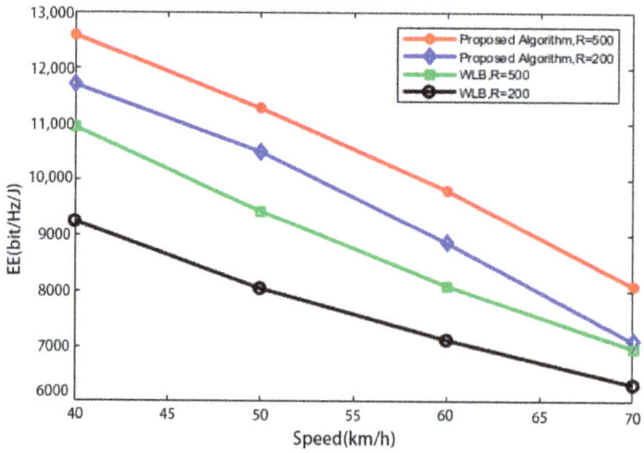

Figure 8. The EE values of VUEs for different radii.

The algorithm in this paper makes rational use of spectrum resources near the overlapping area, and the number of all BS2 primary users will have an impact on the energy efficiency of the vehicle user system in BS1. As shown in Figure 9, with the increase in the number of BS2 primary users, the system energy efficiency declined because the number of vehicles available for the special mode in the overlapping area decreased, and the system energy efficiency decreased to a certain extent. However, under the algorithm without-load balancing, the system energy efficiency was not affected by the main users in BS2.

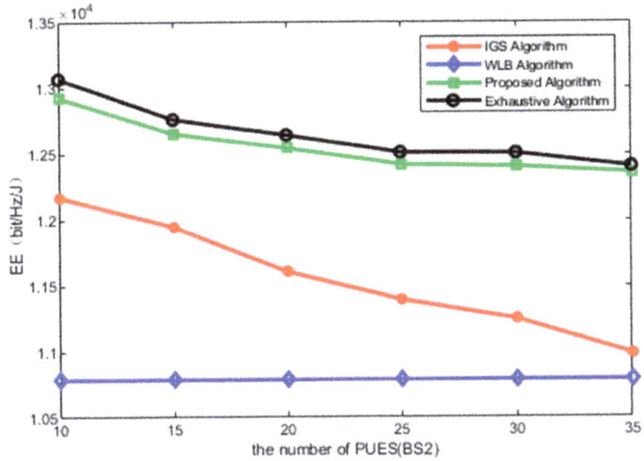

Figure 9. The EE values of VUEs for different numbers of PUEs (BS2).

6. Conclusions

In this paper, a joint scheme of mobile vehicle load balancing was proposed, and a new method of power control and resource allocation mode selection was presented. The joint scheme can significantly improve the system energy efficiency, meet the requirements of vehicle users for spectrum resources and minimal delays, and support centralized/distributed hybrid V2X communication and safety-related V2X services. By considering different resource allocation modes, we proposed a power control strategy to calculate the optimal power allocation value of a single VUE in different modes and at the same time, reduce the reuse of the same channel interference by the VUE and PUE, thereby improving the overall energy efficiency. We proposed two new algorithms, the URRAA and ORRAA, to realize resource allocation in different regions. The simulation results showed that, compared with the other schemes, our proposed scheme can not only significantly improve the energy efficiency of VUEs, but that it can also meet the demands for minimal delays among vehicles.

Author Contributions: Conceptualization, J.B.; methodology, J.B.; software, J.B. and X.Q.; validation, J.B., X.Q.; formal analysis and investigation, J.B.; Investigation, J.B.; resources, J.B. and Z.J.; Data curation, J.B.; writing—original draft preparation, J.B.; writing—review and editing, J.B., X.Q. and Z.J.; All authors have read and agreed to the published version of the manuscript.

Funding: This research was funded by: "Research and Development of Key Technologies and Application Demonstration of Integrated Food Data Supervision" Platform in Xinjiang Region. The funded project number is: 2020A03001-2.

Institutional Review Board Statement: Not applicable.

Informed Consent Statement: Not applicable.

Data Availability Statement: Not applicable.

Conflicts of Interest: The authors declare no conflict of interest.

References

1. Qian, L.P.; Wu, Y.; Zhou, H.; Shen, X. Non-orthogonal multiple access vehicular small cell networks: Architecture and solution. *IEEE Netw.* **2017**, *31*, 15–21. [CrossRef]
2. Li, X.; Ma, L.; Xu, Y.; Shankaran, R. Resource allocation for D2D based V2X communication with imperfect CSI. *IEEE Internet Things J.* **2020**, *7*, 3545–3558. [CrossRef]
3. Sun, G.; Boateng, G.O.; Ayepah-Mensah, D.; Liu, G.; Wei, J. Autonomous Resource Slicing for Virtualized Vehicular Networks With D2D Communications Based on Deep Reinforcement Learning. *IEEE Syst. J.* **2020**, *14*, 4694–4705. [CrossRef]

4. Deng, L.; He, Y.; Zhang, Y.; Chen, M.; Li, Z.; Lee, J.Y.B.; Zhang, Y.J.A.; Song, L. Device-to-device load balancing for cellular networks. *IEEE Trans. Commun.* **2018**, *67*, 3040–3054. [CrossRef]
5. Xu, F.; Li, Y.; Wang, H.; Zhang, P.; Jin, D. Understanding mobile traffic patterns of large scale cellular towers in urban environment. *IEEE/ACM Trans. Netw.* **2017**, *25*, 1147–1161. [CrossRef]
6. Moutsinas, G.; Guo, W. Probabilistic Stability of Traffic Load Balancing on Wireless Complex Networks. *IEEE Syst. J.* **2020**, *14*, 2551–2556. [CrossRef]
7. *TR 36.885 V0.5.0*; Third Generation Partnership Project (3GPP), Technical Specification Group Radio Access Network, Study on LTE-Based V2X Services, Release 14. 3GPP: Sophia Antipolis, France, February 2016.
8. Rafael, M.; Gozalvez, J. LTE-V for Sidelink 5G V2X vehicular communications: A new 5G technology for short-range vehicle-to-everything communications. *IEEE Veh. Technol. Mag.* **2017**, *12*, 30–39.
9. Li, X.; Ma, L.; Shankaran, R.; Xu, Y.; Orgun, M.A. Joint power control and resource allocation mode selection for safety-related v2x communication. *IEEE Trans.* **2019**, *68*, 7970–7986. [CrossRef]
10. Qin, Z.; Yue, X.; Liu, Y.; Ding, Z.; Nallanathan, A. User association and resource allocation in unified NOMA enabled heterogeneous ultra dense networks. *IEEE Commun.* **2018**, *56*, 86–92. [CrossRef]
11. Guan, X.; Huang, Y.; Dong, C.; Wu, Q. User Association and Power Allocation for UAV-Assisted Networks: A Distributed Reinforcement Learning Approach. *Chin. Commun.* **2020**, *17*, 110–122. [CrossRef]
12. Zhang, H.; Song, L.; Zhang, Y. Load balancing for 5G ultra-dense networks using device-to-device communications. *IEEE Trans. Wireless Commun.* **2018**, *17*, 4039–4050. [CrossRef]
13. Hasan, M.M.; Kwon, S. Cluster-based load balancing algorithm for ultra-dense heterogeneous networks. *IEEE Access* **2020**, *8*, 2153–2162. [CrossRef]
14. Xiao, H.; Zhu, D.; Chronopoulos, A.T. Power Allocation With Energy Efficiency Optimization in Cellular D2D-Based V2X Communication Network. *IEEE Trans. Intell. Transp. Syst.* **2020**, *21*, 4947–4957. [CrossRef]
15. Walter, M.; Shutin, D.; Schmidhammer, M.; Matolak, D.W.; Zajic, A. Geometric Analysis of the Doppler Frequency for General Non-Stationary 3D Mobile-to-Mobile Channels Based on Prolate Spheroidal Coordinates. *IEEE Trans. Veh. Technol.* **2020**, *69*, 10419–10434. [CrossRef]
16. Ding, Y.; Huang, Y.; Tang, L.; Qin, X.; Jia, Z. Resource Allocation in V2X Communications Based on Multi-Agent Reinforcement Learning with Attention Mechanism. *Mathematics* **2022**, *10*, 3415. [CrossRef]
17. Chen, C.; Wang, B.; Zhang, R. Interference Hypergraph-Based Resource Allocation (IHG-RA) for NOMA-Integrated V2X Networks. *IEEE Internet Things J.* **2019**, *6*, 161–170. [CrossRef]
18. Abbas, F.; Fan, P.; Khan, Z. A novel low-latency V2V resource allocation scheme based on cellular V2X communications. *IEEE Trans.* **2019**, *20*, 2185–2197. [CrossRef]
19. Zhang, X.; Peng, M.; Yan, S.; Sun, Y. Deep Reinforcement Learning Based Mode Selection and Resource Allocation for Cellular V2X Communications. *IEEE Internet Things J.* **2020**, *7*, 6380–6391. [CrossRef]
20. Zhou., H.; Jiang, K.; He, S.; Min, G.; Wu, J. Distributed Deep Multi-Agent Reinforcement Learning for Cooperative Edge Caching in Internet-of-Vehicles. *IEEE Trans. Wirel. Commun.* **2023**. [CrossRef]
21. Zhou, H.; Wu, T.; Chen, X.; He, S.; Guo, D.; Wu, J. Reverse Auction-Based Computation Offloading and Resource Allocation in Mobile Cloud-Edge Computing. *IEEE Trans. Mob. Comput.* **2022**. [CrossRef]
22. Deng, D.; Wang, C.; Wang, W. Joint Air-to-Ground Scheduling in UAV-Aided Vehicular Communication: A DRL Approach With Partial Observations. *IEEE Commun. Lett.* **2022**, *26*, 1628–1632. [CrossRef]
23. Du, Z.; Wu, C.; Yoshinaga, T.; Yau, K.-L.A.; Ji, Y.; Li, J. Federated Learning for Vehicular Internet of Things: Recent Advances and Open Issues. *IEEE Open J. Comput. Soc.* **2020**, *1*, 45–61. [CrossRef]
24. Xu, W.; Gao, F.; Tao, X.; Zhang, J.; Alkhateeb, A. Computer Vision Aided mmWave Beam Alignment in V2X Communications. *IEEE Trans. Wirel. Commun.* **2023**, *22*, 2699–2714. [CrossRef]
25. Hossain, M.A.; Noor, R.M.; Yau, K.-L.A.; Azzuhri, S.R.; Z'aba, M.R.; Ahmedy, I. Comprehensive Survey of Machine Learning Approaches in Cognitive Radio-Based Vehicular Ad Hoc Networks. *IEEE Access* **2020**, *8*, 78054–78108. [CrossRef]
26. Shen, K.; Yu, W. Fractional Programming for Communication Systems—Part I: Power Control and Beamforming. *IEEE Trans. Signal Process.* **2018**, *66*, 2616–2630. [CrossRef]
27. Zhou, Z.; Xiong, F.; Chen, X.; He, Y.; Mumtaz, S. Energy-efficient vehicular heterogeneous networks for green cities. *IEEE Trans.* **2018**, *14*, 1522–1531. [CrossRef]
28. Li, X.; Shankaran, R.; Orgun, M.A.; Fang, G.; Xu, Y. Resource allocation for underlay D2D communication with proportional fairness. *IEEE Trans.* **2018**, *67*, 6244–6258. [CrossRef]
29. Meng, C.; Wang, G.; Dai, X. Secure energy-efficient transmission for SWIPT intelligent connected vehicles with imperfect CSI. *IEEE Access* **2019**, *7*, 154649–154658. [CrossRef]
30. Liu, Z.; Xie, Y.; Yuan, Y.; Ma, K.; Chan, K.Y.; Guan, X. Robust Power Control for Clustering-Based Vehicle-to-Vehicle Communication. *IEEE Syst. J.* **2020**, *14*, 2557–2568. [CrossRef]

31. Wang, Z.; Liu, L.; Wang, X.; Zhang, J. Resource allocation in ofdma networks with imperfect channel state information. *IEEE Commun. Lett.* **2014**, *18*, 1611–1614. [CrossRef]
32. Dinkelbach, W. On nonlinear fractional programming. *Manag. Sci.* **1967**, *13*, 492–498. [CrossRef]

MDPI AG
Grosspeteranlage 5
4052 Basel
Switzerland
Tel.: +41 61 683 77 34

Mathematics Editorial Office
E-mail: mathematics@mdpi.com
www.mdpi.com/journal/mathematics

www.ingramcontent.com/pod-product-compliance
Lightning Source LLC
Chambersburg PA
CBHW061555220926
51850CB00005B/342
* 9 7 8 3 7 2 5 8 5 6 8 3 1 *